Advances in Digital Speech Transmission

Advances in Digital Speech Transmission

Edited by

Rainer Martin
Ruhr University Bochum, Bochum, Germany

Ulrich Heute
Christian-Albrechts-University, Kiel, Germany

Christiane Antweiler
RWTH Aachen University, Aachen, Germany

John Wiley & Sons, Ltd

Other Wiley Editorial Offices

John Wiley & Sons Inc., 111 River Street, Hoboken, NJ 07030, USA

Jossey-Bass, 989 Market Street, San Francisco, CA 94103-1741, USA

Wiley-VCH Verlag GmbH, Boschstr. 12, D-69469 Weinheim, Germany

John Wiley & Sons Australia Ltd, 42 McDougall Street, Milton, Queensland 4064, Australia

John Wiley & Sons (Asia) Pte Ltd, 2 Clementi Loop #02-01, Jin Xing Distripark, Singapore 129809

John Wiley & Sons Canada Ltd, 6045 Freemont Blvd, Mississauga, Ontario, L5R 4J3, Canada

Wiley also publishes its books in a variety of electronic formats. Some content that appears in print may
not be available in electronic books.

Library of Congress Cataloging-in-publication Data

Martin, Rainer.
    Advances in digital speech transmission / Rainer Martin, Ulrich Heute, Christiane Antweiler.
        p. cm.
    Includes bibliographical references and index.
    ISBN 978-0-470-51739-0 (cloth)
1. Speech processing systems. 2. Signal processing–Digital techniques. I. Heute, Ulrich. II. Antweiler,
Christiane. III. Title.
    TK7882.S65.M53 2008
    621.39′9–dc22

                                                                                        2007039292

British Library Cataloguing in Publication Data

A catalogue record for this book is available from the British Library

ISBN 978-0-470-51739-0 (HB)

Typeset by the authors using LATEX software
Printed and bound in Great Britain by Antony Rowe Ltd, Chippenham, England
This book is printed on acid-free paper responsibly manufactured from sustainable forestry in which at
least two trees are planted for each one used for paper production.

To

Peter Vary

– a great scientist, teacher, and friend –

for sharing his visions and insights
into Digital Speech Transmission

Contents

3 Parametric Quality Assessment of Narrowband Speech in Mobile Communication Systems   51

Marc Werner

II Adaptive Algorithms in Acoustic Signal Processing   77

4 Kalman Filtering in Acoustic Echo Control: A Smooth Ride on a Rocky Road   79

Gerald Enzner

6 Acoustic Source Localization with Microphone Arrays   135

Nilesh Madhu, Rainer Martin

IV Joint Source-Channel Coding 279

10 Parameter Models and Estimators in Soft Decision Source Decoding 281

Tim Fingscheidt

11 Optimal MMSE Estimation for Vector Sources with Spatially and Temporally Correlated Elements   311

Stefan Heinen, Marc Adrat

12 Source Optimized Channel Codes & Source Controlled Channel Decoding   329

Stefan Heinen, Thomas Hindelang

13 Iterative Source-Channel Decoding & Turbo DeCodulation   **365**

Marc Adrat, Thorsten Clevorn, Laurent Schmalen

V Speech Processing in Hearing Instruments   399

14 Binaural Signal Processing in Hearing Aids   401

Volkmar Hamacher, Ulrich Kornagel, Thomas Lotter, Henning Puder

15 Auditory-profile-based Physical Evaluation of Multi-microphone Noise Reduction Techniques in Hearing Instruments

VI Speech Processing for Human–Machine Interfaces

16 Automatic Speech Recognition in Adverse Acoustic Conditions

17 Speaker Classification for Next-Generation Voice-Dialog Systems 497

Felix Burkhardt, Florian Metze, Joachim Stegmann

List of Contributors

Marc Adrat
Research Establishment for Applied Science (FGAN)
53343 Wachtberg, Germany
adrat@fgan.de

Christiane Antweiler
Institute of Communication Systems and Data Processing
RWTH Aachen University, 52056 Aachen, Germany
antweiler@ind.rwth-aachen.de

Colin Breithaupt
Institute of Communication Acoustics
Ruhr University Bochum, 44780 Bochum, Germany
colin.breithaupt@rub.de

Felix Burkhardt
T-Systems Enterprise Services GmbH
Goslarer Ufer 35, 10589 Berlin, Germany
felix.burkhardt@t-systems.com

Thorsten Clevorn
Infineon Technologies AG
47259 Duisburg, Germany
thorsten.clevorn@infineon.com

Simon Doclo
ESAT-SCD
Katholieke Universiteit Leuven, Belgium
simon.doclo@esat.kuleuven.be

Koen Eneman
ExpORL, Dept. Neurosciences
Katholieke Universiteit Leuven, Belgium
koen.eneman@med.kuleuven.be

Gerald Enzner
Institute of Communication Acoustics
Ruhr University Bochum, 44780 Bochum, Germany
gerald.enzner@rub.de

Tim Fingscheidt
Institute for Communications Technology
Braunschweig Technical University, 38106 Braunschweig, Germany
t.fingscheidt@tu-bs.de

Bernd Geiser
Institute of Communication Systems and Data Processing
RWTH Aachen University, 52056 Aachen, Germany
geiser@ind.rwth-aachen.de

Volkmar Hamacher
Siemens Audiologische Technik GmbH
Gebbertstrasse 125, 91058 Erlangen, Germany
volkmar.hamacher@siemens.com

Stefan Heinen
Infineon Technologies AG
47259 Duisburg, Germany
stefan.heinen@infineon.com

Ulrich Heute
Institute for Circuit and System Theory
Christian-Albrechts-University of Kiel, 24143 Kiel, Germany
uh@tf.uni-kiel.de

Thomas Hindelang
Nokia Siemens Networks GmbH & Co. KG
St.-Martin-Str. 76, 81541 Munich, Germany
hindelang@ieee.org

Hans-Günter Hirsch
Niederrhein University of Applied Sciences
47805 Krefeld, Germany
hans-guenter.hirsch@hs-niederrhein.de

Peter Jax
Deutsche Thomson OHG
30625 Hannover, Germany
peter.jax@thomson.net

Ulrich Kornagel
Siemens Audiologische Technik GmbH
Gebbertstrasse 125, 91058 Erlangen, Germany
ulrich.kornagel@siemens.com

Arne Leijon
Sound and Image Processing Lab
KTH Stockholm, Sweden
arne.leijon@ee.kth.se

Thomas Lotter
Siemens Audiologische Technik GmbH
Gebbertstrasse 125, 91058 Erlangen, Germany
thomas.tl.lotter@siemens.com

Nilesh Madhu
Institute of Communication Acoustics
Ruhr University Bochum, 44780 Bochum, Germany
nilesh.madhu@rub.de

Rainer Martin
Institute of Communication Acoustics
Ruhr University Bochum, 44780 Bochum, Germany
rainer.martin@rub.de

Florian Metze
Deutsche Telekom Laboratories
Ernst-Reuter-Platz 7, 10587 Berlin, Germany
florian.metze@telekom.de

Marc Moonen
ESAT-SCD
Katholieke Universiteit Leuven, Belgium
marc.moonen@esat.kuleuven.be

Henning Puder
Siemens Audiologische Technik GmbH
Gebbertstrasse 125, 91058 Erlangen, Germany
henning.puder@siemens.com

Stéphane Ragot
France Télécom R&D/TECH/SSTP
2 av. Pierre Marzin, 22307 Lannion Cedex, France
stephane.ragot@orange-ftgroup.com

Laurent Schmalen
Institute of Communication Systems and Data Processing
RWTH Aachen University, 52056 Aachen, Germany
schmalen@ind.rwth-aachen.de

Ann Spriet
ESAT-SCD
Katholieke Universiteit Leuven, Belgium
ann.spriet@esat.kuleuven.be

Joachim Stegmann
T-Systems Enterprise Services GmbH
Deutsche-Telekom-Allee 7, 64295 Darmstadt, Germany
joachim.stegmann@t-systems.com

Hervé Taddei
Nokia Siemens Networks GmbH & Co. KG
Otto-Hahn-Ring 6, 81739 Munich, Germany
herve.taddei@ieee.org

Marc Werner
QUALCOMM CDMA Technologies GmbH
90411 Nürnberg, Germany
marc.werner@qualcomm.com

Jan Wouters
ExpORL, Dept. Neurosciences
Katholieke Universiteit Leuven, Belgium
jan.wouters@med.kuleuven.be

Preface

When the book *Digital Speech Transmission – Enhancement, Coding and Error Concealment* by Peter Vary and Rainer Martin appeared in 2006, it was clear that a subject of this importance and this range could not be treated in all its details on 600-some pages. Important aspects had to be left out and had to be postponed to a succeeding volume.

The opportunity for such an extension came when friends, colleagues and former doctoral students of Peter Vary decided to launch an edited book on recent developments in this field – in honor of a man who has contributed significantly to the progress of digital signal and, especially, speech processing. The edited book is published on the occasion of his 60th birthday.

The present volume is the result of this effort. It comprises tutorial and research contributions on recent Advances in Digital Speech Transmission – all of them written by known experts in the field. This volume thus presents valuable additions and updates on a broad range of subjects written for graduate students and researchers in speech communications.

We would like to thank all contributing authors for sharing our enthusiasm and for the timely delivery of their manuscripts. We would also like to express our gratitude to the staff of John Wiley & Sons, Ltd, especially to Tiina Ruonamaa, Sarah Hinton, and Brett Wells for supporting this project in any possible way. Last but not least, the staff of the editors' research institutions at the Aachen, Bochum, and Kiel Universities contributed many reviews and helped with the editing work. Their efforts are sincerely appreciated.

Bochum   Rainer Martin
Kiel   Ulrich Heute
Aachen   Christiane Antweiler

Chapter 1

Introduction

Rainer Martin, Ulrich Heute, Christiane Antweiler

In the era of mobile communication networks, of Voice over IP (VoiP), and of hands-free voice interfaces, new opportunities as well as new challenges arise. While traditional voice services are confined to the rather narrow frequency range below 4 kHz, new technologies enable the transition to higher bandwidth and thus better quality speech transmission systems. At the same time, speech communication systems are increasingly used in *adverse acoustic conditions*, i.e., noisy and reverberant environments. It turns out that a convincing end-to-end quality improvement is obtained only when all components of the transmission chain – analog and digital – are optimized to a comparable level of quality. This entails improved methods for acoustic front-end processing as well as for speech coding and transmission. As a consequence, new wideband speech coding systems, new speech enhancement and error concealment algorithms, and new quality assessment methods have emerged.

Another important application in digital speech transmission is hearing instruments that comprise increasingly powerful digital processors. Advanced speech processing algorithms are currently integrated into hearing aids and will result in significant improvements for hearing-impaired people.

Likewise, the recognition performance of automatic speech recognizers (ASR) can also be improved dramatically if models of noise and reverberation are integrated. Furthermore, a satisfying user experience requires these systems to become aware of *who* is actually using them and to adapt to the needs of specific users. Many challenging tasks in the conceptual design and in the signal processing modules of modern speech transmission systems need to be solved before a uniformly pleasant user experience is accomplished.

Advances in Digital Speech Transmission   Edited by R. Martin, U. Heute and C. Antweiler
© 2008 John Wiley & Sons, Ltd

The general theme of this book is the presentation and the analysis of solutions for improved-quality design of speech transmissions under adverse acoustic and heterogeneous network conditions. The book is organized into six parts:

    I.   **Speech Quality Assessment**
   II.   **Adaptive Algorithms in Acoustic Signal Processing**
  III.   **Speech Coding for Heterogeneous Networks**
  IV.   **Joint Source-Channel Coding**
   V.   **Speech Processing in Hearing Instruments**
  VI.   **Speech Processing for Human–Machine Interfaces,**

which will be briefly introduced below. Each chapter comes with an extensive list of references that may serve as a resource for further study.

Intrinsically related to the *Advances in Digital Speech Transmission* is the question of how to measure the quality of voice communication systems. This topic is treated in **Part I, Assessment of Speech Quality**. In his chapter on ***Speech-Transmission Quality: Aspects and Assessment for Wideband vs. Narrowband Signals*** Ulrich Heute discusses the impact of increasing the signal bandwidth first on speech processing algorithms in general and secondly on the methods for quality assessment in particular. He argues convincingly that the traditional total quality assessment approach, for instance, in the form of total quality listening tests and *mean opinion scores* (MOS) should be succeeded by quality assessment procedures with diagnostic abilities. Therefore, the task is to find more or less orthogonal descriptors for wideband speech quality that provide a basis for the computation of total quality scores, and to develop algorithms for the computational assessment of speech signals. In his chapter on ***Parametric Quality Assessment of Narrowband Speech in Mobile Communication Systems***, Marc Werner provides an overview of and insights into automated *non-intrusive* quality monitoring for mobile voice communication channels. His approach is based on mapping measurable system parameters at the receiving end to speech quality measures. Such parameters are, for instance, the signal power or the frame-error and the bit-error rate (FER / BER). The quality measure is then computed via a linear function of these parameters whose coefficients are optimized in the minimum mean-square error sense. Speech transmissions in the GSM and the UMTS systems serve as examples.

Part II is dedicated to **Adaptive Algorithms in Acoustic Signal Processing**. It begins with Gerald Enzner's chapter on ***Kalman Filtering in Acoustic Echo Control: A Smooth Ride on a Rocky Road***, which takes a fresh look at the acoustic echo control problem and develops a model-based approach. By dropping the assumption of a deterministic echo-path model and by replacing it with a statistical model, he arrives at a unifying solution to the acoustic echo-control problem. His solution comprises an echo canceler and a post-filter and turns out to be very robust to echo-path variations, while at the same time it is conceptually elegant and simple to implement. The chapter ***Noise Reduction – Statistical Analysis and Control of Musical Noise*** by Colin Breithaupt and Rainer Martin investigates the statistical fluctuations in the spectral parameters of state-of-the-art noise reduction systems. A careful analytic analysis of the most widely used spectral enhancement approaches explains the emergence of musical noise in these algorithms. It is shown that the histogram of log-spectral amplitudes provides a good indication of audible spectral fluctuations. Furthermore, a solution is presented for controlling these fluctuations without impairing the perceived speech quality. In their chapter *Acoustic Source Localization with Microphone Arrays*, Nilesh Madhu and Rainer Martin provide an overview on time delay of arrival (TDOA) and source localization techniques. This includes the popular generalized cross-correlation (GCC) method as well as multi-microphone techniques such as steered response power (SRP), multiple signal classification (MUSIC), and maximum-likelihood (ML) approaches. The chapter elaborates on the links between these methods and provides simulation examples highlighting their respective performance. The chapter *Multi-Channel System Identification with Perfect Sequences – Theory and Applications* by Christiane Antweiler explains how sequences with perfect correlation properties can be used to identify multiple input – single output (MISO) systems. Perfect sequences are a most elegant tool to be used in conjunction with the normalized least mean-square (NLMS) algorithm. This chapter extends this method to the real-time identification of multiple channels and provides an example in medical technology: the online assessment of the Eustachian tube.

Part III, Speech Coding for Heterogeneous Networks, is opened by an overview on *Embedded Speech Coding: From G.711 to G.729.1* by Bernd Geiser, Stéphane Ragot, and Hervé Taddei. The authors review the general theory and the methods for successive refinement coding of speech. Tree-structured vector quantization (TSVQ), multi-stage vector quantization (MSVQ), and transform domain vector quantization with progressive decoding and the application in state-of-the-art speech coders are discussed in detail. Furthermore, this chapter summarizes the latest developments in embedded wideband coding for VoiP speech transmission systems and the network aspects of such schemes. Bandwidth extension (BWE) of speech signals constitutes another important research area in speech signal processing that currently enjoys renewed and significant interest. In Chap. 9, *Backwards Compatible Wideband Telephony*, Peter Jax discusses the implications of this technology for the migration from narrowband to wideband speech transmission systems and outlines his approach to this task. He explains solutions that do not need side-information (stand-alone BWE) and those that make use of information about the wideband speech envelope. If the side-information is embedded into the speech signal by means of watermarking technology, the result is a speech signal that can be reproduced either on a narrowband or on a wideband terminal.

Part IV, Joint Source-Channel Coding, presents a series of four chapters that deal with the exploitation of redundancies in speech and channel coding schemes for improving the quality of the received signal. Tim Fingscheidt's chapter on *Parameter Models and Estimators in Soft Decision Source Decoding* first presents an overview of soft decision source decoding (SDSD) and of modeling techniques for source parameters. Secondly, it discusses estimators for the recovery of degraded, received speech parameters by means of extrapolation and interpolation and presents simulation results for various transmission scenarios. In Chap. 11, this work is extended towards a general discussion of *Optimal MMSE Estimation for Vector Sources with Spatially and Temporally Correlated Elements* by Stefan Heinen and Marc Adrat. While the general solution of this task would lead to an overwhelming computational complexity, the authors tackle this problem by imposing a Markov property on the speech parameters in the temporal *and* the vector dimensions. Thus, they are capable of deriving MMSE and near-optimal MMSE estimators with significantly reduced computational requirements. Furthermore, they provide simulation results for digital audio broadcast (DAB) and GSM systems. In their chapter on *Source Optimized Channel Codes & Source Controlled Channel Decoding* Stefan Heinen and Thomas Hindelang establish the link between source and channel (de-)coding in two competing approaches. In the source-optimized channel coding (SOCC) system the channel codes are tailored to the source statistics. It thus achieves an efficient utilization of the available bit rate and a low reconstruction error at the receiver. The second approach of source-controlled channel decoding (SCCD) improves the decoding of convolutional codes by exploiting residual redundancies in source parameters. The authors compare and discuss the merits of both approaches on the basis of a single, general source and transmission channel model and are thus able to draw interesting conclusions. Part IV concludes with the chapter on *Iterative Source-Channel Decoding & Turbo DeCodulation* by Marc Adrat, Thorsten Clevorn, and Laurent Schmalen, which builds on the two approaches of the previous chapter. Here, the authors extend soft source and channel decoding techniques towards iterative methods, also known as *turbo*-decoding methods. Besides a review of turbo-techniques, two novel approaches, iterative source-channel decoding (ISCD) and Turbo DeCodulation (TdeC), are developed and analyzed by means of extrinsic information transfer (EXIT) charts. It is shown that ISCD outperforms non-iterative transmission schemes in terms of the signal-to-noise ratio of the reconstructed speech signal.

An interesting area for research in digital speech transmission is the application of **Speech Processing in Hearing Instruments**. **Part V** is dedicated to this topic. The first chapter in this part is authored by Volkmar Hamacher, Ulrich Kornagel, Thomas Lotter, and Henning Puder and discusses *Binaural Signal Processing in Hearing Aids: Technologies and Algorithms*. The realization of a wireless data link between the left ear and the right ear hearing device is a challenge in itself but also enables the employment of binaural processing schemes. The authors first describe the design of the wireless link as implemented in a commercial hearing aid. Then, they discuss the potential that lies in the possibility to exchange data at various bit rates between the left and the right side. They show that considerable potential resides in these processing schemes, especially in terms of user comfort. Advanced signal processing schemes such as binaural beamformers, however, require higher data rates. Thus, many interesting research questions are still to be answered before such methods can be applied. In Chap. 15, *Auditory-profile-based Physical Evaluation of Multi-microphone Noise Reduction Techniques in Hearing Instruments*, by Koen Eneman, Arne Leijon, Simon Doclo, Ann Spriet, Marc Moonen, and Jan Wouters discusses multi-microphone beamforming approaches for hearing devices and introduces an assessment method for noise reduction algorithms that takes the speech perception capabilities of hearing impaired listeners into account. This evaluation is based on the *auditory profile* of the listener, i.e., the characterization of the hearing loss in terms of various measures such as the audiogram. The newly developed instrumental measures provide a prediction of the usefulness of signal enhancement algorithms for people with hearing deficiencies. It can be expected that such measures will reduce the effort that comes with listening tests and thus will facilitate the development of new and more effective algorithms.

Part VI deals with **Speech Processing for Human–Machine Interfaces**. Voice driven human–machine interfaces are typically used in a hands-free mode. Thus, the automatic speech recognizer (ASR) has to cope with significant levels of noise and reverberation. In Chap. 16, Hans-Günter Hirsch discusses *Automatic Speech Recognition in Adverse Acoustic Conditions*, which entail ambient noise and reverberation. He gives an overview on state-of-the-art approaches and presents a new database that was developed to allow the evaluation and comparison of recognition experiments in noisy and reverberant conditions. Furthermore, he presents a new model adaptation approach that dramatically improves the recognition performance in such environments. The topic of the final chapter is *Speaker Classification for Next-Generation Voice-Dialog Systems*. In this chapter, Felix Burkhardt, Florian Metze, and Joachim Stegmann discuss how voice services may be personalized when the speaker can be classified as belonging to a certain target group. Useful classification criteria are, for instance, the age and the gender of the caller, or they may be related to the emotions of the caller when accessing the dialog system. The authors provide an overview on classification methods and present corresponding algorithms. They show how these classification criteria can be used in dialog systems and present evaluation results for their specific approach.

The editors believe that each of these chapters by itself will serve as a valuable resource and reference for students and researchers and that cross links between these topics will trigger new ideas and thus contribute to the progress of the field.

I

Speech Quality Assessment

Chapter 2

Speech-Transmission Quality: Aspects and Assessment for Wideband vs. Narrowband Signals

Ulrich Heute

2.1 Introduction

For decades, users of a telephone connection have expected a speech transmission with a small bandwidth below 4 kHz and a further quality limitation by some disturbances, but sufficient intelligibility. Now, wideband speech transmission up to 7 kHz is being offered in a growing number of services. This yields a more pleasant sound, and even intelligibility advantages can be shown. But wideband transmission, coding, and general processing require new features to be taken into account. An auditory determination of the resulting speech quality follows the same lines as that for narrowband signals. The results, however, need a new interpretation, and instrumental quality measures rely even less on a saturated understanding than those for telephone-band speech. This holds for direct total-quality (i.e., MOS) estimation and for prediction of single, diagnostic quality attributes (like, e.g., "noisiness"), as well as for their integration into overall quality. The state of the art, new approaches, first results of present investigations, and open questions are considered, especially for wideband speech, with an emphasis on differences in comparison with the telephone-band case.

Advances in Digital Speech Transmission   Edited by R. Martin, U. Heute and C. Antweiler
© 2008 John Wiley & Sons, Ltd

2.2 Speech Signals

In the context of this chapter, and as indicated in Fig. 2.1, speech is treated as a continuous acoustical time function, termed $s_o(t)$ or $s_1(t)$. It may either be created by a human speaker, or it may leave a loudspeaker, handset, or earphone. In the latter case, the corresponding electrical signal $y_o(t)$ comes from a digital-to-analog converter (DAC) with succeeding interpolation low-pass filter (Ipo-LP), whose input is a discrete sequence $y(k)$. That signal comes from a digital system. This device transmits or, equivalently, stores and thereby, generally, "somehow influences" the input sequence $x(k)$. These values are samples of the continuous-time signal $x_o(t)$. By $x_o(t)$, we denote a filtered version of the microphone signal $\check{x}_o(t)$. The filter provides a band-limitation and, especially, avoids aliasing in the analog-to-digital conversion (ADC). So, formally, we have

- the microphone signal $\check{x}_o(t) \sim s_o(t)$,

- its filtered version $x_o(t)$,

- the system's input samples $x(k) = x_o(kT_S), \ k \in \mathbb{Z}$,

- the system's output samples $y(k) = y_o(kT_S), \ k \in \mathbb{Z}$, and

- the acoustical output signal $s_1(t) \sim y_o(t)$.

Here, T_S denotes the sampling interval. It is defined via the sampling frequency f_S which, as is well known, has to be chosen such that the sampling theorem is fulfilled:

$$f_S = \frac{1}{T_S} \geq 2 \cdot f_c; \tag{2.1}$$

f_c describes the maximum frequency appearing in $x_o(t)$.

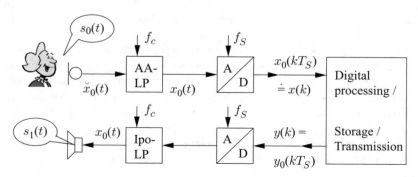

Figure 2.1: Speech signals and their electrical and digital counterparts in a processing / storage / transmission scenario

2.3 Telephone-Band Speech Signals

In our context, the speech signal of interest is transmitted via some telephone connection, with various possibilities of digital encodings. This may be a simple PCM, with basically a transmission of non-linearly quantized values $y(k) = [x(k)]_Q$, or any more or less refined compression system, like an ADPCM (Adaptive Differential Pulse-Code Modulation) or a CELP (Code-Excited Linear Prediction) codec, yielding a more complex modification of $x(k)$. Also, some signal-enhancement techniques, like reduction of echoes and noise, may be included.

In classical digital telephony, the sampling rate is fixed to

$$f_S = 8 \text{ kHz}. \tag{2.2}$$

This would theoretically allow for a bandwidth of $x_o(t)$ with an upper limit at

$$f_c = 4 \text{ kHz}$$

according to (2.1). In fact, however, the telephone-speech band is much more limited: Usually, the pass-band of the telephone band-pass is said to cover frequencies

$$f \in B_n \doteq [0.3, 3.4] \text{ kHz}. \tag{2.3}$$

Equation (2.3) defines[1] so-called *telephone-band* or *narrowband speech*. But even in this narrow frequency range, the spectrum is not too well preserved: Fig. 2.2 shows the tolerance scheme of the input filter for an ISDN transmission system [ITU G.712 2001]. Obviously, quite strong linear distortions are tolerated within B_n.

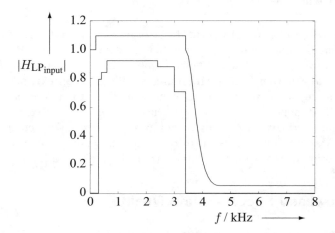

Figure 2.2: Linear tolerance scheme for the ISDN-system input filter

---

[1]The symbol $\doteq$ is used throughout this chapter for definitions.

2.3.1 Narrowband Speech Intelligibility

The above-named bandwidth complies with early investigations of speech intelligibility [Fletcher, Galt 1950]. Figure 2.3 displays the *articulation s* depending on the bandwidth. This term is defined as $s \doteq 1 - e$, and e denotes the probability of errors in phoneme understanding. So, s can be interpreted as the rate of correct phoneme perception. Obviously, a low-pass with upper cut-off frequency $f_c = 3.4\,\text{kHz}$ already allows some 97 % of all sounds to be understood, and the lower frequency limit due to the telephone high-pass at 300 Hz has an even smaller influence.

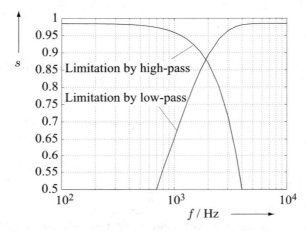

Figure 2.3: Sound intelligibility s depending on the cut-off frequencies of a low-
pass and a high-pass filter (parabola approximations of measurements in
[Fletcher, Galt 1950])

Thus, sound intelligibility usually seems to be of no concern in telephone-band speech. The recognition of whole sentences is even better, namely around 99 %, unless specifically difficult words appear, like unknown names, where the human ability of interpolating missing information from surrounding sounds and context fails. This remains true even when, beyond band limitation, other (small) distortions occur, like quantization noise in PCM, non-linearities, or even aliasing components in certain compression systems (for an overview on present techniques, see, e.g., [Vary et al. 1998], [Heute 2005], [Vary, Martin 2006]. Instead, in our context, quality refers to *sound quality*, i.e., the perception of the acoustic signal $s_1(t)$ in Fig. 2.1 by a human listener.

2.3.2 Narrowband Speech-Sound Quality

Of course, the shape of $s_1(t)$ depends strongly on that of $s_o(t)$, i.e., on possible peculiarities of the speaker in terms of, e.g., the average fundamental frequency f_o (the "height" of the voice), the variation of f_o (monotony vs. melody), the harmonicity and clearness (vs. noisiness or hoarseness) of voiced sounds, and the spectral-shaping abilities due to the speakers vocal-tract features – in short: on the timbre of the original speech signal. The analysis of these "voice-quality" aspects (see, e.g., [Laver 1980]) is, however, beyond the scope of this chapter.

Our concern is the influence of the processing system on the sound quality of $s_1(t)$.

Since the quality attributed to the output of a system is the result of a perception plus a judgment process [Jekosch 2005], quality assessment naturally asks for the involvement of listening subjects. Before addressing some details of such so-called subjective quality tests in Sec. 2.5.1, a few general observations will be mentioned.

The band limitation of (2.3) will, first of all, lead to the well-known, telephone-typical impression of "thin" speech, in contrast to a "full" speech sound especially with the now missing low frequencies. The timbre will also be strongly changed by a band limitation. This "coloration" [Raake 2006] may even falsify the speaker's identity. Beyond, other distortions like losses due to frequency-selective attenuation or added components like noise or echoes are perceived with their own specific distortion timbres.

The long-time average speech spectrum is known to have a low-pass character, also with the very low frequencies being quite weak – which also complies with the choice of the telephone band in (2.3). Figure 2.4 depicts the measurements of [Dunn, White 1940] in a stylized form, with a linear growth of components below some 250 Hz, a band of maximum contributions between 250 Hz and 500 Hz, and a spectral decay of some 9 dB/octave starting at 500 Hz (which is approximately the average first-formant frequency $F1$). Later measurements (e.g., [Blomberg, Elenius 1970], [Serafat, Heute 1996]) and approximations (e.g., [French, Steinberg 1947], [ITU P.50 1999]) give, cum grano salis, the same general impression: Components above 4 kHz are attenuated

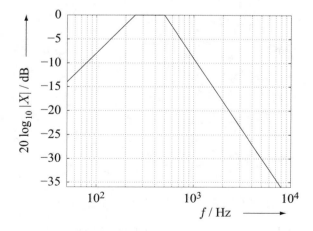

Figure 2.4: Stylized long-time average speech spectrum

by about 30 dB and more. Using the model of Fig. 2.4, a calculation of the power suppressed by an ideal 3.4 kHz low-pass shows that less than 0.9 % (corresponding to a linear-distortion level of -21 dB) is missing.

However, the telephone high-pass cut-off at 300 Hz, realized ideally, would remove more than 22 % of the total power (corresponding to -6.4 dB), and even a softer filtering still would lead to relatively high losses. This contradicts the above observations made from Fig. 2.3: Extending the band B_n of (2.3) *should* enhance the sound fidelity.

2.4 Wideband Speech Signals

Wideband speech is defined to cover frequencies

$$f \in B_w \doteq [50, 7000] \text{ Hz}. \tag{2.4}$$

Then the sampling rate of (2.2) is replaced by

$$f_S = 16 \text{ kHz}. \tag{2.5}$$

Figure 2.5 shows the corresponding input-filter tolerance scheme as defined for the first internationally introduced wideband-coding system [ITU G.722 1993]. Beyond the larger band width, the reduction of allowed in-band variation is obvious, in comparison with Fig. 2.2.

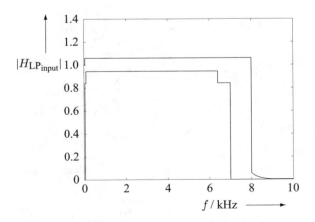

Figure 2.5: Linear tolerance scheme for wideband speech coding with a split-band ADPCM according to the ITU-T recommendation G.722

2.4.1 Wideband-Speech Intelligibility and Sound Quality

Indeed, speech with this more than doubled bandwidth is found to have a much more natural and full sound. A calculation based on the model spectrum of Fig. 2.4 again shows that, here, only 0.2 % of the power is dropped by an ideal 7 kHz low-pass and 0.11 % by an ideal 50 Hz high-pass filter. The equivalent levels of -26.6 dB and -29.4 dB show a much better match. Moreover, a closer look to the results in [Fletcher, Galt 1950] reveals that a larger bandwidth is indeed helpful also in the sense of intelligibility. In their report, they define an *articulation index* as

$$A = -c \cdot \log_{10}(1 - s) = -c \cdot \log_{10}(e).$$

If e_i denotes the understanding-error probability occurring when only the i-th of n narrow bands are passed, then the error probability when using n bands is found to be given by

$$e = e_1 \cdot e_2 \cdot ... \cdot e_n,$$

i.e.,

$$A = -c \cdot \sum_{i=1}^{n} \log_{10}(e_i) = -c \cdot \sum_{i=1}^{n} \log_{10}(1 - s_i).$$

Defining the contribution $\mathrm{d}A$ of a band $\mathrm{d}f$ at frequency f to the articulation index as the *importance* $\mathrm{d}A/\mathrm{d}f \doteq \mathrm{D}(f)$ of a certain frequency, an *accumulated importance* function is found by integration up to frequency f. Per definitionem, a "flat" 8 kHz-bandwidth transmission is assumed to yield $A = 1.0$. For this case, Fig. 2.3 shows that an articulation $s = 0.985$ is achieved. Thereby, the constant $c = 0.55$ is determined experimentally.

Figure 2.6 shows a simplified replica of the accumulated importance. Obviously, the components above 3.4 kHz add about 18 %, those below 300 Hz ca. 4 % important information, whereas not much is lost outside B_w.

The contradiction between the above statement and that of Sec. 2.3.1 may be resolved by a look to *specific*, not just *all possible* sounds. Voiced sounds like vowels have a short-time spectral envelope with much similarity to the long-time average low-pass behavior of Fig. 2.4, though with formants and anti-formants as additional maxima and valleys, respectively; unvoiced sounds tend rather to have a band-pass character, with less pronounced formants but, especially, the absolute maximum at higher frequencies – namely, around $f_c = 3.5$ kHz or even higher (see Fig. 2.7).

Obviously, this essential information is cut off or at least reduced by the narrowband telephone input filter of Fig. 2.2. Since the absolute power of unvoiced sounds is much lower than that of voiced ones (by some 15 dB or more, see Fig. 2.7), the loss of the strongest part is even worse. Thus, the discrimination between two fricatives becomes

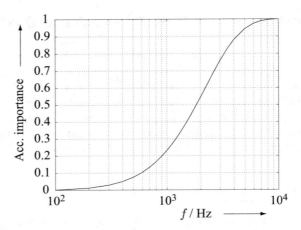

Figure 2.6: Accumulated importance of frequency components to speech articulation (exponential approximation of measurements in [Fletcher, Galt 1950])

difficult. This explains why the corresponding components are important but do not increase the *average* intelligibility too much. An enhanced brightness and clearness are often stated, describing the improvement in other words.

On the other hand, voiced sounds have line spectra with a spacing f_o, where f_o is the fundamental frequency of the short-time periodic sound with period $T_o = 1/f_o$. Especially for male voices, $f_o \in (60, 100)$ Hz may often occur, so that the telephone high-pass removes or strongly attenuates the first 3–5 harmonics; the fullness of the voice is thus lost, which (only) for certain dark vowels also increases the risk of confusion. These effects are, of course, less strong for female voices with their higher f_o values.

The perception of speaker-specific features, beyond f_o, is also considerably augmented by the step from B_n to B_w: The third formant is known to be less sound-typical and variable than the first two, and also to carry some information on the talker identity. The fourth formant is more or less only speaker-related, and the same holds for higher formants. For the neutral vocal tract occurring during the "schwa" sound /ə/, i.e., the acoustic tube between glottis and mouth with an almost uniform cross-section, the formant frequencies are found to be

$$F\nu = (2\nu - 1) \cdot F1 \approx (2\nu - 1) \cdot 500\,\text{Hz}\,.$$

So, $F3 \approx 2500\,\text{Hz}$ is well within B_n, though possibly somewhat attenuated (see Fig. 2.2); but $F4 \approx 3500\,\text{Hz}$ coincides with the edge of the telephone band, and $F5 = 4500\,\text{Hz}$ is outside B_n; these speaker-specific spectral features are strongly attenuated or even removed, while they are well maintained in wideband-speech signals.

In total, as said before, naturalness is enhanced if B_w is allowed. In a formal listening test for this feature, a 4-point increase from 2 to 6 on a 10-point scale was found for

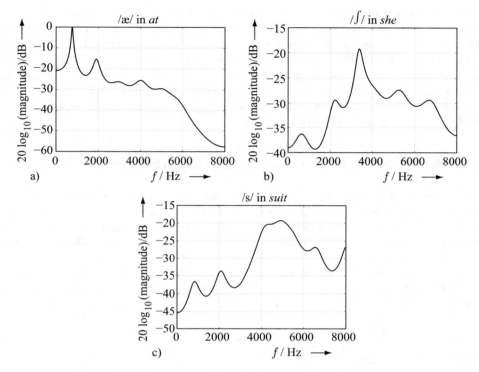

Figure 2.7: Envelopes of short-time spectra of
a) the vowel /æ/
b) the fricative /ʃ/
c) the fricative /s/ spoken by a male speaker

the step from B_n to B_w; even more could be gained for larger bandwidths [Moore, Tan 2003]. Interestingly, not much naturalness improvement is found by reducing the lower *or* increasing the upper cut-off frequency *alone*. Similarly, from a quality test in [Krebber 1995], it can be seen that, on a 5-point scale, a decrease of the lower band edge to 100 Hz alone or the doubling of the upper edge to 7 kHz alone give improvements of only 0.5 points, while the extension of both sides yields a gain of 1.3 points.

This complies with the generally accepted assumption that a balance of bandwidth and band position is needed. Although a band extension of only 300 Hz at the lower end, but some 3500 Hz at the upper end of the frequency axis seems to be quite unbalanced; a look to the non-linear frequency resolution of the ear [Zwicker 1982] is helpful: The use of the warped Bark-frequency scale

$$\Theta/\text{Bark} = 13 \cdot \arctan\left(0.76 \cdot \frac{f}{\text{Hz}}\right) + 3.5 \cdot \arctan\left(\left(\frac{f}{7.5 \text{ kHz}}\right)^2\right) \qquad (2.6)$$

reveals that the critical bands with $\Theta = 1, 2$, and 3 are added on one side, $\Theta = 18, 19, 20$ and (partially) 21 on the other side – so, 3 and 3.5 Bark are quite well balanced.

A comparison test reported in [Voran 1997] says, somewhat in contrast to other results, that mainly the missing lower frequencies should be added to enhance quality; even a shift of B_n by about 100 Hz downward would be helpful. A contradicting result reported in [Raake 2006] says that a band between 200 Hz and 7000 Hz would have a 0.4-point advantage, on a 5-point total quality scale, over B_w as defined in (2.4), so that the *higher* frequencies would be more important. Still, "averaging" all these observations confirms that the choice of B_w in (2.4) is a good one with strong improvements over narrowband transmission.

2.4.2 Wideband Speech Transmission and Processing

Talking about speech *processing* means, of course, that now the sequences $x(k)$ and $y(k)$ in Fig. 2.1 are considered, rather than the continuous acoustical waves $s_o(t)$ and $s_1(t)$. Covering B_w instead of B_n necessitates the step from (2.2) to (2.5), i.e., doubling the sampling rate. This is, however, not the only change: There are often special features to be taken care of, when wideband speech is to be handled. This is essential also because just at the edge of narrowband coverage, between 3 kHz and 5 kHz, the human ear is very sensitive. In parts, this is counteracted by the growth of the critical-band widths at higher frequencies, so that certain effects are less perceivable.

Wideband Speech Coding

An international standard for wideband speech transmission has been available since 1988, in an earlier version of [ITU G.722 1993]. It contains the above input-filter tolerance scheme as well as the description of a split-band ADPCM. The lower half of the spectrum, up to 4 kHz, is transmitted by means of a backward-adaptive differential coding scheme (similar to that of the DECT standard [ITU G.726 1990]), with 8 kHz sampling rate but with 6 instead of 4 bits/sample, while the upper half undergoes another ADPCM coding with 8 kHz sampling frequency but only 2 bits/sample. The increased resolution of the lower, stronger spectral components is needed in order not to spoil the quality gain due to the larger bandwidth by a now more audible quantization noise; as the upper frequency components are much smaller, a lower word-length suffices to cover their dynamic range. The band-splitting and re-synthesis are realized by quadrature-mirror filter (QMF) pairs in transmitter and receiver. The total bit rate

$$f_{B,WB-ADPCM} = 8 \cdot (6 + 2) \text{ kbit/s} = 64 \text{ kbit/s}$$

is identical to that of the narrowband ISDN-standard log-PCM system. It may, however, be reduced by dropping one or two bits of the lower-band signal; this is allowed since the adaptations inside the coder and decoder are based, sub-optimally, on the first four of the maximally six bits in any case. The quality will, of course, suffer in the lower-rate options, but it is still felt to be considerably better than that of the narrowband ISDN transmission.

Nevertheless, there used to be little enthusiasm for the offered "better sound at the same rate". The interest in wideband-speech transmission did not grow until three other developments appeared: the availability of interfaces beyond the classical hand-set, like head-sets or hands-free terminals, varying-rate coding schemes for mobile phones, and the virtually unlimited data volumes possible in the Internet.

For the mobile-telephony development after 1980, the data rates of the narrowband PCM, $f_{B,PCM} = 64$ kbit/s, and ADPCM, $f_{B,ADPCM} = 32$ kbit/s, were too high. Techniques with lower rates and with still similar quality were known from literature, but were unrealizable with available technology. On the other hand, the well-known simple LPC vocoder with its low rate of 4 kbit/s would have too poor a quality. So, a dedicated codec was developed in the ETSI standardization process for GSM, compressing the speech signal to 13 kbit/s and transmitting it, with a large amount of redundancy for error protection, at a gross rate of 22.8 kbit/s [ETSI GSM 06.10 1988]. Soon after the GSM start, it was found that either a lower rate at almost the same quality or a much better quality at 13 kbit/s would be realizable. The latter idea resulted in the so-called *enhanced full-rate* (EFR) coder [ETSI GSM 06.60 1996]; it is based on the principles of an algebraic code-excited linear-predictive (ACELP) system [ITU G.729 1996] developed for speech transmission at 8 kbit/s with a sound quality close to that of narrowband ADPCM. The same principles turned out also to be useful for higher as well as for even lower rates, down to about 4 kbit/s. Together with the realizability of such complex codings, now achievable due to the fast progress of electronics, the old, plausible idea of a varying data rate became feasible, taking care of the strongly fluctuating transmission quality of mobile-radio connections. The net rate for speech and the redundancy added for error protection could be adapted to the channel conditions in the *adaptive multi-rate* (AMR) coding scheme [ETSI GSM 06.90 1998] developed for GSM and UMTS. Since for very good channels high rates of up to more than 20 kbit/s also became available, it was only a logical step to use these high data flows for speech with a larger bandwidth, as included in the so-called *AMR wideband* (AMR-WB) system [ITU G.722.2 2002]. The band up to $f_c = 6.4\,\text{kHz}$ is transmitted via ACELP with different bit allocations for the codec parameters, and at the highest rate of 23.85 kbit/s, an additional information about the band between 6.4 kHz and 7 kHz is transmitted.

In *Voice (transmission) over (the) Internet Protocol* (VoIP), it is not because of fading effects as in mobile telephony, but due to heavily varying traffic load that the AMR ideas are also of interest. As in early VoIP implementations there were some drawbacks concerning link reliability and packet losses, it was attractive to offer a compensation by a high-quality speech sound. So, wideband coding was applied, and nowadays AMR-WB is frequently chosen. It also became a common choice in other applications, like storage of high-quality speech segments for speech-output systems.

Interestingly, the bands in this new scheme are not just those used in the split-band ADPCM, which seem to be quite natural a choice, with a lower-half and an upper-half band, up to and above 4 kHz, respectively. Instead, a coherent wider band with

$f_c = 6.4\,\mathrm{kHz}$ is covered, and only the small remainder is dealt with separately, as mentioned above. The reason can be found in several investigations concerning a *much simplified* handling of the higher frequencies.

Artificial Bandwidth Extension

In the years between 1985 and 1995, no wideband speech transmission was envisaged for the near future. On the other hand, there were obvious problems with narrowband telephone and natural wideband speech occurring together, namely, during broadcast telephone interviews or in radio or TV programs with "phone-in" speakers. There were attempts to overcome the quality switching between "full and natural" and "thin and tinny" by adding higher and lower frequencies *without* transmitting them [Croll 1972], [Patrick, Xydeas 1981], [Patrick 1983], [Patrick et al. 1983]. The lower frequencies were (re-)generated by a non-linearity arousing sub-harmonics of the higher f_o-multiples present in B_n during voiced segments. In order to really improve the perceived quality, however, more refined techniques were found to be needed, based on pitch-detection, i.e., f_o measurements. The availability (that is: transmission) of at least some additional information about the missing components would be helpful. In order to (re-) generate an upper band, noise modulation via parts of the narrowband speech spectrum and band-pass filtering were applied. Also here, some more refined methods were added, again taking care of the harmonic structure in voiced segments.

A later study [Carl 1994], [Carl, Heute 1994] applied narrowband (CELP-) coding techniques to estimate an extrapolated wideband spectral shape from that of the transmitted narrow band. During the narrowband LPC-code-book training phase, wideband LPC-envelope descriptions are also stored and then used to create a "shadow code-book". In the decoding, the full-band envelopes are then combined, within the respective bands, with the transmitted telephone-band excitation, spectral lines from a harmonic-modelling method [Almeida, Silva 1984] in the lowest band (low-frequency regeneration, LFR), and spectral-folding, i.e., aliasing components from a down- and up-sampling *high-frequency regeneration* (HFR) [Makhoul, Berouti 1979] or a refined variant thereof (see Fig. 2.8). While the exact spectral shape in the upper band proved to be less critical, the strength of the excitation was found to be crucial; again, some additional information would be helpful. A very similar approach was followed in [Epps, Holmes 1998].

In [Jax 2002], the problem was revisited, with a different basis. The history of succeeding speech segments was taken into account by means of a hidden Markov model, a statistical estimation was applied, and its aim and limitations were formulated via the information-theoretic concept of the trans-information between the measured narrowband signal description and the wideband counterpart to be concluded from it. A considerable quality improvement was reported [Jax, Vary 2002].

Some of the observations found in these studies lead, in [Paulus 1996], [Paulus, Schnitzler 1996], to the idea of wideband-speech coding in two bands, though with unequal bandwidths. The re-creation of correct spectral lines is not only critical in the lowest

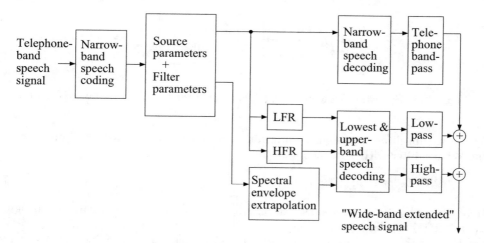

Figure 2.8: Bandwidth expansion within a narrowband-speech transmission system

frequency band below B_n, but also above 4 kHz. The spectral shape becomes less crucial above 6 kHz, where mainly a gain factor is needed, and only unvoiced sounds show components at all. These findings, confirmed by investigations in [Paulus 1996], also comply with the good results of an approach in [Dietrich 1984], where the full band B_w is transmitted and then augmented by artificial expansion up to 12 kHz.

Still, systems with a bandwidth extension from B_n to B_w are useful, e.g., in cases of mixed narrowband and wideband transmission systems, and their quality is of interest.

Wideband Speech Enhancement

Besides a band limitation, non-linear distortions, e.g., due to a codec, echoes due to room reflections in hands-free situations, and additive background noise have an impact on the perceived speech-signal quality. While modern, sophisticated coding systems leave little audible distortion and echoes can be cancelled with a small remainder, noise reduction is critical. On the one hand, it may also be capable of diminishing residual echoes or small distortions; on the other hand, it may itself cause signal deformations and / or artifacts, especially the well-known *musical noise*, a randomly varying multitude of narrowband, short-time spectral components. Good compromise solutions are, however, available and even applied in commercial products [Schmidt 2001].

Mostly, single-channel, i.e., single-microphone techniques have been developed for use in the telephone band B_n. For more details, the reader is referred to [Vary et al. 1998], [Heute 2006], [Vary, Martin 2006]; here, only the basics are outlined in order to then point out the special problems occurring in wideband-speech denoising.

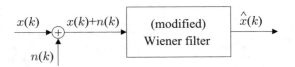

Figure 2.9: Speech enhancement by a Wiener filter modified by an exponent η

The optimum-filter approach, with a minimum mean-square error criterion, leads to the Wiener filter with a frequency response

$$H^{(WF)}(e^{j\Omega}) = \Phi_{xx}(e^{j\Omega})/[\Phi_{xx}(e^{j\Omega}) + \Phi_{nn}(e^{j\Omega})]\,. \tag{2.7}$$

Here, $\Phi_{xx}(e^{j\Omega})$ denotes the power-density spectrum (PDS) of the clean speech signal $x(k)$ (after band-limitation, sampling, and digitization, see Fig. 2.1), $\Phi_{nn}(e^{j\Omega})$ is the PDS of the digital version $n(k)$ of the added noise. Empirically, variants of the Wiener filter have been found useful, replacing $H^{(WF)}(e^{j\Omega})$ by a modified version $[H^{(WF)}(e^{j\Omega})]^{\eta}$ (see Fig. 2.9). In any case, the numerator in (2.7) is unknown, as the clean signal is not available; so, it has to be replaced by

$$\hat{\Phi}_{xx}(e^{j\Omega}) = [\Phi_{xx}(e^{j\Omega}) + \Phi_{nn}(e^{j\Omega})] - \hat{\Phi}_{nn}(e^{j\Omega})\,, \tag{2.8}$$

with an estimated noise PDS $\hat{\Phi}_{nn}(e^{j\Omega})$. The modified-Wiener-filter output PDS is thus

$$\Phi_{\hat{x}\hat{x}}(e^{j\Omega}) = [\Phi_{xx}(e^{j\Omega}) + \Phi_{nn}(e^{j\Omega})] \cdot G(e^{j\Omega})\,, \tag{2.9}$$

with

$$G(e^{j\Omega}) = \{1 - \hat{\Phi}_{nn}(e^{j\Omega})/[\Phi_{xx}(e^{j\Omega}) + \Phi_{nn}(e^{j\Omega})]\}^{(2\eta)}\,. \tag{2.10}$$

Alternatively, "spectral subtraction" is a standard approach, leaving the noisy-signal phase untouched and reducing the magnitude spectrum or the PDS according to the noise contribution. Applied to the power spectrum, it is described by the expression

$$\Phi_{\hat{x}\hat{x}}(e^{j\Omega}) = [\Phi_{xx}(e^{j\Omega}) + \Phi_{nn}(e^{j\Omega})] - \hat{\Phi}_{nn}(e^{j\Omega})\,,$$

where $\hat{\Phi}_{nn}(e^{j\Omega})$ is the same estimated noise PDS as in (2.8). Also this can be written as a spectral weighting. The resulting PDS is found to be

$$\Phi_{\hat{x}\hat{x}}(e^{j\Omega}) = [\Phi_{xx}(e^{j\Omega}) + \Phi_{nn}(e^{j\Omega})] \cdot G(e^{j\Omega}) \tag{2.11}$$

with

$$G(e^{j\Omega}) = 1 - \hat{\Phi}_{nn}(e^{j\Omega})/[\Phi_{xx}(e^{j\Omega}) + \Phi_{nn}(e^{j\Omega})]\,. \tag{2.12}$$

Obviously, with $\eta = 1/2$, Eqs. (2.9) and (2.10) are identical to (2.11) and (2.12): Wiener filtering and spectral subtraction are closely related. The same holds for all variants, possibly including further exponentiations within the above expressions, weighting smaller or larger terms differently [Heute 2006].

So, generally, noise reduction can be explained by a spectral weighting factor

$$G(e^{j\Omega}) = f(\hat{\Phi}_{nn}(e^{j\Omega}), [\Phi_{xx}(e^{j\Omega}) + \Phi_{nn}(e^{j\Omega})]), \qquad (2.13)$$

depending on the measured noisy-speech PDS and the estimated noise PDS. Of course, also "measurement" means estimation; the difference in the above wordings indicates, however, that the *noisy speech* signal is indeed available – in this case, "estimation" concerns only the necessary short-time computation via a periodogram or with some filter-bank; in contrast, the *noise* signal *alone* is *not available* during speech activity but only in speech pauses. Exploiting pauses requires a reliable, noise-robust pause detection and strong noise stationarity over times of speech activity. In particular the latter problem caused the development of proposals for noise-PDS updating *during* speech. They rely mainly on the idea that always in *some* narrow frequency bands (e.g., bins of a sufficiently long FFT) there will be no speech components so that the noise-PDS can be updated in these bands. The decision on local speech absence is based on the assumption that small enough components most probably belong to the disturbance. This technique of "minimum statistics" [Martin 1994], [Martin 2001] has become a standard, with many variants and amendments [Doblinger 1995], [Martin 2006]. Instead of minima tracking, the deviation from a sliding-average spectrum can be applied for a decision, assuming that the short-time speech spectrum varies much faster than that of (even instationary) noise [Arslan et al. 1995], [Hirsch, Ehrlicher 1995]. The latter approach was later refined and modified in [Gülzow 1999], [Gülzow 2001]. A slowly varying noise-PDS shape and an instantaneously varying gain factor were separated, in order to track faster instationarities of the disturbance.

As mentioned above, both required spectra can be found from various spectral analysis methods, be it a simple periodogram with an underlying FFT, a filter-bank system with equally or unequally spaced bands, realized by a wavelet-packet / tree-structure or a polyphase-DFT configuration, or even with adaptive bandwidths [Gülzow et al. 1998], [Gülzow et al. 2003].

The estimations become difficult in the upper frequency region of B_n. Here, the speech components are some 30 dB below the possible maximum (see Fig. 2.4). Either the noise components may become relatively large, in cases of wideband or nearly white disturbances (like wind noise), which makes their separation from speech difficult, or they may become very small in cases of low-pass type noise (like car noise), which asks for the computational differentiation between two small entities. Similarly, at very low frequencies, below the first-formant region, there are small speech components while especially low-pass noise adds strong disturbances. This is particularly bad for fricatives with their band-pass spectrum (see Fig. 2.4). All these problems are strongly aggravated when wideband speech with a bandwidth B_w is dealt with. Novel considerations and techniques are therefore needed. Figure 2.10 shows examples of vowel and fricative spectra with and without additive vehicle, factory, and white noise.

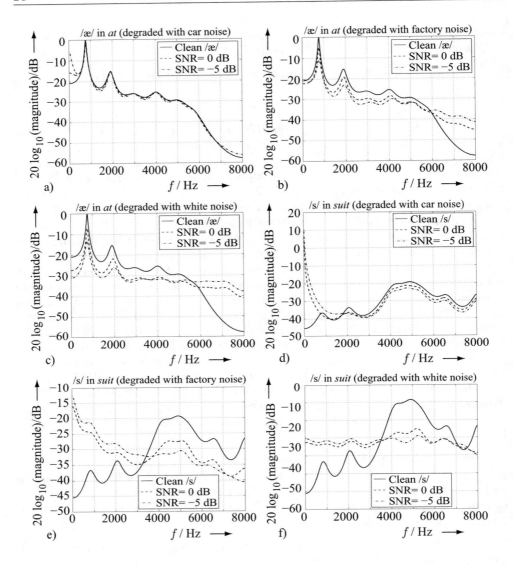

Figure 2.10: Short-time spectra of a vowel /æ/ and a fricative /s/ without disturbance and with additive car, factory, and white noise (SNR = 0 dB and
SNR = 5 dB)

All short-time spectral measurements with speech and / or noise signals yield results
with a certain variance, of course; the relative variation becomes especially large,
however, if the expected "true" values are as low as those of speech above 4 kHz.
The above-mentioned instantaneous gain adaptation within the noise-PDS estimation
[Gülzow 2001] relies on the absolutely lowest spectral values per analysis frame. If
this method is simply applied also to wideband noisy speech, the gain will randomly

fluctuate from frame to frame, causing an unpleasant, fluttering residual noise at the output. In [Janardhanan, Heute 2005b], this was observed and counteracted by finding the smallest values within a few predefined, separate frequency bands, and in [Janardhanan, Heute 2005a], it was shown that an adaptive choice of these sub-bands gives an even better quality.

Besides the fast-varying gain, a slowly adapting noise-PDS shape is needed in the above approach. Here, a (linear or non-linear) smoothing over the frequency axis helps to reduce the spectral variance. This is especially needed for wideband speech. Smoothing is, however, outperformed [Janardhanan, Heute 2006] by applying a spectral measurement technique relying on a DFT / FFT with not just one fixed window, but with several, orthogonal windows, namely, discrete prolate spheroidal sequences and a final averaging [Thomson 1982].

While, for narrowband noise reduction, a broad literature is available, including comparisons of different spectral analysis-synthesis techniques, noise estimations, and rules for the determination of the weighting factor $G(e^{j\Omega})$ in (2.13), similar thorough investigations for the wideband case are still on their way. As indicated above, and as seen in the above sections on coding and bandwidth extension too, particular effects and requirements beyond just a doubled sampling frequency are to be expected.

In this book, more on current enhancement work is to be found in Chap. 5, on coding with variable rate and possible upper-band transmission in Chap. 8, and on bandwidth extensions aspects in Chap. 9.

2.5 Speech-Quality Assessment

As mentioned in the Introduction, the quality of a speech sound can be validly described only by a human being, judging after auditory perception according to expectations and experiences. Yet, instrumental measures, mimicking users' evaluations, are of interest – and they should be possible since "everything that can be heard must also be measurable" [Berger 1998]. Both auditory and instrumental assessments will be briefly explained, as they have been developed and used over years for telephone-band speech, before, again, the wideband-speech case and its peculiarities are dealt with.

2.5.1 Auditory Quality Determination

Quality Tests

Test persons, or "subjects", are asked to "use" a telephone and judge the quality of the transmission system. This is done in a test laboratory, providing the various systems of interest (via hardware or software) and modeling the typical telephone situation more or less well:

- Conversation tests: Two users *talk and listen* to each other via real-time systems.

- Listening-only tests (LOT): Test subjects *listen* to speech signals processed / influenced by the systems (and possibly pre-stored in the laboratory); there are two distinct versions:

  1. LOT with a pair-comparison between two signal variants, namely,

     - between the clean original and the disturbed / processed / influenced signals, rating degradations,

     - between differently processed / influenced signals, rating quality differences.

  2. LOT with single processed / influenced signals rated for their absolute quality in comparison with an *internal reference* of the listeners, i.e., their experience from everyday telephone use.

Integral and Diagnostic Quality Ratings

In all cases, two types of results can be aimed at:

1. Integral quality grades describe an *overall* impression of an average user. A conversation may be good or just fair, a degradation inaudible or annoying, a system slightly better or much worse than another one, and a single system may be excellent or bad compared with an average experience. Also other integral statements may be collected, e.g., concerning the conversational or listening effort.

2. Diagnostic quality features describe *details* of the impressions separately. A conversation may suffer from strong delays while there is only little noise, and degradations or absolute ratings in LOTs may have been evoked by different components in our perception; the relevant features are termed *dimensions* or *attributes*.

The integral-quality impression results from a (generally: nonlinear) superposition of the attributes.

It is obvious that the conclusions to be drawn, by a system evaluator or developer, from the above tests differ largely. A conversation test models the telephone situation as closely as possible, and, combined with an attribute analysis, it would give most detailed information about problems and potentials of a system. An absolute LOT rating, however, will not even be able to include all disturbing effects; it may still suffice to rank several systems, but it will not give further insight into particular weaknesses or strengths. On the other hand, it is clear that the latter test requires much less time, especially from the test persons, than a thorough conversation test.

LOT with Absolute Category Rating (ACR) of the Integral Quality

For cost reasons, total-quality LOTs with a five-point ACR scale are most frequently used: A sufficient number of normal-hearing listeners grade sufficiently many phonemically balanced sentences spoken by several normal-speaking persons and sent through the systems of interest. An internationally standardized scale [ITU P.800 1996] is applied, as displayed in Table 2.1.

Table 2.1: Numerical and verbal descriptions for LOT absolute-category quality ratings

Numerical grade	5	4	3	2	1
Verbal grade	excellent	good	fair	poor	bad

The integer values given by the users are averaged, giving a real number between 1.0 and 5.0 as the *Mean-Opinion Score* (MOS). A standard ISDN / log-PCM telephone connection will reach MOS $\approx 4.1 \ldots 4.3$, if compared with other narrowband-transmission methods.

Attribute-Oriented LOT

Within the same test scenario as sketched above, listeners may also be asked for their impression about certain attributes, marked on another suitable scale. A long list of attributes was proposed by [Voiers 1977], discussed and used in [Quackenbusch et al. 1988], assuming a discrimination between dimensions concerning the speech signal and other components. Examples are "dull" signals or "hissing" background sounds. Other dimensions have been suggested and are a topic of current research (see Secs. 2.5.2, 2.6.2, and [Heute 2007]). The necessary experimental effort grows, of course, with the dimensionality – not necessarily only in a linear way.

2.5.2 Instrumental Quality Determination

Even an integral-quality ACR LOT may be too expensive and time-consuming, especially during a system-development phase, where many parameter variations should be checked. Therefore, *subject-free* measures have been suggested, often termed *objective* in contrast to *subject-ive*, but here referred to as *instrumental*. A computer algorithm acts as a measuring instrument, determines a numerical value from the signal samples, and then maps it to a prediction of a quality term. The prediction should of course be close to the real, i.e., auditory value; the usual figure of merit describing the predictor performance is the correlation coefficient ρ between both results for a large variety of systems and transmission conditions.

Instrumental Integral-Quality Measures

Usually, the simplest – and poorest – model of the telephone situation, namely, the ACR-LOT, is the subject of an instrumental replacement. An estimation $\widehat{MOS}$ is calculated for a MOS value that would be found from listening without any direct comparison. The result $\widehat{MOS}$ is based on a computational evaluation of both the clean and the distorted signal. This means that the "poor" model again is poorly modeled, namely, by a signal comparison.

Many measures have been proposed and investigated. For details, the reader is referred to the literature – an overview in [Heute 2007], an early thorough investigation in [Quackenbusch et al. 1988], in-depth treatments in, e.g., [Hauenstein 1997], [Berger 1998], [Mattila 2001], and derivations of the presently most common technique in [Beerends, Stemerdink 1994], [Rix, Hollier 2000], [Rix et al. 2001], [Rix et al. 2002], [Beerends et al. 2002].

The latter technique – as several similar ones – relies on the following steps:

- inclusion of telephone filtering ("hearing situation"),

- level and time alignment,

- equalization of certain linear distortions;

- comparison between input and output signals (i.e., *not* their difference as a "distortion signal"),

- comparison of signal *segments* (i.e., *short-time* evaluations),

- comparison after *auditory transformation*

    - from frequency in Hz to the inner-ear Bark-scale (see (2.6)),

    - from signal power to loudness

    - including masking effects and

    - an asymmetry in the perception of missing and additional components, and

- averaging of the segmental results at the end.

The international standard PESQ (Perceptual Evaluation of Speech Quality, see [ITU P.862 2001]), based on this scheme and sketched in Fig. 2.11, differs from the other mentioned proposals in terms of details within measurements and perceptual model. The performance of PESQ is expressed by a correlation $\rho = 0.935$ between estimated and real MOS values in a benchmarking experiment.

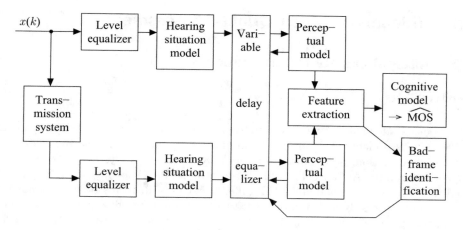

Figure 2.11: Block diagram of PESQ

Instrumental Diagnostic Quality Measures

Assuming the availability of both input and disturbed signals, one may also measure different parts of the disturbances separately. First, linear distortions can be addressed by a measurement of the average frequency response, despite additive and non-linear effects [Schüssler, Dong 1990]. Furthermore, additive-noise information may be extracted – e.g., by means of methods known from speech enhancement (see Sec. 2.4.2), additive artifacts like musical noise can be measured, or missing signal segments may be detected by means of techniques known from frame or packet-loss monitoring [Ludwig et al. 2003] and from error concealment (see Sec. 2.6.2 b). With such information, one may try to predict single auditory attributes, e.g., "dullness" in the case of suppressed higher frequencies or "hissing background" in the case of wide-band noise, depending on the dimensions defined before.

In a recent study, the above approach [Heute et al. 2005] has been investigated in depth. Via a multi-dimensional scaling (MDS) and a succeeding semantic-differential analysis, the attributes *continuity, noisiness,* and *frequency content and directness* were chosen as orthogonal features to be then determined instrumentally [Wältermann et al. 2006b], [Scholz et al. 2006a], [Kühnel 2007]; correlations of $\rho = 0.927 \ldots 0.936$ were found between auditorily and instrumentally predicted attributes; more details will be covered in Sec. 2.6.2.

From the instrumentally described attributes, a total-quality prediction shall be calculated by a suitable combination, so that, finally, both integral and diagnostic informations are available. An essential advantage of this detour to a quality estimation via single dimensions is the fact that the attributes are generic, i.e., independent of the set of systems used for their definition and tool development. A direct total-quality prediction will, in contrast, always be somewhat tuned to the training data used in the derivation of the estimator's parameters.

2.6 Wideband Speech-Quality Assessment

2.6.1 Integral Quality Determination

Auditory Tests

The "methods for subjective determination of transmission quality" defined in [ITU P.800 1996] are identical for narrowband and wideband speech; also the scales to be used in conversation, pair-comparison, or ACR listening-only tests are the same. So, for a total-quality ACR LOT, Table 2.1 is still valid. Of course, the signals to be used in a narrow- or wideband LOT differ – not only in terms of their bandwidths and sampling rates, but also with regard to the type of pre-filtering. For telephone-band speech, the use of the "send-side intermediate-reference system" (IRS$_\text{send}$) filter is recommended for the speech recording. The standard IRS$_\text{send}$ filter [ITU P.48 1993] models the anti-aliasing filter together with the send-side characteristics of the classical hand-sets, while a later, modified version [ITU P.830 1996] disregards the strong upper band limitation. The speech reproduction is assumed to pass the "receive-side intermediate reference system" (IRS$_\text{receive}$), for which again a standard and a modified version are available in the same recommendation. Often, the cascade of both filters is taken care of within the recorded data, where the modified version is preferred, and no further output-signal filtering occurs (see Fig. 2.12). For wideband tests, a "flat" frequency response is to be used for components within B_w.

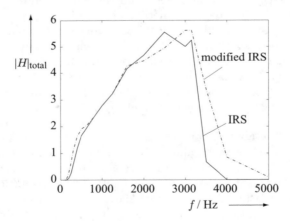

Figure 2.12: Linear frequency response of a cascade of IRS$_\text{send}$ and IRS$_\text{receive}$ filters

MOS Results for Wideband vs. Narrowband Speech

As mentioned in Sec. 2.5.1, a standard narrowband ISDN transmission would receive an average quality grade MOS $\approx 4.1 \ldots 4.3$ in an ACR LOT comprising only narrowband systems. These values indicate a "good" quality with some tendency towards "even better", according to Table 2.1. The remaining space between these scores and the top value $5.0 \doteq$ "excellent" may be explained, partly, by the audible quantization noise of this 8-bit log-PCM coding.

In [Raake 2006], this system is compared to a clean wideband-speech signal within B_w without any compression (although digitized by a linear 16-bit ADC, sampling at a rate $f_S = 16\,\text{kHz}$). The result is not self-evident. One might expect the log-PCM to stay at MOS = 4.14 as found in a preceding narrowband test, and the wideband signal to gain a MOS close to 5.0 then. Instead, the best result for wideband speech (with a band slightly different from B_w) is MOS = 4.16, while the telephone speech drops to MOS = 3.21, i.e., is rated as being only "somewhat better than fair" now.

The reasons are psychological ones. In the narrowband test, the classical handset was the basis, either by its direct use or by an IRS simulation. If, now, the perception of the higher quality of a wideband transmission is to be enabled, a better sound reproduction is needed. In order to still keep the impression of a telephone situation, a "high-fidelity phone" was devised in [Raake 2006], namely, a hand-set-like monaural version of a high-quality headphone. Still, this apparatus looks different and therefore evokes, in a user, the expectation for a sound like that known from a good audio system. This will limit the quality judgment on a signal which is *still band-limited* (though to B_w), and it will even more "de-grade" the quantized narrowband PCM output.

Furthermore, a general observation is known from many tests: Listeners hesitate to use the extreme categories "1.0" or "5.0" in Table 2.1 at all. So, a gap will always appear at both ends. What should be kept in mind, however, is the quality gain of about 1 MOS by doubling the bandwidth, which confirms the slightly more optimistic result in [Krebber 1995] mentioned in Sec. 2.4.1.

Extended Quality Scales

As seen in the preceding section, the inclusion of processed wideband-speech signals into quality comparisons causes that part of the MOS scale which is used for the included narrowband examples to shrink. So, the discrimination becomes less fine.

A natural solution seems to be an extension of the scale at its upper end. There are indeed scales applied in quality assessment with values up to ten or even an open upper end, in order to avoid the above-mentioned hesitation to give really high grades. But a straight-forward extension of the MOS scale in Table 2.1 is not suitable. The value 5.0 is associated with the term "excellent" – so, at least, the verbal descriptions would have to be redefined completely if grades higher than 5.0 should be allowed.

In [Raake 2006], an approach is suggested that stays with the standard 5-point scale while keeping the fine resolution on an intermediate scale. For this purpose, an extended *R-scale* is developed. The term R is the *transmission-rating factor* described in

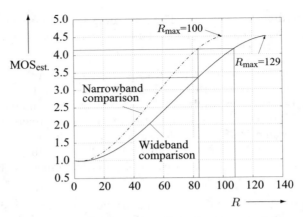

Figure 2.13: Mappings from R to MOS, using the standard scale with $R_{\max} = 100$ and an extended scale with $R_{\max} = 129$

[ITU G.107 2005] for narrowband telephony. A perfect narrowband speech transmission is rated by $R = 100$, a "very satisfying" one under certain default conditions (as to circuit noise, loudness, room noise, etc.) receives $R = 93.2$. From these desirable values, *impairment* terms are subtracted, one each for possible speech-simultaneous, delay, and (coding-) distortion effects. This implies that all impairments have an additive, i.e., linear, impact on the overall rating. It is assumed, then, that the inherent band limitation in telephone speech is also an impairment, so that an ideal wideband transmission would receive a value $R_{\max} > 100$. This new upper limit is found by an interpolation of R values known from a set of narrowband and mixed narrowband / wideband tests with their MOS results. In [ITU G.107 2005], their corresponding narrowband ratings are found, and from the ratios of these numbers, a linear transformation is derived. Alternatively, from a very large set of MOS results, a quadratic interpolation is calculated. From both, the end of the R scale is found to be either $R_{\max} = 112$ or $R_{\max} = 138$, with a compromise figure of $R_{\max} = 129$. Figure 2.13 shows, for the latter choice, the mapping of the standard and the extended R-scale to the estimated MOS values. For instance, a standard narrowband log-PCM transmission would, in a narrowband-system comparison, receive the well-known MOS ≈ 4.2, while in a comparison including wideband systems it would be graded by MOS ≈ 3.3 only. A value MOS ≈ 4.2 would now be achieved by a wideband system with $R > 100$, on the extended scale. The observations of Sec. 2.6.1 are reflected well in this model.

Instrumental Measures

The ITU-standard PESQ was sketched in Sec. 2.5.2 (see Fig. 2.11) as a perceptually based distance measure with an estimation $\widehat{\mathrm{MOS}}$ for MOS values of an ACR-LOT. For use with wideband-speech signals, an extension has also been standardized [ITU P.862.2 2005]. *Wideband PESQ* (WB-PESQ) differs from PESQ only in two details.

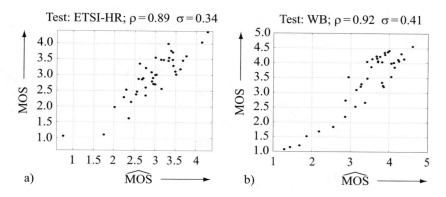

Figure 2.14: Estimated $\widehat{\text{MOS}}$ values vs. auditively found MOS
a) PESQ results from an ETSI test
b) WB-PESQ results from an ITU test

First, the signals are assumed to cover the full band B_w as in (2.4); no IRS filter is employed. Secondly, the mapping from the raw WB-PESQ distance-measurement result to the MOS estimate is newly defined, such that the 5-point scale of Table 2.1 is maintained.

In [Takahashi et al. 2005] a study is reported, showing that WB-PESQ estimates auditory MOS values with a high correlation of $\rho \approx 0.94$ for a large number of transmissions with varied bandwidths, coding distortions, and packet-loss probabilities in some VoIP scenarios. The figures in this report, however, show also that some problems known from PESQ as well as other, similar methods also appear in WB-PESQ – quite naturally, considering that the modifications are minor. The critical effects are the following: Two different systems transmitting speech under varying conditions (as to background noise, bandwidth, frame losses, etc.) may be graded subjectively with (almost) the same MOS, while the estimates cover a range $\Delta\widehat{\text{MOS}} \approx 1.0$. Vice-versa, in other cases, the true MOS may vary over $\Delta\text{MOS} \approx 1.0$, while the estimates are (almost) identical. This "ranking problem" was illustrated in [Heute et al. 2005], and can also be seen in Fig. 2.14, both for a narrowband and a wideband measurement.

Furthermore, systematic tilts or shifts can be observed. Certain systems follow, for varied conditions, the same tendency but with a constant or growing distance; this is indicated in a stylized manner in Fig. 2.15, derived from [Takahashi et al. 2005]. Similar observations are reported in [Côté et al. 2006], where also a different estimation performance was found for male and female voices. From detailed analyses, it was suggested to modify parts of WB-PESQ: The "flat" input filter is, in fact, still a highpass, attenuating frequencies below 200 Hz, but passing all (also: distortion) components between 7 kHz and 8 kHz. While the missing lowest frequencies are rather seen as an advantage, the upper band edge needs a suppression, as recommended for wideband terminals in [ITU P.341 2005] and as assumed in the definition of B_w in (2.4). In addition, the compensation of linear distortions and the emphasis of asym-

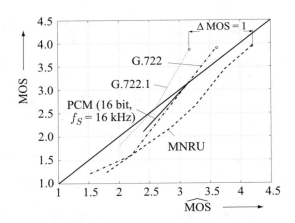

Figure 2.15: Stylized demonstration of shifts and tilts in WB-PESQ results

metric error components in the perceptual model inside WB-PESQ are modified, for an experiment, which does show an amelioration of certain problematic effects but not a removal.

So, enhanced versions of the available total-quality estimations are desirable. Although (WB-)PESQ and similar techniques are built on a psychoacoustic basis, it is necessary to include more knowledge especially on the perception of the upper frequencies. Hints to possible approaches may be found in the special care for peculiarities needed in wideband-speech coding, processing, and enhancement (see Sec. 2.4.2).

Beyond, the general problem of all direct integral-quality estimations remains: There is no diagnostic ability. So, also for the wideband case, the attribute-oriented approach needs to be revisited.

2.6.2 Attribute-Oriented Quality Determination

Auditory Dimension Analysis

As mentioned in Sec. 2.5.1, lists of numerous plausible quality attributes exist [Voiers 1977], [Quackenbusch et al. 1988]. A smaller set is, however, desirable for different reasons. The use of 10 or more dimensions makes overlaps and redundancy unavoidable, whereas few orthogonal attributes would give a clearer and more compact information. Also, the analysis and interpretation effort grows with the size of the set, and especially the necessary number N of system and condition examples ("stimuli") increases. There are two common ways to find a reduced dimensionality, as applied in the project mentioned in Sec. 2.5.2 [Wältermann et al. 2006b].

a) *Multi-Dimensional Scaling* (MDS): A variety of N stimuli are prepared. All $N*(N-1)$ possible stimuli pairs (both A–B and B–A) are presented to a listener group.

Table 2.2: Stress S and covered variance R^2 for narrowband and wideband MDS-based dimensionality reduction

Bandwidth	n	S	R^2
Narrowband B_n	3	0.232	0.74
Narrowband B_n	4	0.195	0.79
Wideband B_w	3	0.240	0.66
Wideband B_w	4	0.190	0.75

The test persons do not grade the speech quality but, rather, rate the (dis-)similarity of the two sounds, on a continuous scale between "very similar" and "not similar at all". After the test and a subject-related normalization, $N * (N - 1)$ distance values result. From these, all stimuli can be represented ideally in an $(N - 1)$-dimensional space (assuming identical distance measurements for A–B and B–A). As the number of test signals is chosen to be as large as possible, while the perceptual effects searched for are hoped to be few, a map with a lower dimensionality $n < (N-1)$ is constructed, with unavoidable inaccuracies. These are expressed by two common figures of merit. The "squeezing" of N points onto n dimensions causes a so-called *stress* term S, which should be made small; dropping dimensions has the consequence that not all variabilities can be covered in the n-dimensional space – the *covered variance* R^2 should be made close to 1.0 [Kruskal, Wish 1978]. The necessary space is found stepwise, observing the decrease of S or / and the growth of R^2 when incrementing n. The MDS within the above-named project was described in [Wältermann et al. 2006b] for the narrowband case, in [Wältermann et al. 2006a] for wideband tests. Dimensionalities of $n = 3 \ldots 4$ appeared to be appropriate (see Table 2.2).

The dimensions are, however, abstract in the sense that they can not yet be interpreted as "named attributes". Such names can now be found with the help of (system) experts knowing how some processing creates a certain effect on the signal sound. In a succeeding attribute-oriented LOT (see Sec. 2.5.1), the chosen dimensionality and attributes have to be verified. Such a test can, however, also be directly applied for the definition of the reduced perceptual space, as outlined in the next paragraph.

b) *Semantic Differential* (SD): Listeners are asked to grade the single stimuli on scales of a highly redundant, large set of predefined descriptors. The list may also be found in a pre-experiment, asking listeners for their own intuitive descriptions. In [Wältermann et al. 2006b] and [Wältermann et al. 2006a], as many as 217 and 135 candidate names were found, respectively. They were reduced by inspection to 13 and 28 antonym pairs, respectively, with which the actual LOTs were carried out. Finally, a principal-component analysis was performed with these reduced groups, leading to n final attributes.

In the first, narrowband, case, $n = 3$ still nameless factors $F_{1,2,3}$ carry a variance $R^2 = 0.935$. In the second, wideband, case, $n = 4$ is needed to cover $R^2 = 0.933$; here, the fourth axis still adds $\Delta R^2 = 0.172$.

Table 2.3: Attributes for narrowband and wideband quality analysis from [Wälter-mann et al. 2006a], [Wältermann et al. 2006b]

Band-width	n	F_1	ΔR_1^2	F_2	ΔR_2^2	F_3	ΔR_3^2	F_4	ΔR_4^2
Narrow-band B_n	3	directness/ frequency content	0.427	conti-nuity	0.342	noisi-ness	0.166	–	–
Wide-band B_w	4	continuity	0.331	dis-tance	0.218	lisping	0.212	noisi-ness	0.172

Names are then found by rotating the n axes such that high correlations with n of the pre-defined antonyms appear, and / or by observing the expert interpretations of the reduced MDS-space results. The names are – naturally – not unique, due to, e.g., the redundancy in the predetermined set.

After the above study, the following choices were made (see Table 2.3). As the first of three narrowband dimensions was linked, by the listeners, with frequency-content descriptions both in terms of pairs like "dark / bright" and "distant / close", the ambiguous term *directness and frequency content* was selected. The second factor – as well as the first one for wideband speech – was clearly related to short-time effects in the signals, appearing either as interruptions or as instantaneous sound insertions; it was termed *continuity*. The third attribute in the narrowband, the fourth one in the wideband case, were found to describe hissing distortions and noisy components; so the term *noisiness* is appropriate.

In the wideband-signal results, *frequency content and directness* seems to have van-ished. It is, however, actually split up into two well related factors: *Distance* (vs. nearness) summarizes the perception of a sound that had to travel some way from the speaker's mouth to the microphone, as happens in handsfree telephony. In such applications, multi-path, i.e., echo and reverberation, effects appear, changing the spectral shape by comb-filter-type frequency responses; so, "(in-) directness" is al-most synonymous. *Lisping* may be explained as an incorrect reproduction of the higher frequencies, causing, especially in the fricative /s/, a virtual unpleasant ar-ticulation shift. A relation to "frequency content" can therefore be seen. The am-biguous double descriptor of the narrowband case is thus split into two separate terms.

A further analysis will reveal whether such sub-dimensions are helpful, possibly also for other attributes. Frequency content has to do, on the one hand, with the linearly transmitted input spectrum, thereby, with the system's average frequency response and especially the band-limitations. On the other hand, it also describes further components, which may be artificially added on purpose or by an incorrectness. Con-tinuity was said to be affected by real interruptions as well as by short insertions – at least two sub-attributes may be appropriate. Noisiness may hint to a hissing speech reproduction, e.g., due to an overemphasis of higher frequencies in fricatives, as well as to added noise, which may itself have quite different characteristics in the wideband case (see Sec. 2.4.2).

Instrumental Attribute Measurement

Following the ideas outlined in Sec. 2.5.2 for narrowband assessment, measurements are searched for which now characterize the systems' influences on the attributes distance (or directness), continuity (or short-time artifacts), lisping (or frequency content), and noisiness (or hissing sound and added noise). The requirements are now defined more specifically.

The measurements should yield *few* parameters. These variables should be chosen such that they are *orthogonal* and, beyond, can be directly used to parameterize a *model system.* This idealized distorting system should then influence just one single dimension such that the measured quantities as well as the attribute-LOT result can be reproduced as closely as possible.

a) *Directness and Frequency Content* (DFC): For telephone-band speech, it was shown in [Scholz et al. 2006b] that width and position of the averaged passband of the system alone are able to predict DFC-LOT results quite well by a simple linear formula:

$$\widehat{\mathrm{DFC}} = -20.5865 + 0.2466\ \mathrm{ERB/Bark} + 1.873\ \Theta_G/\ \mathrm{Bark}\,. \tag{2.14}$$

The passband is described by the equivalent rectangular bandwidth (ERB) and the position by the center of gravity of the frequency response, after its limitation to a range (Θ_1, Θ_2) above relevant thresholds and its transformation to the Bark scale using (2.6). Denoting this modified response by $G(\Theta)$, we have

$$\mathrm{ERB} = \frac{\int\limits_{\Theta_1}^{\Theta_2} G(\Theta) \cdot \mathrm{d}\Theta}{\max\limits_{\Theta \in [\Theta_1, \Theta_2]} \{G(\Theta)\}} \tag{2.15}$$

$$\Theta_G = \frac{\int\limits_{\Theta_1}^{\Theta_2} G(\Theta) \cdot \Theta \cdot \mathrm{d}\Theta}{\int\limits_{\Theta_1}^{\Theta_2} G(\Theta) \cdot \mathrm{d}\Theta}\,. \tag{2.16}$$

The correlation between $\widehat{\mathrm{DFC}}$ and DFC values is $\rho \approx 0.9635$. Although this seems to say that the estimation (2.14) suffices, it is admitted in [Scholz et al. 2006b] that, in general, further parameters have to be included – the high correlation being due to too small a number of systems in the test.

In a wideband scenario, as addressed in Sec. 2.6.2, these additional features are definitely needed: ERB and Θ_G may be able to model a "correct frequency content" (vs.

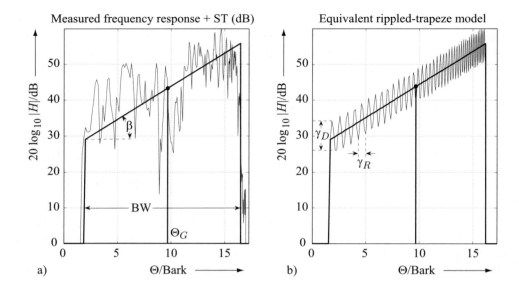

Figure 2.16: a) Measured and b) idealized frequency response representations of a transmission system with bandwidth B_W (to be expressed by ERB), position Θ_G, slope β, and a ripple with rate γ_R and depth γ_D

lisping); still, an average-slope parameter β of the frequency response inside the pass-band, i.e., a pre- or de-emphasis behavior, should be explicitly introduced. Directness needs quantities reflecting distance effects as mentioned in the preceding subsection. A comb-filter impact is known to create a "rippled" frequency response; so, in [Scholz et al. 2006b], [Scholz et al. 2006a], a *ripple rate* γ_R and a *ripple depth* γ_D are proposed to take care of this.

Figure 2.16 displays a measured frequency response and its idealized representation by the above parameters. The latter can be used immediately for the realization of an "idealized system" creating distortions which are equivalent to those of the original system [Huo et al. 2006].

b) *Continuity*: Both for narrowband and wideband speech, discontinuities have a strong impact on the perceived quality. This comes as no surprise. Front clipping (e.g., by a misadjusted voice-activity driven switch) or, even more, gaps inside the signal flow (due to lost data blocks, in a frame or packet-based transmission) create the feeling of an unreliable connection. They may even lead to intelligibility problems and, especially then, to strong dissatisfaction. A detection and characterization is needed.

For such drop-outs of longer signal sequences, techniques can be used which were developed for online monitoring of services by so-called *in-service non-intrusive measurement devices* (INMD) [ITU P.561 1996], [ITU P.562 2000].

Lost frames in a block-coding system, with time durations of $T_F \approx 10 \ldots 20 \ldots$ msec, may be concealed – in the simplest case by a repetition of the last completely received frame. This leads to a high correlation between samples at frame distance, which can be exploited for detection. Erroneous alarms can be avoided by a check of the fundamental frequency f_0, which may, incidentally, lead to similar correlations if $f_0 = \lambda * 1/T_F$, $\lambda \in \mathbb{Z}$ [Ludwig 2003], [Ludwig et al. 2003]. Similarly, simple packet-loss concealments may be detected.

Unconcealed losses of data blocks, with simply a longer zero sequence inserted into the signal, can be found from an observation of an *energy gradient* $\Delta E(i)$ [Ludwig 2003]. It is defined as the change of energy between succeeding block numbers $(i-1)$ and i:

$$\triangle E(i) \doteq \frac{E(i)}{E(i-1)} - 1 \,. \tag{2.17}$$

The block energy is calculated from the squared signal samples after appropriate band limitation to B_n or B_w, or by summation / integration over squared spectral values within B_n or B_w. Occurrence of $E(i) = 0$ indicates a lost block. Then, $\Delta E(i) = -1.0$ appears. If the following block is transmitted again, $E(i-1) = 0$ and $E(i) > 0$ lead to $\Delta E(i) \to \infty$; a limitation to $E_{\max} = +1$ is useful. Then, a sequence $\Delta E(i) = -1$ and $\Delta E(i+1) = +1$ indicates a single block loss. If more losses happen in a sequel, $E(i) = E(i+1) = 0$ are found, $\Delta E(i)$ is set to zero. The first value $\Delta E(i) = +1$ thereafter will indicate the recovered transmission. Figure 2.17 shows an example of a signal with several single frame losses visible both in the corresponding spectrogram and in the sequence of energy gradients. The detection potential was evaluated in [Ludwig 2003], and a reliability of 95 % correctly identified losses was stated. A further refinement was proposed which is also able to deal with a combination of slowly muted block repetitions and zero sequences.

More carefully concealed losses are less easily identified [Ludwig 2003]. However, the better the concealment, the less important is its detection. The quality impact will be correspondingly smaller.

From the detection results, a parametric description can be derived. It should consist of a loss-probability estimate, first, but also an indication of a (simple) concealment if applied, and of a statement on the type of losses. They may occur as independent, single random events as well as in longer bursts of drop-outs, with a certain probability of burst lengths. Beyond, these descriptors may be constant or time-varying during a connection. Once suitable parameters are found, an idealized model for this type of disturbance can be realized relatively simply, dropping or possibly replacing blocks with appropriate probability. All these approaches are applicable both to narrowband and to wideband speech.

Other short-time effects include pulse-like disturbances, e.g., those due to bit errors, which are instantaneous and therefore occupy the whole frequency band, or blocks of narrowband distortions, like the musical noise created by simple noise-reduction techniques. Figure 2.18 depicts a spectrogram of a corresponding case, with small, rectangular spots indicating the insertion of artificial narrowband, almost tonal sounds.

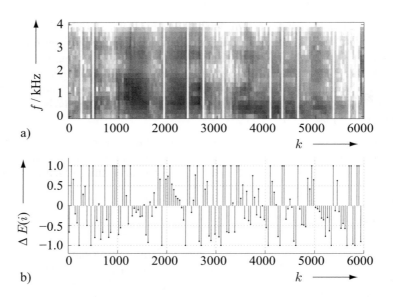

Figure 2.17: Spectrogram and corresponding energy gradient sequence for a signal with frame losses without concealment

They may be found obviously by a technical observation of the spectrogram. The so-called "relative approach" of [Genuit 1996] is a good candidate. It detects "unexpected" short-time components by a comparison with an averaged Bark-scale and loudness-transformed spectrogram, the latter describing the "expected" behavior. This method was also applied to packet-loss concealment [Kettler et al. 2003]; it could therefore be devised for a combined lost-data and tonal-noise detection.

A proposal due to [Goh et al. 1998] for post-processing of spectral-subtraction results can also be exploited for the characterization of added short-time spectral contributions. It monitors the variation over a set of straight lines with different angles in a time-frequency plane.

Both approaches are worth an investigation for narrowband and wideband signals. In the latter case, however, additional problems are to be expected due to the small, often noise-like speech components at higher frequencies (see Sec. 2.4.2).

Even if it is assumed that all discontinuous effects can be reliably found and described, further evaluation remains difficult. An estimator $\hat{C}$ for a true continuity C may be derived – once C and its relation to the perceived quality are clear. Hints may be found from the thorough investigations of relations between quality and loss statistics in [Raake 2006], especially on the use of impairment terms as intermediate quantities.

c) *Noisiness*: From inspection of the examples which are included in the MDS and SD analysis, it becomes clear that systems with considerable noisiness values on the negative part of the related dimension scale do not simply suffer from additive white

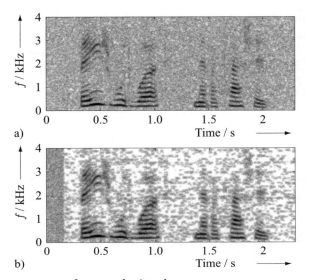

Figure 2.18: Spectrogram of a speech signal
    a) with additive white noise and
    b) after application of a simple de-noising technique showing musical-noise artifacts

noise. Colored background noise, broadband decoding and output-terminal noise, and signal-correlated noise appear as well as a noise-like roughness or hoarseness, which can be identified by a careful signal analysis as very short interruptions (below 5 ms) – so short that continuity is not concerned. A systematic investigation for narrowband signals is reported in [Kühnel 2007]. In this case, it is known that background noise must have passed both the send and receive handset filters (see Fig. 2.12). The system noise is filtered much less heavily by the receiver only, whereby the signal-correlated part is limited to $f_c = 4\,\text{kHz}$.

A possible approach is then to determine background noise N_{BG} inside B_n during speech pauses, evaluate the band between $3\,\text{kHz}$ and $4\,\text{kHz}$ in speech periods for correlated noise N_{corr}, check for higher-frequency system noise N_{HF} above $4\,\text{kHz}$, and find hoarseness by a suitably modified energy-gradient observation (see (2.17)) of the *short-interruption* rate SI. A measurement of higher-frequency parts requires, of course, that the signal between the actual transmission / processing and the acoustical output, i.e., $y(k)$ before becoming $y_o(t)$ in Fig. 2.1, is sampled with a rate $f_S > 8\,\text{kHz}$. From the measured powers within the above bands plus the term SI, a linear superposition formula can be devised [Kühnel 2007]. It is found to be well correlated with auditory noisiness ratings, according to $\rho \approx 0.8$, but with a quite strong speaker-gender dependency. An equivalent formula with modified parameters is claimed to yield $\rho \approx 0.93$; however, this is still a topic for deeper investigations. The corresponding idealized distortion system is defined easily, on the other hand, inserting correspondingly filtered noise types and short gaps.

Table 2.4: Correlations between predicted noisiness and auditory attributes

Dimension	Female speakers	Male speakers
Noisiness	0.8839	0.7215
Lisping / Frequency Content	0.2075	0.4075
Continuity	0.4715	0.1531
Distance / Directness	0.2772	0.7034

Assuming that the approach, in general, is reasonable, the extension to wideband speech is straight-forward: Interruptions are found in the same way, background noise covers 50 Hz up to 7 kHz, signal-correlated noise can be separately seen between 7 kHz and 8 kHz, and the terminal signal has to be sampled with $f_S > 16$ kHz and analyzed above 8 kHz to find high-frequency noise; the idealized system follows the above descriptions with these modifications.

Only preliminary results are available. They indicate that other descriptors as well as varied total-noisiness models have to be investigated, especially in the wideband case. The sub-dimension idea, however, appears to be very appropriate anyhow again. Table 2.4 shows that even for wideband signals the correlations between auditory and instrumental results are not too poor for the present formula with $\rho \approx 0.72 \ldots 0.88$, for male and female speakers, respectively, but the orthogonality with other dimensions is partly violated.

Total-Quality Calculation

As said in the statement about total and diagnostic quality in Sec. 2.5.1 and in the final paragraph of Sec. 2.5.2, an integral-quality prediction may be constructed from the single instrumental attribute predictions. If sufficiently verified estimations $\hat{C}$ for the continuity C, $\widehat{FC}$ for the frequency content FC (covering lisping in Table 2.3), $\hat{D}$ for the distance (or directness) D, and $\hat{N}$ for the noisiness N are given, it is an obvious idea to step back to the MDS and SD analysis in the beginning of this section: The auditively identified four dimensions $\dim_1 = C$, $\dim_2 = FC$, $\dim_3 = D$, and $\dim_4 = N$ are able to describe the difference between the systems and conditions and, thereby, their perceived qualities. In the simplest approach, a linear combination is used [Wältermann et al. 2006a]:

$$\text{MOS} = \text{const.} + \sum_{i=1}^{4} b_i \cdot \dim_i . \tag{2.18}$$

For $\widehat{\text{MOS}}$, all dimensions in (2.18) have to be replaced by their estimates $\hat{C}, \widehat{FC}, \hat{D}$, and $\hat{N}$. Numerical values for b_i are also given in the above report, but they are certainly preliminary, at least because of the very limited number of stimuli included.

2.6.3 Combined Direct and Attribute-Based Total Quality Determination

For years, the direct integral-quality estimation on the basis of a Bark-spectral loudness distance by PESQ and similar systems, on the one hand, and, on the other hand, the search for attributes, their instrumental estimation and their combination into an integral-quality prediction have been dealt with separately. Since 2005, a convergence can be observed. Within ITU-T, the enhancement of total-quality estimators by dimension analysis is investigated.

In the present overall measures, *some* knowledge about specific effects from certain systems was already exploited to broaden the validity. For example, the "Telecommunications Objective Speech-Quality Assessment" (TOSQA) program of [Berger 1998], [ITU COM 12-34-E 1997], [ITU COM 12-19-E 2000] contains in its later version TOSQA2001, beyond a wideband-speech mode, separate modules detecting and evaluating noise, interruptions, or non-linearities. Now, decomposed and hopefully orthogonal perceptual effects are to be addressed more generally in a new study. This should lead to improved predictions $\widehat{\text{MOS}}$, but, more important, also deliver a quality diagnosis [Berger 2005].

On the other side, in the present overall-algorithms, some quantities are measured that are used to emphasize or diminish certain perceptually more or less important features; the ideas behind these algorithms may be used in a dimension-based analysis to develop alternative attribute descriptors.

2.7 Concluding Remarks

For telephone-band speech quality, PESQ is claimed to be a world standard [Beerends, van Vugt 2004] for instrumental quality measurements, predicting the MOS results of well-defined ACR-LOTs reliably under numerous conditions. Even in the same context, it is admitted, however, that amendments would be helpful, reducing the sensitivity towards unknown types of distorting systems, but also adding diagnostic abilities by a "degradation decomposition". The way via decomposed attributes first and a derivation of an overall quality estimation thereafter is also accepted, but no standard is currently available.

For wideband speech, the auditory analysis is not really different, in contrast to wideband-speech processing which needs additional care. Both total and diagnostic instrumental approaches, however, pose more and new questions, to which only partial answers are known. The recent investigations of dimensionality as such and of "named" attributes or sub-dimensions require a persistent continuation, with a large variety of stimuli included. Modeling both attribute-specific disturbance generation and total-quality perception needs more integration of psychoacoustical and system-theoretical knowledge. A cross-exploitation of internal features from both approaches is promising.

For wideband as well as for narrowband speech, other tasks have also to do with quality assessments. The continuous QoS monitoring of signal features by an INMD was already mentioned [ITU P.561 1996], [Ludwig 2003]. This is beyond the scope of this chapter, as well as quality estimation from tabulated, expected impairments in a compound-network planning phase [ITU G.107 2005], or quality monitoring by means of in-service transmission-channel observations. The latter field is, however, dealt with in Chap. 3.

The thoughts and results reported in this chapter reflect, mainly, the author's personal view, and the work of the speech-processing group in Kiel is in the focus. Other work has, naturally, been observed as closely as possible, but necessarily not completely. An apology is appropriate for this limitation. On the other hand, much of the contents comes from close and fruitful cooperation with friends and colleagues elsewhere – too many to list all of them here. As a representative of many, the team of S. Möller at TUB / T-Labs Berlin is named. The author wishes to express his sincere gratitude to him and his co-workers as well as, especially, to the LNS members in Kiel.

Bibliography

Almeida, L. B.; Silva, F. M. (1984). Variable- Frequency Synthesis: An Improved Harmonic-Coding Scheme, *Proceedings of the IEEE International Conference on Acoustics, Speech, and Signal Processing (ICASSP)*, pp. 27.5.1–4.

Arslan, L.; McCree, A.; Viswanathan, V. (1995). New Methods for Adaptive noise Suppression, *Proceedings of the IEEE International Conference on Acoustics, Speech, and Signal Processing (ICASSP)*, Detroit, USA, pp. 812–815.

Beerends, J. G.; Hekstra, A. P.; Rix, A. W.; Hollier, M. P. (2002). Perceptual Evaluation of Speech Quality (PESQ), the New ITU Standard for End-to-End Speech-Quality Assessment, Part II - Psychoacoustic Model, *Journal of the Audio Engineering Society*, vol. 50, pp. 765–778.

Beerends, J. G.; Stemerdink, J. A. (1994). A Perceptual Speech-Quality Measure Based on a Psychoacoustic Sound Representation, *Journal of the Audio Engineering Society*, vol. 42, pp. 115–123.

Beerends, J. G.; van Vugt, J. M. (2004). *Speech Quality-Degradation Decomposition*, 4th IEEE Benelux Signal Processing Symposium, Hilvarenbeek, The Netherlands.

Berger, J. (1998). *Instrumental Methods for Speech-Quality Estimation*, PhD thesis. Arbeiten über Digitale Signalverarbeitung, vol. 13, U. Heute (ed.), University Kiel (in German).

Berger, J. (2005). *Requirements of a New Model for Objective Speech-Quality Assessment P.OLQA.*, ITU-T, Geneva, Switzerland.

Blomberg, M.; Elenius, K. (1970). *Statistical Analysis of Speech Signals*, Quarterly Progress and Status Report, Deptartment for Speech, Music, and Hearing, vol. 11, no. 4, KTH Stockholm, Sweden.

Carl, H. (1994). *Investigation of Different Speech-Coding Methods and an Application to Bandwidth Extension of Narrowband Speech Signals.*, PhD thesis. Arbeiten über Digitale Signalverarbeitung, vol. 4, U. Heute (ed.), Ruhr-Universität Bochum (in German).

Carl, H.; Heute, U. (1994). Bandwidth Enhancement of Narrowband Speech Signals, *Proceedings of the European Signal Processing Conference (EUSIPCO)*, Edinburgh, U.K., pp. 1178–1181.

Côté, N.; Gautier-Turbin, V.; Raake, A.; Möller, S. (2006). *Analysis of a Quality Prediction Model for Wideband-Speech Quality, the WB-PESQ.*, 2nd ISCA-DEGA Workshop on Perceptual Quality of Systems, Berlin, Germany.

Croll, M. (1972). *Sound-Quality Improvement of Broadcast Telephone Calls*, BBC Research Report RD 1972 /26, British Broadcasting Corporation, U.K.

Dietrich, M. (1984). Performance and Implementation of a Robust ADPCM Algorithm for Wideband Speech Coding with 64 kbit/s, *International Zuerich Seminar on Digital Communications*, Zürich, Switzerland.

Doblinger, G. (1995). Computationally Efficient Speech Enhancement by Spectral Minima Tracking, *Proceedings of the European Conference on Speech Communication and Technology (EUROSPEECH)*, Madrid, Spain, pp. 1513–1516.

Dunn, H. K.; White, S. D. (1940). Statistical Measurements on Conversational Speech, *Journal of the Acoustical Society of America*, vol. 11, pp. 278–288.

Epps, J.; Holmes, W. H. (1998). Speech Enhancement Using STC-Based Bandwidth Extension., *Proceedings of the International Conference on Spoken Language Processing (ICSLP)*, paper no. 0711.

ETSI GSM 06.10 (1988). *GSM Full-Rate Transcoding.*, Recommendation, Sophia-Antipolis, France.

ETSI GSM 06.60 (1996). *Digital Cellular Telecommunications System: Enhanced Full-Rate (EFR) Speech Transcoding.*, Recommendation, Sophia-Antipolis, France.

ETSI GSM 06.90 (1998). *Digital Cellular Telecommunications System: Adaptive Multi-Rate (AMR) Speech Coding.*, Recommendation, Sophia-Antipolis, France.

Fletcher, H.; Galt, H. (1950). The Perception of Speech and its Relation to Telephony, *Journal of the Acoustical Society of America*, vol. 22, pp. 89–151.

French, N. R.; Steinberg, J. C. (1947). Factors Governing the Intelligibility of Speech Sounds, *Journal of the Acoustical Society of America*, vol. 19, pp. 90–119.

Genuit, K. (1996). Objective Evaluation of Acoustic Quality Based on a Relative Approach, *Internoise*, Liverpool, UK, pp. 3233–3238.

Goh, Z.; Tan, K. C.; Tan, B. T. G. (1998). Postprocessing Method for Suppressing Musical Noise Generated by Spectral Subtraction, *IEEE Transactions on Speech and Audio Processing*, vol. 6, pp. 287–292.

Gülzow, T. (1999). Spectral-Subtraction-Based Speech Enhancement Using a New Estimation Technique for Non-Stationary Noise, *Proceedings of the International Workshop on Acoustic Echo and Noise Control (IWAENC)*, Pocono Manor, USA, pp. 76–79.

Gülzow, T. (2001). *Quality Enhancement for Heavily Disturbed Speech Signals – Detection of a Carrier Mismatch and Suppression of Additive Noise*, PhD thesis. Arbeiten über Digitale Signalverarbeitung, no. 20, U. Heute (ed.), Shaker, vol. 20, U. Heute (ed.), University Kiel (in German).

Gülzow, T.; Engelsberg, A.; Heute, U. (1998). Comparison of a Discrete Wavelet Transformation and a Non-Uniform Poly-Phase Filter-Bank Applied to Spectral-Subtraction Speech Enhancement, *Signal Processing*, vol. 64, pp. 5–19.

Gülzow, T.; Ludwig, T.; Heute, U. (2003). Spectral-Subtraction Speech Enhancement in Multi-Rate Systems with and without Non-Uniform and Adaptive Bandwidths, *Signal Processing*, vol. 83, pp. 1613–1631.

Hauenstein, M. (1997). *Psychoacoustically Motivated Measures for Instrumental Speech Quality Assessment*, PhD thesis. Arbeiten über Digitale Signalverarbeitung, vol. 10, U. Heute (ed.), University Kiel (in German).

Heute, U. (2005). Speech and Audio Coding – Aiming at High Quality and Low Data Rates, *in* J. Blauert (ed.), *Communication Acoustics*, Springer, Berlin, Germany.

Heute, U. (2006). Noise Reduction, *in* E. Hänsler; G. Schmidt (eds.), *Topics in Acoustic Echo and Noise Control*, Springer, Berlin, Germany.

Heute, U. (2007). Telephone-Speech Quality, *in* E. Hänsler; G. Schmidt (eds.), *Speech and Audio Processing in Adverse Environments*, Springer, Berlin, Germany.

Heute, U.; Möller, S.; Raake, A.; Scholz, K.; Wältermann, M. (2005). Integral and Diagnostic Speech-Quality Measurement: State of the Art, Problems, and New Approaches, *Proceedings of the Forum Acusticum*, Budapest, Hungary, pp. 1695–1700.

Hirsch, H. G.; Ehrlicher, C. (1995). Noise Estimation Techniques for Robust Speech Recognition, *Proceedings of the IEEE International Conference on Acoustics, Speech, and Signal Processing (ICASSP)*, Detroit, USA, pp. 153–156.

Huo, L.; Scholz, K.; Heute, U. (2006). Idealized System for Studying the Speech-Quality Dimension "Directness / Frequency Content", *2nd ISCA-DEGA Workshop Perceptual Quality of Systems*, Berlin, Germany, pp. 109–114.

ITU COM 12-19-E (2000). *Results of Objective Speech-Quality Assessment of Wideband Speech Using the Advanced TOSQA2001*, Study Group 12 edn, Contrib. 19, ITU-T, Geneva, Switzerland.

ITU COM 12-34-E (1997). *TOSQA – Telecommunication Objective Speech-Quality Assessment*, study group 12 edn, Contrib. 34, ITU-T, Geneva, Switzerland.

ITU G.107 (2005). *The E-Model, a Computational Model for Use in Transmission Planning*, Recommendation ITU-T, Geneva, Switzerland.

ITU G.712 (2001). *Transmission Performance Characteristics of Pulse Code Modulation Channels*, Recommendation ITU-T, Geneva, Switzerland.

ITU G.722 (1993). *7 kHz Audio Coding within 64 kbit/s*, Recommendation ITU-T, Geneva, Switzerland.

ITU G.722.2 (2002). *Wideband Coding of Speech at Around 16 kbit/s Using Adaptive Multi-rate Wideband (AMR-WB)*, Recommendation ITU-T, Geneva, Switzerland.

ITU G.726 (1990). *40, 32, 24, 16 kbit/s Adaptive Differential Pulse-Code Modulation (AD-PCM)*, Recommendation ITU-T, Geneva, Switzerland.

ITU G.729 (1996). *Coding of Speech at 8 kbit/s Using Conjugate-Structure Algebraic Code-Excited Linear Prediction.*, Recommendation ITU-T, Geneva, Switzerland.

ITU P.341 (2005). *Transmission Characteristics for Wideband (150–7000 Hz) Digital Hands-Free Telephony Terminals*, Recommendation ITU-T, Geneva, Switzerland.

ITU P.48 (1993). *Specification of an Intermediate Reference System*, Recommendation ITU-T, Geneva, Switzerland.

ITU P.50 (1999). *Artificial Voices*, Recommendation ITU-T, Geneva, Switzerland.

ITU P.561 (1996). *In Service, Non-Intrusive Measurement Device – Voice-Service Measurements*, Recommendation ITU-T, Geneva, Switzerland.

ITU P.562 (2000). *Analysis and Interpretation of INMD Voice-Service Measurements*, Recommendation ITU-T, Geneva, Switzerland.

ITU P.800 (1996). *Methods for Subjective Determination of Transmission Quality*, Recommendation ITU-T, Geneva, Switzerland.

ITU P.830 (1996). *Subjective Performance Assessment of Telephone-Band and Wideband Digital Codecs*, Recommendation ITU-T, Geneva, Switzerland.

ITU P.862 (2001). *Perceptual Evaluation of Speech Quality (PESQ): An Objective Method for End-to End Speech Quality Assessment of Narrowband Telephone Networks and Speech Codecs*, Recommendation ITU-T, Geneva, Switzerland.

ITU P.862.2 (2005). *Wideband Extension of Recommendation P.862 for the Assessment of Wideband Telephone Networks and Speech Codecs*, Recommendation ITU-T, Geneva, Switzerland.

Janardhanan, D.; Heute, U. (2005a). Wideband Speech Enhancement Using a Modified Noise Estimation, *Electronic Speech Signal Processing Conference (ESSV)*, Prague, Czech Republic, pp. 143–150.

Janardhanan, D.; Heute, U. (2005b). Wideband Speech Enhancement Using a Robust Noise Estimation, *Proceedings of German Annual Conference on Acoustics (DAGA)*, Munich, Germany.

Janardhanan, D.; Heute, U. (2006). Wideband Speech Enhancement with a Multi-Window Method for Spectrum Estimation, *ITG Conference Sprachkommunikation*, Kiel, Germany.

Jax, P. (2002). *Enhancement of Band-Limited Speech Signals: Algorithms and Theoretical Bounds*, PhD thesis. Aachener Beiträge zu digitalen Nachrichtensystemen, vol. 15, P. Vary (ed.), RWTH Aachen University.

Jax, P.; Vary, P. (2002). An Upper Bound on the Quality of Artificial Bandwidth Extension of Narrowband Speech Signals, *Proceedings of the IEEE International Conference on Acoustics, Speech, and Signal Processing (ICASSP)*, Orlando, USA, pp. 237–240.

Jekosch, U. (2005). *Voice and Speech Quality Perception*, Springer, Berlin, Germany.

Kettler, F.; Gierlich, H. W.; Rosenberger, F. (2003). Application of the Relative Approach to Optimize Packet-Loss Concealment Implementations, *Proceedings of German Annual Conference on Acoustics (DAGA)*, Aachen, Germany, pp. 662–663.

Krebber, W. (1995). *Speech-Transmission Quality of Telephone Handsets*, PhD thesis. Fortschritts Berichte VDI, vol. 10, no. 357, H.-D. Lüke (ed.), RWTH Aachen University (in German).

Kruskal, J.; Wish, M. (1978). Multidimensional Scaling, *in* E. M. Uslaner (ed.), *Quantitative Applications in the Social Sciences*, Sage, Newbury Park, USA.

Kühnel, C. (2007). *Investigation of Instrumental Measurement and Systematic Variation of the Speech-Quality Dimension "Noisiness"*, Diploma Thesis, University Kiel (in German).

Laver, J. (1980). *The Phonetic Description of Voice Quality*, Cambridge University Press, UK.

Ludwig, T. (2003). *Measurement of Speech Characteristics for Reference-Free Quality Evaluation of Telefone-Band Speech*, PhD thesis. Arbeiten über Digitale Signalverarbeitung, vol. 23, U. Heute (ed.), University Kiel (in German).

Ludwig, T.; Scholz, K.; Heute, U. (2003). Speech-Quality Evaluation in Telephone Networks, *DAGA*, Aachen Germany, pp. 718–719.

Makhoul, J.; Berouti, M. (1979). High-Frequency Regeneration in Speech-Coding Systems, *Proceedings of the IEEE International Conference on Acoustics, Speech, and Signal Processing (ICASSP)*, pp. 428–431.

Martin, R. (1994). Spectral Subtraction Based on Minimum Statistics, *Proceedings of the European Signal Processing Conference (EUSIPCO)*, Edinburgh, UK, pp. 1182–1185.

Martin, R. (2001). Noise Power-Spectral Density Estimation Based on Optimal Smoothing and Minimum Statistics, *IEEE Transactions on Speech and Audio Processing*, vol. 9, pp. 504–512.

Martin, R. (2006). Bias Compensation Methods for Minimum Statistics Power Spectral Density Estimation, *Signal Processing*, vol. 86, pp. 1215–1229.

Mattila, V. (2001). *Perceptual Analysis of Speech Quality in Mobile Communication*, PhD thesis, Tampere University of Technology.

Moore, B. C. J.; Tan, C. T. (2003). Perceived Naturalness of Spectrally Distorted Speech and Music, *Journal of the Acoustical Society of America*, vol. 114, pp. 408–419.

Patrick, P. J. (1983). *Enhancement of Band-Limited Speech Signals*, Diss., Loughborough Univ. of Technology, U.K.

Patrick, P. J.; Steele, R.; Xydeas, C. S. (1983). Frequency Compression of 7.6 kHz Speech into 3.3 kHz Bandwidth, *Proceedings of the IEEE International Conference on Acoustics, Speech, and Signal Processing (ICASSP)*, Boston, USA, pp. 1304–1307.

Patrick, P. J.; Xydeas, C. (1981). Speech-Quality Enhancement by High-Frequency Band Generation, *International Conference on Digital Processing of Signals in Communications*, pp. 365–373.

Paulus, J. (1996). *Coding of Wideband Speech Signals at Low Data Rate*, PhD thesis. Aachener Beiträge zu digitalen Nachrichtensystemen, vol. 6, P. Vary (ed.), RWTH Aachen University (in German).

Paulus, J.; Schnitzler, J. (1996). 16 kbit/s Wideband Speech Coding Based on Unequal Sub-Bands, *Proceedings of the IEEE International Conference on Acoustics, Speech, and Signal Processing (ICASSP)*, Atlanta, USA, pp. 255–258.

Quackenbusch, S. R.; Barnwell, T. P.; Clemens, M. A. (1988). *Objective Measures of Speech Quality*, Prentice Hall, Englewood Cliffs, USA.

Raake, A. (2006). *Speech Quality of VOIP – Assessment and Prediction*, John Wiley & Sons, Ltd, Chichester, UK.

Rix, A. W.; Beerends, J. G.; Hollier, M. P.; Hekstra, A. P. (2001). Perceptual Evaluation of Speech Quality (PESQ) – a New Method for Speech-Quality Assessment of Telephone Networks and Codecs, *Proceedings of the IEEE International Conference on Acoustics, Speech, and Signal Processing (ICASSP)*, Salt Lake City, USA, pp. 749–752.

Rix, A. W.; Hollier, M. P. (2000). The Perceptual Analysis Measurement System for Robust End-to-End Speech Quality Assessment, *Proceedings of the IEEE International Conference on Acoustics, Speech, and Signal Processing (ICASSP)*, Istanbul, Turkey, pp. 1515–1518.

Rix, A. W.; Hollier, M. P.; Hekstra, A. P.; Beerends, J. G. (2002). Perceptual Evaluation of Speech Quality (PESQ), the New ITU Standard for End-to-End Speech-Quality Assessment, Part I – Time-Delay Compensation, *Journal of the Audio Engineering Society*, vol. 50, pp. 755–764.

Schmidt, G. U. (2001). *Design and Realisation of a Multi-Rate System for Hands-Free Telephony*, PhD thesis, Fortschritts Berichte VDI, vol. 10, no. 674, Technical University Darmstadt.

Scholz, K.; Wältermann, M.; Huo, L.; Raake, A.; Möller, S.; Heute, U. (2006a). Comparison of the Instrumental Description of the Quality Dimension "Directness / Frequency Content" for Narrowband and Wideband Speech, *ITG Conference Sprachkommunikation*, Kiel, Germany.

Scholz, K.; Wältermann, M.; Huo, L.; Raake, A.; Möller, S.; Heute, U. (2006b). Estimation of the Quality Dimension "Directness / Frequency Content" for the Instrumental Assessment of Speech Quality, *Proceedings of the International Conference on Spoken Language Processing (ICSLP)*, Pittsburgh, USA, pp. 1523–1526.

Schüssler, H. W.; Dong, Y. (1990). A New Method for Measuring the Performance of Weakly Non-Linear Systems, *Frequenz*, vol. 44, pp. 82–87.

Serafat, R.; Heute, U. (1996). A wideband speech-model process as a test signal, *Proceedings of the European Signal Processing Conference (EUSIPCO)*, pp. 487–490.

Takahashi, A.; Kurashima, A.; Morioka, C.; Yoshino, H. (2005). Objective Quality Assessment of Wideband Speech by an Extension of ITU-T Recommendation P.862, *Interspeech*, Lisbon, Portugal, pp. 3153–3156.

Thomson, D. J. (1982). Spectrum Estimation and Harmonic Analysis, *Proceedings of the IEEE*, vol. 70, pp. 1055–1096.

Vary, P.; Heute, U.; Hess, W. (1998). *Digitale Sprachsignalverarbeitung*, Teubner, Stuttgart, Germany.

Vary, P.; Martin, R. (2006). *Digital Speech Transmission*, John Wiley & Sons, Ltd., Chichester.

Voiers, W. D. (1977). Diagnostic Acceptability Measure for Speech Communication Systems, *Proceedings of the IEEE International Conference on Acoustics, Speech, and Signal Processing (ICASSP)*, Hartfort, USA, pp. 204–207.

Voran, S. (1997). Listener Ratings of Speech Passbands, *IEEE Workshop Speech Coding Telecomm.*, Pocono Manor, USA.

Wältermann, M.; Raake, A.; Möller, S. (2006a). Perceptual Dimensions of Wideband-Transmitted Speech, *2nd ISCA-DEGA Workshop Perceptual Quality of Systems*, Berlin, Germany, pp. 103–108.

Wältermann, M.; Scholz, K.; Raake, A.; Heute, U.; Möller, S. (2006b). Underlying Quality Dimensions of Modern Telephone Connections, *Proceedings of the International Conference on Spoken Language Processing (ICSLP)*, Pittsburgh, USA, pp. 2170–2173.

Zwicker, E. (1982). *Psychoakustik*, Springer, Berlin, Germany.

Chapter 3

Parametric Quality Assessment of Narrowband Speech in Mobile Communication Systems

Marc Werner

3.1 Introduction

Modern speech and audio transmission systems often employ dedicated digital signal processing algorithms like advanced speech coding, echo cancellation, and noise reduction. The effect of these non-linear operations on the resulting speech quality, especially in combination with transmission errors on the communication link, is difficult to estimate. However, the perceived speech quality is for many applications and services the most important criterion for user satisfaction. It is therefore a primary interest of system designers and operators to assess and predict the quality of delivered speech signals under different conditions. In the domain of digital mobile communication networks, narrowband speech telephony is still by far the most frequently used service. In the competitive environment of the cellular business, operators need to optimize the transmission performance of their networks and to monitor constantly the resulting speech quality.

This chapter provides a classification of methods for narrowband speech quality assessment, as well as a short overview of the evolution of these techniques. It then focuses on a specific approach to how efficient automated quality monitoring procedures for mobile communication systems can be designed. The developed methodology is based on the evaluation of measurement parameters that are available at the receiver, and can be applied to any mobile communication system providing such measurements. As

Advances in Digital Speech Transmission Edited by R. Martin, U. Heute and C. Antweiler
© 2008 John Wiley & Sons, Ltd

examples, dedicated speech quality measures are derived for use in GSM and UMTS networks. The performance of these parametric measures is quantified by comparison against benchmark quality assessment methods.

3.1.1 Subjective Listening Tests and Classes of Objective Measures

Classically, speech quality assessments have been carried out in subjective tests. These tests provide excellent accuracy for many assessment scenarios if sufficient effort is taken in obtaining judgments by a large group of human subjects under standardized laboratory conditions.

On the other hand, objective speech quality measures have been developed mostly in the form of computer programs, which allow quite reliable speech quality assessments for certain applications without expensive subjective tests.

Subjective Listening Tests

The auditive quality of narrowband speech (and thereby the quality of the underlying processing algorithm or transmission system) can be determined by subjective listening tests following the procedures standardized by ITU-T in its recommendations P.800 and P.830 [ITU-T 1996a], [ITU-T 1996c]. High-quality original speech signals are used to produce a large amount of speech material to be assessed, reflecting the various transmission or processing conditions that are expected for the system under test. For telephony applications, the speech material is first processed by a filter representing a typical handset transmit path.

The processed speech signals are usually presented to the test subjects in speech clips with a duration of 5 s to 10 s. Additionally, speech clips exhibiting certain reference degradations, as well as the original speech, should be included in the test material. Speech clips with two male and two female speakers should be used for each testing condition.

The listening quality of each speech clip is rated by a large number of test persons. Different types of rating include the ACR (Absolute Category Rating), DCR (Degradation Category Rating), and CCR (Comparison Category Rating). ACR tests are the most common and the quickest to conduct. For these tests, the listening quality (LQ) rating is given on a five-point opinion scale reproduced in Table 3.1 [ITU-T 1996a].

The mean rating of all test persons for a certain speech clip (or for all clips describing a certain speech processing system under test), called the Mean Opinion Score (MOS), is then calculated to characterize the speech quality of the clip or system. Because external factors such as the speech laboratory environment can influence the results of listening tests, the MOS scores are normalized with the help of reference speech clips exhibiting well-defined distortions (e.g., MNRU, Modulated Noise Reference Unit) [ITU-T 1996b].

Table 3.1: Opinion scale used in subjective listening tests

Score	Listening quality
5	Excellent
4	Good
3	Fair
2	Poor
1	Bad

Apart from the standardized ACR-LQ subjective testing method recommended in [ITU-T 1996a], other subjective quality assessment models have been proposed and used, like the Diagnostic Acceptability Measure (DAM), which aims at categorizing the perceived distortion according to a larger number of labels [Quackenbush et al. 1988].

Because telephony services are characterized by the interaction of two or more speakers, the listening quality assessments described above reflect only one part of the overall user's quality perception. Additional quality attributes of the so-called Conversational Quality (CQ) are the talking quality (describing the perception of a user's own voice) and the interaction quality (characterizing impairments like delay and double-talk distortions) [Rix et al. 2006]. Conversational tests are slower and more expensive than listening tests.

In the field of audio processing and transmission, subjective listening tests like the recommended ITU-R BS.1116 focus on the assessment of relatively small signal degradations by comparison tests [ITU-R 1998].

Subjective listening tests are time-consuming and expensive. They are therefore mostly carried out to classify the quality of single components of a speech transmission system such as a newly developed speech codec or noise reduction filter, in a final selection or decision state.

During the development phase of speech processing algorithms, quicker and cheaper methods of speech quality assessment are desirable. Other applications, like the automated monitoring of end-to-end speech quality in cellular communication systems, also call for more efficient assessment techniques and may in turn allow a larger tolerance regarding the accuracy of the quality judgments. For these purposes, the listening quality can be estimated using objective speech quality measurement algorithms, introduced in subsequent sections.

Objective Speech Quality Measures

Instrumental or objective speech quality measures are based on algorithms that aim at the prediction of listening test results in a reproducible way and irrespective of external factors.

Intrusive objective measures analyze information from the transmitter and receiver of the system under test, mostly the transmitted and received speech signals themselves. They usually contain models of the human auditory and cognitive perception of sound, and address certain psychoacoustic effects, such as [Zwicker, Fastl 1999], [Beerends, Stemerdink 1994], [Rix et al. 2001]:

- non-linear transformation of sound pressure level to perceived loudness,

- non-linear transformation of frequency to perceived pitch,

- spectral and temporal masking,

- unequal weighting of new and deleted signal components,

- unequal weighting of distortions depending on their position within the clip (recency effects).

A comparison of the original and distorted speech signals after a transformation into the perceptual domain then yields a typically one-dimensional quality score that can be mapped to MOS values of subjective ACR-LQ tests using some monotonic mapping function. The objective measurement of speech quality in speech communication systems using intrusive methods is only possible by setting up test connections with well-defined speech material and by thus generating additional network load.

Non-intrusive models, on the other hand, allow an estimation of the perceived speech quality by exploiting general system properties and information taken only from the receiver side of the communication path. This makes them especially useful for quality assessment during the normal operation of a speech communication system, where the original speech signals are not available.

Some of these non-intrusive methods are *signal-based*, and thus analyze the distorted speech signal itself, making use of the above-mentioned perception models and psychoacoustic effects. The absence of a reference signal, however, makes this a difficult task, and the accuracy of signal-based non-intrusive measures usually falls behind intrusive models.

Other non-intrusive quality measures do not analyze speech signals but certain parameters of the underlying network. Such *parametric* models have been used for network planning, where an estimate of the expected general speech quality is made based on network properties (E-Model, [ITU-T 2002]). But most non-intrusive parametric measures are designed to assess the quality of individual speech transmissions by analyzing transmission parameters that are available at the receiver. These parameters usually originate from measurements that are taken at a sufficiently high frequency (at least several times a second) and reported back to the transmitter for purposes like link adaptation, radio resource management, or handover. These measures, once trained to their specific measurement environment, exhibit a low complexity and can be used for the purpose of automated, efficient and low-cost speech quality monitoring without creating additional network load.

The performance of objective speech quality measures is usually characterized by their correlation with subjective tests. For this purpose, the absolute value of the correlation coefficient $-1 \leq \rho \leq 1$ is calculated. ρ is a measure of how well the objective quality scores can be mapped to the reference subjective MOS scores.

According to a recommended classification scheme [ITU-T 2003a], the reference MOS values from subjective ACR-LQ tests are termed MOS-LQS (listening quality – subjective), while estimated MOS values delivered by objective quality measures have to be denoted as MOS-LQO (listening quality – objective). MOS estimates stemming from the parametric E-Model [ITU-T 2002] for network planning are referred to as MOS-LQE.

Current state-of-the-art objective measures exhibit very good correlations between MOS-LQS and MOS-LQO of $|\rho| \geq 0.9$ if applied to the measurement conditions for which they have been designed.

3.1.2 Overview of Objective Speech Quality Measures

In the following sections, a short synopsis of some objective speech quality assessment models is given. For complementary information, the reader is referred to a recent overview paper [Rix et al. 2006].

Intrusive Models

The development of intrusive objective speech quality measures was primarily driven by advances in speech coding where models of speech production are exploited to obtain high-quality speech compression at low bit rates. For these and other speech processing applications, the quality of decoded or received speech cannot be characterized sufficiently by simple measures such as the signal-to-noise ratio (SNR), but the auditory perception of distortions must be accounted for. Modern speech codecs often exhibit a coding noise that is shaped with the spectrum of the transmitted signal and therefore reduced in its audibility.

An early quality assessment approach addressing this issue was made 1987 by Brandenburg with the Noise-to-Mask Ratio (NMR) [Brandenburg 1987], which quantifies the degree of psychoacoustic masking of noise by the original signal waveform.

The transformation of waveforms into the perceptual domain using auditory and cognitive models was proposed by Karjalainen [Karjalainen 1985]. This concept is now widely used in speech quality assessment methods [Quackenbush et al. 1988], [Wang et al. 1992], [Park et al. 2000], [Beerends, Stemerdink 1994], [Rix et al. 2001]. Usually, a handset-filtered version of the original and distorted speech signal is transformed into the frequency domain, taking into account the non-linear perception of frequency and loudness. Effects like spectral and temporal masking and unequal weighting of loud and silent signal intervals, are then used in a cognitive model to assess the perceptual difference of the two signals. These models must be trained by a large amount of test data from ACR-LQ tests and validated by unknown data.

The Perceptual Speech Quality Measure (PSQM) [Beerends, Stemerdink 1994] was the first intrusive measure for the quality of narrowband speech to be standardized by the ITU-T in the recommendation P.861 [ITU-T 1998]. It was based largely on the above described principles and was especially optimized for the assessment of speech codecs. However, for the task of estimating distortions found in transmission networks, certain limitations were identified.

Other recent speech quality assessment algorithms include TOSQA [ITU-T 1997] and PACE [Juric 1998]. The new and current ITU-T standard P.862 [ITU-T 2001] recommends the PESQ (Perceptual Evaluation of Speech Quality) algorithm [Rix et al. 2001] for use in a wide range of applications, including communication systems. PESQ is based on an improved version of PSQM and on the PAMS algorithm (Perceptual Analysis Measurement System) [Rix, Hollier 2000] and includes, e.g., a dedicated time alignment procedure to address variable delays found in speech transmission systems. For the calculation of MOS-LQO values from the PESQ output scores, a dedicated mapping function was developed [ITU-T 2003b].

An extended version of PESQ was recently introduced for the assessment of wideband speech [ITU-T 2005].

Signal-based Non-intrusive Models

Signal-based non-intrusive speech quality measures are a relatively new development. Out of a competition including the ANIQUE model [Kim 2005], the so-called Single-Ended Assessment Model (SEAM) [Gray et al. 2000] was adopted as ITU-T recommendation P.563 in 2004 [ITU-T 2004]. It includes a speech production model to identify signal components that cannot be produced by the human vocal tract and are therefore classified as distortions, as well as an identification of additive noise, clipping and muting.

Owing to the absence of a reference signal, the performance of signal-based non-intrusive assessment models is slightly inferior to that of intrusive methods. However, the current results are promising and this field remains an interesting area of research.

Parametric Non-intrusive Models

Parametric non-intrusive quality assessment models have been derived to characterize the speech quality in communication networks. In wire-line networks, Broom evaluated Voice-over-IP network characteristics to estimate the resulting speech quality [Broom 2006]. For wireless systems, an approach was made by Gaspard who analyzed several statistical parameters, all derived from the channel bit error rate measurement

(RXQUAL) in GSM systems [Gaspard 1994]. The mapping of the parameter space to MOS-LQS scores from subjective tests was performed using a multidimensional MMSE fitting technique.

A similar approach was taken by Karlsson et al. who, apart from the bit error rate, included parameters reflecting decoding erasures of speech frames in the GSM full-rate codec, and fitted a linear combination of average parameter values to the MOS-LQS scores [Karlsson et al. 1999]. This approach was extended by Wän-stedt et al. for the Adaptive Multi-Rate (AMR) speech codec [Wänstedt et al. 2002].

Further refinements in parametric speech quality assessment were proposed with re-spect to the averaging and mapping procedures of radio transmission parameters of cellular communication networks to reference quality scores for GSM and UMTS sys-tems employing the Enhanced Full-Rate (EFR) codec and the AMR codec [Werner et al. 2003], [Werner et al. 2004], [Werner, Vary 2005].

3.1.3 Development of Parametric Models

The radio transmission is the most critical part of cellular voice communication. Therefore, an analysis of radio transmission parameters is well suited to describe the experienced end-to-end speech quality. The development of non-intrusive speech quality measures for cellular communication systems, based on these parameters, will be described in detail in the remainder of this chapter.

The presented development procedure maximizes the correlation of quality assessment results with reference speech quality scores. It is applied to two different cellular networks: GSM, employing the EFR or AMR codec, and UMTS, using the AMR codec. The speech quality measurement method can be easily adapted to other digital cellular systems as long as appropriate transmission measurement parameters are available. In this chapter, only the listening quality of narrowband (30–3400 Hz) speech transmissions is considered. Aspects of wideband-speech quality are covered in Chap. 2.

The optimization of these speech quality measures is based on measurement data provided by a GSM and UMTS network operator as well as on link-level simulations of the GSM and UMTS speech transmissions, to be described in Sec. 3.2. These simulations allow an analysis of the correlation of certain parameters with the resulting speech quality in terms of MOS-LQO, obtained by the intrusive PESQ measure [Rix et al. 2001]. Furthermore, an algorithm for the switching of AMR codec modes is presented.

Section 3.3 covers the techniques of mapping transmission parameter values to speech quality scores. Methods for improving the quality correlation are described. Then, the procedures for optimizing parameter-based speech quality measures are applied to the GSM-EFR, GSM-AMR, and UMTS-AMR systems.

3.2 Simulations of GSM and UMTS Speech Transmissions

In all digital cellular communication systems, the network performance is optimized by dedicated algorithms that are employed to maximize the transmission quality in speech channels while consuming only a reasonable necessary amount of radio resources such as bandwidth, transmit power, or spreading codes. These algorithms include speech processing functions like speech coding and error concealment but also radio-related mechanisms like the so-called radio resource management (RRM).

Owing to the highly variable nature of the radio channel, transmission impairments cannot be avoided. Fast fading, for instance, can result in signal level variations in the range 30 to 40 dB within milliseconds. Even with fast power control mechanisms as employed in UMTS, residual transmission errors remain after de-interleaving and channel decoding. The measurement of these transmission errors is vital for the RRM, which allocates, e.g., the transmit powers, user data rates, and coding/modulation schemes to the individual users in an economical way to maximize the system capacity and the overall transmission quality [Werner 2005]. Transmission parameters based on the received signal and interference power levels, as well as on the frame and bit error rates (FER/BER) are defined for the RRM and the handover operation in most current cellular communication systems such as the GSM and the third-generation UMTS.

For most transmission parameters, an immediate evaluation is of limited suitability for the purpose of speech quality monitoring. The most common technical parameters of current cellular systems disregard special features of speech processing elements like the speech decoder's unequal sensitivity to different bit error positions, and the effects of error concealment on the received speech clip.

The practice of many operators to use thresholds of transmission parameters for statistical speech quality monitoring is therefore inaccurate and does not allow a detailed analysis of customer satisfaction. For the development of refined parametric speech quality assessment models, an extensive analysis of transmitted speech material, together with corresponding transmission parameter progressions, is necessary.

3.2.1 Simulation Environment

To produce speech clips that reflect real-world radio transmission conditions, link-level simulations of the GSM and UMTS downlink physical speech transmissions can be used. In the following, the simulation models that were set up within the System Studio software environment [Syn 2004] to derive the parametric speech quality measures in Sec. 3.3 are described.

Elements of the Transmission Chain

The simulation models for GSM and UMTS include EFR or AMR speech coding, and all necessary elements of the equivalent physical baseband transmission chain, e.g., multiplexing, channel coding, and interleaving.

In the UMTS simulations, the rate matching, spreading and scrambling procedures according to the 3GPP standard [3GPP 2000a] are implemented. At the receiver, the respective inverse algorithms, de-scrambling, de-spreading, and inverse rate matching, are applied.

In the GSM simulations, a binary error channel is employed that is dynamically controlled by channel BER measurements recorded in the real network (see Fig. 3.1). Within each speech frame (20 ms), it is assumed that the channel BER is constant and the bit errors are randomly distributed. The validity of these assumptions was checked with a simulation of the Typical Urban (TU) fading channel [ETSI 1999a], producing typical burst errors that are then broken up by de-interleaving. Figure 3.2 depicts the distribution of bit error burst lengths after de-interleaving, for different terminal velocities and an exemplary overall BER of 4.6%. It is shown that, compared with a purely random and memoryless distribution of bit errors, no ($v = 30$ km/h) or only a slight ($v = 10$ km/h, $v = 3$ km/h) increase of longer burst errors due to fast fading remains. However, for a constant overall BER, only a large shift in the burst length distribution towards long error bursts produces significantly more channel decoding failures, which may then reduce the perceived speech quality. The approximation with randomly distributed bit errors per speech frame can thus be regarded as valid for the purpose of speech quality measurement.

The UMTS simulations are carried out using a channel model that reflects both intracell and intercell interference as well as a slow fading process. This slow fading may result, e.g., from variations in the assignment of radio resources, and is modeled by a first-order Markov process reflecting different kinds of channel conditions from

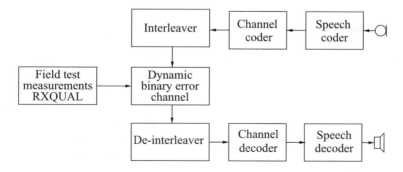

Figure 3.1: Binary error channel of GSM simulations, controlled by field test measurements

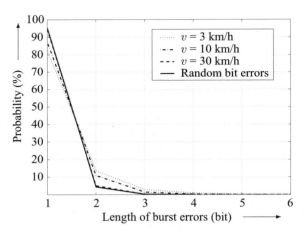

Figure 3.2: Distribution of error burst lengths in GSM TU channels, after de-
interleaving, for different terminal velocities v (overall BER: 4.6%)

nearly error-free to heavily disturbed transmissions. Taking the correlation properties of UMTS scrambling codes into account, the intercell interference is modeled as
simple Additive White Gaussian Noise (AWGN), whereas the orthogonality imperfec-
tions of the downlink spreading codes due to multipath propagation are modeled as
Orthogonal Code Noise Sequences (OCNS) [3GPP 2005].

For channel decoding, a Soft-Output Viterbi Algorithm (SOVA) [Hagenauer, Hoeher
1989] is used in both GSM and UMTS simulations. The implemented EFR and AMR
speech decoders employ error concealment: speech frames that did not pass a CRC
check after convolutional decoding are marked by a Bad Frame Indication (BFI). The
BFI flag controls the concealment process, which is based on a state machine with
seven states. Depending on the state, certain parameters of the codec are replaced
by attenuated counterparts from the previous frame or by averaged values [ETSI
1999d].

AMR Mode Switching

The AMR codec can be interpreted as an example of an RRM procedure for speech
channels. It consists of eight independent speech codecs (modes 0...7) with data rates
ranging from 4.75 kbit/s to 12.2 kbit/s. The 12.2 kbit/s mode (identical to the EFR
codec) is usually the default codec employed for good transmission channel conditions.
At the expense of a slightly reduced baseline quality of lower modes, the data rate
can be decreased to allow for an increased transmission robustness due to a higher
bit energy and/or better channel coding, while maintaining the same level of other
consumed radio resources.

In GSM, the channel bit rate per user is constant at 22.8 kbit/s. Different channel
coding schemes are used for each AMR mode, filling up the individual net rate to the

constant gross channel rate. In UMTS, the channel bit rate per user is determined by the chosen spreading factor and the rate matching algorithm. A constant channel encoder of rate $r = 1/3$ is employed for the reference UMTS-AMR speech channel [3GPP 2002]. Note that the gain in transmission reliability for lower AMR modes is realized by a higher coding gain in the GSM system but by a larger channel bit energy in the UMTS system.

It is assumed that in UMTS a preselection of four AMR modes takes place as in GSM [ETSI 1998] (the selection of modes is not specified in the UMTS standard). As the characteristics of neighboring AMR modes differ only slightly, the quality loss induced by the reduction of the number of possible modes from eight to four is small. On the other hand, a selection of a mode subset allows for a quicker link adaptation as the step size of the net data rates becomes larger, and also reduces the number of switching instants. In the UMTS link-level simulations, modes 0, 2, 5, and 7, with the data rates 4.75, 5.9, 7.95, and 12.2 kbit/s, are selected for an optimum resulting average speech quality under an exemplary, uniformly distributed range of transmission conditions. For GSM, the modes 0, 5, 6 (10.2 kbit/s), and 7 represent the best choice under these conditions. The difference in mode selections can be explained by the different channel coding in UMTS and GSM. The mode preselection in real networks will also consider the individual expected occurrence of different channel quality conditions.

The adaptation of AMR modes to changing transmission conditions is crucial for the performance of the adaptive speech coding. A dedicated AMR mode switching scheme is introduced in the link-level simulations which optimizes the resulting speech quality perception by providing a compromise between inherent speech quality and robustness against interference.

In the downlink, channel degradations must be quickly estimated at the mobile station and reported back to the base station, which selects the AMR mode to be used for subsequent speech frames. These estimates must be taken at a frequency well above that of the standardized measurement parameters described in Sec. 3.2.2 to allow for a frame-by-frame switching. In GSM, an SIR estimation method on the basis of non-binary real (soft) channel values is recommended for this purpose [ETSI 1999c]. A sliding-window FIR lowpass filter with 100 coefficients is applied to these values to facilitate a sufficiently reliable estimation. The further mapping procedure of filtered channel values to certain AMR switching thresholds is left open for optimization by manufacturers; an example would be to take the mean value of the 20 most unreliable filtered samples per speech frame and to define appropriate switching thresholds.

However, the GSM approach cannot be applied to UMTS, where settings of the radio transmission like the spreading factor and rate matching scheme vary between different AMR modes or even within one mode, depending on the dynamic RRM procedures. The relation between soft channel values and transmission quality is variable in this case.

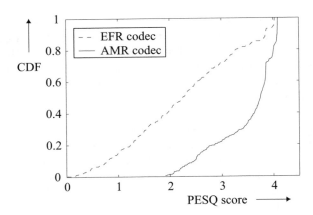

Figure 3.3: CDF of PESQ scores for GSM transmission simulations with channel BER uniformly distributed over [0, 18%]

Therefore, the dedicated AMR mode switching method employed in the described GSM and UMTS link-level simulations is based on an analysis of the soft output bits of the SOVA [Hagenauer, Hoeher 1989] channel decoder. The evaluation of the SOVA softbits is carried out immediately before the speech decoding and can therefore correlate better with the resulting speech quality than an evaluation of channel values.

The dedicated AMR mode switching method offers a significant quality improvement, compared with the standard method recommended by the GSM specification [ETSI 1999c], especially for bad and medium channels. It reaches up to 0.6 PESQ score points for certain clips transmitted over a highly variable channel. A more detailed description of the dedicated AMR mode switching method can be found in [Werner et al. 2004] and [Werner, Vary 2005].

The general improvement in speech quality by the introduction of the AMR codec, compared with the EFR codec, is depicted in Fig. 3.3. A selection of GSM transmission simulations is chosen, which represents a uniform distribution over the speech clips' mean channel BER, ranging from 0 to 18%. The quality advantage of the AMR codec under these conditions is clearly visible in the offset of the cumulative probability density function (CDF) of the resulting PESQ scores.

3.2.2 GSM Transmission Parameters

The analysis of GSM transmission parameters regarding their correlation with the resulting speech quality is based on field test measurement data collected from approximately 150 hours of GSM-1800 downlink test calls. The measurement data covers the progression of several physical layer measurement parameters for a variety of radio propagation conditions:

RXQUAL: The channel BER is averaged over intervals of 480 ms and mapped to the logarithmic RXQUAL parameter [ETSI 1999b] with eight BER levels from 0 (BER < 0.2%) to 7 (BER > 12.8%). RXQUAL serves as an estimate of the current channel quality during an active call, and controls the binary error channel in the GSM simulations performed. In the GSM system, values below four are desirable, because at a gross BER of less than 1.6%, nearly all bit errors within the most important class-I-bits can be corrected by the channel decoder.

RXLEV: The received power level at the mobile station is measured in dBm (relative to 1 mW) and mapped linearly to an RXLEV index [ETSI 1999b] ranging from 0 (< −110 dBm) to 63 (> −48 dBm) in 1 dBm steps. The minimum required value specified in the GSM standard [ETSI 1999a] ranges from −104 to −100 dBm (RXLEV > 6 ... 10). Measurements are reported every 480 ms. Among other factors, the received power level depends on the radio channel in terms of path loss and slow fading. However, it is not a measure of the signal-to-interference power ratio (SIR), but only an estimate of the sum of the desired signal plus interference.

Using the RXQUAL progressions as input, numerous EFR and AMR transmission simulations are carried out. The speech source contains a male and a female voice speaking a German sentence. As a result, several thousands of male and female speech clips are generated from the measurement data, each having a duration of approximately 9 s.

Additional transmission parameters that had not been included in the original field test measurement database are recorded in the simulations:

AMRmode (in the case of AMR transmissions): The four preselected AMR modes are indexed as $\{0, 1, 2, 3\}$. A system with dynamic adaptation of these modes to the current channel quality as described above is assumed. The presented models for speech quality assessment in Sec. 3.3 can easily be adapted to a de-activated mode switching.

The BFI rate, or Frame Erasure Rate (FER), and the BFI distribution within the speech clip, are of great relevance to the speech quality. Therefore the FER and some derivations are recorded as transmission parameters as well:

FER: Frame erasures for speech frames,

LFER: Sequence lengths of consecutively erased speech frames in the speech clip,

MxLFER: Maximum sequence length of consecutively erased frames,

MnMxLFER: Mean of four local maximum sequence lengths of erased speech frames for the first, second, third and fourth quarters of the speech clip. The mean of these four local maxima will be calculated using L_P norms, see (3.1). The maximization over short clip periods is regarded to be similar to the human perception of severe signal distortions.

Although the FER is not part of the standard GSM downlink measurement report, FER values for the uplink are usually stored within the Operation and Maintenance

Center (OMC) and an OMC function estimating the downlink FER exists in most cases. This feature depends on the OMC manufacturer.

Note that the progressions of all described parameters are to be averaged by an appropriate L_P norm per speech clip.

3.2.3 UMTS Transmission Parameters

The UMTS transmission parameters available at the receiver and their reporting frequency depend on the manufacturer's hard- and software implementation and on the network configuration. Some parameters specified for measurement reports of the mobile station and for measurements at the base station are given in Table 3.2 [3GPP 2000b].

Unfortunately, a BER measurement resembling the GSM RXQUAL is available only in the UMTS uplink. The UMTS outer loop power control measures the uplink received SIR and compares the measurements with a target value. On the downlink, the block error rate (BLER) is measured instead. The uplink BLER can also be derived from BER estimations [Heck et al. 2002].

The UMTS BLER is equivalent to the GSM FER parameter if one AMR speech frame is transmitted within one TTI (Transmission Time Interval), which is the specified procedure for the UMTS reference channels of net rate 12.2 kbit/s [3GPP 2002]. The derived parameters LFER, MnMxLFER and MxLFER are available in UMTS as well. Additionally, the AMR mode itself is taken into account as a transmission parameter. A system with dynamic adaptation of AMR modes according to Sec. 3.2.1 is assumed as in the GSM-AMR case.

Table 3.2: Some UMTS measurement parameters

Quantity	Downlink parameter	Uplink parameter
Power	Received total power, Received CPICH power, Transmitted power	Received total power, Transmitted carrier power, Transmitted code power
SIR	CPICH-E_c/N_0	SIR, SIR-error
BER/FER	Transport channel BLER	Transport channel BER, Physical channel BER
Data rate	AMR mode	AMR mode
Physical channel parameters	Spreading factor, Scrambling code, Coding/Puncturing schemes	Spreading factor, Scrambling code, Coding/Puncturing schemes

(CPICH: UMTS Common Pilot Channel)

3.3 Speech Quality Measures based on Transmission Parameters

The parametric speech quality assessment methods described in this chapter are non-intrusive as they are based only on the above received transmission parameters, usually reported a number of times per second in current cellular communication systems.

Earlier approaches [Gaspard 1994], [Karlsson et al. 1999], [Wänstedt et al. 2002] to use simple combination functions of transmission parameters for the prediction of the listening quality in cellular networks are extended by additionally including individually selected, psychoacoustically motivated normalization functions for each parameter.

In the process of calculating optimized quality prediction functions (see Sec. 3.3), the PESQ scores M_i, with $M_i \in [0,5]$, are taken as reference speech quality figures. A transformation function exists [ITU-T 2003b] that maps the PESQ scores to MOS values. While this procedure poses some restriction on the accuracy of the developed measures, the intrusive signal-based MOS-LQO quality assessments by PESQ have been proven as a reliably close approximation of subjective test results in various experiments [Rix et al. 2001]. Nevertheless, MOS-LQS scores from listening tests should be preferred as reference if available.

The recorded transmission parameters from field tests and simulations are now analyzed with respect to their correlation with the resulting speech quality in terms of PESQ scores. For this purpose, a dedicated L_P averaging procedure of parameter progressions per speech clip, as well as a linearization function, are applied. Speech quality measures based on single transmission parameters and on parameter combinations are then derived in Sec. 3.3.2.

3.3.1 Correlation Analysis

The parameter progressions $\zeta_i(k)$ with measurement time index $k = 1 \ldots N$ of the parameter $\zeta \in \{$RXQUAL, FER, LFER, MxLFER, MnMxLFER, AMRmode$\}$ are evaluated for each speech clip i. To compare these parameter progressions with the single PESQ score M_i, L_P norms of $\zeta_i(k)$ are calculated:

$$L_P(\zeta_i(k)) = \left[\frac{1}{N} \sum_{k=1}^{N} (\zeta_i(k))^P \right]^{1/P}, \quad P > 0. \tag{3.1}$$

The L_1 norm corresponds to the arithmetic mean and the L_2 norm is equivalent to the quadratic mean of $\zeta_i(k)$. The reason for calculating different L_P norms for each parameter progression is the different psychoacoustic perception of parameter outliers. Large values of P amplify parameter variations whereas small P-values have a smoothing effect. The values $P \in \{1/20, 1/19, \ldots, 1/2, 1, 2, \ldots, 19, 20\}$ are tested in the present procedure.

For each parameter and each value of P, the mapping function that fits the L_P norms to the objective PESQ scores M_i, over all speech clips i, is approximated with respect to a minimum mean-square error using a monotonic polynomial $f_{\zeta,P}$ of degree $m \leq 4$:

$$M_i \approx f_{\zeta,P}(L_P(\zeta_i(k))) \quad \forall i \tag{3.2}$$

(the dependency of f on ζ and P will be omitted in the following expressions).

The correlation coefficient $\rho(f(L_P(\zeta_i(k))), M_i)$ is calculated after a transformation of the L_P norms to the PESQ domain by the polynomial f. The optimum P-value $\hat{P}$, which maximizes the correlation, is identified along with the corresponding linearization polynomial $\hat{f}$ for each parameter ζ.

An example of the polynomial fitting and selection of optimal values for P and m is depicted in Figs. 3.4 and 3.5.

In Fig. 3.4, the effect of linearization is shown. The relation between the resulting speech quality (PESQ) and the L_6 norm of RXQUAL is depicted for a subset of

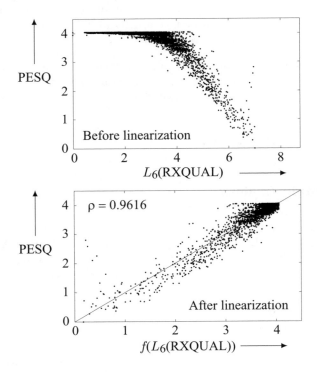

Figure 3.4: RXQUAL–PESQ correlation and polynomial linearization: Transformation of RXQUAL (L_6 norms) for GSM-EFR transmissions. © 2003 IEEE

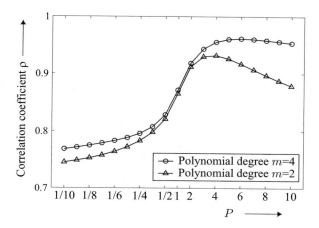

Figure 3.5: RXQUAL–PESQ correlations for different L_P norms and polynomial degrees m for GSM-EFR transmissions. © 2003 IEEE

GSM-EFR speech clips as a scatter-plot. The distribution of points in the upper diagram indicates a non-linear dependency and therefore a low linear correlation of the parameter norms with the PESQ scores.

After the transformation of the L_6 norms by the mapping polynomial f (horizontal axis in the lower diagram of Fig. 3.4), the linear correlation coefficient ρ can be properly calculated.

Figure 3.5 illustrates the dependency of the resulting correlation $\rho(f(L_P(\zeta_i(k))), M_i)$ on the P-value of the L_P norm and on the polynomial degree m, for the parameter RXQUAL. It can be observed that the highest correlation is obtained for $\hat{P} = 6$ and $\hat{m} = 4$. A higher-order polynomial up to $m = 6$ could not further improve the resulting correlation under the constraint of monotonicity.

It should be noted that the deterministic linearization function f does not change the degree of dependency between speech quality and parameter value itself but only improves the linear correlation measure. On the other hand, the optimization of P offers a real correlation gain.

The procedure of L_P averaging and linearization is applied to the GSM-EFR, GSM-AMR and UMTS-AMR transmission parameters. The correlation results are given in Table 3.3. The large optimum P-values for RXQUAL indicate that outliers are perceived more strongly than is suggested by the numerical value of this parameter. Note that for the FER parameter, $L_1(\text{FER}) = \text{FER}$, $L_2(\text{FER}) = \text{FER}^{1/2}$, and $L_4(\text{FER}) = \text{FER}^{1/4}$, because the constituent elements are taken from the binary set $\{0, 1\}$ (frame either received correctly or erased).

Table 3.3: Correlation coefficients and optimum L_P norms of GSM and UMTS transmission parameters

Parameter ζ	GSM-EFR		GSM-AMR		UMTS-AMR	
	$\hat{P}$	$\rho(\hat{f}, M_i)$	$\hat{P}$	$\rho(\hat{f}, M_i)$	$\hat{P}$	$\rho(\hat{f}, M_i)$
RXQUAL	6	0.9419	14	0.9317	–	–
RXLEV	any	< 0.7	any	< 0.7	–	–
FER	2	0.9633	1	0.9556	4	0.9752
LFER	6	0.8864	20	0.8682	10	0.9239
MxLFER	–	0.9088	–	0.8747	–	0.9331
MnMxLFER	1	0.9383	1	0.9282	1	0.9525
AMRmode	–	–	0.2	0.9259	2	0.8761

3.3.2 Parametric Speech Quality Measures

FER-based Measures

For GSM-EFR, GSM-AMR and UMTS-AMR transmissions, the Frame Erasure Rate (FER) parameter exhibits an outstanding speech quality correlation of $\rho = 0.9633$, $\rho = 0.9556$, or $\rho = 0.9752$, respectively. Therefore, an evaluation of this parameter alone can serve as a good speech quality estimation rule. For the GSM case, the correlation is depicted in Figs. 3.6 (EFR) and 3.7 (AMR).

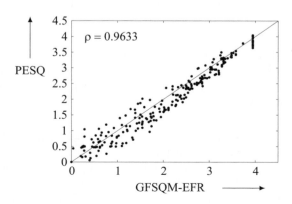

Figure 3.6: GSM: correlation of GFSQM-EFR and PESQ scores

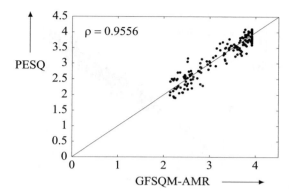

Figure 3.7: GSM: correlation of GFSQM-AMR and PESQ scores

The GSM FER-based speech quality measures (GFSQM) for EFR and AMR transmissions are given by

$$\text{GFSQM-EFR} = f_{\text{GE0}}(L_2(\text{FER})) \tag{3.3}$$

and

$$\text{GFSQM-AMR} = f_{\text{GA0}}(L_1(\text{FER})), \tag{3.4}$$

with optimum linearization polynomials $f_{\text{GE0}}(x)$ and $f_{\text{GA0}}(x)$. These polynomials are identified using a random subset covering half of the available simulation data. The given correlation coefficients are then calculated using the remaining part of the data.

By comparing Figs. 3.6 and 3.7, the quality advantage of the AMR over the EFR transmission, which was already illustrated in Fig. 3.3, can be observed in the distribution of data points. For the examined range of channel conditions, the unsatisfactory speech quality of some EFR clips (PESQ score below 2) is improved in the AMR transmission (all quality scores above 2).

The FER-based speech quality measure (UFSQM) for UMTS-AMR is given by

$$\text{UFSQM-AMR} = f_{\text{UA0}}(L_4(\text{FER})), \tag{3.5}$$

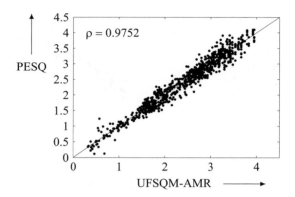

Figure 3.8: UMTS: correlation of UFSQM-AMR and PESQ scores

where f_{UA0} represents the optimum linearization polynomial. The correlation of the UFSQM-AMR measure with PESQ scores is indicated in Fig. 3.8.

Measures Based on Parameter Combinations

To enhance the robustness of the parameter-based quality assessment against measurement errors of single parameters, multiple parameters can be combined into a single speech quality measure. To find a suitable combination rule, the MSECT (Minimum Mean Square Error Coordinate Transformation) [Zahorian, Jagharghi 1992] procedure is employed. In the MSECT procedure, multidimensional data in predefined categories within a source space of dimension D are mapped onto target positions in a target space of dimension $Q < D$. The mapping function is optimized with respect to the minimum mean-square error between the mapping points and the specified target positions of training datasets. The optimal mapping function is of the form

$$\mathbf{c} = \mathbf{T} \cdot \mathbf{v} + \mathbf{o}. \tag{3.6}$$

Source vectors $\mathbf{v}$ are mapped in a linear way to target vectors $\mathbf{c}$, i.e., an optimal mapping matrix $\mathbf{T}$ and offset vector $\mathbf{o}$ are identified by the algorithm. This procedure is based on training datasets for which the target positions are already known.

The MSECT method is applied to the given task of mapping parameter vectors to PESQ scores. In this application, parameter groups resulting in different speech quality levels are regarded as the categories of the source space. Distinct PESQ values serve as target positions in the one-dimensional target space. L_P norms of the chosen parameters ζ are linearized by their according polynomial function f before serving as input vectors.

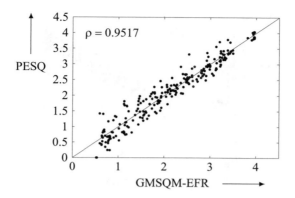

Figure 3.9: GSM: correlation of GMSQM-EFR and PESQ scores. © 2004 IEEE

With regard to a combination of the three GSM-EFR parameters with the highest individual PESQ correlations, the resulting speech quality measure is

$$\mathbf{v}_{\mathrm{GE}} = \begin{pmatrix} f_{\mathrm{GE1}}(L_6(\mathrm{RXQUAL})) \\ f_{\mathrm{GE2}}(L_2(\mathrm{FER})) \\ f_{\mathrm{GE3}}(L_1(\mathrm{MnMxLFER})) \end{pmatrix},$$

$$\mathrm{GMSQM\text{-}EFR} = \mathbf{T}_{\mathrm{GE}} \cdot \mathbf{v}_{\mathrm{GE}} + o_{\mathrm{GE}}, \tag{3.7}$$

with optimized values for $\mathbf{T}_{\mathrm{GE}}$ and o_{GE} and coefficients of $f_{\mathrm{GE}n}$.

For the GSM-AMR case, a combination of the above parameters, expanded by the AMR mode, yields

$$\mathbf{v}_{\mathrm{GA}} = \begin{pmatrix} f_{\mathrm{GA1}}(L_{14}(\mathrm{RXQUAL})) \\ f_{\mathrm{GA2}}(L_1(\mathrm{FER})) \\ f_{\mathrm{GA3}}(L_1(\mathrm{MnMxLFER})) \\ f_{\mathrm{GA4}}(L_{0.2}(\mathrm{AMRmode})) \end{pmatrix},$$

$$\mathrm{GMSQM\text{-}AMR} = \mathbf{T}_{\mathrm{GA}} \cdot \mathbf{v}_{\mathrm{GA}} + o_{\mathrm{GA}}. \tag{3.8}$$

Fifty percent of the available speech clips and PESQ scores are chosen as training data for the MSECT algorithm. The correlation coefficients of GMSQM-EFR and GMSQM-AMR are $\rho = 0.9517$ or $\rho = 0.9516$, respectively, based only on the datasets excluding the training data. These correlations are depicted as scatter-plots in Figs. 3.9 and 3.10.

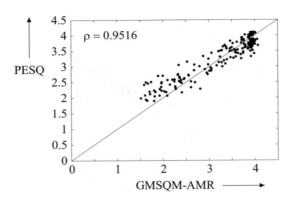

Figure 3.10: GSM: correlation of GMSQM-AMR and PESQ scores. © 2004 IEEE

For the UMTS-AMR combination measure, the two parameters FER and Mn-MxLFER, which exhibit the highest PESQ correlation are combined with the AMR mode:

$$\mathbf{v}_{\text{UA}} = \begin{pmatrix} f_{\text{UA1}}(L_4(\text{FER})) \\ f_{\text{UA2}}(L_1(\text{MnMxLFER})) \\ f_{\text{UA3}}(L_2(\text{AMRmode})) \end{pmatrix} ,$$

$$\text{UMSQM-AMR} = \mathbf{T}_{\text{UA}} \cdot \mathbf{v}_{\text{UA}} + o_{\text{UA}} , \tag{3.9}$$

with a mapping vector $\mathbf{T}_{\text{UA}}$ and offset value o_{UA} generated by the MSECT algorithm. The correlation of the UMSQM-AMR measure with PESQ scores is $\rho = 0.9805$, illustrated in Fig. 3.11.

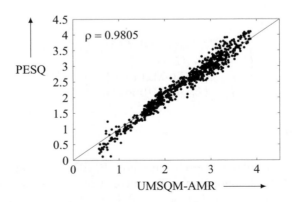

Figure 3.11: UMTS: correlation of UMSQM and PESQ scores

3.4 Discussion and Conclusions

It should be noted that the specific coefficients of the linearization polynomials f as well as the weighting factors in $\mathbf{T}$ and offset values in o of the derived measures depend on the configuration of the cellular network, e.g., the RRM settings and the employed AMR switching procedure. For each cellular network, they are to be determined individually, e.g., using the described procedure, separately for the uplink and downlink directions. All measures developed here as examples were established for the downlink direction.

In general, the approach to expressing the speech quality of voice transmissions in digital cellular communication networks by evaluating transmission parameters produces encouraging results. The correlation coefficients of the six presented parameter-based non-intrusive speech quality measures with reference PESQ scores are all in the range $0.95 \leq \rho \leq 0.98$ for GSM-EFR, GSM-AMR, and UMTS-AMR systems. These results motivate the utilization of parameter-based speech quality measures for automated quality monitoring in any cellular communication network that delivers suitable measurement parameters. It is shown that even the evaluation of one single transmission parameter (FER) can lead to a reliable speech quality prediction if a dedicated perception-based averaging method (L_P norm) as well as a linearization function are applied.

Bibliography

3GPP (2000a). *TS 25.213: Universal Mobile Telecommunications System (UMTS); Spreading and modulation (FDD)*. Ver. 3.2.0.

3GPP (2000b). *TS 25.215: Universal Mobile Telecommunications System (UMTS); Physical layer – Measurements (FDD)*. Ver. 3.4.0.

3GPP (2002). *TS 34.121: Universal Mobile Telecommunications System (UMTS); Terminal Conformance Specification; Radio Transmission and Reception (FDD)*. Ver. 3.9.0.

3GPP (2005). *TS 25.101: Universal Mobile Telecommunications System (UMTS); User Equipment (UE) radio transmission and reception (FDD)*. Ver. 6.8.0.

Beerends, J.; Stemerdink, J. (1994). A Perceptual Speech-Quality Measure Based on a Psychoacoustic Sound Representation, *Journal of the Audio Engineering Society*, vol. 42, pp. 115–123.

Brandenburg, K. (1987). Evaluation of Quality for Audio Encoding at Low Bit Rates, *Proceedings 82nd Audio Engineering Society Conv., preprint 2433*.

Broom, S. (2006). VoIP Quality Assessment: Taking Account of the Edge-Device, *IEEE Transactions on Audio, Speech and Language Processing*, vol. 14, no. 6, pp. 1977–1983.

ETSI (1998). *Rec. GSM 06.90: Adaptive Multi-Rate (AMR) Speech Transcoding*. Ver. 7.2.1.

ETSI (1999a). *Rec. GSM 05.05: Radio Transmission and Reception*. Ver. 8.14.0.

ETSI (1999b). *Rec. GSM 05.08: Radio Sub-System Link Control*. Ver. 8.4.0.

ETSI (1999c). *Rec. GSM 05.09: Link Adaptation.* Ver. 8.1.0.

ETSI (1999d). *Rec. GSM 06.61: Substitution and Muting of Lost Frames for Full Rate Speech Channels.* Ver. 8.0.1.

Gaspard, I. (1994). Efficient Methods for Evaluation and Prediction of Subjective Speech Quality in GSM Mobile Networks, *Proceedings of the IEEE Vehicular Technology Conference (VTC)*, vol. 1, pp. 334–337.

Gray, P.; Hollier, M.; Massara, R. (2000). Non-intrusive Speech Quality Assessment Using Vocal Tract Models, *IEE Proceedings on Vision, Image and Signal Processing*, vol. 147, no. 6, pp. 493–501.

Hagenauer, J.; Hoeher, P. (1989). A Viterbi Algorithm with Soft-Decision Outputs and its Applications, *Proceedings IEEE Globecom*, pp. 1680–1686.

Heck, K.; Staehle, D.; Leibnitz, K. (2002). Diversity Effects on the Soft Handover Gain in UMTS Networks, *Proceedings IEEE Vehicular Technology Conference (VTC) Fall*, vol. 2, pp. 1269–1273.

ITU-R (1998). *Rec. BS.1116: Methods for the Subjective Assessment of Small Impairments in Audio Systems Including Multichannel Sound Systems.*

ITU-T (1996a). *Rec. P.800: Methods for Subjective Determination of Transmission Quality.*

ITU-T (1996b). *Rec. P.810; Modulated Noise Reference Unit (MNRU).*

ITU-T (1996c). *Rec. P.830: Subjective Performance Assessment of Telephone-band and Wideband Codecs.*

ITU-T (1997). *Study Group 12: TOSQA – Telecommunications Objective Speech Quality Assessment.* COM12-34-E.

ITU-T (1998). *Rec. P.861: Objective Quality Measurement of Telephone-band (300–3400 Hz) Speech Codecs.*

ITU-T (2001). *Rec. P.862: Perceptual Evaluation of Speech Quality (PESQ), an Objective Method for End-to-End Speech Quality Assessment of Narrowband Telephone Networks and Speech Codecs.*

ITU-T (2002). *Rec. G.107: The E-model, a Computational Model for Use in Transmission Planning.*

ITU-T (2003a). *Rec. P.800.1: Mean Opinion Score (MOS) Terminology.*

ITU-T (2003b). *Rec. P.862.1: Mapping Function for Transforming P.862 Raw Result Scores to MOS-LQO.*

ITU-T (2004). *Rec. P.563: Single-ended Method for Objective Speech Quality Assessment in Narrow-band Telephony Applications.*

ITU-T (2005). *Rec. P.862.2: Wideband Extension to Recommendation for the Assessment of Wideband Telephone Networks and Speech Codecs.*

Juric, P. (1998). An Objective Speech Quality Measurement in the QVoice, *Proceedings of the IEEE 5th International Workshop on Systems, Signals and Image Processing (IWSSIP)*, pp. 156–163.

Karjalainen, M. (1985). A New Auditory Model for the Evaluation of Sound Quality of Audio System, *Proceedings of the IEEE International Conference on Acoustics, Speech, and Signal Processing (ICASSP)*, pp. 608–611.

Karlsson, A.; Heikkila, G.; Minde, T.; Nordlund, M.; Timus, B. (1999). Radio Link Parameter based Speech Quality Index – SQI, *Proceedings of the IEEE Workshop on Speech Coding*, pp. 147–149.

Kim, D.-S. (2005). ANIQUE: An Auditory Model for Single-ended Speech Quality Estimation, *SAP*, vol. 13, no. 5, pp. 821–831.

Park, S.-W.; Ryu, S.-K.; Park, Y.-C.; Youn, D.-H. (2000). A Bark Coherence Function for Perceived Speech Quality Estimation, *Proceedings of the International Conference on Spoken Language Processing (ICSLP)*, vol. 2, pp. 218–221.

Quackenbush, S.; Barnwell, T.; Clements, M. (1988). *Objective Measures of Speech Quality*, Englewood Cliffs, NJ: Prentice-Hall.

Rix, A.; Beerends, J.; Hollier, M.; Hekstra, A. (2001). Perceptual Evaluation of Speech Quality (PESQ) – A New Method for Speech Quality Assessment of Telephone Networks and Codecs, *Proceedings of the IEEE International Conference on Acoustics, Speech, and Signal Processing (ICASSP)*, vol. 2, pp. 749–752.

Rix, A.; Beerends, J.; Kim, D.-S.; Kroon, P.; Ghitza, O. (2006). Objective Assessment of Speech and Audio Quality – Technology and Applications, *IEEE Transactions on Audio, Speech and Language Processing*, vol. 14, no. 6, pp. 1890–1901.

Rix, A.; Hollier, M. (2000). The Perceptual Analysis Measurement System for Robust End-to-End Speech Quality Assessment, *Proceedings of the IEEE International Conference on Acoustics, Speech, and Signal Processing (ICASSP)*, vol. 3, pp. 1515–1518.

Syn (2004). *CoCentric System Studio User Guide*, 2004.09 edn.

Wang, S.; Sekey, A.; Gersho, A. (1992). An Objective Measure for Predicting Subjective Quality of Speech Coders, *IEEE Journal on Selected Areas of Communication*, vol. 10, no. 5, pp. 819–829.

Wänstedt, S.; Pettersson, J.; Xiangchun, T.; Heikkilä, G. (2002). Development of an Objective Speech Quality Measurement Model for the AMR Codec, *Proceedings Workshop Measurement of Speech and Audio Quality in Networks (MESAQIN)*, pp. 77–82.

Werner, M. (2005). *Methods for Quality and Capacity Enhancements in UMTS Mobile Radio Networks*, PhD thesis, Aachen University (in German).

Werner, M.; Junge, T.; Vary, P. (2004). Quality Control for AMR Speech Channels in GSM Networks, *Proceedings of the IEEE International Conference on Acoustics, Speech, and Signal Processing (ICASSP)*, vol. 3, pp. 1076–1079.

Werner, M.; Kamps, K.; Tuisel, U.; Beerends, J.; Vary, P. (2003). Parameter-based Speech Quality Measures for GSM, *IEEE International Symposium on Personal, Indoor and Mobile Radio Communications (PIMRC)*, vol. 3, pp. 2611–2615.

Werner, M.; Vary, P. (2005). Quality Control for UMTS-AMR Speech Channels, *Proceedings 9th European Conference on Speech Communication and Technology (Interspeech)*.

Zahorian, A.; Jagharghi, A. (1992). Minimum Mean Square Error Transformations of Categorical Data to Target Positions, *IEEE Transactions on Signal Processing*, vol. 40, no. 1, pp. 13–23.

Zwicker, E.; Fastl, H. (1999). *Psychoacoustics*, 2nd edn, Springer, Berlin.

II

Adaptive Algorithms in Acoustic Signal Processing

Chapter 4

Kalman Filtering in Acoustic Echo Control: A Smooth Ride on a Rocky Road

Gerald Enzner

4.1 Introduction

Acoustic echo occurs in all modern voice communication systems with hands-free acoustic transducers. It has been recognized that the precise separation of the disturbing echo from the desired near-end speech is a difficult task in adaptive signal processing. Thus, a lot of work has been devoted to the development of adaptive filters for acoustic echo cancellation and suppression and to the design of sophisticated control mechanisms to ensure their robustness in adverse conditions. Because of the heuristic combination of adaptive filters and control mechanisms, however, the available systems may not satisfy a given optimization criterion. The difficulty in the design of hands-free voice communication systems has been subsumed in the metaphorical statement "From Algorithms to Systems - It's a Rocky Road" [Hänsler 1997]. In this chapter, we will show that the Kalman filter is key to a most elegant and yet efficient unification of adaptive filtering and adaptation control.

Our presentation of the mainstream in algorithm development for single channel acoustic echo control is based on the block diagram in Fig. 4.1. The available system concepts mostly use a combination of adaptive filtering to adjust an echo canceler, suboptimal control mechanisms, and some kind of post-processing for residual echo suppression. Nowadays, there is often little or no signal processing at all applied in receiving direction. Choosing the building blocks of a complete system successfully

Advances in Digital Speech Transmission   Edited by R. Martin, U. Heute and C. Antweiler
© 2008 John Wiley & Sons, Ltd

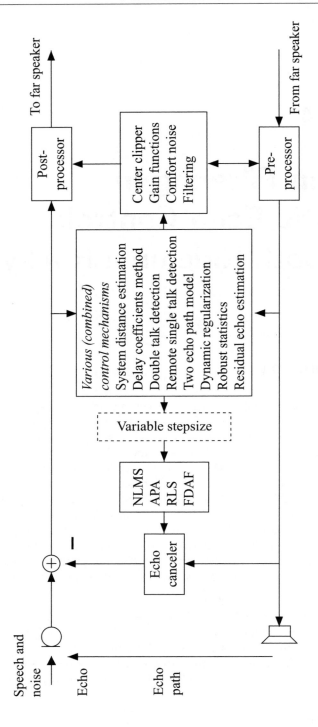

Figure 4.1: Typical building blocks of an acoustic echo controller

and optimizing their parameters is a time-consuming procedure and requires extensive practical experience.

Because many specific variants were proposed in literature, this overview section cannot be complete, but it will help to clarify open problems and the aims of this chapter. For further reading and extensive bibliography on the history of acoustic echo control, the following articles and textbooks are recommended: [Hänsler 1992], [Hänsler 1994], [Breining et al. 1999], [Gay, Benesty 2000], [Benesty et al. 2001], [Hänsler, Schmidt 2004], [Vary, Martin 2006].

4.1.1 Adaptive Filter Structures for Acoustic Echo Control

Voice controlled switching was developed in the 1960s and is still used in many products to suppress the acoustic echo of the far speaker. In most implementations, the hands-free microphone signal is strongly attenuated, whenever a received signal is detected from the far speaker side. Alternatively, if near-end speech activity is predominant, the loudspeaker signal can be attenuated. Voice controlled switches can be implemented in a very simple way in analog or digital form, but the fundamental problem is that switching effectively leads to an unacceptable half-duplex transmission between both ends of the communication system. In particular, the perception of background noise is very unnatural in this case. Therefore, voice controlled switches (and other gain functions in sending or receiving direction of the system) are nowadays implemented in conjunction with comfort noise injection.

In the 1970s and 1980s, the *adaptive echo canceler* in parallel to the electroacoustic echo path was identified as a seemingly ideal solution for acoustic echo control [Hänsler 1992], [Hänsler 1994]. The working assumption (to this day) is that the adaptive filter uses the known loudspeaker signal to generate an exact replica of the acoustic echo. This echo replica is then subtracted from the microphone signal in order to obtain the undistorted near-end speech. It has been realized, however, that the fast and robust tracking of the true time-varying echo path of real acoustic environments is an extremely difficult issue. Residual echo always remains after the echo canceler. As a result, it is now widely accepted that an echo canceler alone will not be able to deliver sufficient echo attenuation [Hänsler, Schmidt 2004], [Vary, Martin 2006].

The most prominent *adaptive algorithms* for adjusting the echo canceler coefficients are the NLMS (normalized least mean-square), APA (affine projection), RLS (recursive least-squares), and FDAF (frequency-domain adaptive filter) algorithm. All of them are based on an iterative update of the filter coefficients, which is controlled by a stepsize parameter. The NLMS algorithm is definitely the simplest variant, but the underlying model assumptions are "over-conservative" [Haykin 2002] and therefore the algorithm is slow in following the true echo path. If the stepsize is chosen to be large, the NLMS will diverge in the presence of observation noise (near-end speech and background noise). Unfortunately, the other adaptive algorithms basically suffer from the same problems. Although APA, RLS, and FDAF incorporate received signal statistics (far-end speech properties) into the learning process, they still do not consider the observation noise and echo path properties, which is a prerequisite for fast and robust adaptation. The latter deficiency must be attributed to the fact that NLMS, APA, RLS, FDAF, and their derivatives are all obtained from a similar *deterministic* optimization criterion, i.e., least-squares, constrained least-squares, or certain variants of the least-squares criterion [Haykin 2002].

Combinations of echo cancelers and voice controlled switches are often implemented to improve an insufficient echo attenuation, but the related distortion of the near-end speech is annoying. Therefore, in the 1990s, the *frequency-selective adaptive postfilter* in the sending path of the communication system was proposed to reduce the residual echo after the echo canceler [Martin, Altenhöner 1995], [Martin 1995], [Martin, Gustafsson 1996], [Gustafsson et al. 1998]. In principle, the operation of the postfilter for residual echo suppression is very similar to that of a noise suppression filter and it was reported that both functionalities can be combined efficiently [Gustafsson et al. 1998], [Le Bouquin Jeannès et al. 2001]. The key to an effective postfilter for residual echo attenuation is the availability of the power spectral density (PSD) of the residual echo. However, the residual echo is not a directly measurable signal in the presence of observation noise and, therefore, the calculation (estimation) of the residual echo PSD is not trivial [Enzner et al. 2002a], [Enzner et al. 2002b].

It should be noted that, to some extent, even the postfilter for residual echo suppression will cause distortion of the near-end speech, either by unwanted speech attenuation or by remaining residual echo fragments. Thus, regarding the overall quality, the adaptive echo canceler should still remove as much echo as possible in order to minimize the distortion related to postfiltering [Enzner, Vary 2005].

4.1.2 Control of Adaptive Filters

Adaptive filters have one fundamental problem in common. On the one hand, the adaptation must be fast enough to track time-varying acoustic echo paths, on the other hand, the adaptation must be robust against interfering near-end speech and background noise. Both requirements are conflicting and, for this reason, sophisticated control mechanisms were designed to enable the fast and robust adaptation of echo cancelers using the NLMS, APA, RLS, or FDAF algorithm [Mader et al. 2000], [Benesty et al. 2001], [Haykin 2002], [Hänsler, Schmidt 2004]. In the attempt to satisfy both requirements, many systems utilize a time-varying adaptive stepsize (or adaptive memory in case of the RLS algorithm). However, a perfect solution is not available. In the following, we summarize the most important concepts:

The optimum stepsize for the NLMS algorithm (in the MMSE sense) has been derived several times, e.g., [Meissner et al. 1980], [Yamamoto, Kitayama 1982], [Frenzel 1992], [Mader et al. 2000], [Haykin 2002], [Hänsler, Schmidt 2004]. The optimum time- and frequency-dependent stepsize function for the FDAF and PBFDAF (partitioned block frequency-domain adaptive filter) has been deduced in [Nitsch 2000]. Unfortunately, these optimal stepsizes cannot be implemented directly, because the required time-varying and potentially frequency-dependent system distance (square error) between echo canceler and true echo path is hardly accessible in practice. Thus, suboptimal control mechanisms have been developed to approximate the optimum stepsize. Some methods have been designed explicitly to estimate the system distance, others can be employed either to control the adaptation directly or to facilitate the estimation of the system distance.

- The popular *delay coefficients method* computes the system distance from the leading coefficients of the adaptive filter, provided that the echo path has natural or artificial zeros at the corresponding index positions [Yamamoto, Kitayama 1982]. It is important to note that the delay coefficients method alone is not able to deliver a reliable estimate of the system distance. An additional detector for echo path changes is required to avoid stalling (freezing) of the adaptation [Frenzel 1992], [Haykin 2002], [Hänsler, Schmidt 2004].

- The famous concept of *double talk detection* can be used to directly halt the adaptation of the echo canceler in the presence of near-end speech at the microphone [Benesty et al. 2001] or, alternatively, to control the estimation of the system distance [Hänsler, Schmidt 2004]. Double talk detectors are based on cross-correlation measures, e.g., [Benesty et al. 2000], or simply on the comparison of signal powers [Duttweiler 1978]. *Remote single talk detection* as described by [Hänsler, Schmidt 2004] is closely related to double talk detection.

- Also the *two echo path model* [Ochiai et al. 1977] can be utilized in different ways to control the adaptation (either directly or indirectly through the stepsize). Basically, the approach models a fast and a slowly changing echo path by a background and a foreground adaptive filter, respectively. A power comparison of the two error signals determines the actual control strategy for the foreground adaptive filter. This technique finds widespread application in cases where its computational complexity is acceptable.

- *Dynamic regularization* controls gradient adaptive filters by means of a time-varying additive quantity in the denominator of the gradient [Haykin 2002], [Buchner, Kellermann 2002], [Myllylä, Schmidt 2002]. It has been shown in [Hänsler, Schmidt 2003] that the optimum regularization of the NLMS algorithm is, theoretically, equivalent to its optimum stepsize control.

Regarding the post-processor or postfilter of the system, there are also several options how to control a fixed or adaptive, linear or nonlinear, scalar or frequency-dependent echo attenuation. Post-processing techniques are, however, not so well documented in the literature as adaptive echo cancellation filters. Byproducts of the control mechanisms for the echo canceler are sometimes used to control the post-processor. Some practical hints on post-processing can be found, for example, in [Eneroth et al. 2000], [Benesty et al. 2001, Chap. 7], and [Hänsler, Schmidt 2004].

A mathematical proof of the tight relationship between the optimum statistical adaptation of echo canceler and postfilter coefficients has just recently appeared in the literature [Hänsler, Schmidt 2000], [Schmidt 2001], [Enzner et al. 2002a], [Enzner 2003]. This relationship helps to realize an intelligent interaction of both filters, leading to an improved output signal quality of the system. Thanks to the synergy, the joint control of echo canceler and postfilter can be even simpler than their individual control. This has been demonstrated by the simple and robust acoustic echo controller proposed by [Enzner, Vary 2003], which provides the required echo attenuation and also preserves the full-duplex ability of the system [Enzner et al. 2004].

4.1.3 Open Problems / Organization of this Chapter

The previous two sections could only give hints towards the large variety of system options for designing acoustic echo controllers. A closer look reveals that each system component (adaptive echo canceler, control mechanism, or post-processor) is based on several tuning parameters, e.g., time constants, thresholds, frame lengths, regularization parameters, etc., which have to be optimized by the system developer. Extensive experience is required in order to create a reliable system with full duplex ability and persistent echo attenuation. In any case, a precise statement regarding the optimality of the resulting system is not possible.

The variety of system options results from the fact that most researchers have focused on either the design of adaptive filters, control mechanisms, or post-processing. Thus, the main objective of this chapter is to create a unified understanding of adaptive filtering and adaptation control. This requires, first, the definition of an adequate statistical model of the time-varying acoustic environment of hands-free telephones and, secondly, the rigorous mathematical derivation of the signal processing solution that satisfies a given optimization criterion – subject to the model.

In Sec. 4.2.1, we introduce our system model for acoustic echo control. In contrast to the traditional approach, the acoustic echo path is characterized as a random process with statistical mean and covariance. The non-zero covariance of the echo path reflects the uncertainty about the true echo path coefficients. Conversely, the known echo path input (far-end speech) is modeled as a deterministic signal.

Based on the new system model, Sec. 4.2.2 derives the general nonlinear MMSE (minimum mean-square error) estimator for the near-end speech components in the microphone signal. The derivation demonstrates, for the first time, that the general conditional mean estimator for the speech signal basically maps onto a conditional mean estimator for the unknown echo path. Moreover, it is proven rigorously that the general MMSE estimator for the speech signal decomposes into an acoustic echo canceler (based on the conditional echo path mean) and an MMSE post-processor (postfilter) for residual echo suppression.

Under the assumption of Gaussianity of the involved random processes, Sec. 4.2.3 explains that the MMSE post-processor is given by a linear Wiener filter, the key parameter of which is the estimation error covariance of the echo path.

In Sec. 4.3, we will introduce a refinement of the system model. The echo generation model is formulated in the block frequency-domain in order to facilitate efficient signal processing and the time evolution of the true echo path is modeled as a first-order statistical Markov process. Since the echo signal at the microphone is always recorded in the presence of near-end speech or background noise, the echo path therefore obeys a general stochastic state-space model. That creates the perspective of utilizing the Kalman filter for conditional echo path mean and covariance calculation.

Despite the simplicity of the underlying system model, Sec. 4.4 shows that the performance of the resulting adaptive algorithm clearly satisfies the requirements of advanced hands-free voice terminals with full-duplex ability, though sophisticated adaptation control or cumbersome parameter tuning are not needed anymore.

In summary, we shall see in this chapter that:

- the proposed statistical echo path model (stochastic state-space model) is in line with the real boundary conditions of the acoustic echo control problem;

- the derivation of the MMSE solution for acoustic echo control, based on the new echo path model, proves the principal coexistence of echo canceler and postfilter in hands-free voice communication systems;

- the Kalman filter jointly and recursively adjusts the echo canceler and postfilter coefficients and therefore constitutes an outstandingly compact and robust signal processing solution for acoustic echo control;

- the unification of adaptive filtering and adaptation control through the Kalman filter simplifies derivation, design, and realization of acoustic echo controllers.

4.2 A Comprehensive Theory of Acoustic Echo Control

The following estimation framework is introduced in the context of the acoustic echo control problem. However, the basic concept is formulated generally enough, so that other applications in acoustic signal processing or in communication theory might profit, especially where channel estimation, channel equalization, or interference cancellation is of fundamental interest.

4.2.1 Stochastic Modeling of the Echo Path

We consider the generic signal model of hands-free voice communication systems at discrete time k as shown in Fig. 4.2. In receiving direction, a possibly processed version $x(k)$ of the received signal $x'(k)$ is played back by the loudspeaker. In sending direction, the microphone captures near-end speech $s(k)$ as well as the room reflections

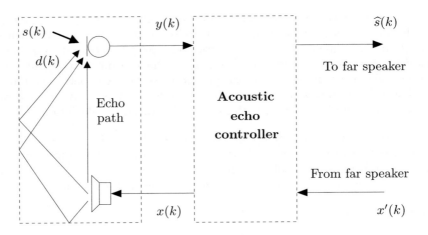

Figure 4.2: Acoustic front-end of a hands-free voice communication system

(echoes) of the loudspeaker signal $x(k)$. The microphone signal $y(k)$ is thus considered as an additive mixture of speech $s(k)$ and echo $d(k)$:

$$y(k) = s(k) + d(k) \tag{4.1}$$

$$= s(k) + \sum_{n=0}^{N-1} w_n \, x(k-n) \ . \tag{4.2}$$

If sufficiently linear electroacoustic transducers are used in the system and if the significant memory of the echo path is not longer than N samples, then the convolution based on the impulse response coefficients w_n is generally accepted as a realistic model of the echo generation. Owing to transmission delay, e.g., in mobile networks and Internet telephony, the acoustic echo $d(k)$ is not tolerated by the far speaker.

In the traditional theory of acoustic echo control, the speech signal $s(k)$ and the loudspeaker signal $x(k)$ are both modeled as independent *random processes*, while the echo path coefficients w_n are treated as unknown *deterministic* parameters. For these model assumptions, an MMSE optimization of the AEC leads to the classical Wiener solution, in which the echo canceler duplicates the true echo path in order to compensate the acoustic echo in the microphone signal.

In contrast to the traditional approach, we consider the echo as the linear convolution of a *measurable*, i.e., *deterministic* loudspeaker signal $x(k)$ with the unknown echo path coefficients w_n. Because of the uncertainty about the fluctuating acoustic echo path, the coefficients w_n are now modeled as independent *random processes* with statistical expectations $w_{1,n}$ at index positions n and covariances p_{ij}:

$$w_{1,n} = \mathrm{E}\{w_n\} \tag{4.3}$$

$$p_{ij} = \mathrm{E}\{w_{r,i} \, w_{r,j}\} \ . \tag{4.4}$$

Here, the mean $w_{1,n}$ represents a systematic (i.e., deterministic) component of the echo path, while the residual $w_{r,n} = w_n - w_{1,n}$ denotes an unpredictable (i.e., purely random) component. This new system model matches the practical boundary conditions of the echo control problem, in which the echo path coefficients w_n are not directly measurable, while there is indeed no uncertainty about the echo path input signal $x(k)$. The effect of the proposed system model regarding the MMSE solution for acoustic echo control will be demonstrated in the following section.

4.2.2 Minimum Mean-Square Error (MMSE) Solution

In the acoustic echo control literature, as mentioned before, the echo canceler is often considered as the ideal solution to remove the echo of hands-free telephones, while the widely used post-processor is treated only as an auxiliary part of the system. Based on the new system model, we will develop an alternative view, which is more appropriate for the design of acoustic echo controllers.

The full-duplex operation of the hands-free telephone in Fig. 4.2 requires a strong attenuation of the acoustic echo signal $d(k)$ by the acoustic echo controller and, ideally, an undistorted reproduction of the speech $s(k)$ at the system output $\widehat{s}(k)$. Mathematically, this signal processing conflict can be expressed as a statistical optimization problem which aims, e.g., at the MMSE between $s(k)$ and $\widehat{s}(k)$:

$$\epsilon^2 = \mathrm{E}\big\{\big(s(k) - \widehat{s}(k)\big)^2\big\} \;\to\; \min . \tag{4.5}$$

Not making assumptions on the statistics of the involved signals and not imposing linearity of the estimator, the system output $\widehat{s}(k)$ according to (4.5) is obtained as the conditional mean of $s(k) = s_k$, given the observed data $y(k) = y_k$ up to and including the current time instant [Papoulis 1984], [Scharf 1991]:

$$\widehat{s}(k) = \mathrm{E}\{s_k \,|\, \mathbf{y}^k_{-\infty}\} \tag{4.6}$$

$$= \int\limits_{-\infty}^{\infty} s_k \, p_{s|y}(s_k \,|\, \mathbf{y}^k_{-\infty}) \, \mathrm{d}s_k , \tag{4.7}$$

where $\mathbf{y}^k_{-\infty} = (y_k, y_{k-1}, \ldots, y_{-\infty})^T$ is an example of our vector notation of time-domain signals and $p(\cdot)$ denotes probability density functions.

With Bayes rule, we can reformulate the integrand in (4.7), obtaining

$$\widehat{s}(k) = \int\limits_{-\infty}^{\infty} s_k \, \frac{p_{s,y|y}(s_k, \mathbf{y}^k_{k-R+1} \,|\, \mathbf{y}^{k-R}_{-\infty})}{p_{y|y}(\mathbf{y}^k_{k-R+1} \,|\, \mathbf{y}^{k-R}_{-\infty})} \, \mathrm{d}s_k \tag{4.8}$$

$$= \int\limits_{-\infty}^{\infty} s_k \, \frac{p_{y|y,s}(\mathbf{y}^k_{k-R+1} \,|\, \mathbf{y}^{k-R}_{-\infty}, s_k) \, p_{s|y}(s_k \,|\, \mathbf{y}^{k-R}_{-\infty})}{p_{y|y}(\mathbf{y}^k_{k-R+1} \,|\, \mathbf{y}^{k-R}_{-\infty})} \, \mathrm{d}s_k , \tag{4.9}$$

for any $R \in \mathbb{N}$. Assuming that R is larger than or equal to the span of correlation of $s(k)$, we have $p_{s|y}(s_k \mid \mathbf{y}_{-\infty}^{k-R}) = p_s(s_k)$, and thus

$$\widehat{s}(k) = \int\limits_{-\infty}^{\infty} s_k \, \frac{p_{y|y,s}(\mathbf{y}_{k-R+1}^k \mid \mathbf{y}_{-\infty}^{k-R}, s_k) \, p_s(s_k)}{p_{y|y}(\mathbf{y}_{k-R+1}^k \mid \mathbf{y}_{-\infty}^{k-R})} \, \mathrm{d}s_k \,. \tag{4.10}$$

According to the principal of orthogonality of mean-square estimation [Papoulis 1984], we can drop the condition $\mathbf{y}_{-\infty}^{k-R}$, if we rewrite this integral in terms of the mean-removed signal $\mathbf{e}_{k-R+1}^k = \mathbf{y}_{k-R+1}^k - \mathrm{E}\{\mathbf{y}_{k-R+1}^k \mid \mathbf{y}_{-\infty}^{k-R}\}$ using the corresponding probability densities $p_e(\cdot)$ and $p_{e|s}(\cdot)$:

$$\widehat{s}(k) = \int\limits_{-\infty}^{\infty} s_k \, \frac{p_{e|s}(\mathbf{e}_{k-R+1}^k \mid s_k) \, p_s(s_k)}{p_e(\mathbf{e}_{k-R+1}^k)} \, \mathrm{d}s_k \,. \tag{4.11}$$

Using Bayes rule again, the result is obviously equivalent to a conditional mean estimator based on the finite data set $\mathbf{e}_{k-R+1}^k$, i.e.,

$$\widehat{s}(k) = \mathrm{E}\{s_k \mid \mathbf{e}_{k-R+1}^k\} \,. \tag{4.12}$$

Hence, we have a two-stage estimation procedure. At first we have to determine the vector of mean-removed samples, $\mathbf{e}_{k-R+1}^k$, given the history of observations $\mathbf{y}_{-\infty}^{k-R}$, and subsequently we have to determine s_k, given this vector $\mathbf{e}_{k-R+1}^k$. In accordance with (4.2), the mean removal obeys the following expressions:

$$\mathbf{e}_{k-R+1}^k = \mathbf{y}_{k-R+1}^k - \mathrm{E}\{\mathbf{y}_{k-R+1}^k \mid \mathbf{y}_{-\infty}^{k-R}\} \tag{4.13}$$

$$= \mathbf{y}_{k-R+1}^k - \mathrm{E}\{\mathbf{s}_{k-R+1}^k \mid \mathbf{y}_{-\infty}^{k-R}\} - \mathrm{E}\{\mathbf{X}_{k-N-R+1}^{k^T}\mathbf{w} \mid \mathbf{y}_{-\infty}^{k-R}\} \tag{4.14}$$

$$\approx \mathbf{y}_{k-R+1}^k - \mathbf{X}_{k-N-R+1}^{k^T} \mathrm{E}\{\mathbf{w} \mid \mathbf{y}_{-\infty}^{k-R}\} \tag{4.15}$$

$$= \mathbf{s}_{k-R+1}^k + \mathbf{X}_{k-N-R+1}^{k^T}\mathbf{w} - \mathbf{X}_{k-N-R+1}^{k^T} \mathrm{E}\{\mathbf{w} \mid \mathbf{y}_{-\infty}^{k-R}\} \tag{4.16}$$

$$= \mathbf{s}_{k-R+1}^k + \mathbf{X}_{k-N-R+1}^{k^T}\mathbf{w}_r \tag{4.17}$$

$$= \mathbf{s}_{k-R+1}^k + \mathbf{b}_{k-R+1}^k \,, \tag{4.18}$$

where $\mathbf{w} = (w_0, \ldots, w_{N-1})^T$ denotes the vector of unknown echo path coefficients, $\mathbf{w}_r = \mathbf{w} - \mathrm{E}\{\mathbf{w} \mid \mathbf{y}_{-\infty}^{k-R}\}$ the unpredictable, i.e., non-systematic component of the echo path, $\mathbf{X}_{k-N-R+1}^k = (\mathbf{x}_{k-N+1}^k, \ldots, \mathbf{x}_{k-N-R+1}^{k-R})$ the matrix of shifted echo path input vectors, and $\mathbf{b}_{k-R+1}^k = \mathbf{X}_{k-N-R+1}^{k^T}\mathbf{w}_r$ the residual echo vector after mean removal. The approximation in (4.15) is well justified by the fact that the vector $\mathbf{s}_{k-R+1}^k$ is hardly predictable from the generally noisy observations $\mathbf{y}_{-\infty}^{k-R}$ of the past.

Drawing the conclusion from this section, we have proven the separability of the echo control problem into a system identification task and an optimum filtering problem.

- The system identification task consists of determining a systematic (mean) echo path component $\mathbf{w}_1 = \mathrm{E}\{\mathbf{w} \mid \mathbf{y}_{-\infty}^{k-R}\}$, given the observations $\mathbf{y}_{-\infty}^{k-R}$. The quantity $\mathbf{w}_1$ is then used for creating an echo replica in order to partially compensate the echo in the microphone signal according to (4.15). The latter step is referred to as *echo cancellation*, e.g., [Sondhi 1967], [Hänsler, Schmidt 2004].

- An MMSE post-processor according to (4.12) has to be applied to attenuate the unpredictable residual echo $b(k)$ that is still present in the error signal $e(k) = s(k) + b(k)$ after echo cancellation. Echo attenuation by post-processing is termed *echo suppression* [Hänsler, Schmidt 2004], [Vary, Martin 2006].

This separability holds for any probability distribution and is indeed optimal in the MMSE sense. In contrast to earlier work, the optimal strategies for echo cancellation and post-processing were derived jointly from the MMSE criterion.

4.2.3 MMSE Processor in the Gaussian Case

Provided that the unknown system (*here:* the acoustic echo path $\mathbf{w}$) obeys a first-order Gauss–Markov model, it is known that the Kalman filter performs a recursive calculation of the conditional mean $\mathbf{w}_1 = \mathrm{E}\{\mathbf{w}|\mathbf{y}_{-\infty}^{k-R}\}$ of the unknown system, given the noisy observations $y(k)$ of the past [Papoulis 1984], [Scharf 1991].

In the case of Gaussianity of the involved random processes, it is further known that the conditional mean estimator in (4.12) is equivalent to a minimum phase Wiener filter $\mathbf{w}_2$ with $\mathbf{e}_{k-R+1}^{k}$, obtained from (4.15), as the input signal vector and $\widehat{s}(k)$ as the output signal, e.g., [Papoulis 1984], [Scharf 1991]:

$$\widehat{s}(k) = \mathbf{w}_2^T \mathbf{e}_{k-R+1}^{k} \tag{4.19}$$

$$= \mathbf{w}_2^T \mathbf{s}_{k-R+1}^{k} + \mathbf{w}_2^T \mathbf{b}_{k-R+1}^{k} \tag{4.20}$$

$$= \widetilde{s}(k) + \widetilde{b}(k) , \tag{4.21}$$

where the symbols $\widetilde{s}(k)$ and $\widetilde{b}(k)$ are used to denote post-processed speech and residual echo components, respectively. Just note, by introducing an appropriate algorithmic delay into the estimation problem posed in (4.6), we could also have a linear phase Wiener filter in order to avoid phase distortions.

According to [Papoulis 1984], [Proakis, Manolakis 1996], the Wiener filter is determined by the normal equations,

$$\mathbf{R}_{ee}\mathbf{w}_2 = \boldsymbol{\varphi}_{se} , \tag{4.22}$$

where $\mathbf{R}_{ee} = \mathrm{E}\{\mathbf{e}_{k-R+1}^{k} \mathbf{e}_{k-R+1}^{k\,T}\}$ and $\boldsymbol{\varphi}_{se} = \mathrm{E}\{s_k \mathbf{e}_{k-R+1}^{k}\}$ are the correlation matrix of $e(k)$ and the cross-correlation vector of $s(k)$ with $e(k)$, respectively. With (4.17) and by invoking the independence of $s(k)$ and $\mathbf{w}_r$, we arrive at $\boldsymbol{\varphi}_{se} = \boldsymbol{\varphi}_{ss}$ and

$$\mathbf{R}_{ee} = \mathbf{R}_{ss} + \mathbf{X}_{k-N-R+1}^{k}{}^{T} \mathbf{p}\, \mathbf{X}_{k-N-R+1}^{k} \cdot \tag{4.23}$$

Suppose that the estimation error covariance matrix $\mathbf{p} = \mathrm{E}\{\mathbf{w}_r\mathbf{w}_r^T\}$ is available. Then we could calculate the correlation matrix $\mathbf{R}_{ss}$ and thus vector $\mathbf{r}_{ss}$ from (4.23), since $\mathbf{X}_{k-N-R+1}^{k}$ is known and since $\mathbf{R}_{ee}$ is measurable from $e(k)$. Hence, the key problem of postfiltering is to determine the quantity $\mathbf{p}$.

In contrast to deterministic adaptive algorithms, e.g., LMS and RLS, the Kalman filter calculates the estimation error covariance $\mathbf{p} = \mathrm{E}\{\mathbf{w}_r\mathbf{w}_r^T\}$ as a byproduct of the statistical estimation procedure for $\mathbf{w}_1 = \mathrm{E}\{\mathbf{w}|\mathbf{y}_{-\infty}^{k-R}\}$. In that sense, we have to consider the Kalman filter as a joint conditional mean and covariance estimator that perfectly fits the requirements of the acoustic echo control problem.

4.3 The Kalman Filter for Conditional Mean and Covariance Estimation

Despite its potential for optimality, the attempt to use the Kalman filter in acoustic echo cancellation hardly ever appears in literature. In an early paper [Meissner et al. 1980], the authors compared the NLMS algorithm with a degenerated version of the Kalman filter (assuming a time-invariant echo path and other significant statistical simplifications). In [Lippuner, Kälin 1999], the Kalman filter was examined for white noise input only and in [Lippuner 2002], it was just utilized to derive a model-based scalar stepsize for the NLMS algorithm.

It seems that the general form of the Kalman filter has been completely avoided in acoustic echo control. This can be attributed to the high computational complexity and to the possibility of numerical instability of the exact high-dimensional Kalman filter. Moreover, an appropriate signal model for the Kalman filter (in the form of observation and process noise covariance matrices of the underlying state-space model) has not existed up to now [Hänsler, Schmidt 2004].

In this section, we shall demonstrate that the Kalman filter is indeed the adequate signal processing tool for acoustic echo control according to the MMSE criterion. The approach that we pursue uses a blockwise reformulation of the linear convolution echo path model (4.2) in the DFT domain (Sec. 4.3.1). The time evolution of the echo path will be described by a first-order statistical Markov model in the transform domain (Sec. 4.3.2). In this way, we obtain a mathematically tractable stochastic state-space model of the echo path, which provides at least a minimum of reliable *a priori* information for echo path estimation. The model formulated in the block frequency-domain naturally considers a possible correlation of all input signals.

Based on the system model, we can easily write down the exact Kalman filter for echo path estimation in the DFT domain (Sec. 4.3.3). Since we have nearly diagonal covariance matrices in the transform domain, the aforementioned drawbacks of the exact Kalman filter can be circumvented at the cost of mild approximations. We obtain a diagonalized version of the exact Kalman filter, which represents a self-contained, near optimum, and yet workable solution for acoustic echo control (Sec. 4.3.4).

Interestingly, we can show that the diagonalized Kalman filter decomposes into frequently used standard components of adaptive signal processing: gradient-based adaptive filtering, optimum stepsize control, and system distance (convergence state) estimation. Conversely, the diagonalized Kalman filter can be understood as a unification of the classical concepts of adaptive filtering and adaptation control (Sec. 4.3.5).

4.3.1 Linear Echo Path Model in DFT-Matrix Form

Block-processing generally aims at the efficient calculation of output signals from short-time stationary frames of the input signals. The derivation of a block-processing algorithm for echo path estimation clearly requires a blockwise formulation of the linear measurement equation (4.2) for the echo path. Here, this happens by means of the overlap-save method using the discrete Fourier transform (DFT):

Consider the vector $\mathbf{x}(\kappa) = \mathbf{x}_{\kappa R - M + 1}^{\kappa R}$ at frame-time index $\kappa \in \mathbb{Z}$, which contains the M latest samples of the loudspeaker signal $x(k)$ in the time-domain, i.e.,

$$\mathbf{x}(\kappa) = \big(x(\kappa R - M + 1), x(\kappa R - M + 2), \ldots, x(\kappa R)\big)^H , \tag{4.24}$$

where R is the frame-shift and superscript ^{H} denotes Hermitian transposition. From $\mathbf{x}(\kappa)$, we construct a complex excitation matrix $\mathbf{X}(\kappa)$ in the DFT domain, i.e.,

$$\mathbf{X}(\kappa) = \text{diag}\{\mathbf{F}_M \mathbf{x}(\kappa)\} , \tag{4.25}$$

where $\mathbf{F}_M$ is the Fourier matrix of size $M \times M$ and $\text{diag}\{\cdot\}$ produces a diagonal matrix from its input vector. Note that the elements of $\mathbf{X}(\kappa)$ can be calculated efficiently from the loudspeaker signal by the fast Fourier transform (FFT).

Next, we define the echo path vector $\mathbf{W}(\kappa)$ in the DFT domain, based on the zero-padded time-domain coefficients $\mathbf{w}(\kappa)$, assuming that the finite number of model taps $w_n(\kappa)$ at frame-time κ will cover the length of the echo path ($N = M - R$):

$$\mathbf{W}(\kappa) = \mathbf{F}_M \begin{pmatrix} \mathbf{w}(\kappa) \\ \mathbf{0} \end{pmatrix} \tag{4.26}$$

$$\mathbf{w}(\kappa) = \big(w_0(\kappa), w_1(\kappa), \ldots, w_{M-R-1}(\kappa)\big)^H . \tag{4.27}$$

Now we can express the linear convolution model in (4.2) by a compact matrix form using the overlap-save method [Proakis, Manolakis 1996]. The vector of the R latest samples of the microphone signal, $\mathbf{y}(\kappa) = \mathbf{y}_{(\kappa-1)R+1}^{\kappa R}$, is additively composed of the near-end speech vector $\mathbf{s}(\kappa) = \mathbf{s}_{(\kappa-1)R+1}^{\kappa R}$ and the echo signal vector $\mathbf{d}(\kappa) = \mathbf{d}_{(\kappa-1)R+1}^{\kappa R}$, which in turn stems from linear filtering of the loudspeaker signal:

$$\mathbf{y}(\kappa) = \mathbf{s}(\kappa) + \mathbf{d}(\kappa) \tag{4.28}$$

$$= \mathbf{s}(\kappa) + \mathbf{Q}^H \mathbf{F}_M^{-1} \mathbf{X}(\kappa) \mathbf{W}(\kappa) . \tag{4.29}$$

The projection matrix $\mathbf{Q}^H = (\ \mathbf{0} \quad \mathbf{I}_R\ )$ of size $R \times M$ is responsible for the linearization of the cyclic convolution in the DFT domain. $\mathbf{I}_R$ is the identity matrix of size $R \times R$. The compact matrix notation implies that $\mathbf{W}(\kappa)$ defines a constant echo path at frame-time index κ, i.e., for the duration of R samples. That does not mean a fundamental limitation of the proposed theory, since the R samples usually correspond to a duration of only a few milliseconds.

A system model entirely in the frequency-domain is obtained when (4.29) is pre-multiplied by the zero-padding matrix $\mathbf{Q} = (\ \mathbf{0} \quad \mathbf{I}_R\ )^H$ and the Fourier matrix $\mathbf{F}_M$:

$$\mathbf{Y}(\kappa) = \mathbf{F}_M \mathbf{Q}\, \mathbf{y}(\kappa) \tag{4.30}$$

$$= \mathbf{F}_M \mathbf{Q}\, \mathbf{s}(\kappa) + \mathbf{F}_M \mathbf{Q} \mathbf{Q}^H \mathbf{F}_M^{-1} \mathbf{X}(\kappa) \mathbf{W}(\kappa) \tag{4.31}$$

$$= \mathbf{S}(\kappa) + \mathbf{C}(\kappa) \mathbf{W}(\kappa) . \tag{4.32}$$

The abbreviation $\mathbf{C}(\kappa) = \mathbf{F}_M \mathbf{Q} \mathbf{Q}^H \mathbf{F}_M^{-1} \mathbf{X}(\kappa)$ represents the influence of the known loudspeaker signal $x(k)$. The sequence of transformed signal vectors $\mathbf{S}(\kappa) = \mathbf{F}_M \mathbf{Q}\, \mathbf{s}(\kappa)$ is assumed to be independent, zero-mean, uncorrelated, and fully characterized by the time-varying covariance matrix $\mathbf{\Psi}_{ss}(\kappa)$, i.e.,

$$\mathrm{E}\left\{\mathbf{S}(\kappa)\right\} = \mathbf{0} \tag{4.33}$$

$$\mathrm{E}\left\{\mathbf{S}(\kappa)\mathbf{S}^H(\lambda)\right\} = \mathbf{\Psi}_{ss}(\kappa)\, \delta(\kappa - \lambda) . \tag{4.34}$$

Here, we used the unit pulse $\delta(\kappa)$ to express the uncorrelatedness, i.e., $\delta(\kappa - \lambda) = 1$ for $\kappa = \lambda$ and $\delta(\kappa - \lambda) = 0$ otherwise. Because of the good decorrelation properties of the DFT [Gray 2002], the $M \times M$ covariance matrix $\mathbf{\Psi}_{ss}(\kappa)$ is closely related to a diagonal matrix that contains the time-varying power spectral density (PSD) $\Phi_{ss}(\ell, \kappa)$ of the speech signal $s(k)$ at discrete frequencies $\Omega_\ell = 2\pi\ell/M$, $\ell = 0, 1, \ldots, M - 1$, i.e.,

$$\mathbf{\Psi}_{ss}(\kappa) \approx R \cdot \mathrm{diag}\left\{\mathbf{\Phi}_{ss}(\kappa)\right\} \tag{4.35}$$

$$\mathbf{\Phi}_{ss}(\kappa) = \left(\Phi_{ss}(0, \kappa), \Phi_{ss}(1, \kappa), \ldots, \Phi_{ss}(M - 1, \kappa)\right)^H . \tag{4.36}$$

The normalization factor R has to be applied to relate magnitude-squared DFT coefficients to the definition of power spectral densities [Proakis, Manolakis 1996].

4.3.2 Markov Model of the Time-Varying Echo Path

In order to describe the smooth transition between successive realizations $\mathbf{W}(\kappa)$ and $\mathbf{W}(\kappa + 1)$ of the time-varying echo path, we consider a first-order Gauss–Markov model [Haykin 2002] as proposed in [Enzner 2006]:

$$\mathbf{W}(\kappa + 1) = A \cdot \mathbf{W}(\kappa) + \Delta \mathbf{W}(\kappa) . \tag{4.37}$$

It is assumed that the forgetting factor A is close to but smaller than unity. The unpredictability of acoustic echo path changes is taken into account by a stationary sequence of independent and uncorrelated process noise vectors $\Delta \mathbf{W}(\kappa)$ with zero-mean and covariance matrix $\mathbf{\Psi}_{\Delta\Delta}$, i.e.,

$$\mathrm{E}\{\Delta \mathbf{W}(\kappa)\} = \mathbf{0} \tag{4.38}$$

$$\mathrm{E}\{\Delta \mathbf{W}(\kappa)\Delta \mathbf{W}^H(\lambda)\} = \mathbf{\Psi}_{\Delta\Delta} \, \delta(\kappa - \lambda) . \tag{4.39}$$

Owing to the formulation in the DFT domain, the covariance matrix $\mathbf{\Psi}_{\Delta\Delta}$ is also close to a diagonal matrix:

$$\mathbf{\Psi}_{\Delta\Delta} \approx M \cdot \mathrm{diag}\{\mathbf{\Phi}_{\Delta\Delta}\} . \tag{4.40}$$

Based on the model equation (4.37) and using the stationarity (time invariance) of $\mathbf{\Psi}_{\Delta\Delta}$ as shown by (4.39), we can evaluate the echo path covariance matrix as follows:

$$\mathbf{\Psi}_{ww} = \mathrm{E}\{\mathbf{W}(\kappa)\mathbf{W}^H(\kappa)\} \tag{4.41}$$

$$= A^2 \mathrm{E}\{\mathbf{W}(\kappa - 1)\mathbf{W}^H(\kappa - 1)\} + \mathrm{E}\{\Delta \mathbf{W}(\kappa - 1)\Delta \mathbf{W}^H(\kappa - 1)\} \tag{4.42}$$

$$= A^2 \mathbf{\Psi}_{ww} + \mathbf{\Psi}_{\Delta\Delta} . \tag{4.43}$$

This result can be rearranged to obtain an interesting proportionality between the covariances of echo path changes and echo path:

$$\mathbf{\Psi}_{\Delta\Delta} = (1 - A^2)\mathbf{\Psi}_{ww} . \tag{4.44}$$

This simple relation can be very useful to determine the usually unknown covariance of the echo path changes, $\mathbf{\Psi}_{\Delta\Delta}$, from the easily measurable echo path covariance $\mathbf{\Psi}_{ww}$, given the estimated echo path.

The results of an experimental verification of the Markov model for acoustic echo control have been reported in [Enzner 2006]. The experiments have shown that the simple model in (4.37) based on the time- and frequency-independent parameter A is indeed suitable to approximate the behavior of the echo path in realistic acoustic environments. For a frame-shift of $R = 64$ at 8 kHz sampling frequency, it turns out that a reasonable choice of the parameter A lies in the range $0.99 < A < 0.9999$, depending on the intensity of the echo path variation caused by the near speaker. To illustrate the corresponding degree of echo path variability, consider a Markov model with the typical parameter $A = 0.999$. It can be verified that, in this case, the echo attenuation of a perfectly adjusted echo canceler would drop to about 0 dB within 2–3 seconds after the adaptation of the echo canceler is halted.

4.3.3 Exact Kalman Filter for the Conditional Mean and Covariance

Combining the statistical Markov model in (4.37) and the linear observation model in (4.32), we obtain a general stochastic state-space model for the unknown echo path $\mathbf{W}(\kappa)$ in the DFT domain. The two model equations are reproduced here for the convenience of presentation:

$$\mathbf{W}(\kappa + 1) = A \cdot \mathbf{W}(\kappa) + \Delta\mathbf{W}(\kappa) \tag{4.45}$$

$$\mathbf{Y}(\kappa) = \mathbf{S}(\kappa) + \mathbf{C}(\kappa)\mathbf{W}(\kappa) . \tag{4.46}$$

In the language of state-space modeling of linear dynamical systems [Haykin 2002], Eqs. (4.45) and (4.46) are often referred to as the *state equation* and the *measurement equation*, respectively. A block diagram of the entire state-space model is depicted in Fig. 4.3.

According to the state equation (4.45), the echo path $\mathbf{W}(\kappa)$ is regarded as the state of a linear recursive system. The purely stochastic system input $\Delta\mathbf{W}(\kappa)$ is fully characterized by the time-invariant covariance matrix $\mathbf{\Psi}_{\Delta\Delta}$.

The measurement equation (4.46) linearly relates the unknown state $\mathbf{W}(\kappa)$ to the observations $\mathbf{Y}(\kappa)$ by the time-varying observation matrix $\mathbf{C}(\kappa)$. Regarding the communication, the additive vector $\mathbf{S}(\kappa)$ represents the desired speech signal. From the viewpoint of system identification, however, the unknown vector $\mathbf{S}(\kappa)$ with time-varying covariance matrix $\mathbf{\Psi}_{ss}(\kappa)$ is an undesirable observation noise. It has been reported in [Enzner 2006] that an approximation of $\mathbf{\Psi}_{ss}(\kappa)$ can be calculated conveniently using a *decision-directed* approach based on the output of the echo canceler or postfilter.

Given the state-space model of the unknown echo path, it becomes intuitively clear that the Kalman filter must be *the* tool for acoustic echo path estimation. In [Haykin 2002], the Kalman filter is derived as the *linear* MMSE estimator of the state of a linear dynamical system. In [Scharf 1991], the Kalman filter is developed as the nonlinear MMSE state estimator under the assumption of *Gaussianity* of independent

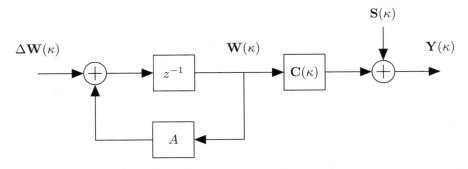

Figure 4.3: State-space model of the unknown echo path $\mathbf{W}(\kappa)$ in the DFT domain

process and observation noises. A very intuitive presentation of the Kalman filter equations can be found, for example, in [Unbehauen 1997]. The original work of Kalman is documented in [Kalman 1960].

As shown in the time-domain by Sec. 4.2.2 and Sec. 4.2.3, the conditional mean $\mathbf{w}_1(k) = \mathrm{E}\{\mathbf{w}|\mathbf{y}_{-\infty}^{k-R}\}$ and the estimation error covariance $\mathbf{p} = \mathrm{E}\{\mathbf{w}_r\mathbf{w}_r^T\}$ of the echo path are the key parameters of echo canceler and postfilter (the two components of the MMSE processor for acoustic echo control). The respective conditional mean and co-variance quantities in the DFT domain can be defined as:

$$\mathbf{W}_1(\kappa) = \mathrm{E}\{\mathbf{W}(\kappa)\,|\,\mathbf{Y}(\kappa-1), \mathbf{Y}(\kappa-2), \ldots, \mathbf{Y}(-\infty)\} \tag{4.47}$$

$$\mathbf{P}(\kappa) = \mathrm{E}\{\mathbf{W}_r(\kappa)\mathbf{W}_r^H(\kappa)\}\,, \tag{4.48}$$

where $\mathbf{W}_r(\kappa) = \mathbf{W}(\kappa) - \mathbf{W}_1(\kappa)$. The relation between time- and frequency-domain quantities is given by the discrete Fourier transform as $\mathbf{W}_1 \approx \mathbf{F}_M(\mathbf{w}_1^T\ \mathbf{0})^T$ and $\mathbf{P} \approx \mathbf{F}_M\,\mathbf{p}\,\mathbf{F}_M^H$. The approximation here is merely due to the frame-oriented organi-zation and processing of the data in the DFT domain compared with the sample-based use of the same data in the time-domain.

The Kalman filter *jointly* and *recursively* computes the parameters $\mathbf{W}_1(\kappa)$ and $\mathbf{P}(\kappa)$ from the past observations $\mathbf{Y}(\kappa-1), \mathbf{Y}(\kappa-2), \ldots, \mathbf{Y}(-\infty)$. The efficiency that is related to the recursiveness of the Kalman filter is of great advantage for the realtime application of acoustic echo control. Following the approach in [Unbehauen 1997], the computation is accomplished by a set of coupled matrix iteration formulas.

Two equations realize an extrapolation (prediction) step according to the state-equation (4.45) of the unknown system:

$$\mathbf{W}_1(\kappa+1) = A \cdot \mathbf{W}_1^+(\kappa) \tag{4.49}$$

$$\mathbf{P}(\kappa+1) = A^2 \cdot \mathbf{P}^+(\kappa) + \mathbf{\Psi}_{\Delta\Delta}\,. \tag{4.50}$$

A statistical correction of predicted parameters $\mathbf{W}_1(\kappa)$ and $\mathbf{P}(\kappa)$ is performed using the noisy input data $\mathbf{Y}(\kappa)$:

$$\mathbf{W}_1^+(\kappa) = \mathbf{W}_1(\kappa) + \mathbf{K}(\kappa)\Big(\mathbf{Y}(\kappa) - \mathbf{C}(\kappa)\mathbf{W}_1(\kappa)\Big) \tag{4.51}$$

$$\mathbf{P}^+(\kappa) = \Big(\mathbf{I}_M - \mathbf{K}(\kappa)\mathbf{C}(\kappa)\Big)\mathbf{P}(\kappa)\,, \tag{4.52}$$

where the Kalman gain $\mathbf{K}(\kappa)$ is defined as:

$$\mathbf{K}(\kappa) = \mathbf{P}(\kappa)\mathbf{C}^H(\kappa)\Big(\mathbf{C}(\kappa)\mathbf{P}(\kappa)\mathbf{C}^H(\kappa) + \mathbf{\Psi}_{ss}(\kappa)\Big)^{-1}\,. \tag{4.53}$$

At time κ, the order of the execution of the iteration formulas is the following: as soon as a new data vector $\mathbf{Y}(\kappa)$ becomes available, first the Kalman gain is computed according to (4.53), then the mean and covariance update is performed according to (4.51) and (4.52), and, finally, the predictions of the conditional mean and covariance for the next frame indexed $\kappa + 1$ are computed by (4.49) and (4.50). For the optimal initialization of this iterative procedure, the following unconditional expectations have to be chosen at time $k = 0$:

$$\mathbf{W}_1(0) = \mathrm{E}\{\mathbf{W}(0)\} \tag{4.54}$$

$$\mathbf{P}(0) = \mathrm{E}\{(\mathbf{W}(0) - \mathbf{W}_1(0))\,(\mathbf{W}(0) - \mathbf{W}_1(0))^H\}\,. \tag{4.55}$$

In the typical case without *a priori* knowledge of the initial echo path $\mathbf{W}(0)$, we may simply choose $\mathbf{W}_1(0) = \mathbf{0}$. However, to start the recursion, at least some *a priori* information must be available about the corresponding estimation error covariance $\mathbf{P}(0) = \mathrm{E}\{\mathbf{W}(0)\mathbf{W}^H(0)\} = \mathbf{\Psi}_{ww}$. In the typical situation with strong acoustic coupling between loudspeaker and microphone, we may choose $\mathbf{P}(0) = \mathbf{I}_M$.

Despite the recursiveness, the exact Kalman filter is not yet suitable for an implementation. Computational complexity and memory requirements are extremely high due to the presence of non-diagonal $M \times M$ matrices. In acoustic echo control, the matrix dimension M could easily range from a few hundred to a few thousand. Furthermore, the processing of large-scale non-diagonal matrices is subject to a potential numerical instability, especially the matrix inversion in the Kalman gain. On the other hand, the covariance matrices in the DFT domain are nearly diagonal, which means that the signal processing of the exact DFT-based Kalman filter is highly redundant.

4.3.4 Diagonalization of the Kalman Filter

A series of mild approximations will help to rewrite the exact Kalman filter in an efficient diagonalized form. The approximations are mainly justified by the good decorrelation and diagonalization properties of the DFT [Gray 2002]. If we make sure that vector lengths and matrix dimensions cover the span of correlation of the involved signals [Brillinger 1981], we can consider the following approximations as "mild" in the sense that the related algorithm degradation will be small compared with the enormous simplification that is achieved.

- Eqs. (4.53) and (4.50) of the exact Kalman filter can be simplified by exploiting the diagonal approximation of observation and process noise covariance matrices according to (4.35) and (4.40), respectively.

- In Eqs. (4.52) and (4.53), the observation matrix $\mathbf{C}(\kappa) = \mathbf{F}_M\mathbf{Q}\mathbf{Q}^H\mathbf{F}_M^{-1}\mathbf{X}(\kappa)$ can be approximated by $\mathbf{C}(\kappa) \approx R/M \cdot \mathbf{X}(\kappa)$. This approximation is justified by the fact that the main diagonal of the projection matrix $\mathbf{G} = \mathbf{F}_M\mathbf{Q}\mathbf{Q}^H\mathbf{F}_M^{-1}$ is dominant and that the off-diagonals rapidly decay, so that $\mathbf{G}$ is in fact close to a scaled version of the identity [Benesty et al. 2001, Chap. 8], i.e., $\mathbf{G} \approx R/M \cdot \mathbf{I}_M$.

- The expression $\mathbf{C}(\kappa)\mathbf{P}(\kappa)\mathbf{C}^H(\kappa)$ in the Kalman gain can be replaced using the observation that $\mathbf{G}\mathbf{D}\mathbf{G}^H \approx R/M \cdot \mathbf{D}$, when $\mathbf{D}$ is a diagonal matrix. For a better understanding of this, note that the concatenation of two projections will not cause more than just one: $\mathbf{G}\mathbf{G}^H = \mathbf{G}$. Assuming the estimation error covariance $\mathbf{P}(\kappa)$ is diagonal, we thus have $\mathbf{C}(\kappa)\mathbf{P}(\kappa)\mathbf{C}^H(\kappa) \approx R/M \cdot \mathbf{X}(\kappa)\mathbf{P}(\kappa)\mathbf{X}^H(\kappa)$.

We do not apply any approximation to Eq. (4.51) of the exact Kalman filter, because the correct phase information is very important in the computation of the estimation error $\mathbf{Y}(\kappa) - \mathbf{C}(\kappa)\mathbf{W}_1(\kappa)$. Indeed, we have to substitute the exact definition of the matrix $\mathbf{C}(\kappa)$ into (4.51) and *not* a previously shown approximation. Otherwise, the strictly linear convolution would be replaced by a circular convolution, which might have a disastrous impact on the echo path identification [Benesty et al. 2001, Chap. 8]. In (4.51), we replace only the noisy observation $\mathbf{Y}(\kappa)$ using its previous definition by the Fourier transform: $\mathbf{Y}(\kappa) = \mathbf{F}_M \mathbf{Q} \mathbf{y}(\kappa)$. In this way, we can incorporate the data vector $\mathbf{y}(\kappa)$, which contains the original samples of the microphone signal in the time-domain.

As a result of approximations and modifications, we obtain a simplified version of the exact Kalman filter, in which the statistical extrapolation step is given by

$$\mathbf{W}_1(\kappa + 1) = A \cdot \mathbf{W}_1^+(\kappa) \tag{4.56}$$

$$\mathbf{P}(\kappa + 1) = A^2 \cdot \mathbf{P}^+(\kappa) + M \cdot \mathrm{diag}\{\mathbf{\Phi}_{\Delta\Delta}\}\,, \tag{4.57}$$

the correction step on the basis of the input data $\mathbf{y}(\kappa)$ is performed as

$$\mathbf{W}_1^+(\kappa) = \mathbf{W}_1(\kappa) + \mathbf{K}(\kappa)\mathbf{F}_M \mathbf{Q}\Big(\mathbf{y}(\kappa) - \mathbf{Q}^H \mathbf{F}_M^{-1}\mathbf{X}(\kappa)\mathbf{W}_1(\kappa)\Big) \tag{4.58}$$

$$\mathbf{P}^+(\kappa) = \Big(\mathbf{I}_M - \frac{R}{M}\mathbf{K}(\kappa)\mathbf{X}(\kappa)\Big)\mathbf{P}(\kappa)\,, \tag{4.59}$$

and the Kalman gain $\mathbf{K}(\kappa)$ is approximated through

$$\mathbf{K}(\kappa) = \mathbf{P}(\kappa)\mathbf{X}^H(\kappa)\Big(\mathbf{X}(\kappa)\mathbf{P}(\kappa)\mathbf{X}^H(\kappa) + M\mathrm{diag}\left\{\mathbf{\Phi}_{ss}(\kappa)\right\}\Big)^{-1}. \tag{4.60}$$

If the initialization $\mathbf{P}(0)$ of the iteration is diagonal, then the covariance matrices of *a priori* and *a posteriori* estimation errors, $\mathbf{P}(\kappa)$ and $\mathbf{P}^+(\kappa)$, as well as the Kalman gain, $\mathbf{K}(\kappa)$, automatically become diagonal matrices. So without any further assumptions, the signal processing of the whole Kalman filter now consists of diagonal matrices – the only exception being Eq. (4.58), which still represents a strictly linear convolution.

Because of the diagonalized structure of the algorithm, a realization of the proposed concept merely requires basic vector arithmetics $(+/-/\cdot/\div)$ and FFT/IFFT. This is a striking feature of our diagonalized Kalman filter, since its computational complexity, memory requirements, and numerical properties will therefore be comparable to the known and feasible concepts for acoustic echo control.

Nevertheless, our modifications have *not* substantially reduced the key advantage of the Kalman filter, i.e., the fundamental algorithm structure for joint conditional mean and covariance estimation has been preserved.

4.3.5 Unification of Adaptive Filtering and Adaptation Control

We have demonstrated that the Kalman filter unifies the two worlds of adaptive echo cancellation and statistical postfiltering. In this section, we show that the diagonalized Kalman filter also establishes the optimal relationship between the following (partly well known) components of advanced systems for acoustic echo control:

- the frequency-domain adaptive filter (FDAF) [Ferrara 1985];

- the optimum stepsize for the FDAF in the MMSE sense [Nitsch 2000]; and

- a simple and robust, model-based statistical estimator of the *convergence state* of the adaptive filter, i.e., of the time- and frequency- dependent system distance between echo canceler and echo path [Enzner, Vary 2003].

Frequency-Domain Adaptive Filter (FDAF)

Based on the diagonal structure of the Kalman gain, we may introduce the following abbreviation, which defines a vector $\boldsymbol{\mu}(\kappa)$ in accordance with (4.60):

$$\mathbf{K}(\kappa) = \text{diag}\{\boldsymbol{\mu}(\kappa)\}\mathbf{X}^H(\kappa) \ . \tag{4.61}$$

Then, we substitute (4.58) and (4.61) into (4.56) and rewrite (4.56) in the form of a conventional gradient update of $\mathbf{W}_1(\kappa)$. The vector $\mathbf{e}(\kappa)$ in the following set of equations obviously takes the meaning of an error signal in the time-domain:

$$\mathbf{W}_1(\kappa + 1) = A \cdot \left(\mathbf{W}_1(\kappa) + \Delta\mathbf{W}_1(\kappa)\right) \tag{4.62}$$

$$\Delta\mathbf{W}_1(\kappa) = \text{diag}\{\boldsymbol{\mu}(\kappa)\}\mathbf{X}^H(\kappa)\mathbf{F}_M\mathbf{Q}\mathbf{e}(\kappa) \tag{4.63}$$

$$\mathbf{e}(\kappa) = \mathbf{y}(\kappa) - \mathbf{Q}^H\mathbf{F}_M^{-1}\mathbf{X}(\kappa)\mathbf{W}_1(\kappa) \ . \tag{4.64}$$

For $A \to 1$, this algorithm is identical to the *unconstrained frequency-domain adaptive filter* proposed by [Mansour, Gray 1982]. The vector $\boldsymbol{\mu}(\kappa)$ represents a possibly time- and frequency-dependent stepsize factor. The reason for the qualification *unconstrained* is that the resulting filter vector $\mathbf{W}_1(\kappa)$ will not necessarily satisfy the overlap-save constraint in (4.26). A *constrained* version of the algorithm is obtained when the update in (4.62) is pre-multiplied by $\mathbf{F}_M(\mathbf{I}_M - \mathbf{Q}^H\mathbf{Q})\mathbf{F}_M^{-1}$. That results in the following *constrained* update in place of (4.62):

$$\mathbf{W}_1^c(\kappa + 1) = A \cdot \left(\mathbf{W}_1^c(\kappa) + \mathbf{F}_M(\mathbf{I}_M - \mathbf{Q}^H\mathbf{Q})\mathbf{F}_M^{-1}\Delta\mathbf{W}_1(\kappa)\right) \ . \tag{4.65}$$

We consider $\mathbf{W}_1^c(\kappa) = \mathbf{F}_M(\mathbf{I}_M - \mathbf{Q}^H\mathbf{Q})\mathbf{F}_M^{-1}\mathbf{W}_1(\kappa)$ as a constrained conditional expectation of the echo path $\mathbf{W}(\kappa)$ at time κ. For $A \to 1$, the resulting adaptive algorithm is identical to the *constrained FDAF* proposed in [Ferrara 1980], [Ferrara 1985]. There, it was shown that it will converge to the Wiener solution for $\mathbf{W}_1(\kappa)$. Note that the "constraining" of the unconstrained FDAF will not exactly compensate for approximations used in the derivation of the diagonalized Kalman filter.

FDAF has become a first choice in acoustic echo cancellation [Gay, Benesty 2000], [Benesty et al. 2001], because of its ability to realize high-order adaptive filters with good convergence properties and moderate computational complexity [Haykin 2002].

Optimum Stepsize for the FDAF

Now we consider Eqs. (4.60) and (4.61). Recalling that all the involved matrices are diagonal, it can easily be verified that the Kalman filter delivers the following expression for the stepsize, $\text{diag}\{\boldsymbol{\mu}(\kappa)\}$, of the FDAF:

$$\text{diag}\{\boldsymbol{\mu}(\kappa)\} = \mathbf{P}(\kappa)\Big(\mathbf{P}(\kappa)\mathbf{X}(\kappa)\mathbf{X}^H(\kappa) + M\text{diag}\{\boldsymbol{\Phi}_{ss}(\kappa)\}\Big)^{-1}. \qquad (4.66)$$

A frequency-selective optimal stepsize, similar to the Kalman stepsize in (4.66), has been derived in [Nitsch 2000] by a lengthy mathematical procedure that minimizes the convergence state $\mathbf{P}(\kappa+1)$ at time $\kappa+1$, given the previous value $\mathbf{P}(\kappa)$ at time κ. One important difference, however, should be noted. Equation (4.66) utilizes the instantaneous power spectrum $\mathbf{X}(\kappa)\mathbf{X}^H(\kappa)$ of the loudspeaker signal $x(k)$, while the stepsize in [Nitsch 2000] relies on the PSD of $x(k)$ in this place. The difference in our result can be traced back to the improved system model introduced in Sec. 4.2.1, in which the loudspeaker input $x(k)$ is modeled as a measurable, i.e., deterministic signal and *not* as a random process.

The FDAF in (4.62) and (4.63) combined with the Kalman stepsize in (4.66) can be considered as a *normalized* frequency-domain adaptive filter with optimum *regularization* in the denominator of the gradient $\Delta\mathbf{W}_1(\kappa)$. Indeed, the observation noise covariance $\boldsymbol{\Phi}_{ss}(\kappa)$ represents the optimum regularization quantity according to the Kalman filter. The balance between the regularization $\boldsymbol{\Phi}_{ss}(\kappa)$ and the convergence state $\mathbf{P}(\kappa)$ controls the speed of adaptation and avoids a divergence of the adaptive filter in the presence of observation noise.

The critical issue regarding the implementation of the optimal stepsize in (4.66) and the postfilter according to (4.22) is always the uncertainty about the convergence state $\mathbf{P}(\kappa)$ of the adaptive filter. The convergence state (echo path estimation error covariance) is not directly measurable and therefore has to be determined, somehow, from the available signals. We have now reached the key problem of acoustic echo control and the next section shows that Kalman filtering has an answer for this, too.

A Recursive Statistical Convergence State Estimator

We simply combine Eqs. (4.59) and (4.57) of the Kalman filter and then replace the Kalman gain $\mathbf{K}(\kappa)$ according to (4.61), obtaining the *Riccati difference-equation* for the convergence state $\mathbf{P}(\kappa)$ of the adaptive echo canceler:

$$\mathbf{P}(\kappa+1) = A^2 \cdot \Big(\mathbf{I}_M - \frac{R}{M}\text{diag}\{\boldsymbol{\mu}(\kappa)\}\mathbf{X}^H(\kappa)\mathbf{X}(\kappa)\Big)\mathbf{P}(\kappa) + M \cdot \text{diag}\{\boldsymbol{\Phi}_{\Delta\Delta}\}. \quad (4.67)$$

This difference-equation fully describes the dynamic behavior of the FDAF. In the literature, the convergence behavior of adaptive filters is usually determined by an explicit convergence analysis, e.g., [Haykin 2002] and references therein. Here, the difference-equation (4.67) was obtained directly from the Kalman filter.

We observe that the estimation error covariance $\mathbf{P}(\kappa)$ depends on both the echo path characteristics and the adaptive filter properties. Essentially, (4.67) calculates a prediction $\mathbf{P}(\kappa + 1)$ of the estimation error covariance on the basis of the current stepsize of the adaptive filter, $\boldsymbol{\mu}(\kappa)$, the current estimation error covariance, $\mathbf{P}(\kappa)$, and the process noise covariance $\boldsymbol{\Phi}_{\Delta\Delta}$, which stands for the degree of time-variability of the echo path. According to (4.66), the stepsize is controlled by the balance of observation noise and estimation error covariance. Using the prediction $\mathbf{P}(\kappa + 1)$, the optimal stepsize $\boldsymbol{\mu}(\kappa + 1)$ and a prediction $\mathbf{P}(\kappa + 2)$ can be calculated at time $\kappa + 1$.

4.4 AEC Performance of the Frequency-Domain Adaptive Kalman Filter

We measure the *echo attenuation* after echo cancellation and postfiltering in terms of the echo return loss enhancement $\mathrm{ERLE}_{W_1} = \sigma_d^2/\sigma_b^2$ and $\mathrm{ERLE}_{W_{12}} = \sigma_d^2/\sigma_{\hat{b}}^2$, respectively. The *speech quality* is evaluated by means of the resulting signal-to-echo ratio $\mathrm{SER}_e = \sigma_s^2/\sigma_{s-e}^2$ after the echo canceler and $\mathrm{SER}_{\hat{s}} = \sigma_s^2/\sigma_{s-\hat{s}}^2$ after the postfilter. The compound of ERLE and SER is a suitable measure to reflect the overall performance of an acoustic echo controller – including the tracking ability and the robustness of the adaptive algorithm.

For the performance analysis of the adaptive algorithm, we first consider a *time-varying* echo path, which is generated by the Markov model in (4.37). We use the realistic transition factor $A = 0.999$ (cf. Sec. 4.3.2), the frame-shift $R = 64$, the DFT length $M = 512$, and we have 8 kHz sampling frequency. The analysis is based on real speech input in a wide range of input signal-to-echo ratios $\mathrm{SER}_y = \sigma_s^2/\sigma_d^2$. $\mathrm{SER}_y = 0\,\mathrm{dB}$ simulates hard double talk, $\mathrm{SER}_y = -40\,\mathrm{dB}$ corresponds to remote single talk, and $\mathrm{SER}_y = 40\,\mathrm{dB}$ basically represents near-end single talk. The background noise level is adjusted such that the signal-to-noise ratio of the near-end speech is 10 dB, while the received signal from the far-end can be considered as clean speech – a situation that stands, for example, for the car acoustic environment.

We obtain the results in Fig. 4.4, if the Kalman filter is matched to the realistic time-varying echo path model. The ERLE_{W_1} by the echo canceler ranges from 0 to 20 dB depending on the input SER_y. In the remote single talk case, we have a saturation of ERLE_{W_1} since the echo canceler at time κ is determined by means of the "incomplete" data up to time $\kappa - 1$. The total echo attenuation $\mathrm{ERLE}_{W_{12}}$ by echo canceler and postfilter ranges from 0 to 50 dB. That complies with the practical requirements for acoustic echo controllers. During remote single talk, more than 40 dB ERLE is indeed

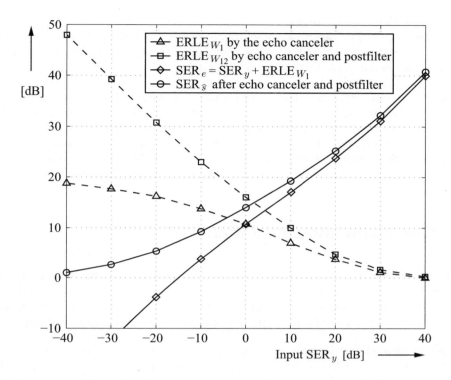

Figure 4.4: ERLE and output SER for various input SER. Kalman filter matches the simulated, realistic time-varying echo path: $A = 0.999$, $R = 64$, $M = 512$. Speech material consists of eight phonetically balanced sentences

recommended [ITU-T Rec. P.342 2000], [Verband der Automobilindustrie 2002], while in the other cases less echo attenuation is sufficient.

In addition to that, it can be seen from Fig. 4.4 that the echo attenuation by the echo canceler results in a direct speech quality improvement. This is shown by the fact that $\mathrm{SER}_e = \mathrm{SER}_y + \mathrm{ERLE}_{W_1}$ (in dB). Postfiltering consistently adds improvement to the speech quality as shown by $\mathrm{SER}_{\widehat{s}} > \mathrm{SER}_e$. In the prominent double talk situation, i.e., for $\mathrm{SER}_y = 0\,\mathrm{dB}$, we have an instrumental speech quality of $\mathrm{SER}_{\widehat{s}} \approx 14\,\mathrm{dB}$ at the system output. However, due to the psychoacoustic effect of masking, the perceived quality is even better than indicated by the $\mathrm{SER}_{\widehat{s}}$. This has been confirmed by informal subjective listening tests based on a realtime prototype system that builds on the proposed algorithms [Enzner, Vary 2006].

Concerning a possible model mismatch between the Kalman filter and the true echo path, we have to distinguish two cases. At first, let us assume that the Kalman filter is set up for realistic echo path variability, but the echo path remains constant in time. Compared with Fig. 4.4, the echo attenuation by echo canceler and postfilter will actually increase, because the time-invariant echo path can be "tracked" easier and because the Kalman filter over-estimates the echo path uncertainty (estimation error covariance). The resulting speech quality after the echo canceler will increase too, but, since the postfilter attenuation overshoots, the speech quality at the system output is just comparable to the one in Fig. 4.4.

In contrast to this "good-natured" model mismatch, we also have to expect an echo path variability stronger than that presumed by the realistic Markov model. This may happen in the case of untypically strong movements of the near speaker. In this case, the mismatched Kalman filter might fail in adjusting the echo canceler and postfilter coefficients fast enough and the echo signal could become audible. Luckily, strong echo path variations only occur sporadically in realworld applications.

4.5 Discussion and Conclusions

To this day, the optimal Kalman filter for the tracking of time-varying systems has not received sufficient attention in acoustic echo control. This can be attributed to its high complexity and its potential for numerical instability. Furthermore, a comprehensive system model was not available for the Kalman filter [Hänsler, Schmidt 2004].

The solutions deployed for acoustic echo control were mostly based on deterministic adaptive filters, e.g., LMS or RLS, although the acoustic environment of hands-free telephones is obviously characterized by statistical uncertainties: the presence of noise and non-stationary near-end speech, the randomly fluctuating echo path, and its non-persistent excitation. As a result of the statistical under-modeling of the echo control problem, the development of sophisticated control mechanisms for adaptive filters, e.g., double talk detection, has become an art of its own. Nonetheless, the design of fast and robust algorithms for realistic acoustic environments remained difficult.

It was demonstrated in this chapter, that the echo control problem can be modeled in a very compact and comprehensive way by a stochastic state-space model of the time-varying echo path in the DFT domain. It was shown that the corresponding Kalman filter can be diagonalized efficiently, practically meaning that computational and numerical drawbacks of the standard Kalman filter are circumvented. Basic vector arithmetics and FFT/IFFT are indeed sufficient for the implementation of the proposed concept. Furthermore, we have pointed out the direction for determining the model parameters of the Kalman filter from the available signals [Enzner 2006].

From the strict mathematical approach in this chapter, we obtained an outstandingly compact and robust signal processing solution for acoustic echo control. The adaptive algorithm is inherently robust and does not require additional control mechanisms. In fact, we observed that Kalman filtering can be understood as a unification of classical adaptive filtering and adaptation control. Moreover, the Kalman filter is *the* tool for jointly and recursively adjusting acoustic echo canceler and postfilter, the two indispensable components of the MMSE processor for acoustic echo control.

Bibliography

Benesty, J.; Gänsler, T.; Morgan, D.; Sondhi, M.; Gay, S. (2001). *Advances in Network and Acoustic Echo Cancellation*, Springer.

Benesty, J.; Morgan, D.; Cho, J. (2000). A New Class of Double Talk Detectors based on Cross-Correlation, *IEEE Transactions on Speech and Audio Processing*, vol. 8, March, pp. 168–172.

Breining, C.; Dreiseitel, P.; Hänsler, E.; Mader, A.; Nitsch, B.; Puder, H.; Schertler, T.; Schmidt, G.; Tilp, J. (1999). Acoustic Echo Control, An Application of Very-High-Order Adaptive Filters, *IEEE Signal Processing Magazine*, vol. 16, no. 4, pp. 42–69.

Brillinger, D. (1981). *Time Series: Data Analysis and Theory*, McGraw-Hill, Inc., New York.

Buchner, H.; Kellermann, W. (2002). Improved Kalman Gain Computation for Multichannel Frequency-Domain Adaptive Filtering and Application to Acoustic Echo Cancellation, *Proceedings of International Conference on Acoustics, Speech, and Signal Processing (ICASSP)*, Orlando, Florida, pp. 1909–1912.

Duttweiler, D. (1978). A Twelve-Channel Digital Echo Canceler, *IEEE Transactions on Communications*, vol. 26, May, pp. 647–653.

Eneroth, P.; Gay, S.; Gänsler, T.; Benesty, J. (2000). A Real-Time Stereophonic Acoustic Subband Echo Canceler, *in* S. Gay; J. Benesty (eds.), *Acoustic Signal Processing for Telecommunications*, Kluwer Academic Publishers, pp. 135–152.

Enzner, G. (2003). Hands-Free Communication: A Unified Concept of Acoustic Echo Cancellation and Residual Echo Suppression, *Proceedings of Deutsche Jahrestagung für Akustik (DAGA)*, Aachen, Germany.

Enzner, G. (2006). *A Model-Based Optimum Filtering Approach to Acoustic Echo Control: Theory and Practice*, PhD thesis, Aachener Beiträge zu digitalen Nachrichtensystemen, vol. 22, P. Vary (ed.), Wissenschaftsverlag Mainz in Aachen, RWTH Aachen University.

Enzner, G.; Martin, R.; Vary, P. (2002a). Partitioned Residual Echo Power Estimation for Frequency-Domain Acoustic Echo Cancellation and Postfiltering, *European Transactions on Telecommunications*, vol. 13, no. 2, pp. 103–114.

Enzner, G.; Martin, R.; Vary, P. (2002b). Unbiased Residual Echo Power Estimation for Hands-Free Telephony, *Proceedings of International Conference on Acoustics, Speech, and Signal Processing (ICASSP)*, Orlando, Florida, pp. 1893–1869.

Enzner, G.; Mauler, D.; Vary, P. (2004). Realtime Performance of Acoustic Echo Canceler and Postfilter for Residual Echo Suppression in the Car Environment, *Proceedings of Congrès Français d'Acoustique (CFA) and Deutsche Jahrestagung für Akustik (DAGA)*, Strasbourg, France.

Enzner, G.; Vary, P. (2003). Robust and Elegant, Purely Statistical Adaptation of Acoustic Echo Canceler and Postfilter, *Proceedings of International Workshop on Acoustic Echo and Noise Control (IWAENC)*, Kyoto, Japan, pp. 43–46.

Enzner, G.; Vary, P. (2005). New Insights into the Statistical Signal Model and the Performance Bounds of Acoustic Echo Control, *Proceedings of European Signal Processing Conf. (EUSIPCO)*, Antalya, Turkey.

Enzner, G.; Vary, P. (2006). Frequency-Domain Adaptive Kalman Filter for Acoustic Echo Control in Hands-free Telephones, *Signal Processing, Elsevier*, vol. 86, no. 6, pp. 1140–1156.

Ferrara, E. (1980). Fast Implementation of LMS Adaptive Filters, *IEEE Transactions on Acoustics, Speech, and Signal Processing*, vol. 28, August, pp. 474–475.

Ferrara, E. (1985). Frequency-Domain Adaptive Filtering, *in* C. Cowan; P. Grant (eds.), *Adaptive Filters*, Prentice Hall, Englewood Cliffs (NJ), pp. 145–179.

Frenzel, R. (1992). *Freisprechen in gestörter Umgebung*, PhD thesis, Fortschritt-Berichte VDI, Reihe 10, Nr. 228, Düsseldorf: VDI Verlag, Technical University of Darmstadt.

Gay, S. L.; Benesty, J. (eds.) (2000). *Acoustic Signal Processing for Telecommunications*, Kluwer Academic Publishers.

Gray, R. M. (2002). Toeplitz and Circulant Matrices: A Review, *Technical report*, Information Systems Laboratory, Stanford University, Stanford (California). http://ee.stanford.edu/~gray/toeplitz.pdf.

Gustafsson, S.; Martin, R.; Vary, P. (1998). Combined Acoustic Echo Control and Noise Reduction for Hands-Free Telephony, *Signal Processing, Elsevier*, vol. 64, no. 1, pp. 21–32.

Hänsler, E. (1992). The Hands-Free Telephone Problem – An Annotated Bibliography, *Signal Processing, Elsevier*, vol. 27, no. 3, pp. 259–271.

Hänsler, E. (1994). The Hands-Free Telephone Problem: An Annotated Bibliography Update, *Annales des Télécommunications*, vol. 49, no. 7–8, pp. 360–367.

Hänsler, E. (1997). From Algorithms to Systems - It's a Rocky Road, *Proceedings of International Workshop on Acoustic Echo and Noise Control (IWAENC)*, London, UK.

Hänsler, E.; Schmidt, G. (2000). Hands-Free Telephones – Joint Control of Echo Cancellation and Postfiltering, *Signal Processing, Elsevier*, vol. 80, no. 11, pp. 2295–2305.

Hänsler, E.; Schmidt, G. (2003). Control of LMS-Type Adaptive Filters, *in* S. Haykin; B. Widrow (eds.), *Least-Mean-Square Adaptive Filters*, John Wiley & Sons, Ltd., New York, pp. 175–240.

Hänsler, E.; Schmidt, G. (2004). *Acoustic Echo and Noise Control: A Practical Approach*, John Wiley & Sons, Ltd., New York.

Haykin, S. (2002). *Adaptive Filter Theory*, 4th edn, Prentice-Hall, Upper Saddle River, NJ.

ITU-T Rec. P.342 (2000). *Transmission Characteristics for Telephone Band (300–3400 Hz) Digital Loudspeaking and Hands-Free Telephony Terminals*.

Kalman, R. (1960). A New Approach to Linear Filtering and Prediction Problems, *Transactions ASME, Journal of Basic Engineering*, vol. 82, March, pp. 35–45.

Le Bouquin Jeannès, R.; Scalart, P.; Faucon, G.; Beaugeant, C. (2001). Combined Noise and Echo Reduction in Hands-Free Systems: A Survey, *IEEE Transactions on Speech and Audio Processing*, vol. 9, no. 8, pp. 808–820.

Lippuner, D. (2002). *Model-Based Step-Size Control for Adaptive Filters*, PhD thesis, Series in Signal and Information Processing, Volume 8, Hans-Andrea Loeliger (ed.), Hartung-Gorre Verlag, Konstanz, ETH Zürich (Diss. No. 14461).

Lippuner, D.; Kälin, A. N. (1999). Tracking Behavior of Model-Based Adaptive FIR Filters with Noise Variance Estimation, *Proceedings of International Workshop on Acoustic Echo and Noise Control (IWAENC)*, Pocono Manor, Pennsylvania, pp. 156–159.

Mader, A.; Puder, H.; Schmidt, G. (2000). Step-Size Control for Acoustic Echo Cancellation Filters – An Overview, *Signal Processing, Elsevier*, vol. 80, no. 9, pp. 1697–1719.

Mansour, D.; Gray, A. (1982). Unconstrained Frequency-Domain Adaptive Filters, *IEEE Transactions on Acoustics, Speech, and Signal Processing*, vol. 30, no. 5, pp. 726–734.

Martin, R. (1995). *Freisprecheinrichtungen mit mehrkanaliger Echokompensation und Störgeräuschreduktion*, PhD thesis, Aachener Beiträge zu digitalen Nachrichtensystemen, vol. 3, P. Vary (ed.), Verlag der Augustinus Buchhandlung, RWTH Aachen University.

Martin, R.; Altenhöner, J. (1995). Coupled Adaptive Filters for Acoustic Echo Control and Noise Reduction, *Proceedings of International Conference on Acoustics, Speech, and Signal Processing (ICASSP)*, pp. 3043–3046.

Martin, R.; Gustafsson, S. (1996). The Echo Shaping Approach to Acoustic Echo Control, *Speech Communication, Elsevier*, vol. 20, no. 3-4, pp. 181–190.

Meissner, P.; Wehrmann, R.; van der List, J. (1980). A Comparative Analysis of Kalman and Gradient Methods for Adaptive Echo Cancellation, *AEÜ. International Journal of Electronics and Communications*, vol. 34, no. 12, pp. 485–492.

Myllylä, V.; Schmidt, G. (2002). Pseudo-Optimal Regularization for Affine Projection Algorithms, *Proceedings of International Conference on Acoustics, Speech, and Signal Processing (ICASSP)*, Orlando, Florida, pp. 1917–1920.

Nitsch, B. (2000). A Frequency-Selective Stepfactor Control for an Adaptive Filter Algorithm Working in the Frequency-Domain, *Signal Processing, Elsevier*, vol. 80, no. 9, pp. 1733–1745.

Ochiai, K.; Araseki, T.; Ogihara, T. (1977). Echo Canceler with Two Echo Path Models, *IEEE Transactions on Communications*, vol. 25, June, pp. 589–595.

Papoulis, A. (1984). *Probability, Random Variables, and Stochastic Processes*, 2nd edn, McGraw-Hill, New York.

Proakis, J. G.; Manolakis, D. G. (1996). *Digital Signal Processing: Principles, Algorithms, and Applications*, Prentice-Hall, Upper Saddle River, New Jersey.

Scharf, L. L. (1991). *Statistical Signal Processing*, Addison-Wesley Publishing Company, Reading, Massachusetts.

Schmidt, G. (2001). *Entwurf und Realisierung eines Multiratensystems zum Freisprechen*, PhD thesis, Fortschritt-Berichte VDI, Reihe 10, Nr. 674, Düsseldorf: VDI Verlag, Technical University of Darmstadt.

Sondhi, M. M. (1967). An Adaptive Echo Canceler, *Bell Systems Technical Journal*, vol. 46, March, pp. 497–511.

Unbehauen, R. (1997). *Systemtheorie 1*, 7th edn, R. Oldenburg Verlag, München Wien.

Vary, P.; Martin, R. (2006). *Digital Speech Transmission - Enhancement, Coding, and Error Concealment*, John Wiley & Sons, Ltd., Chichester.

Verband der Automobilindustrie (2002). *VDA Specification for Car Hands-free Terminals (Version 1.4)*, Verband der Automobilindustrie, Postfach 17 05 63, 60079 Frankfurt.

Yamamoto, S.; Kitayama, S. (1982). An Adaptive Echo Canceller with Variable Step Gain Method, *Transactions IECE Japan*, vol. E65, no. 1, pp. 1–8.

Chapter 5

Noise Reduction - Statistical Analysis and Control of Musical Noise

Colin Breithaupt, Rainer Martin

5.1 Introduction

A key issue of adaptive speech enhancement schemes, especially of those acting in the spectral domain, is the prevention of unnatural fluctuations in the processed signal. These fluctuations are perceived as non-stationary artifacts and are often described as *musical noise*. In the case of adaptive spectral filters, the annoying fluctuations in the residual noise are caused by single spectral outliers in the spectral gain function. This spectral gain is multiplied with the short-term spectrum of noisy speech so that the product results in an estimate of the clean speech spectrum. Outliers in the spectral gain result in a clean speech estimate with isolated spectral peaks that excite the synthesis branch of a spectral analysis–synthesis system thus causing *musical noise*. The aim of overcoming *musical noise* has been the subject of research for over twenty years. Once this problem is solved, speech enhancement systems based on adaptive modifications of the short-term signal spectrum provide effective means for noise reduction.

In the following overview, we give a short description of the different approaches available in the literature that aim at a reduction of *musical noise*. Many proposals deal with the issue of *avoiding musical noise* in the first place. The basic idea is to reduce the variability of the adaptive spectral gain function during low SNR conditions. One common technique is essentially to replace the estimated spectral gain by a constant

Advances in Digital Speech Transmission   Edited by R. Martin, U. Heute and C. Antweiler
© 2008 John Wiley & Sons, Ltd

attenuation whenever the estimated SNR is below a threshold. This is frequently termed as spectral flooring. A drawback of this method is that it limits the amount of noise reduction [Berouti et al. 1979], [Cappé 1994]. Another way of reducing the variability of the estimated spectral gain is to smooth the spectral gain function or its parameters. The smoothing can be done over time [Gustafsson et al. 2001], [Hasan et al. 2004] or frequency [Fingscheidt et al. 2005]. In [Gülzow et al. 2003] temporal and spectral smoothing is implicitly combined by reducing the resolution of the spectral analysis in spectral regions with low SNR. A difficulty that comes with smoothing methods in general is noise shaping during the presence of speech. Therefore, the degree of smoothing has to be adapted carefully to the temporal evolution of the speech signal. Finally, the most successful approach of reducing noise with a low amount of *musical noise* and a relatively low distortion of the speech signal are the non-linear estimators presented in [Ephraim, Malah 1984] and [Ephraim, Malah 1985]. They are extra-ordinarily robust in Gaussian noise conditions. However, their performance heavily depends on a specific estimation procedure of the *a priori* SNR. The estimation algorithm proposed in [Ephraim, Malah 1984] avoids *musical noise* by trading off fluctuations in the estimate of the *a priori* SNR against distortions of the speech signal, especially of the onsets of speech.

While methods of avoiding *musical noise* work well in quasi-stationary Gaussian noise, their efficiency in non-stationary noise is limited. As estimation errors are generally inevitable, it is desirable to design algorithms such that they do not produce *musical noise* even if errors occur in the estimation of their controlling parameters. In [Goh et al. 1998] and [Hansen 1991] post-processing methods are presented for *suppressing musical noise* that could not be avoided. In order to suppress spectral fluctuations, they make use of the difference in the spectro-temporal structure of speech and *musical noise*. Like the smoothing techniques, these methods attempt to distinguish between speech components and spectral outliers of noise, which is difficult to accomplish, for example, in babble noise. In [Gustafsson et al. 1998] and [Virag 1999] the spectral masking properties of the human ear have also been exploited. A difficulty that comes with this approach is that an estimate of the clean speech signal itself is necessary in order to calculate the masking thresholds. As algorithms for spectral masking distinguish between noise-like and tone-like maskers, fluctuations in the preliminary estimate of the clean speech can be misinterpreted as tonal masker which results in wrong spectral masks.

Summing up this overview, it can be observed that, so far, noise reduction without *musical noise* has not been achieved in general. As to the structure of *musical noise* it can be said that, if adaptive spectral gain functions are used and if a high noise reduction is attempted, *musical noise* appears almost independently of the type of noise. The subjective annoyance of these artifacts depends on the statistical distribution of outliers versus time and frequency on one side and on the properties of the human auditory system on the other.

In this chapter, an analysis and an interpretation of the statistics of the residual noise is presented as it leads to new insights about the origin and counter-measures against *musical noise*. First, in order for us to be able to link the empirical distributions of processed noise to the enhancement method used, an overview of non-linear clean speech spectral estimators is given by means of their input–output characteristics. The focus is on these estimators, because they form the basis of today's most successful noise reduction systems. Together with the spectral estimators, the algorithm for estimating the *a priori* SNR [Ephraim, Malah 1984] is also analyzed, as this estimate in conjunction with the non-linear estimators plays an important role in the naturalness of the residual noise. The discussion of empirical histograms of processed noise then follows. It demonstrates how specific input–output characteristics lead to spectral outliers in processed noise. Finally, a post-processing method is described that suppresses outliers in the spectral gain function in such a way that spectral components of speech are implicitly protected. This method is thus able to suppress *musical noise* without the need to adapt controlling parameters to the temporal evolution of the speech signal.

5.2 Speech Enhancement in the DFT Domain

In this chapter we consider noisy speech signals $y(k) = s(k) + n(k)$, $k \in \mathbb{Z}$, which are a sum of a speech signal $s(k)$ and a statistically independent noise signal $n(k)$. The noisy signal $y(k)$ is analyzed by a frame-wise short-term discrete Fourier transform (DFT) of length M. This results in spectral coefficients $Y_\mu(\lambda) = S_\mu(\lambda) + N_\mu(\lambda)$, where $\mu = 0 \ldots M - 1$ denotes the frequency bin index and where $\lambda \in \mathbb{Z}$ is the frame index. Typically, tapered window functions $w(\tau)$, $\tau = 0 \ldots M - 1$, such as the Hann window are used in the analysis process.

In order to achieve a high noise reduction and, at the same time, a preservation of speech components, the analysis–synthesis system has to meet several requirements. A high spectral resolution that resolves the pitch harmonics allows for suppression of noise during speech presence. The processing of the highly dynamic speech signal also requires a high temporal resolution for the proper reproduction of plosives and speech onsets. In contrast to this quick responsiveness of the filters to temporal changes in the speech energy, the variance of estimated control quantities, however, must be low for an artifact-free reproduction of enhanced speech. As a consequence the design of noise reduction algorithms can be improved, if the trade-off between temporal and spectral resolution and the variance of spectral parameters is properly controlled.

These partly conflicting requirements can only be satisfied with a signal-adaptive processing scheme. Processing in the short-term Fourier domain provides much flexibility in this sense. However, controlling the temporal dynamics of spectral parameters independently in all frequency bins is difficult and often leads to the unwanted spectral artifacts. This is especially true when the disturbing noise is non-stationary.

5.2.1 Optimal Speech Estimators

Typical DFT-based noise reduction algorithms employ a multiplicative gain function in the DFT-domain [Vary, Martin 2006]. For an estimate $\widehat{S}_\mu(\lambda)$ of the clean speech spectral coefficient, an adaptive spectral filter gain $G_\mu(\lambda)$ is calculated that is then applied to the observed spectral coefficients $Y_\mu(\lambda)$:

$$\widehat{S}_\mu(\lambda) = G_\mu(\lambda) \, Y_\mu(\lambda) \, . \tag{5.1}$$

The result is transformed back into the time domain via the inverse DFT and the enhanced time signal is synthesized using the overlap-add method and a synthesis window $w_{\text{synth}}(\tau)$, $\tau = 0 \ldots M - 1$. If the analysis window is the Hann window $w(\tau) = w_{\text{hann}}(\tau)$ and if adjacent frames are half-overlapping, no synthesis window is needed, i.e., $w_{\text{synth}}(\tau) = 1$, to achieve perfect reconstruction of the time signal.

The filter gain $G_\mu(\lambda)$ can be derived for various optimization criteria. Next to the well-known Wiener filter, we will analyze the short-time spectral amplitude (STSA) estimator of [Ephraim, Malah 1984] and the log-spectral amplitude (LSA) estimator of [Ephraim, Malah 1985]. Additionally, the filter of [Martin 2005a] is considered. It is referred to as the LG filter. Further estimators are presented in [Breithaupt, Martin 2003], [Martin 2002], [Accardi, Cox 1999], [Lotter, Vary 2004], and [You et al. 2005]. All subsequently considered filter gain functions are a function of the *a priori* SNR $\xi_\mu(\lambda)$ and the *a posteriori* SNR $\gamma_\mu(\lambda)$ which are defined as

$$\text{\textit{a priori} SNR:} \qquad \xi_\mu(\lambda) = \frac{\Phi_{ss,\mu}(\lambda)}{\Phi_{nn,\mu}(\lambda)},$$

$$\text{\textit{a posteriori} SNR:} \qquad \gamma_\mu(\lambda) = \frac{|Y_\mu(\lambda)|^2}{\Phi_{nn,\mu}(\lambda)},$$

where $\Phi_{ss,\mu}(\lambda) = E\left\{|S_\mu(\lambda)|^2\right\}$ and $\Phi_{nn,\mu}(\lambda) = E\left\{|N_\mu(\lambda)|^2\right\}$ are the speech power and the noise power in frequency bin μ, respectively. The filter gains are denoted here as $G_\mu(\lambda) = G^{\text{filter}}[\xi_\mu(\lambda), \gamma_\mu(\lambda)]$. Note that G^{filter} is calculated in each bin μ and for each frame λ independently. We therefore omit the indices μ and λ whenever quantities are considered that do not depend on previous frames or on adjacent spectral bins.

The gain functions that are analyzed in this chapter are:

$$G^{\text{wiener}}[\xi, \gamma] = G^{\text{wiener}}[\xi] = \frac{\xi}{1 + \xi}, \tag{5.2}$$

$$G^{\text{stsa}}[\xi, \gamma] = \frac{\sqrt{\pi}}{2} \frac{\sqrt{\nu}}{\gamma} e^{-\frac{\nu}{2}} \left((1 + \nu) \, \text{I}_0\left(\frac{\nu}{2}\right) + \nu \, \text{I}_1\left(\frac{\nu}{2}\right) \right), \tag{5.3}$$

$$\text{with } \nu = \frac{\xi}{1 + \xi} \, \gamma,$$

$$G^{\text{lsa}}[\xi, \gamma] = \frac{\xi}{1 + \xi} \, \exp\left(\frac{1}{2} \, \text{Expint}\left\{\frac{\xi}{1 + \xi} \gamma\right\}\right), \tag{5.4}$$

$$G^{\text{lg}}[\xi, \gamma] = \frac{1}{\sqrt{\gamma}} \frac{L^+ \exp(E^+) \, \text{erfc}(L^+) - L^- \exp(E^-) \, \text{erfc}(L^-)}{\exp(E^+) \, \text{erfc}(L^+) + \exp(E^-) \, \text{erfc}(L^-)}, \tag{5.5}$$

$$\text{with } L^\pm = 1/\sqrt{\xi} \pm \sqrt{\gamma} \text{ and } E^\pm = \pm 2\sqrt{\gamma/\xi}.$$

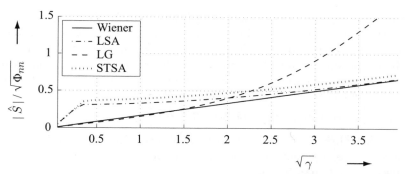

Figure 5.1: Filter input–output characteristics for the theoretic filter gains G^{filter} during speech pauses, with $\xi = 0.2 = \text{const}$. The graphs show the normalized filtered spectral magnitude $|\widehat{S}|/\sqrt{\Phi_{nn}} = G^{\text{filter}}[\xi, \gamma]\sqrt{\gamma}$

In (5.3), the functions $I_n(\cdot)$ are the modified Bessel functions of the first kind and order n. Expint$\{\cdot\}$ in (5.4) is the exponential integral function with Expint$\{x\} = -\text{Ei}\{-x\}$, see [Gradshteyn, Ryzhik 2000, (8.211.1)]. In (5.5), erfc$(\cdot)$ is the complementary error function.

While the Wiener filter G^{wiener}, the STSA estimator G^{stsa}, and the LSA estimator G^{lsa} model $S_\mu(\lambda)$ and $N_\mu(\lambda)$ as complex Gaussians, G^{lg} results from the minimum mean-square error (MMSE) estimator of the complex clean-speech spectral coefficients for Laplacian speech and Gaussian noise models [Martin 2005a], [Martin 2005b].

In Fig. 5.1 the normalized value for the filtered spectral magnitude (5.1) is plotted for filter gains (5.2) to (5.5) with $\xi = \xi_{\text{const}} = \text{const}$, i.e.,

$$\frac{|\widehat{S}|}{\sqrt{\Phi_{nn}}} = G^{\text{filter}}[\xi_{\text{const}}, \gamma]\frac{|Y|}{\sqrt{\Phi_{nn}}} = G^{\text{filter}}[\xi_{\text{const}}, \gamma]\sqrt{\gamma}. \tag{5.6}$$

We refer to this type of plot as the input–output characteristics of the filter. For the chosen small value of $\xi_{\text{const}} = 0.2$, the input–output characteristics for low SNR conditions and for speech absence can be analyzed. In this first analysis, it is assumed that Φ_{nn} and ξ were known, i.e., they do not have to be estimated from the observed signal. For $\xi = \text{const}$, the Wiener gain function $G^{\text{wiener}}[\xi] = \xi/(1 + \xi)$ results in a constant multiplicative factor. Thus, the input–output characteristics of the Wiener filter has a constant slope $G^{\text{wiener}}[\xi_{\text{const}}]$ in Fig. 5.1. While the LSA estimator approaches the Wiener solution asymptotically for larger values of $\sqrt{\gamma}$, output values $\widehat{S} \to 0$ theoretically do not occur. The minimum value is (see [Ephraim, Malah 1985, eqn. (19)]):

$$\frac{|\widehat{S}|}{\sqrt{\Phi_{nn}}}\bigg|_{\gamma=0} = G^{\text{lsa}}[\xi, \gamma]\sqrt{\gamma}\big|_{\gamma=0} = \left(\frac{\xi}{1 + \xi}\right)^{\frac{1}{2}} e^{-c/2}, \tag{5.7}$$

where $c = 0.5772\ldots$ is Euler's constant [Gradshteyn, Ryzhik 2000, eqn. (9.73)]. Note that the minimum value (5.7) implies $G^{\mathrm{lsa}}[\xi, \gamma] \to \infty$ for $\gamma \to 0$ and $\xi > 0$. The same applies to the gain functions G^{stsa} and the one described in [Lotter, Vary 2004]. These gain functions have a singularity for zero input $Y_\mu(\lambda) = 0$. In practical implementations, an upper limit $G_{\max}$ is therefore used for G^{filter}. The limit is usually set to $G_{\max} = \lim_{\xi \to \infty} G^{\mathrm{filter}}[\xi, \gamma] = 1$ (see [Lotter, Vary 2004]). In Fig. 5.1, the influence of $G_{\max}$ is visible for values of $\sqrt{\gamma}$ below 0.3.

Finally, the LG filter, as compared with the Wiener filter, emphasizes large input values, because the supergaussian speech model of this filter attributes large values of γ to speech rather than to noise.

Note that the filter gain functions G^{filter} are all derived under the assumption that speech is present in the current signal. In order to account for the possibility that no speech is present in a given bin μ, the speech presence uncertainty is often considered as the *a posteriori* probability $P(\mathcal{H}_1|Y_\mu(\lambda))$ of speech being present in the current spectral bin μ [Malah et al. 1999], [Cohen 2001]. Here, speech presence in a spectral bin is denoted as hypothesis $\mathcal{H}_1$, speech absence as $\mathcal{H}_0$. Depending on the gain function, this *soft-gain* modification of the gain function G^{filter} may appear as an additional factor or as an exponent [Cohen 2001] to G^{filter}. The resulting weighted gain function is additionally limited towards lower values by a constant $G_{\min}$. This leaves a noise floor in the processed signal that is used to cover annoying artifacts. If not stated otherwise, the *soft-gain* method is not used in this chapter, because in non-stationary noises – like babble noise – estimators of $P(\mathcal{H}_1|Y_\mu(\lambda))$ do not provide a notable benefit.

Parameter Estimation

As the true values of the parameters Φ_{nn}, γ, and ξ are not available in real environments, they have to be replaced by their estimates $\widehat{\Phi}_{nn}$, $\widehat{\gamma}$, and $\widehat{\xi}$. For the estimate $\widehat{\Phi}_{nn,\mu}(\lambda)$ of the noise power in each frequency bin μ for each frame λ, methods like the minimum statistics estimator [Martin 2001] or the IMCRA method [Cohen 2003] are available.

In order to compensate for estimation errors in $\widehat{\Phi}_{nn,\mu}(\lambda)$, an over-estimation factor $o_n \geq 1$ is often considered in the estimate of the *a posteriori* SNR:

$$\widehat{\gamma}_\mu(\lambda) = \frac{|Y_\mu(\lambda)|^2}{o_n \, \widehat{\Phi}_{nn,\mu}(\lambda)}. \tag{5.8}$$

In [Malah et al. 1999] it is recommended that one chooses a range $o_n = 1.2\ldots 1.4$. Although an over-estimation $o_n > 1$ is helpful to reduce *musical noise*, it also causes clipping of low-energy speech components. This effect results from the fact that $\gamma_\mu(\lambda)$ will be under-estimated by $\widehat{\gamma}_\mu(\lambda)$, which leads to low values of the gain function even in the presence of low-energy speech components. As low-energy speech components mainly appear in higher frequency bins, the processed speech signal therefore sounds unnecessarily muffled, if an over-estimation $o_n \geq 1$ is used. For the remainder of this chapter we therefore use $o_n = 1$.

The decision-directed approach for the a priori SNR estimation

The *decision-directed* approach by Ephraim and Malah is an established estimator
$\widehat{\xi}$ for the *a priori* SNR. Originally, it was proposed in the form [Ephraim, Malah
1984]

$$\widehat{\xi}_\mu(\lambda) = \alpha \, G_\mu(\lambda - 1)^2 \, \widehat{\gamma}_\mu(\lambda - 1) + (1 - \alpha) \, \max\left(0, \widehat{\gamma}_\mu(\lambda) - 1\right) . \tag{5.9}$$

The parameter α controls the trade-off between speech distortion and noise fluctuations [Cappé 1994], [Malah et al. 1999], [Ephraim, Cohen 2005]. A higher value of α
suppresses more *musical noise*, but it also leads to more clipping of low energy speech
components and speech onsets, resulting in muffled speech, especially in white noise.
Owing to the similarity of this estimator to a recursive averaging system, the constant
α is generally described as the smoothing constant [Cappé 1994]. Typical values of α
are in the range 0.92 to 0.98.

Note that the filter gain function is an integral part of this recursion. As we can see
in Fig. 5.2 it has a profound effect on the smoothing during a speech pause. A strong
smoothing can only be observed for the STSA estimator [Ephraim, Malah 1984] that
was originally proposed in conjunction with (5.9).

In [Ephraim, Malah 1985], it was reported that (5.9) in conjunction with G^{lsa} results in a lower noise level compared with (5.9) and G^{stsa}. While some of the increased noise attenuation can be attributed to the gain function itself (see Fig. 5.1),
some is due to the averaging recursion (5.9). As can be seen in Fig. 5.2, the mean
value of $\widehat{\xi}$ is lower for the LSA estimator than in the case of the STSA during
speech absence (frames 0 to 24). In order to quantify this effect, the temporal mean
$\overline{\xi}_\mu^{\mathrm{filter}} = E\{\widehat{\xi}_\mu(\lambda) \,|\, \mathcal{H}_0, G^{\mathrm{filter}}\}$ during speech pauses can be compared. Note that we

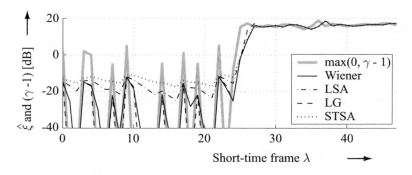

Figure 5.2: Estimate $\widehat{\xi}_\mu(\lambda)$ of the *a priori* SNR as obtained by the *decision-directed*
approach (5.9) with different filter functions G^{filter} and $\alpha = 0.98$. A
sufficiently smooth trajectory results only for the STSA estimator. The
LSA estimator results in less smoothing. For the Wiener and the LG filter,
the estimate $\widehat{\xi}_\mu(\lambda)$ is not a smoothed version of the *a posteriori* SNR γ

can write $\overline{\xi}^{\text{filter}}$ without indices, because the estimates are obtained independently for each spectral bin μ.

During speech absence, an approximation of the mean estimate $\overline{\xi}^{\text{filter}}$ of the *a priori* SNR in the case of Gaussian noise is (see Appendix 5.8.1)

$$\overline{\xi}^{\text{stsa}} \approx \frac{(1-\alpha)e^{-1}}{1-\alpha\frac{\pi}{4}}, \tag{5.10}$$

$$\overline{\xi}^{\text{lsa}} \approx \frac{(1-\alpha)e^{-1}}{1-\alpha e^{-c}} < \overline{\xi}^{\text{stsa}}, \tag{5.11}$$

$$\overline{\xi}^{\text{wiener}} \approx (1-\alpha)e^{-1} < \overline{\xi}^{\text{lsa}}, \tag{5.12}$$

$$\overline{\xi}^{\text{lg}} \approx \overline{\xi}^{\text{wiener}}. \tag{5.13}$$

As all filter gain functions attenuate the signal more for lower ξ, the mean $\overline{\xi}$ can be used to define a noise reduction figure similar to the one introduced in [Vary 1985] in order to quantify the maximum theoretically possible noise suppression. In order to relate $\overline{\xi}^{\text{filter}}$ to the noise reduction, the mean value of γ during speech pauses, $E\{\gamma \mid \mathcal{H}_0\} = \overline{\gamma}_0 = 1$, is used to estimate the average gain $G^{\text{filter}}\left[\overline{\xi}^{\text{filter}}, \overline{\gamma}_0\right]$ in spectral bins with low SNR. For $\alpha = 0.98$, Table 5.1 lists the theoretic values for $\overline{\xi}^{\text{filter}}$ and $G^{\text{filter}}\left[\overline{\xi}^{\text{filter}}, \overline{\gamma}_0\right]$ as well as the mean values from a prototypical filter implementation that uses (5.9). The theoretic approximations are all slightly under-estimating the real values. As $\overline{\xi}^{\text{stsa}}$ and $\overline{\xi}^{\text{lsa}}$ are lower limits (see Appendix 5.8.1), this can be expected. For the Wiener and the LG filter the mean of $\widehat{\xi}$ is slightly higher than the theoretic values, because their input–output characteristics have a relatively large slope around $\overline{\gamma}_0$ compared with the STSA and the LSA estimator, which makes the approximation of their mean by $G^{\text{filter}}\left[\overline{\xi}^{\text{filter}}, \overline{\gamma}_0\right]$ inaccurate.

As will become clear in the investigation of the influence of (5.9) on the input–output characteristics that will be the subject of Sec. 5.3.1, the input–output characteristics

Table 5.1: While the mean value of $\widehat{\xi}$ for $\alpha = 0.98$ in a prototypical filter implementation is very close to the theoretic approximation $\overline{\xi}^{\text{filter}}$, the average noise reduction is significantly over-estimated for the Wiener and the LG filter. All values are expressed in decibel

	STSA	LSA	Wiener	LG
$\overline{\xi}^{\text{filter}}$	-15.0	-17.9	-21.3	≈ -21
Filter implementation: mean of $\widehat{\xi}$	-13.7	-16.9	-20.1	-19.5
$G^{\text{filter}}\left[\overline{\xi}^{\text{filter}}, E\{\gamma \mid \mathcal{H}_0\}\right]$	-16.0	-20.4	-42.7	-42.1
Real noise reduction $G^{\text{filter}}\left[\widehat{\xi}, \widehat{\gamma}\right]$	-14.8	-19.4	-28.6	-26.1

of the Wiener filter and the LG estimator are severely distorted for values around $\overline{\gamma}_0$ by using $\widehat{\xi}$ of (5.9). Therefore, for the Wiener and the LG gain function, the substitution $\gamma \to \overline{\gamma}_0$ is not admissible for an estimate of the average noise reduction during low SNR conditions, as it results in large deviations of the theoretical gain based on the averaged parameters from the empirical noise suppression. This shows that it is important to incorporate both the filter gain function and the parameter estimators in an analysis of a noise reduction system.

Regarding the STSA and the LSA estimator, it can be summarized that this analysis explains the observations made in [Ephraim, Malah 1985] that the LSA estimator produces the lower residual noise when the *a priori* SNR estimator (5.9) is used. This is also easily verified in a listening experiment.

However, the preceding evaluations do not consider the naturalness of the residual noise. In [Ephraim, Malah 1985] it was also reported that the combination of (5.9) and G^{lsa} leads to more fluctuations in the residual noise than (5.9) with G^{stsa}. This observation is confirmed in Fig. 5.2 where the parameter $\widehat{\xi}$ for G^{lsa} fluctuates more than for G^{stsa}. In the case of the Wiener and the LG filter, $\widehat{\xi}_\mu(\lambda)$ has even more peaks of short duration. Here, the *a priori* SNR estimate $\widehat{\xi}_\mu(\lambda)$ basically directly follows $\widehat{\gamma}_\mu(\lambda)$ (see Appendix 5.8.1). It is important to notice that the smoothing property generally attributed to the *decision-directed* approach is negligible for these two filters.

To lower the volatility of $\widehat{\xi}_\mu(\lambda)$ further for the STSA estimator, a lower limit $\xi_{\min} \geq \overline{\xi}^{\text{filter}}$ has been introduced in [Cappé 1994]. This gives rise to reformulate the *decision-directed* approach [Cappé 1994], [Ephraim, Cohen 2005]. Including the noise over-estimation factor o_n, it can be stated in the form

$$\widehat{\xi}_\mu(\lambda) = \max\left(\alpha \, G_\mu(\lambda-1)^2 \, \widehat{\gamma}_\mu(\lambda-1) + (1-\alpha)\left(\widehat{\gamma}_\mu(\lambda) - \frac{1}{o_n}\right), \xi_{\min}\right). \quad (5.14)$$

Note that with a sufficiently high value of the lower limit $\xi_{\min}$, i.e., $\xi_{\min} \gg \overline{\xi}^{\text{filter}}$, the estimate $\widehat{\xi}$ according to (5.14) appears to be similarly smooth for all filter gain functions in Fig. 5.2, because the resulting estimate will be $\widehat{\xi} = \xi_{\min}$ most of the time during low SNR conditions.

5.3 Measurement and Assessment of Unnatural Fluctuations

In this section we outline an approach that can be used to assess fluctuations in the residual noise. We found that this evaluation method shows a strong correlation with the auditory perception of spectral outliers. We derive an approximation of the filter input–output characteristics in combination with the parameter estimator (5.14). As *musical noise* is a phenomenon that occurs most frequently in processed noise, it is especially interesting to analyze the behavior of a filter in spectral regions

with low SNR conditions. By means of the approximated input–output characteristics the statistical distribution of spectral amplitudes of processed noise can be
explained.

5.3.1 Filter Analysis via Approximated Filter Input–Output Characteristics

In a first step, the filter input–output characteristics of Fig. 5.1 are discussed again,
but under the condition that the *a priori* SNR ξ is available only as an estimate $\widehat{\xi}$ from
(5.14). In order to reduce the number of random variables that have an influence on
the estimate, (5.14) is approximated by (see Appendix 5.8.2):

$$\widetilde{\xi}_\mu(\lambda) = \max \left(b_\alpha + (1 - \alpha)\widehat{\gamma}_\mu(\lambda), \xi_{\min} \right), \tag{5.15}$$

with constant bias $b_\alpha = \alpha \left[\left(G^{\text{filter}} \left[\xi_{\min}, \overline{\gamma}_0 \right] \right)^2 + 1 \right] - 1 .$

This approximation of the *decision-directed* approach (5.14) for $o_n = 1$ is valid for
low SNR conditions. In Fig. 5.3 the approximated input–output characteristics are
given that result from substituting ξ_{const} by $\widetilde{\xi}$ in (5.6):

$$\frac{|\widehat{S}|}{\sqrt{\Phi_{nn}}} = G^{\text{filter}} \left[\widetilde{\xi}, \gamma \right] \sqrt{\gamma} . \tag{5.16}$$

Owing to the flooring in (5.14), $\widehat{\xi}_\mu(\lambda)$ takes the constant value $\xi_{\min} > \overline{\xi}^{\text{filter}}$ for most
of the time in the case of speech absence. Thus, for $\gamma < (\xi_{\min} - b_\alpha)/(1 - \alpha)$, i.e.,
when $\widetilde{\xi} = \xi_{\min}$, the input–output characteristics of Figs. 5.1 and 5.3 are identical.

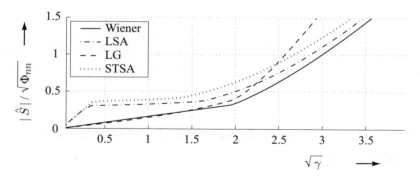

Figure 5.3: The approximated input–output characteristics for $\widetilde{\xi}$ according to (5.15)
and for $\alpha = 0.94$. The lower bound is set to $\xi_{\min} = 0.2 > \overline{\xi}^{\text{filter}}$. As
a consequence of estimating ξ with (5.15) or (5.14), the curvature of the
filter functions is increased for $\gamma > (\xi_{\min} - b_\alpha)/(1 - \alpha)$ as compared with
Fig. 5.1

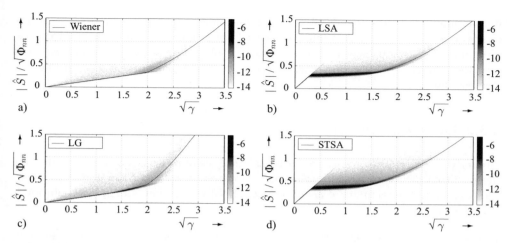

Figure 5.4: Bivariate log-histograms of the spectral amplitudes of Gaussian noise that were processed by prototypical filter implementations using (5.14) with $\xi_{min} = 0.2$ and $\alpha = 0.94$. Their approximation (5.16) (solid line) describes the principal shapes well

But in the case of outliers in noise and hence outliers in $|Y_\mu(\lambda)| = |N_\mu(\lambda)|$, the estimate of the *a priori* SNR is $\widetilde{\xi}_\mu(\lambda) > \xi_{min}$ and is thus a direct function of $|Y_\mu(\lambda)|^2$ according to (5.15). This increases the curvature of the input–output characteristics. As could already be seen in Table 5.1 for the Wiener and the LG filter, bending the filter function towards higher output values limits the noise reduction, as more input values are mapped to higher output values. Raising the lower limit ξ_{min} extends the region where $\widehat{\xi}$ stays constant and where the input–output characteristics are not distorted.

In order to verify that $\widetilde{\xi}_\mu(\lambda)$ of (5.15) is a valid approximation of $\widehat{\xi}_\mu(\lambda)$ from (5.14), the bivariate log-histogram of the spectral amplitudes of noise processed by a prototypical filter implementation is depicted in Fig. 5.4. The grey-scale of this diagram shows the log-probability for the pairs consisting of the input values $\sqrt{\gamma}$ and the corresponding normalized filter outputs $\widehat{S}/\sqrt{\Phi_{nn}}$. Ideally, the shape of this histogram should be identical to the input–output characteristics (5.16). A comparison with the graphs of (5.16), which are also given in the figures, shows that the principal shapes are identical. Furthermore, it can be verified in listening experiments that the noise signal processed by the approximated filter (5.16) leads to the same amount of perceived *musical noise* as the noise filtered by the prototypical filter implementation that is analyzed in Fig. 5.4. The *musical noise* is a result of bending the input–output characteristics towards higher values by making $\widehat{\xi}$ a function of the observed signal $\widehat{\gamma}$ via the *decision-directed* approach. Note that this is the only property that (5.15) and (5.14) have in common. The scatter that is visible in Fig. 5.4 results from the fact that $\gamma_\mu(\lambda - 1)$ in (5.14) is not equal to $\overline{\gamma}_0$ most of the time as is assumed in (5.15). As *musical noise* is also audible with (5.15), the scatter does *not* relate to the audible fluctuations. The dependency of both the estimate of ξ and the gain G^{filter} on the observation $\widehat{\gamma}$, that both the approximated filter using (5.15) and the prototypical filter implementation have in common, is the sole origin of the fluctuations. It is therefore sufficient to use (5.16) with (5.15) for an evaluation of the naturalness of the residual noise.

5.3.2 Outlier Statistics

Similarly to [Vary 1985], we use a statistical analysis of processed noise for evaluating the resulting degree of noise fluctuations. However, because of the non-linear characteristics and the recursive *decision-directed* SNR estimation, an analytic solution is not possible.

Figure 5.5 depicts the log-histogram of $|\widehat{S}_\mu(\lambda)|$ in the case of white Gaussian noise processed by the theoretic filter function (5.6). No speech is present in the signal in order to simulate low SNR conditions as they occur between pitch harmonics and in spectral bands of low speech energy. For the gain functions G^{filter} the parameter ξ was set to a constant values of $\xi_{\text{const}} = 0.2$ as in Fig. 5.1. The log-histogram considers all spectral values of all frames λ excluding the DC and Nyquist frequency bin. The duration of the noise sample is $30\,\text{s}$. The histograms are calculated by dividing the range of observed magnitudes into 400 equally wide histogram bins. These histograms are the marginal distribution of bivariate histograms like those shown in Fig. 5.4. For comparison, the Rayleigh probability density function (pdf) of the scaled magnitude of Gaussian noise, i.e., $|\widehat{S}| = G^{\text{wiener}}[\xi_{\text{const}}]\,|N|$ is also given. In the graph, the filtered spectrum is normalized by the true mean input value $\sqrt{\Phi_{nn}}$. For the mean value of the unfiltered noise, whose pdf is not shown in Fig. 5.5, this would give $E\{|N|\}/\sqrt{\Phi_{nn}} = \sqrt{\pi/4}$ in the case of Rayleigh distributed magnitudes $|N|$. As the Wiener filter G^{wiener} is a constant multiplicative factor for $\xi = \text{const}$, the magnitude of the processed noise is Rayleigh distributed just like that of the scaled signal. In

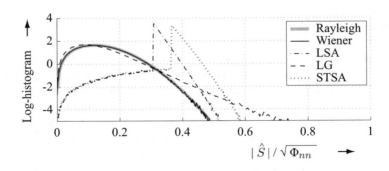

Figure 5.5: Log-histogram of processed white Gaussian noise for the filter functions depicted in Fig. 5.1. The Rayleigh distribution of the scaled input signal, i.e., $|\widehat{S}| = G^{\text{wiener}}[\xi_{\text{const}}]\,|Y|$, coincides with the result for the Wiener filter

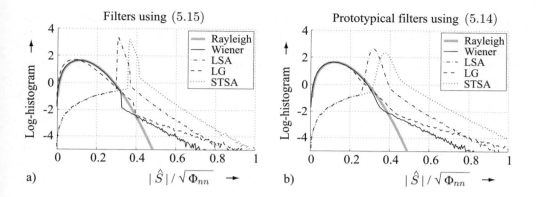

Figure 5.6: Log-histogram of processed Gaussian noise using the approximated *a priori* SNR estimate (5.15) and a prototypical filter implementation that uses (5.14). The histograms exhibit similar tails of outliers that are perceived as unnatural fluctuations. Only the peaks are less pronounced for the prototypical implementation due to the scatter as seen in Fig. 5.4

the case of the LSA estimator G^{lsa}, input values $\sqrt{\gamma} < 2$ in Fig. 5.1 are mapped to an almost constant output value. In Fig. 5.5 the noise processed by the LSA estimator therefore exhibits a histogram with a peak close to the lower limit (5.7). Note that the mean value of this histogram is higher than in the case of the Wiener and the LG filter. For a given value of ξ_{min} the LSA estimator therefore leaves a comparably high noise floor. The noise processed by the LG filter contains an increased number of magnitudes greater than the mean value that form a tail in the histogram. They are caused by the comparably high positive curvature of the filter's input–output characteristic (see Fig. 5.1). The observed magnitudes are spread over a wider range of output values than in the case of the Wiener filter or the LSA estimator. In the resynthesized time signal, these outliers are perceived as fluctuations in the processed noise. For the STSA estimator similar conclusions as for the LSA estimator can be made.

In Fig. 5.6-a the histogram of $|\widehat{S}|/\sqrt{\Phi_{nn}}$ for white Gaussian noise is given for the filter input–output characteristics (5.16) that use the approximated *a priori* SNR estimate (5.15) and $\alpha = 0.94$. Compared with Fig. 5.5, the tails of the histograms are heavier. The processed noise contains more outliers due to the increased curvature of the mapping function (5.16). From the perceptual point of view, this is perceived as an increased amount of audible *musical noise*. Thus, the comparison of the curvature of the filter functions and the statistical analysis as seen in the histograms give an indication for *musical noise*. In addition to listening tests the evaluation of log-histograms of processed stationary signals is a useful way to assess the amount of *musical noise*. *Musical noise* is a consequence of the principal shape of the input–output characteristics.

The practical relevance is demonstrated in Fig. 5.6-b, where the log-histograms are given as they result from the prototypical filter implementation. These histograms exhibit heavy tails similar to those of the approximated filters in Fig. 5.6-a. Only their modes are less pronounced due to the scatter that was observed in Fig. 5.4.

As the perception of the fluctuations in processed noise also depends on the overall shape of the histogram, it is difficult to compare different filters merely by the size of the histogram tails. So far, only a comparison of histograms belonging to one certain filter and different values of its parameters allows for an assessment of the degree of *musical noise*. In the following, the LSA estimator is selected for further analysis. The insights obtained from the LSA case also apply to the other filters introduced in this chapter.

5.4 Avoidance of Processing Artifacts

With the above analysis of the approximated filter input–output characteristics and the log-histograms it is now possible to assess different measures for the suppression of *musical noise*. An exemplary analysis of the *decision-directed* approach (5.14) for estimating the *a priori* SNR will be given here. It is a well known fact that this estimator directly links the amount of *musical noise* and the clipping of low-energy spectral components of speech through the parameters α and ξ_{min} [Malah et al. 1999], [Cappé 1994].

Figure 5.7-a depicts the influence of α on the approximated input–output characteristics (5.16). The corresponding log-histograms of the processed noise are shown in

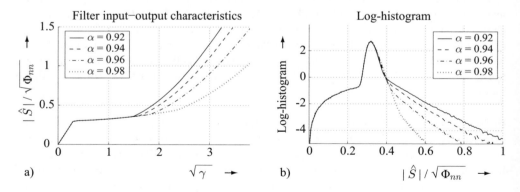

Figure 5.7: Filter input–output characteristics of the LSA estimator for different values of α (Fig. 5.7-a). The lower the value of α, the larger the curvature. The outliers in the noise are spread over a wider range in the processed signal. The amount of audible *musical noise* increases. Accordingly, the amount of spectral outliers increases in Fig. 5.7-b

Fig. 5.7-b. For lower values of α, the input–output characteristic is bent upwards (see Fig. 5.7-a). Therefore, outliers in noise are mapped to higher output values giving more pronounced fluctuations. At the same time, lowering α also raises the mean estimate $\overline{\xi}^{\text{filter}}$ (see (5.2) to (5.5)) so that the condition $\widehat{\xi} > \xi_{\min}$ in (5.14) occurs more often during low SNR conditions. This additionally increases the amount of outliers in the processed noise. The increase in *musical noise* for lower values of α is reflected in the heavier tails in Fig. 5.7-b.

While increasing α reduces the amount of spectral outliers in the processed noise, at the same time, low-energy spectral components of speech are also more attenuated than in the case of smaller α. This is a result of the flatter input–output characteristics for increased values of α. Herein lies the trade-off between speech distortion and *musical noise* [Cappé 1994]. For the preservation of speech, the LG filter shows the best performance. This becomes apparent in Fig. 5.3, where for $\alpha = 0.94$ spectral components $\sqrt{\gamma} > 2.5$ are attenuated least by this filter. In Fig. 5.2 the LG filter consequently is the one that responds fastest to the signal onset. Nevertheless, in noise this property of the LG filter also produces the outliers that have the largest distance from the mean value of the processed noise. This can be seen in Fig. 5.6. There, the outliers form a relatively heavy tail in the log-histogram of the LG filter. The analysis of the input–output characteristics and the log-histograms makes clear that speech distortion and *musical noise* are interdependent when using the estimator (5.14).

A way of *masking* unnatural fluctuations in the residual noise is a higher threshold $\xi_{\min}$. This is achieved at the price of a reduced noise suppression. Figure 5.8-a depicts the input–output characteristics of the LSA estimator for $\alpha = 0.94$ in (5.15) and different values of $\xi_{\min}$. The resulting log-histograms are shown in Fig. 5.8-b. In the figures, the values of $\xi_{\min}$ are expressed in decibel, i.e., $\xi_{\min,\text{dB}} = 10 \log_{10}(\xi_{\min})$.

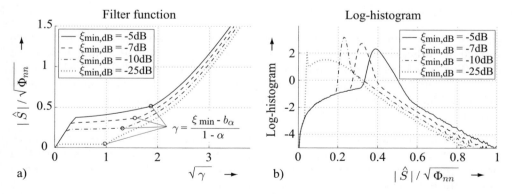

Figure 5.8: Analysis of the LSA estimator for different values of $\xi_{\min}$. The lower the value of $\xi_{\min}$, the more noise can be suppressed. At the same time, the tails in Fig. 5.8-b become more pronounced. The amount of fluctuations in the noise increases

In the log-histogram of Fig. 5.8-b the mean value is lower for smaller values of ξ_{min} indicating a better noise suppression. At the same time, outliers are more pronounced, which is reflected by the fact that the tails are heavier in relation to the mean value for lower values of ξ_{min}. The processed signal consequently contains more audible *musical noise*, if ξ_{min} is lowered. In Fig. 5.8-a the pronunciation of outliers is indicated by the fact that the region $\gamma \geq (\xi_{min} - b_\alpha)/(1 - \alpha)$ begins at lower values of $\sqrt{\gamma}$. The lower border of this region is marked in Fig. 5.8-a. For values of $\sqrt{\gamma}$ greater than the marked border, $\widehat{\xi}$ in (5.15) is a function of the input signal γ and the slope of the input–output characteristics is increased compared with the input–output characteristics of Fig. 5.1. Furthermore, the increased noise reduction observable in Fig. 5.8-b can be deduced from the lower output values $|\widehat{S}|$ for low values of ξ_{min}.

Note that if ξ_{min} corresponds to $-25\,\mathrm{dB}$, the log-histogram in Fig. 5.8-b exhibits a second mode at $|\widehat{S}|/\sqrt{\Phi_{nn}} \approx 0.15$. For a value of $\alpha = 0.94$ the mean *a priori* SNR estimate $\overline{\xi}^{\,\mathrm{lsa}}$ corresponds to $-13.3\,\mathrm{dB}$, and thus we have $\overline{\xi}^{\,\mathrm{lsa}} > \xi_{min}$. Since the estimate $\widehat{\xi}$ fluctuates around the mean value $\overline{\xi}^{\,\mathrm{lsa}}$, the hard limit ξ_{min} that causes an almost constant output $|\widehat{S}|$ comes into effect less often. As a result, the distribution of the processed noise contains less spectral values of similar magnitude that form a peak in the histograms as is the case for ξ_{min} corresponding to $-5\,\mathrm{dB}$ to $-10\,\mathrm{dB}$. Consequently, the signal sounds more natural. Therefore, the variance of the estimate $\widehat{\xi}$ that can be observed in Fig. 5.2 during speech absence actually is an advantage for the perceptual quality of the LSA estimator. It widens the mode of the histogram around the value related to $\overline{\xi}^{\,\mathrm{lsa}}$. It has therefore become common practice in recent implementations to choose a low value of ξ_{min}, for example, a value corresponding to $-25\,\mathrm{dB}$, in order to avoid the accumulation of similar magnitudes in the statistics of the processed noise that would result in a peaked histogram. The masking of *musical noise* that was originally intended by the introduction of ξ_{min} is achieved by applying an overall limit to the final spectral gain instead:

$$G_\mu(\lambda) = \max\left(G^{\mathrm{filter}}\left[\xi_\mu(\lambda), \gamma_\mu(\lambda)\right], G_{min}\right). \tag{5.17}$$

As this change mainly has an effect on small values of G^{filter}, the tails in the histograms formed by outliers in noise are almost the same as for filters limited by higher values of ξ_{min}. The amount of the remaining *musical noise* is comparable whether the limitation of G^{filter} is achieved by a high value of ξ_{min} or by (5.17). However, the naturalness of the residual noise is improved with (5.17) in the case of the STSA and the LSA estimator.

Summarizing this section, it can be said that an optimal choice of the three controlling parameters α, ξ_{min} and G_{min} necessarily results in a trade-off between the distortion of the speech signal, a higher residual noise level, and fluctuations in the residual noise. Since in most applications a high speech quality is required, it is advisable to allow for a fast responsiveness of $\widehat{\xi}$ and G^{filter} to changes in the speech signal in the first place. As a consequence, a low value of α needs to be chosen. The degree of noise reduction is then controlled by ξ_{min} and G_{min}. In order to increase the effectiveness of the filters, low values of ξ_{min} and G_{min} are desirable. This combination of parameters inevitably results in *musical noise*. Therefore, the suppression of fluctuating peaks in the gain function $G_\mu(\lambda)$ can finally be attempted in a post-processing step. The following section deals with such a post-processing technique. It exploits the fact that the peaks caused by outliers in noise are spectrally narrow and have a duration much shorter than the salient spectral features of speech.

5.5 Control of Spectral Fluctuations in the Cepstral Domain

From the previous discussion it becomes clear that spectral filter gain estimation based on the *decision-directed* approach has the tendency to emphasize single bins randomly due to the increased curvature of the input–output characteristics. For Gaussian noise these outliers result in *musical noise*. The duration of such a random outlier is only one frame. Its spectral width is determined by the spectral resolution of the analysis–synthesis system and the windowing function $w(\tau)$. In the case of correlated noise, such as babble noise, however, outliers can be more pronounced in duration and spectral width.

While single *musical tones* may appear as spontaneously as speech onsets, their energy is much lower. Additionally, only a narrow band is affected. In a cepstral representation of the gain function $G_\mu(\lambda)$, this abrupt change in the fine structure of $G_\mu(\lambda)$ is reflected by a change of the coefficients corresponding to higher quefrencies. Smoothing these higher coefficients flattens out the short spectral peaks of *musical noise*. A temporal smoothing of the cepstrum of the gain function $G_\mu(\lambda)$ can therefore be used to amend a peaked shape of $G_\mu(\lambda)$ caused by outliers in $\hat{\xi}_\mu(\lambda)$. Cepstral smoothing has the advantage that different cepstral bins describe different degrees of detail in the spectral structure of the gain function. Smoothing the higher cepstral coefficients affects the fine structure of $G_\mu(\lambda)$. Its temporal dynamics is slowed down. As the narrow spectral peaks of single *musical tones* appear only for a duration of a single frame, they are strongly affected by such a cepstral smoothing.

A cepstral representation of $G_\mu(\lambda)$ from (5.1) is calculated for each frame λ as

$$G_{\mu'}^{\text{cepst}}(\lambda) = \text{IDFT}\left\{\log_e(G_\mu(\lambda))\right\}_M , \quad \mu = 0\ldots(M-1), \tag{5.18}$$

where IDFT is the inverse DFT of length M resulting in quefrency bins $\mu' = 0\ldots(M-1)$.

Note that cepstral smoothing would also affect narrow-band quasi-stationary speech components like pitch harmonics. However, as voiced sounds have high energy, the cepstral bins corresponding to the pitch frequency are very pronounced and can be reliably detected. Less smoothing of these cepstral bins can be applied so that pitch quefrency bins can follow rapid changes. Thus, the formation of the spectral fine structure for voiced speech onsets is almost unaffected. The spectral contrast of voiced speech is able to develop within a few frames.

In the smoothing procedure described in this chapter [Breithaupt et al. 2008], the smoothing is not applied to low cepstral coefficients at all. This preserves the temporal evolution of the spectral envelope of the gain function in the case of speech presence. Thus, speech onsets and broad spectral structures like fricatives and plosives are not distorted. A smoothed version $G^{\text{cepst}}_{\mu',\text{smooth}}(\lambda)$ is calculated as

$$G^{\text{cepst}}_{\mu',\text{smooth}}(\lambda) = \beta_{\mu'}\, G^{\text{cepst}}_{\mu',\text{smooth}}(\lambda - 1) + (1 - \beta_{\mu'})\, G^{\text{cepst}}_{\mu'}(\lambda). \qquad (5.19)$$

Owing to the symmetry of $G^{\text{cepst}}_{\mu'}(\lambda)$, the number of relevant cepstral bins is $D = M/2 + 1$. The smoothing is applied to cepstral bins $\mu' \in \{\mu'_{\text{low}}, \dots, D-1\}\backslash\mathbb{P}'$, where $\mathbb{P}'$ denotes a set of cepstral coefficient indices that is excluded from smoothing. It consists of the cepstral index μ'_{pitch} of the pitch and its two neighbors $(\mu'_{\text{pitch}} - 1)$ and $(\mu'_{\text{pitch}} + 1)$. μ'_{pitch} is found by taking the quefrency bin with the maximum value of $G^{\text{cepst}}_{\mu'}(\lambda)$ in a range of possible pitch related quefrencies:

$$\mu'_{\text{pitch}} = \underset{\mu'}{\text{argmax}} \left\{ G^{\text{cepst}}_{\mu'}(\lambda) \Big| \mu' \in \{\mu'_{\text{pitch,lower}}, \dots, \mu'_{\text{pitch,upper}}\}, \mathcal{H}_1 \right\}. \qquad (5.20)$$

The search for μ'_{pitch} is only meaningful if speech is present. The pitch frequency is assumed to be in the range $f_{\text{pitch}} \in (70\,\text{Hz} \dots 500\,\text{Hz})$. The search interval in terms of quefrencies is then obtained by considering that a pitch frequency f_{pitch} corresponds to a pitch quefrency bin $\mu'_{\text{pitch}} = f_S/f_{\text{pitch}}$, with f_S being the sampling rate. For the cepstral coefficients $\mu' \in \mathbb{P}'$ a smoothing similar to (5.19) is used, but with a smoothing constant $\beta_{\mu',\text{pitch}} < \beta_{\mu'}$. Possible choices are $\beta_{\mu'} = 0.8$ and $\beta_{\mu',\text{pitch}} = 0.4$ for a spectral analysis with $16\,\text{ms}$ frame-shift. $\mathbb{P}'$ is the empty set, when speech pauses are signaled by a voice activity detector (VAD) [Breithaupt, Martin 2006]. For $\mu' \in \{0 \dots \mu'_{\text{low}} - 1\}$ no smoothing is applied at all, giving $G^{\text{cepst}}_{\mu',\text{smooth}}(\lambda) = G^{\text{cepst}}_{\mu'}(\lambda)$. A sufficient protection of the speech envelope is achieved with $\mu'_{\text{low}} = 4$ for $M = 512$ and $f_S = 16\,\text{kHz}$.

Note that the log-function in (5.18) is not essential for the selective smoothing procedure just described. Nevertheless, this non-linear transform of $G_\mu(\lambda)$ considerably reduces noise shaping effects caused by (5.19) in stationary Gaussian noise.

The final smoothed spectral gain function is computed as

$$G_{\mu,\text{smooth}}(\lambda) = \min\left(\exp\left(\text{DFT}\left\{G^{\text{cepst}}_{\mu',\text{smooth}}(\lambda)\right\}_M\right), G_{\max}\right), \quad \mu' = 0 \dots (M-1). \tag{5.21}$$

The cepstral bins $\mu' \in \{D \dots M - 1\}$ needed for this transform are available from the symmetry condition $G^{\text{cepst}}_{\mu'} = G^{\text{cepst}}_{M-\mu'}$. The smoothed spectral gain $G_{\mu,\text{smooth}}(\lambda)$ can be applied according to (5.1).

Although a VAD is used for finding μ'_{pitch}, false alarms do not have a large effect. For background noises or unvoiced sounds, the cepstral bin μ'_{pitch}, as determined according

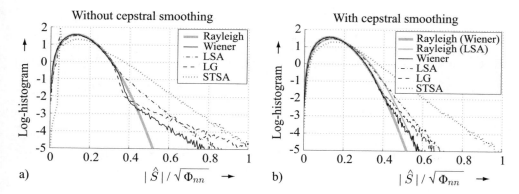

Figure 5.9: Gaussian noise processed by a prototypical filter without and with cepstral smoothing. For the *a priori* SNR estimation, α is 0.94 and ξ_{min} corresponds to $-25\,dB$. G_{min} corresponds to $-15\,dB$. For comparison, Rayleigh distributions with means ρ^{wiener} (graphs "Rayleigh", "Rayleigh (Wiener)") and ρ^{lsa} (graph "Rayleigh (LSA)") are given. Cepstral smoothing gives almost Rayleigh distributed output signals resulting in natural sounding residual noise

to (5.20), does not contribute as significantly to the filter output as in the case of voiced speech. In fact, it is possible to leave out the VAD and do the search (5.20) for every frame λ without significantly affecting the signal quality.

With the outlier statistics of Sec. 5.3.2 the effectiveness of cepstral smoothing in avoiding noise fluctuations can be assessed. Figure 5.9 depicts the log-histograms of Gaussian noise processed by $G_\mu(\lambda)$ and $G_{\mu,\text{smooth}}(\lambda)$, respectively. The *a priori* SNR estimation uses $\alpha = 0.94$. In order to demonstrate the effect of (5.17), the limit ξ_{min} corresponds to $-25\,dB$ in this experiment. The resulting gain is limited to a value of G_{min} corresponding to $-15\,dB$. Note that choosing a higher limit ξ_{min}, as was demonstrated in the previous sections, gives similar results in terms of the content of *musical noise* in the processed signal.

From the log-histograms it is apparent that for all filters except the STSA estimator the amount of outliers is reduced when cepstral smoothing is applied. Accordingly, the processed noise sounds like scaled Gaussian noise without *musical noise*. For the STSA estimator the smoothing has no effect, because the time constants of the cepstral smoothing are similar to those of the *decision-directed* approach for the *a priori* SNR estimate $\widehat{\xi}$ (see Fig. 5.2). Thus, fluctuations in the estimate $|\widehat{S}|$ due to outliers in noise maintain large amplitudes long enough to overcome the smoothing effect of (5.19). Note that this is different if the *soft-gain* method is applied to the STSA estimator [Ephraim, Malah 1984].

In order to get an impression of the naturalness of the residual noise, the log-histogram of the scaled undistorted noise is given for comparison. In Fig. 5.9 the scaling was chosen so that the mean of the scaled noise is equal to the mean of the noise processed

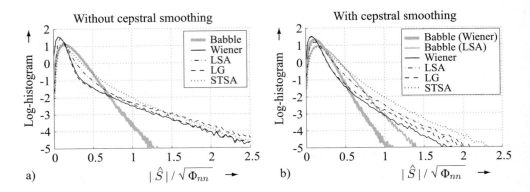

Figure 5.10: Babble noise processed without and with cepstral smoothing as it results from a prototypical filter with parameters as in Fig. 5.9. As gray lines, the log-histograms of scaled undistorted babble noise are given with mean values identical to the Wiener filter result and the LSA estimator result, respectively. A comparison shows the high naturalness of the processed babble noise in case of cepstral smoothing

by the Wiener filter (graphs "Rayleigh" and "Rayleigh (Wiener)"). This mean value is denoted ρ^{wiener}. Correspondingly, for the LSA estimator the mean ρ^{lsa} after processing with the LSA estimator was chosen for the graph "Rayleigh (LSA)" in Fig. 5.9-b. The processed magnitude of the Gaussian noise has a distribution that is very similar to the Rayleigh distribution, if cepstral smoothing is used. Listening experiments confirm the naturalness of the processed noise achieved by cepstral smoothing. In fact, single *musical tones* no longer occur.

Note that an analysis of the approximated filter input–output characteristics as in Sec. 5.3.1 is not possible here, as cepstral smoothing interlinks the gains of all spectral bins μ.

In order to analyze the suppression of *musical noise* in non-stationary noise, log-histograms for babble noise are given in Fig. 5.10. These histograms consider a spectral band of 160 Hz to 1 kHz only, as voiced babble bursts are most numerous in this frequency range. The reduction of outliers due to cepstral smoothing is clearly visible. The naturalness of the residual noise is apparent from the comparison of the histograms of the processed and the scaled noise. Again the scaled noise is given for ρ^{wiener} and ρ^{lsa}. Listening experiments confirm that processing babble noise with cepstrally smoothed gain functions results in a residual noise that sounds very similar to attenuated babble without salient tonal artifacts.

In Fig. 5.11 the spectrogram of a speech sample is depicted. The noisy speech signal is disturbed by babble noise at a segmental SNR of 0 dB. In this experiment the LSA estimator G^{lsa} is used. In the spectrogram 5.11-b, regions have been marked where babble bursts occur during speech presence. As babble bursts are narrow-band signals

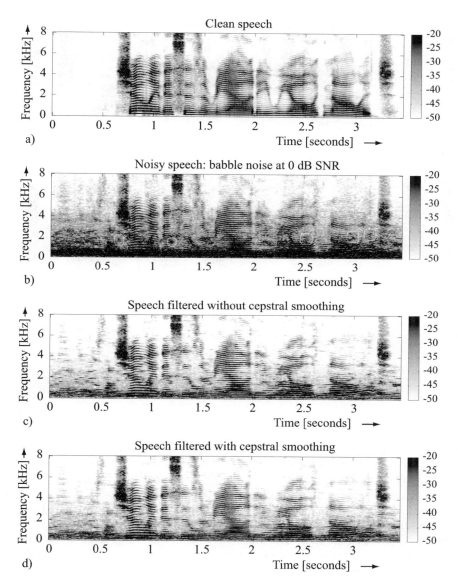

Figure 5.11: Spectrograms of clean, noisy, and filtered signals. The sentence is "Surely this is a reality we all acknowledge". The noise is babble noise at 0 dB segmental SNR. With cepstral smoothing many narrow-band fluctuations of the babble noise are smoothed out. At the same time, cepstral smoothing does not clip speech onsets. Critical babble bursts during speech presence are marked in the noisy spectrogram. In case of cepstral smoothing, the residual babble noise sounds more natural, while the speech quality is not affected

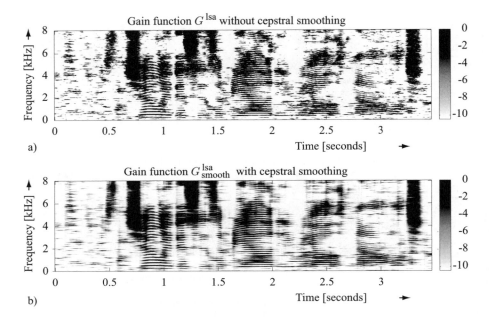

Figure 5.12: Filter gains $G_\mu(\lambda)$ and $G_{\mu,\text{smooth}}(\lambda)$ (in dB) of the LSA estimator for the filtered sample of Fig. 5.11. The parameters are set to $\alpha = 0.94$ and $\xi_{\min} = 0.1$. Although cepstral smoothing results in a temporally "smeared" gain, the speech onsets are not clipped. The harmonic fine structure of voiced speech is also preserved

of short duration they can be effectively suppressed by $G_{\text{smooth}}^{\text{lsa}}$ of (5.21) without affecting the speech. If no cepstral smoothing is applied, the residual noise sounds unnatural, as single babble noise bursts are more salient than before the filtering. In the case of the cepstrally smoothed gain function $G_{\text{smooth}}^{\text{lsa}}$ the residual babble noise sounds much more natural while the speech signal sounds identical to the speech processed without cepstral smoothing. The effect of cepstral smoothing on the gain function is shown in Fig. 5.12. The variability of G^{lsa} is diminished as the dynamics of the fine structure is reduced. Nevertheless, broad-band speech onsets and the pitch harmonics are very well preserved.

5.6 Discussion and Conclusions

In the first part of this chapter the performance of noise reduction filters was analyzed and compared by means of theoretical approximations as well as empirical outlier distributions of processed noise. For state-of-the-art noise reduction systems using the *decision-directed a priori* SNR estimator, the limits of noise reduction were derived as a function of the *decision-directed* smoothing parameter α. For low SNR conditions, the theoretical approximation of the filter input–output characteristics describes measurable and perceived properties very well. Moreover, the approximate input–output characteristics are linked to the empirical distributions as the evaluation of log-histograms reveals. A relation to the amount of perceived *musical noise* is also indicated. In state-of-the-art systems, the residual noise level, speech distortions, and the appearance of spectral fluctuations are intrinsically linked and not easily minimized. Besides other effects, it could be shown that the intrinsic smoothing property of the STSA estimator when combined with the *decision-directed* approach is much less effective for other frequently used gain functions. It can therefore be concluded that the excellent perceptual performance of this system in Gaussian noise must be attributed to the input–output characteristics of the STSA estimator and only to a lesser extend to the *decision-directed* approach.

As a result of the linkage between speech distortions and the appearance of spectral fluctuations we suggest that one selects the smoothing parameters of the *decision-directed* approach such that speech is well reproduced. Additional measures such as smoothing the gain function in the cepstral domain can then be used to control the residual noise statistics. Smoothing in the cepstral domain is ideally suited for this purpose as it allows one to apply less smoothing to the salient features of speech while effectively suppressing random outliers. Furthermore, it does not introduce additional latency in the signal processing chain.

5.7 Acknowledgements

This work was funded in part by the German Research Foundation **DFG**. The authors would also like to thank Dirk Mauler for proof-reading the text and for his valuable comments.

5.8 Appendix

5.8.1 Mean a priori SNR for different filter types and low SNR

For the statistical analysis of the original *decision-directed* approach (5.9), we reinterpret $\widehat{\xi}_\mu(\lambda)$ as the weighted sum of the two random variables $\widehat{\gamma}_\mu(\lambda - 1)$ and $z_\mu(\lambda) = \max(0, \widehat{\gamma}_\mu(\lambda) - 1)$. For the mean values in spectral bins without speech, we have $E\{\widehat{\gamma} \mid \mathcal{H}_0\} = \overline{\gamma}_0 = 1$ and $E\{z \mid \mathcal{H}_0\} = \overline{z}_0 = e^{-1}$, if $N_\mu(\lambda)$ is assumed to be complex Gaussian distributed. For a frame-shift of at least half the frame-length, the two random variables can be assumed to be independent. The mean *a priori* SNR during speech pauses therefore is

$$\overline{\xi}^{\,\text{filter}} = E\left\{\widehat{\xi}_\mu(\lambda)\,\middle|\,\mathcal{H}_0, G^{\text{filter}}\right\}$$

$$= \alpha\, E\left\{\left|G^{\text{filter}}\left[\widehat{\xi}_\mu(\lambda-1),\widehat{\gamma}_\mu(\lambda-1)\right]\right|^2 \widehat{\gamma}_\mu(\lambda-1)\,\middle|\,\mathcal{H}_0\right\}$$
$$+ (1-\alpha)\, E\left\{z_\mu(\lambda)\,|\,\mathcal{H}_0\right\}. \tag{5.22}$$

For the STSA and the LSA estimator, an approximation of the first term of (5.22) can be found by considering that $G^{\text{filter}}\left[\widehat{\xi}_\mu(\lambda),\widehat{\gamma}_\mu(\lambda)\right]\sqrt{\widehat{\gamma}_\mu(\lambda)}$ is almost constant for values of $\widehat{\gamma}$ around $\overline{\gamma}_0 = 1$ and given $\widehat{\xi}$ (see Fig. 5.1). For the STSA estimator , this constant is the lower bound of the output that is obtained for $\widehat{\gamma} = 0$:

$$\left.\left|G^{\text{stsa}}\left[\widehat{\xi}_\mu(\lambda),\widehat{\gamma}_\mu(\lambda)\right]\right|^2 \widehat{\gamma}_\mu(\lambda)\right|_{\widehat{\gamma}=0} = \frac{\pi}{4}\frac{\widehat{\xi}}{1+\widehat{\xi}} \approx \frac{\pi}{4}\widehat{\xi},$$

where the approximation is possible for low SNR conditions, when $\widehat{\xi} \ll 1$. This lower bound can be used to replace the expected value in the first term of (5.22).

The expected value $\overline{\xi}$ is the stationary value of the recursion (5.22), where $z_\mu(\lambda)$ is considered as the innovation:

$$\overline{\xi} = \alpha\,\frac{\pi}{4}\,\overline{\xi} + (1-\alpha)\,\overline{z}_0. \tag{5.23}$$

For the STSA estimator we thus obtain

$$\overline{\xi}^{\,\text{stsa}} \approx \frac{(1-\alpha)\,e^{-1}}{1-\alpha\,\pi/4}. \tag{5.24}$$

For $\alpha = 0.98$ this gives a value $\overline{\xi}^{\,\text{stsa}}$ equivalent to $-14.95\,\text{dB}$, which corresponds to the value observed in [Cappé 1994].

With (5.22) the analysis can be extended to other filter types. The LSA estimator gives

$$\left.\left|G^{\text{lsa}}\left[\widehat{\xi}_\mu(\lambda),\widehat{\gamma}_\mu(\lambda)\right]\right|^2 \widehat{\gamma}_\mu(\lambda)\right|_{\widehat{\gamma}=0} = \frac{\widehat{\xi}}{1+\widehat{\xi}}\,e^{-c} \approx \widehat{\xi}\,e^{-c}$$

and, hence,

$$\overline{\xi}^{\,\text{lsa}} \approx \frac{(1-\alpha)\,e^{-1}}{1-\alpha\,e^{-c}},$$

where c is Euler's constant. For $\alpha = 0.98$, $\overline{\xi}^{\,\text{lsa}}$ corresponds to $-17.86\,\text{dB}$, thus yielding more noise reduction.

The Wiener filter gain G^{wiener} is not constant for small $\widehat{\xi}$ (see Fig. 5.1). As it is not a function of γ either (see (5.2)), substituting G^{wiener} in (5.22) gives

$$\overline{\xi} \approx \alpha\,|\overline{\xi}|^2\,E\left\{\widehat{\gamma}_\mu(\lambda-1)\,|\,\mathcal{H}_0\right\} + (1-\alpha)\,E\left\{z_\mu(\lambda)\,|\,\mathcal{H}_0\right\}$$

Instead of solving this quadratic equation of $\overline{\xi}$, an additional approximation can be made by considering that $\alpha\overline{\xi}^2\,\overline{\gamma}_0 \ll (1 - \alpha)\,\overline{z}_0$ for small $\overline{\xi}$. Therefore, in the case of the Wiener filter and similarly the LG filter, the recursive term in (5.22) does not contribute significantly. In Fig. 5.2 the estimate $\widehat{\xi}_\mu(\lambda)$ thus is a scaled version of $z_\mu(\lambda)$: $\widehat{\xi}_\mu^{\text{wiener}}(\lambda) \approx (1 - \alpha)\,z_\mu(\lambda)$. The mean value is

$$\overline{\xi}^{\text{ wiener}} \approx (1 - \alpha)e^{-1},$$

which corresponds to $-21.3\,\text{dB}$ for $\alpha = 0.98$. From Figs. 5.1 and 5.2 it can be concluded that $\overline{\xi}^{\text{ lg}} \approx \overline{\xi}^{\text{ wiener}}$ during low SNR conditions. Note that it is difficult to estimate a mean value from Fig. 5.2 as the ordinate is in logarithmic scale. Nevertheless, the numerical evaluation shown in Table 5.1 confirms the above approximations.

5.8.2 Approximation of the decision-directed approach for low SNR

The *decision-directed a priori* SNR estimator (5.14) can be re-written as

$$\widehat{\xi}_\mu(\lambda) = \max\left(\alpha\,\left|G^{\text{filter}}\left[\widehat{\xi}_\mu(\lambda - 1), \widehat{\gamma}_\mu(\lambda - 1)\right]\right|^2\widehat{\gamma}_\mu(\lambda - 1)\right.$$

$$\left. + (1 - \alpha)\,(\widehat{\gamma}_\mu(\lambda) - 1),\, \xi_{\min}\right). \qquad (5.25)$$

The overall dependency of $|\widehat{\xi}_\mu(\lambda)|$ from the current observation $\widehat{\gamma}_\mu(\lambda)$ for a low SNR can be shown by combining all those terms that are not a function of the current observation $\widehat{\gamma}_\mu(\lambda)$. The first term in (5.25) depends only on values of the previous frame $\lambda - 1$. The term $\widehat{\gamma}_\mu(\lambda - 1)$ in (5.25) is additionally substituted by its expected value. For speech pauses $(|Y_\mu(\lambda)|^2 = |N_\mu(\lambda)|^2)$ this is $\widehat{\gamma}_\mu(\lambda - 1) \approx E\{\gamma\,|\,\mathcal{H}_0\} = \overline{\gamma}_0 = 1$. If no estimation error occurred, the estimate of the *a priori* SNR is $\widehat{\xi}_\mu(\lambda - 1) = \xi_{\min} > \overline{\xi}^{\text{ filter}}$. This additionally gives for the first term:

$$\alpha\,\left|G^{\text{filter}}\left[\widehat{\xi}_\mu(\lambda - 1), \widehat{\gamma}_\mu(\lambda - 1)\right]\right|^2 \approx \alpha\,\left(G^{\text{filter}}\,[\xi_{\min}, \overline{\gamma}_0]\right)^2,$$

which can be used in (5.25) yielding (5.15).

Bibliography

Accardi, A. J.; Cox, R. V. (1999). A Modular Approach to Speech Enhancement with an Application to Speech Coding, *Proceedings of the IEEE International Conference on Acoustics, Speech, and Signal Processing (ICASSP)*, pp. 201–204.

Berouti, M.; Schwartz, R.; Makhoul, J. (1979). Enhancement of Speech Corrupted by Acoustic Noise, *Proceedings of the IEEE International Conference on Acoustics, Speech, and Signal Processing (ICASSP)*, pp. 208–211.

Breithaupt, C.; Gerkmann, T.; Martin, R. (2008). Cepstral Smoothing of Spectral Filter Gains for Speech Enhancement without Musical Noise, *IEEE Signal Processing Letters*, vol. 15, no. 2, in press.

Breithaupt, C.; Martin, R. (2003). MMSE Estimation of Magnitude-squared DFT Coefficients with Supergaussian Priors, *Proceedings of the IEEE International Conference on Acoustics, Speech, and Signal Processing (ICASSP)*, vol. I, pp. 848–851.

Breithaupt, C.; Martin, R. (2006). Voice Activity Detection in the DFT Domain Based on a Parametric Noise Model, *International Workshop on Acoustic Echo and Noise Control (IWAENC)*.

Cappé, O. (1994). Elimination of the Musical Noise Phenomenon with the Ephraim and Malah Noise Suppressor, *IEEE Transactions on Speech and Audio Processing*, vol. 2, no. 2, pp. 345–349.

Cohen, I. (2001). On Speech Enhancement under Signal Presence Uncertainty, *Proceedings of the IEEE International Conference on Acoustics, Speech, and Signal Processing (ICASSP)*, pp. 661–664.

Cohen, I. (2003). Noise Spectrum Estimation in Adverse Environments: Improved Minima Controlled Recursive Averaging, *IEEE Transactions on Speech and Audio Processing*, vol. 11, no. 5, pp. 466–475.

Ephraim, Y.; Cohen, I. (2005). Recent Advancements in Speech Enhancement, *in* R. C. Dorf (ed.), *The Electrical Engineering Handbook*, CRC Press.

Ephraim, Y.; Malah, D. (1984). Speech Enhancement Using a Minimum Mean-Square Error Short-Time Spectral Amplitude Estimator, *IEEE Transactions on Acoustics, Speech and Signal Processing*, vol. 32, no. 6, pp. 1109–1121.

Ephraim, Y.; Malah, D. (1985). Speech Enhancement Using a Minimum Mean-Square Error Log-Spectral Amplitude Estimator, *IEEE Transactions on Acoustics, Speech and Signal Processing*, vol. 33, no. 2, pp. 443–445.

Fingscheidt, T.; Beaugeant, C.; Suhadi, S. (2005). Overcoming the Statistical Independence Assumption w.r.t. Frequency in Speech Enhancement, *Proceedings of the IEEE International Conference on Acoustics, Speech, and Signal Processing (ICASSP)*, vol. 1, pp. 1081–1084.

Goh, Z.; Tan, K.-C.; Tan, B. (1998). Postprocessing Method for Suppressing Musical Noise generated by Spectral Subtraction, *IEEE Transactions on Speech and Audio Processing*, vol. 6, no. 3, pp. 287–292.

Gradshteyn, I.; Ryzhik, I. (2000). *Table of Integrals, Series, and Products*, 6th edn, Academic Press.

Gülzow, T.; Ludwig, T.; Heute, U. (2003). Spectral-Subtraction Speech Enhancement in Multirate Systems with and without Non-uniform and Adaptive Bandwidths, *EURASIP Signal Processing*, vol. 83, no. 8, pp. 1613–1631.

Gustafsson, H.; Nordholm, S. E.; Claesson, I. (2001). Spectral Subtraction Using Reduced Delay Convolution and Adaptive Averaging, *IEEE Transactions on Speech and Audio Processing*, vol. 9, no. 8, pp. 799–807.

Gustafsson, S.; Jax, P.; Vary, P. (1998). A Novel Psychoacoustically Motivated Audio Enhancement Algorithm Preserving Background Noise Characteristics, *Proceedings of the IEEE International Conference on Acoustics, Speech, and Signal Processing (ICASSP)*, pp. 397–400.

Hansen, J. H. L. (1991). Speech Enhancement Employing Adaptive Boundary Detection and Morphological Based Spectral Constraints, *Proceedings of the IEEE International Conference on Acoustics, Speech, and Signal Processing (ICASSP)*, pp. 901–904.

Hasan, K.; Salahuddin, S.; Khan, M. R. (2004). A Modified A Priori SNR for Speech Enhancement Using Spectral Subtraction Rules, *IEEE Signal Processing Letters*, vol. 11, no. 4, pp. 450–453.

Lotter, T.; Vary, P. (2004). Noise Reduction by Joint Maximum A Posteriori Spectral Amplitude and Phase Estimation with Super-Gaussian Speech Modelling, *Proceedings of the European Signal Processing Conference (EUSIPCO)*, pp. 1457–1460.

Malah, D.; Cox, R. V.; Accardi, A. J. (1999). Tracking Speech-Presence Uncertainty to Improve Speech Enhancement in Non-Stationary Noise Environments, *Proceedings of the IEEE International Conference on Acoustics, Speech, and Signal Processing (ICASSP)*, vol. 2, pp. 789–792.

Martin, R. (2001). Noise Power Spectral Density Estimation Based on Optimal Smoothing and Minimum Statistics, *IEEE Transactions on Speech and Audio Processing*, vol. 9, no. 5, pp. 504–512.

Martin, R. (2002). Speech Enhancement Using MMSE Short Time Spectral Estimation with Gamma Distributed Speech Priors, *Proceedings of the IEEE International Conference on Acoustics, Speech, and Signal Processing (ICASSP)*, vol. 1, pp. 253–256.

Martin, R. (2005a). Speech Enhancement Based on Minimun Mean Square Error Estimation and Supergaussian Priors, *IEEE Transactions on Speech and Audio Processing*, vol. 13, no. 5, pp. 845–856.

Martin, R. (2005b). Statistical Methods for the Enhancement of Noisy Speech, *in* J. Benesty; S. Makino; J. Chen (eds.), *Speech Enhancement*, Springer.

Vary, P. (1985). Noise Suppression by Spectral Magnitude Estimation – Mechanism and Theoretical Limits, *EURASIP Signal Processing*, vol. 8, pp. 387–400.

Vary, P.; Martin, R. (2006). *Digital Speech Transmission*, John Wiley & Sons, Ltd.

Virag, N. (1999). Single Channel Speech Enhancement Based on Masking Properties of the Human Auditory System, *IEEE Transactions on Speech and Audio Processing*, vol. 7, pp. 126–137.

You, C. H.; Koh, S. N.; Rahardja, S. (2005). β-order MMSE Spectral Amplitude Estimation for Speech Enhancement, *IEEE Transactions on Speech and Audio Processing*, vol. 13, pp. 475–486.

Chapter 6

Acoustic Source Localization with Microphone Arrays

Nilesh Madhu, Rainer Martin

6.1 Introduction

Acoustic source localization using microphone arrays is of paramount importance in many speech processing applications such as video conferencing, hands-free speech acquisition in cars, and processors for digital hearing aids. Additionally, acoustic source localization is used in non-speech applications, for example, remote surveillance, fault analysis of machinery, automotive acoustics, and autonomous robots. It comes as no surprise, therefore, that this topic has been the subject of significant research activity for a long time and still enjoys considerable interest in the signal processing community.

This chapter aims to give an overview of contemporary localization algorithms. It will be shown that the most common algorithms depend only on the second order statistics of the microphone signals and fit into a unifying framework that exploits just the *cross-correlation* between the signals of the various microphone pairs. We shall start with a description of the signal model and associated concepts for source localization, followed by the overview of the localization approaches. To provide additional insight, we conclude this chapter with an illustration of the performance of representative algorithms under various conditions of reverberation and noise.

Advances in Digital Speech Transmission   Edited by R. Martin, U. Heute and C. Antweiler
© 2008 John Wiley & Sons, Ltd

6.2 Signal Model

As we are concerned with multichannel approaches to source localization, we shall be dealing with microphone or sensor[1] *arrays* for sound pickup. Figure 6.1 illustrates the general situation in two spatial dimensions (spanned by unit vectors $\mathbf{e}_x$ and $\mathbf{e}_y$). We consider one such array consisting of M microphones at positions $\mathbf{r}_m$ capturing the signal emitted from a source at position $\mathbf{r}_s$. These signals recorded by the microphones may then be expressed in the continuous or the discrete time domain, considering the contribution of all the paths from the source to the individual microphones of the array.

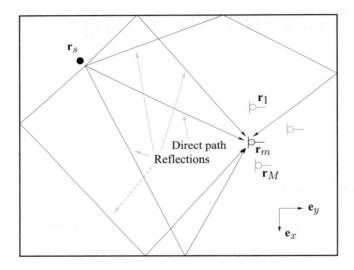

Figure 6.1: Acoustic signal paths for a particular microphone in the x–y plane. $\mathbf{r}_s$ and $\mathbf{r}_m$ denote the locations of the source and the mth microphone, respectively. Each path from the source to the microphone m may be represented by an attenuation and a delay of the source signal. The direct path possesses the least delay. The sum of all the paths constitutes the impulse response of the room for the particular source and microphone position

6.2.1 Continuous Time Model

The signal $\tilde{y}_m(t)$ received at the mth microphone of the array, located at $\mathbf{r}_m = (x_m, y_m, z_m)^T$, due to a source located at $\mathbf{r}_s = (x_s, y_s, z_s)^T$ may be written, in

---

[1]Note that the terms channels, microphones, and sensors will be used synonymously throughout the text.

continuous time t, as

$$\tilde{y}_m(t) = \tilde{a}_m(t) * \tilde{s}_0(t) + \tilde{v}_m(t), \tag{6.1}$$

where $\tilde{s}_0(t)$ represents the source waveform, $\tilde{a}_m(t)$ represents the room impulse response from the source position to the microphone m, the $*$ represents the convolution operator, and $\tilde{v}_m(t)$ represents the noise at the microphone. This may be extended to the general case of Q sources as

$$\tilde{y}_m(t) = \sum_{q=1}^{Q} \tilde{a}_{mq}(t) * \tilde{s}_{0q}(t) + \tilde{v}_m(t), \tag{6.2}$$

where $\tilde{a}_{mq}(t)$ now represents the room impulse response from the qth source to the mth microphone.

6.2.2 Discrete Time Representation

Since we shall mostly deal with the digital representations of the microphone and source signals, the concept of continuous time serves only to clarify some basic ideas. Consequently, relations (6.1) and (6.2) will now be extended to the discrete time case. To simplify the discussion, we approximate the room impulse responses by finite impulse response (FIR) filters of order $L-1$. For the single source case, we now have an impulse response vector

$$\mathbf{a}_m = (a_m(0), a_m(1), \ldots, a_m(L-1))^T \tag{6.3}$$

and the signal at microphone m as

$$y_m(k) = \mathbf{a}_m^T \mathbf{s}_0(k) + v_m(k), \tag{6.4}$$

where k is the discrete time index, $\mathbf{s}_0(k) = (s_0(k), s_0(k-1), \ldots, s_0(k-L+1))^T$, and $v_m(k)$ is the sampled noise signal. For the multi-source scenario we define the impulse response vectors from source q to microphone m as

$$\mathbf{a}_{mq} = (a_{mq}(0), a_{mq}(1), \ldots, a_{mq}(L-1))^T \tag{6.5}$$

and obtain

$$\begin{pmatrix} y_1(k) \\ \vdots \\ y_M(k) \end{pmatrix} = \begin{pmatrix} \mathbf{a}_{11}^T & \cdots & \mathbf{a}_{1Q}^T \\ \vdots & \ddots & \vdots \\ \mathbf{a}_{M1}^T & \cdots & \mathbf{a}_{MQ}^T \end{pmatrix} \begin{pmatrix} s_{01}(k) \\ \vdots \\ s_{0Q}(k) \end{pmatrix} + \mathbf{v}(k), \tag{6.6}$$

where $\mathbf{v}(k) = (v_1(k), v_2(k), \ldots, v_M(k))^T$ is the vector of noise signals.

6.2.3 Formulation in the Frequency Domain

Equations (6.4) and (6.6) may also be formulated in the frequency domain using the Fourier transform of discrete signals (FTDS) [Oppenheim, Schafer 1975], [Vary, Martin 2006, Chap. 3]. Provided that the Fourier transforms of all signals under consideration exist, we obtain the frequency domain equivalent of (6.4) as

$$Y_m(\Omega) = A_m(\Omega)S_0(\Omega) + V_m(\Omega), \tag{6.7}$$

where $\Omega = 2\pi f/f_S$ denotes the normalized frequency variable and f_S is the sampling rate. In the multiple source case we obtain with $\mathbf{Y}(\Omega) = (Y_1(\Omega), Y_2(\Omega), \ldots, Y_M(\Omega))^T$ and $\mathbf{V}(\Omega) = (V_1(\Omega), V_2(\Omega), \ldots, V_M(\Omega))^T$

$$\mathbf{Y}(\Omega) = \begin{pmatrix} A_{11}(\Omega) & \cdots & A_{1Q}(\Omega) \\ \vdots & \ddots & \vdots \\ A_{M1}(\Omega) & \cdots & A_{MQ}(\Omega) \end{pmatrix} \begin{pmatrix} S_{01}(\Omega) \\ \vdots \\ S_{0Q}(\Omega) \end{pmatrix} + \begin{pmatrix} V_1(\Omega) \\ \vdots \\ V_M(\Omega) \end{pmatrix} \tag{6.8}$$

$$= \mathbf{A}(\Omega)\mathbf{S}_0(\Omega) + \mathbf{V}(\Omega).$$

6.2.4 Simplified Model for Localization

The localization algorithms to be considered in the next section assume a dominance of the direct path. Consequently, each $A_{mq}(\Omega)$ may be written as

$$A_{mq}(\Omega) = \alpha'_{mq}(\Omega)e^{-\jmath\Omega\tau_{mq}} + a''_{mq}(\Omega), \tag{6.9}$$

where $|\alpha'_{mq}| >> |a''_{mq}|$, $\alpha'_{mq} \in \mathbb{R}$ represents the attenuation along the direct path and $a''_{mq} \in \mathbb{C}$ indicates the net attenuation and phase smearing caused by the reflections along the indirect paths. τ_{mq} represents the *absolute* time delay of the signal from source q to the microphone m along the direct path. Then, (6.8) takes the form

$$\mathbf{Y}(\Omega) = \begin{pmatrix} \alpha'_{11}(\Omega)e^{-\jmath\Omega\tau_{11}} & \cdots & \alpha'_{1Q}(\Omega)e^{-\jmath\Omega\tau_{1Q}} \\ \vdots & \ddots & \vdots \\ \alpha'_{M1}(\Omega)e^{-\jmath\Omega\tau_{M1}} & \cdots & \alpha'_{MQ}(\Omega)e^{-\jmath\Omega\tau_{MQ}} \end{pmatrix} \mathbf{S}_0(\Omega)$$

$$+ \begin{pmatrix} a''_{11}(\Omega) & \cdots & a''_{1Q}(\Omega) \\ \vdots & \ddots & \vdots \\ a''_{M1}(\Omega) & \cdots & a''_{MQ}(\Omega) \end{pmatrix} \mathbf{S}_0(\Omega) + \mathbf{V}(\Omega) \tag{6.10}$$

$$= \mathbf{A}'(\Omega)\mathbf{S}_0(\Omega) + \mathbf{A}''(\Omega)\mathbf{S}_0(\Omega) + \mathbf{V}(\Omega).$$

The *propagation matrix* $\mathbf{A}'(e^{\jmath\Omega})$ is directly related to the geometric arrangement of the sources and the sensors and thus is key to solving the localization problem. The vectors $\mathbf{A}''(\Omega)\mathbf{S}_0(\Omega)$ and $\mathbf{V}(\Omega)$ constitute disturbances. While the former is obviously correlated with the source signals, the latter is typically modelled as being statistically independent from the source signals. To simplify the discussion we will frequently neglect the contribution $\mathbf{A}''(\Omega)\mathbf{S}_0(\Omega)$ of the indirect paths. In this case we have $\mathbf{A}(\Omega) = \mathbf{A}'(\Omega)$.

For convenience, we introduce a reference point $\mathbf{r}_0$ that may coincide, for example, with one of the microphone locations. Then, we define the source signal spectra $S_q(\Omega)$ that are received at this reference point when only the direct path is considered as

$$S_q(\Omega) = \alpha'_{0q}(\Omega)e^{-\jmath\Omega\tau_{0q}}S_{0q}(\Omega) \tag{6.11}$$

and rewrite (6.10) in terms of the source signal vector

$$\mathbf{S}(\Omega) = (S_1(\Omega),\ldots,S_Q(\Omega))^T \tag{6.12}$$

as

$$
\mathbf{Y}(\Omega) = \begin{pmatrix} \frac{\alpha'_{11}(\Omega)}{\alpha'_{01}(\Omega)}e^{\jmath\Omega\tau_{01}-\jmath\Omega\tau_{11}} & \cdots & \frac{\alpha'_{1Q}(\Omega)}{\alpha'_{0Q}(\Omega)}e^{\jmath\Omega\tau_{0Q}-\jmath\Omega\tau_{1Q}} \\ \vdots & \ddots & \vdots \\ \frac{\alpha'_{M1}(\Omega)}{\alpha'_{01}(\Omega)}e^{\jmath\Omega\tau_{01}-\jmath\Omega\tau_{M1}} & \cdots & \frac{\alpha'_{MQ}(\Omega)}{\alpha'_{0Q}(\Omega)}e^{\jmath\Omega\tau_{0Q}-\jmath\Omega\tau_{MQ}} \end{pmatrix}\mathbf{S}(\Omega)
$$
$$
+ \begin{pmatrix} \gamma_{11}(\Omega) & \cdots & \gamma_{1Q}(\Omega) \\ \vdots & \ddots & \vdots \\ \gamma_{M1}(\Omega) & \cdots & \gamma_{MQ}(\Omega) \end{pmatrix}\mathbf{S}(\Omega) + \mathbf{V}(\Omega)
$$
$$
= \begin{pmatrix} \frac{\alpha'_{11}(\Omega)}{\alpha'_{01}(\Omega)}e^{\jmath\Omega\Delta\tau_{11}} & \cdots & \frac{\alpha'_{1Q}(\Omega)}{\alpha'_{0Q}(\Omega)}e^{\jmath\Omega\Delta\tau_{1Q}} \\ \vdots & \ddots & \vdots \\ \frac{\alpha'_{M1}(\Omega)}{\alpha'_{01}(\Omega)}e^{\jmath\Omega\Delta\tau_{M1}} & \cdots & \frac{\alpha'_{MQ}(\Omega)}{\alpha'_{0Q}(\Omega)}e^{\jmath\Omega\Delta\tau_{MQ}} \end{pmatrix}\mathbf{S}(\Omega)
$$
$$
+ \begin{pmatrix} \gamma_{11}(\Omega) & \cdots & \gamma_{1Q}(\Omega) \\ \vdots & \ddots & \vdots \\ \gamma_{M1}(\Omega) & \cdots & \gamma_{MQ}(\Omega) \end{pmatrix}\mathbf{S}(\Omega) + \mathbf{V}(\Omega), \tag{6.13}
$$

where $\gamma_{mq} = \frac{a''_{mq}}{\alpha'_{0q}e^{-\jmath\Omega\tau_{0q}}}$ represents the normalized indirect components. $\Delta\tau_{mq} = \tau_{0q} - \tau_{mq}$ is the *relative* time delay or *time delay of arrival* (TDOA) with respect to the reference point. If the reference point is close to the array and both, reference point and array, are in the farfield of the sources we may further simplify (6.13) and obtain [Vary, Martin 2006, Chap. 12]

$$\mathbf{Y}(\Omega) = \begin{pmatrix} e^{\jmath\Omega\Delta\tau_{11}} & \cdots & e^{\jmath\Omega\Delta\tau_{1Q}} \\ \vdots & \ddots & \vdots \\ e^{\jmath\Omega\Delta\tau_{M1}} & \cdots & e^{\jmath\Omega\Delta\tau_{MQ}} \end{pmatrix} \mathbf{S}(\Omega)$$

$$+ \begin{pmatrix} \gamma_{11}(\Omega) & \cdots & \gamma_{1Q}(\Omega) \\ \vdots & \ddots & \vdots \\ \gamma_{M1}(\Omega) & \cdots & \gamma_{MQ}(\Omega) \end{pmatrix} \mathbf{S}(\Omega) + \mathbf{V}(\Omega) . \tag{6.14}$$

The direct path contributions in (6.14) encode the spatial positions of the q sources in terms of TDOAs. The indirect paths and noise contributions constitute a disturbance. When we consider the qth source only and assume farfield and anechoic conditions the matrix $\mathbf{A}(\Omega)$ is a vector, the *propagation vector* [Vary, Martin 2006, Chap. 12], which might be parametrized by the source location of the qth source as

$$\mathbf{A}(\mathbf{r}_q, \Omega) = \left(e^{\jmath\Omega\Delta\tau_{1q}(\mathbf{r_q})}, \ldots, e^{\jmath\Omega\Delta\tau_{Mq}(\mathbf{r_q})} \right)^T . \tag{6.15}$$

In the case of a single source we may omit the source index q and obtain

$$\mathbf{A}(\mathbf{r}, \Omega) = \left(e^{\jmath\Omega\Delta\tau_1(\mathbf{r})}, \ldots, e^{\jmath\Omega\Delta\tau_M(\mathbf{r})} \right)^T . \tag{6.16}$$

The farfield scenario is illustrated in two dimensions in Fig. 6.2 for the simple case of a single microphone pair and anechoic, noiseless transmission. In this case the difference of TDOAs allows us to infer the direction of arrival (DOA): the angle of incidence θ and the delay difference T are related as

$$T = \Delta\tau_1 - \Delta\tau_2 = \frac{d\cos\theta}{c} , \tag{6.17}$$

where d is the microphone distance and c denotes the speed of sound.

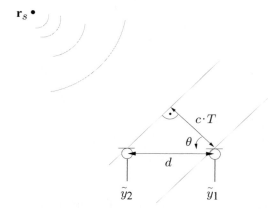

Figure 6.2: TDOA for a microphone pair in the farfield of a source. It can be assumed that the source signal arrives as plane waves at the microphone array

6.3 Localization Approach Taxonomy

The multi-channel approaches to acoustic source localization may broadly be divided into two main classes: indirect and direct. Indirect approaches to source localization are usually two-step methods: first, the relative time delays $\Delta\tau_{mq}$ for the various microphone pairs are evaluated and then the source location is found as the intersection of the corresponding half-hyperboloids centered around the respective microphone pairs. Direct approaches, on the other hand, generally scan the so-called *candidate* source positions and pick the Q most likely candidates – thus performing the localization in a single step.

In the following discussion, we shall start by considering the simple case of two microphones and a single source. Further, without loss of generality, we shall assume the source to be in the farfield, with the source wavefront propagating as plane waves and impinging upon the microphone pair with the corresponding delay T (Fig. 6.2). The extension to microphone arrays with more than two microphones is the subject of the later sections. We shall first present the indirect approaches, followed by the direct ones.

6.4 Indirect Localization Approaches

Indirect approaches explicitly estimate the time delays of arrival (TDOA) before performing the actual localization task. Early approaches [Knapp, Carter 1976], [Etter, Stearns 1981], [Hertz 1986] to time delay of arrival estimation consider an anechoic, farfield signal model:

$$\tilde{y}_1(t) = \tilde{s}(t + \Delta\tau_1) + \tilde{v}_1(t) \qquad (6.18)$$
$$\tilde{y}_2(t) = \tilde{s}(t + \Delta\tau_2) + \tilde{v}_2(t)\,.$$

By means of an LMS-type algorithm the approaches of [Etter, Stearns 1981], [So et al. 1994] then explicitly adapt a time delay T to minimize the mean square error (MSE), also termed *cost function*, between the microphone signals, i.e.,

$$\hat{T} = \underset{T}{\mathrm{argmin}}\ \mathrm{E}\left\{(\tilde{y}_1(t) - \tilde{y}_2(t+T))^2\right\} \qquad (6.19)$$
$$= \underset{T}{\mathrm{argmin}}\ \mathrm{E}\left\{\tilde{y}_1^2(t)\right\} + \mathrm{E}\left\{\tilde{y}_2^2(t+T)\right\} - 2\mathrm{E}\left\{\tilde{y}_1(t)\tilde{y}_2(t+T)\right\}. \qquad (6.20)$$

Note that the first two terms of (6.20) represent the signal power at the two channels and, for stationary input signals, are independent of T. Therefore, they do not contribute to the cost function, simplifying the expression to

$$\hat{T} = \underset{T}{\mathrm{argmax}}\ \mathrm{E}\left\{\tilde{y}_1(t)\tilde{y}_2(t+T)\right\}. \qquad (6.21)$$

Thus, *minimizing* the mean square error may be seen to be equivalent to *maximizing* the cross-correlation between the microphone signals.

However, approaches that explicitly search the optimal time delay in the time domain suffer from two drawbacks: first, in discrete time systems, the time delay could contain fractional sample shifts that requires some form of interpolation and leads to more complicated estimation procedures [Chan et al. 1981]; secondly, the approaches are based on an overly simplistic signal model.

Further improvements along this direction lead to LMS-type algorithms [Reed et al. 1981] where, instead of a delay parameter T, the microphone signal $\tilde{x}_2(t)$ is *filtered* by a filter $\tilde{h}(t)$ such that an approximate solution to

$$\tilde{h}_{\mathrm{opt}}(t) = \underset{\tilde{h}(t)}{\operatorname{argmin}}\ \mathrm{E}\left\{(\tilde{y}_1(t) - \tilde{h}(t) * \tilde{y}_2(t))^2\right\} \tag{6.22}$$

is found. In discrete time the update equations for the normalized LMS (see, e.g., [Vary, Martin 2006, Chap. 13]) approach may be written as

$$e(k) = y_1(k - T_B) - \mathbf{h}^T(k)\mathbf{y}_2(k) \tag{6.23}$$

$$\mathbf{h}(k+1) = \mathbf{h}(k) + \varsigma e(k)\frac{\mathbf{y}_2(k)}{\|\mathbf{y}_2(k)\|^2},$$

where ς is the step size and $T_B > \lceil f_S d/c \rceil$ is an integral sample delay[2] required in order to preserve causality in the case of negative TDOA values. When the direct path is dominant, an estimate $\hat{T}$ of the time delay can be obtained as the abscissa of the largest peak of $\tilde{h}_{\mathrm{opt}}(t)$.

Besides MSE, other optimization criteria may be used to compute the optimal filter impulse response $\tilde{h}(t)$. These lead to a family of TDOA estimation algorithms that fit into the general framework presented below.

6.4.1 Generalized Cross-Correlation (GCC)

Generalized cross-correlation (GCC) [Knapp, Carter 1976] is the term given to the framework that encompasses a wide range of approaches to TDOA estimation. The block diagram of a generalized cross-correlator is shown in Fig. 6.3.

The optimal time delay estimate $\hat{T}$ is obtained as

$$\hat{T} = \underset{T}{\operatorname{argmax}}\ \mathrm{E}\left\{(\tilde{y}_1(t) * \tilde{h}_1(t))(\tilde{y}_2(t+T) * \tilde{h}_2(t))\right\} \tag{6.24}$$

$$= \underset{T}{\operatorname{argmax}}\ \varphi_{\tilde{y}_1 \tilde{y}_2}^g(T),$$

---

[2]The operator $\lceil a \rceil$ rounds a to the nearest upper integer value.

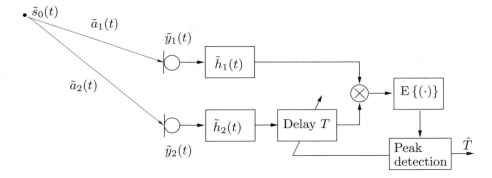

Figure 6.3: Block diagram of the generalized cross-correlation (GCC) method for the estimation of the TDOA

where $\varphi^g_{\tilde{y}_1\tilde{y}_2}(T)$ denotes the *generalized cross-correlation* (GCC) function. The filters $\tilde{h}_m(t)$ are chosen according to the particular optimization criterion considered.

We may also write the generalized cross-correlation function $\varphi^g_{\tilde{y}_1\tilde{y}_2}(T)$ in the frequency domain as

$$\varphi^g_{\tilde{y}_1\tilde{y}_2}(T) = \frac{1}{2\pi} \int\limits_{-\infty}^{\infty} H_1(\omega)H_2^*(\omega)\Phi_{\tilde{y}_1\tilde{y}_2}(\omega)e^{\jmath\omega T}\,\mathrm{d}\omega\,, \tag{6.25}$$

where $\Phi_{\tilde{y}_1\tilde{y}_2}(\omega)$ represents the cross-power spectral density of signals $\tilde{y}_1(t)$ and $\tilde{y}_2(t)$ and ω represents the continuous frequency variable. Further, defining $G(\omega) = H_1(\omega)H_2^*(\omega)$ and rewriting (6.25), we obtain:

$$\varphi^g_{\tilde{y}_1\tilde{y}_2}(T) = \frac{1}{2\pi} \int\limits_{-\infty}^{\infty} G(\omega)\Phi_{\tilde{y}_1\tilde{y}_2}(\omega)e^{\jmath\omega T}\,\mathrm{d}\omega\,. \tag{6.26}$$

The GCC-based approaches to TDOA estimation may be summarized in this framework [Knapp, Carter 1976] as in Table 6.1 below. The term $\Gamma(\omega)$ in Table 6.1 represents the *coherence* between the microphone signals at frequency ω

$$\Gamma(\omega) = \frac{\Phi_{\tilde{y}_1\tilde{y}_2}(\omega)}{\sqrt{\Phi_{\tilde{y}_1\tilde{y}_1}(\omega)\Phi_{\tilde{y}_2\tilde{y}_2}(\omega)}}\,. \tag{6.27}$$

Note that the development so far has assumed perfect knowledge of the cross- and auto-power spectral density of the source signals and the noise. However, in practice, these quantities have to be estimated from a fixed time record of observations.

Table 6.1: Generalized Cross-Correlation (GCC) weighting functions

Weighting function $G(\omega)$	Approach
1	Regular Cross-Correlation (CC)
$\dfrac{1}{\sqrt{\Phi_{\tilde{y}_1\tilde{y}_1}(\omega)\Phi_{\tilde{y}_2\tilde{y}_2}(\omega)}}$	Smoothed Coherence Transform (SCOT)
$\dfrac{1}{\Phi_{\tilde{y}_2\tilde{y}_2}(\omega)}$	Roth (Wiener–Hopf weighting)
$\dfrac{|\Gamma(\omega)|^2}{1-|\Gamma(\omega)|^2}$	Hannan–Thomson (Maximum Likelihood estimate)
$\dfrac{\Phi_{\tilde{s}\tilde{s}}(\omega)}{\Phi_{\tilde{v}_1\tilde{v}_1}(\omega)\Phi_{\tilde{v}_2\tilde{v}_2}(\omega)}$	Eckart weighting
$\dfrac{1}{|\Phi_{\tilde{y}_1\tilde{y}_2}(\omega)|}$	Phase Transform (PHAT)

6.4.2 Adaptive Eigenvalue Decomposition (AED)

The adaptive eigenvalue decomposition (AED) and its variants [Benesty 2000], [Doclo, Moonen 2003] are fairly recent approaches to TDOA estimation. The block diagram of the basic AED algorithm is shown in Fig. 6.4. In the noiseless case, the signal model of (6.1) reduces to

$$\tilde{y}_1(t) = \tilde{a}_1(t) * \tilde{s}(t)$$
$$\tilde{y}_2(t) = \tilde{a}_2(t) * \tilde{s}(t)\,.$$

$$(6.28)$$

The aim, then, is to find optimal, energy constrained filters $\tilde{h}_m(t)$, $(m \in \{1,2\})$ that minimize

$$\mathrm{E}\left\{\tilde{e}^2(t)\right\} = \mathrm{E}\left\{(\tilde{h}_1(t) * \tilde{y}_1(t) - \tilde{h}_2(t) * \tilde{y}_2(t))^2\right\}\,.$$

$$(6.29)$$

In the light of Wiener–Hopf filtering [Haykin 1996], this can be seen as an attempt to match the two microphone signals. Under certain conditions [Xu et al. 1995] and using the commutative property of linear convolution, the signals can be exactly matched when

$$\tilde{h}_1(t) = \beta\tilde{a}_2(t) \quad \text{and}$$
$$\tilde{h}_2(t) = \beta\tilde{a}_1(t)\,,$$

$$(6.30)$$

where β is a scaling factor. The TDOA may then be computed as the difference in the abscissae of the largest values of the respective optimal filters.

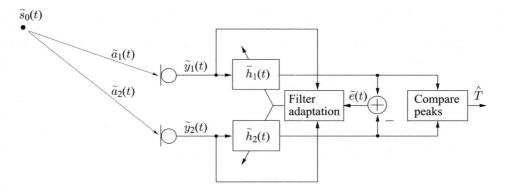

Figure 6.4: Block diagram of the AED algorithm (in continuous time)

Formulating (6.28) and (6.29) in the discrete time domain as in (6.4), we get

$$y_1(k) = \mathbf{a}_1^T \mathbf{s}(k) \tag{6.31}$$
$$y_2(k) = \mathbf{a}_2^T \mathbf{s}(k) \,,$$

where the room impulse responses $\mathbf{a}_m$ have been modeled as FIR filters of length L. Rewriting equation (6.29) using (6.31), we obtain

$$\mathrm{E}\left\{e^2(k)\right\} = \mathrm{E}\left\{(\mathbf{h}_1^T \mathbf{y}_1(k) - \mathbf{h}_2^T \mathbf{y}_2(k))^2\right\} \,, \tag{6.32}$$

where

$$\mathbf{h}_m = (h_m(0),\ h_m(1),\dots,h_m(P-1))^T \tag{6.33}$$
$$\mathbf{y}_m(k) = (y_m(k),\ y_m(k-1),\dots,y_m(k-P+1))^T \,. \tag{6.34}$$

We shall rewrite equation (6.32) more compactly as

$$\mathrm{E}\left\{e^2(k)\right\} = \mathbf{h}^T \mathbf{R_{yy}} \mathbf{h} \,, \tag{6.35}$$

where

$$\mathbf{h} = (\mathbf{h}_1^T,\ -\mathbf{h}_2^T)^T \tag{6.36}$$
$$\mathbf{R_{yy}} = \begin{pmatrix} \mathrm{E}\left\{\mathbf{y}_1 \mathbf{y}_1^T\right\} & \mathrm{E}\left\{\mathbf{y}_1 \mathbf{y}_2^T\right\} \\ \mathrm{E}\left\{\mathbf{y}_2 \mathbf{y}_1^T\right\} & \mathrm{E}\left\{\mathbf{y}_2 \mathbf{y}_2^T\right\} \end{pmatrix} \,. \tag{6.37}$$

It is now easy to see that, when the filter vector $\mathbf{h}$ is energy constrained, the optimal solution $\mathbf{h}_{\mathrm{opt}}$ to the minimization problem is the eigenvector corresponding to the zero eigenvalue of $\mathbf{R_{yy}}$, in the noiseless case, or the eigenvector corresponding to the smallest eigenvalue, when the microphone noises are spatially and temporally uncorrelated and independent of the source signals [Strang 1988]. When the filters $\mathbf{a}_m$ do not have any common zeros, we obtain

$$\mathbf{h}_1 = \beta \mathbf{a}_2 \quad \text{and} \tag{6.38}$$
$$\mathbf{h}_2 = \beta \mathbf{a}_1 \,,$$

provided $P = L$. The TDOA may then be obtained as explained above. For convenience, the energy of the filter vector is constrained to unity. Thus, the scale factor is determined and the following adaptive algorithm for the iterative update of the filter vector $\mathbf{h}(k)$ results

$$\mathbf{h}(k+1) = \frac{\mathbf{h}(k) - \varsigma \mathbf{y}(k)\mathbf{y}(\lambda)^T \mathbf{h}(k)}{\|\mathbf{h}(k) - \varsigma \mathbf{y}(k)\mathbf{y}^T(k)\mathbf{h}(k)\|}, \tag{6.39}$$

where ς denotes the stepsize parameter. Note that if we fix any one filter in (6.29) by, say, a delta function, we obtain the mean square error criterion of (6.22), which, in the frequency domain, corresponds to the GCC-Roth weighting.

6.4.3 Information Theoretic Approach to TDOA Estimation

As shown by [Talantzis et al. 2005], the principle of information maximization can be also used to approach the problem of TDOA estimation. The signal model considered for this method is the noiseless variant of that in (6.18). The idea is to find the time delay T that maximizes the mutual information $\mathcal{I}(\tilde{y}_1; \tilde{y}_2)$ between the microphone signals $\tilde{y}_1(t)$ and $\tilde{y}_2(t + T)$

$$\hat{T} = \underset{T}{\operatorname{argmax}} \, \mathcal{I}\left(\tilde{y}_1(t); \tilde{y}_2(t + T)\right). \tag{6.40}$$

The mutual information between two stochastic random variables a and b is given by [Cover, Thomas 1991]

$$\mathcal{I}(a; b) = \mathcal{H}(a) + \mathcal{H}(b) - \mathcal{H}(a, b), \tag{6.41}$$

where $\mathcal{H}(a)$ represents the *entropy* of the random variable a and $\mathcal{H}(a, b)$ the joint entropy of a and b. Assuming that the source signal (and, consequently, the microphone signals) may be modeled by a zero mean, stationary, stochastic process, we may consider the time evolution of the signals to be realizations of that process. Therefore, the mutual information between the two microphone signals for a time shift T may be written as

$$\mathcal{I}\left(\tilde{y}_1(t); \tilde{y}_2(t + T)\right) = \mathcal{H}\left(\tilde{y}_1(t)\right) + \mathcal{H}\left(\tilde{y}_2(t + T)\right) - \mathcal{H}\left(\tilde{y}_1(t), \tilde{y}_2(t + T)\right). \tag{6.42}$$

The entropy for Gaussian random variables is well known [Cover, Thomas 1991] to be proportional to its variance. Thus, under the assumptions that the signals have a Gaussian density,

$$\mathcal{H}\left(\tilde{y}_1(t)\right) \propto \log_e\left(\varphi_{\tilde{y}_1 \tilde{y}_1}(0)\right) \tag{6.43}$$
$$\mathcal{H}\left(\tilde{y}_2(t + T)\right) \propto \log_e\left(\varphi_{\tilde{y}_2 \tilde{y}_2}(0)\right)$$
$$\mathcal{H}\left(\tilde{y}_1(t), \tilde{y}_2(t + T)\right) \propto \log_e\left(\det\left(\mathbf{R}_{\tilde{\mathbf{y}}\tilde{\mathbf{y}}}\right)\right),$$

where $\tilde{\mathbf{y}}(t) = (\tilde{y}_1(t), \; \tilde{y}_2(t+T))^T$ and $\mathbf{R}_{\tilde{\mathbf{y}}\tilde{\mathbf{y}}}$ is given by

$$\mathbf{R}_{\tilde{\mathbf{y}}\tilde{\mathbf{y}}} = \mathrm{E}\left\{\tilde{\mathbf{y}}(t)\tilde{\mathbf{y}}^T(t)\right\} \tag{6.44}$$

$$= \begin{pmatrix} \mathrm{E}\left\{\tilde{y}_1^2(t)\right\} & \mathrm{E}\left\{\tilde{y}_1(t)\tilde{y}_2(t+T)\right\} \\ \mathrm{E}\left\{\tilde{y}_1(t)\tilde{y}_2(t+T)\right\} & \mathrm{E}\left\{\tilde{y}_2^2(t+T)\right\} \end{pmatrix}$$

$$= \begin{pmatrix} \varphi_{\tilde{y}_1\tilde{y}_1}(0) & \varphi_{\tilde{y}_1\tilde{y}_2}(T) \\ \varphi_{\tilde{y}_1\tilde{y}_2}(T) & \varphi_{\tilde{y}_2\tilde{y}_2}(0) \end{pmatrix}.$$

Consequently,

$$\hat{T} = \underset{T}{\mathrm{argmax}} \; \mathcal{I}\left(\tilde{y}_1(t); \tilde{y}_2(t+T)\right) \tag{6.45}$$

$$= \underset{T}{\mathrm{argmin}} \; \log_e\left(\det\left(\mathbf{R}_{\tilde{\mathbf{y}}\tilde{\mathbf{y}}}\right)\right)$$

$$= \underset{T}{\mathrm{argmax}} \; \varphi_{\tilde{y}_1\tilde{y}_2}(T),$$

where the simplifications above follow as the first two terms in (6.42) are independent of T according to (6.43). Thus, when the signals are Gaussian distributed, maximizing the mutual information is equivalent to maximizing the *cross-correlation* between the microphone signals. In contrast to GCC, however, the optimization criterion in its general form (6.40) can exploit the non-Gaussian structure of the source signals.

6.4.4 Extension to Multiple Microphone Pairs

The above sections described various approaches to estimate the TDOA using a pair of sensors. Obviously, using more than one pair of microphones increases the spatial diversity afforded to the localization system and, consequently, may be exploited to localize the source in more than one spatial dimension and to improve the localization accuracy. The integration of multiple microphone pairs into the existing framework shall briefly be discussed in this section.

The simplest way to extend the two-channel method to an M channel $(M > 2)$ array is to obtain the TDOA estimate T_p for *all* $M(M-1)/2$ microphone pairs, using any of the GCC approaches, e.g., the Roth weighted GCC estimate [Dvorkind, Gannot 2005], simple cross-correlation [Birchfield 2004], the PHAT estimate [Brutti et al. 2005], or the multi-channel AED approach, where an estimate of the impulse response from the source to each microphone is first obtained, from which the TDOA between all microphone pairs may be computed as in the AED approach, etc. Once this is done, one method of obtaining the source position is by solving the non-linear equation relating the vector of obtained TDOA estimates, the geometry of the array, and the source location [Chan, Ho 1994], [Drews 1995], [Huang et al. 2001]

$$\hat{\mathbf{r}}_s = \mathcal{F}(T_1, T_2, \ldots, T_{M(M-1)/2}). \tag{6.46}$$

The methods of [Madhu, Martin 2005], [Bechler, Kroschel 2002], [Bechler, Kroschel 2004] additionally consider weighting the contributions of the TDOA estimates from the various pairs according to some optimality criteria, when computing the 'averaged' source location – thus improving the location estimate.

Another interesting approach to source localization using multiple microphones is that suggested in [Chen et al. 2003b], [Chen et al. 2003a], where the concept of linear prediction is extended to the spatial case. For the two microphone case, the cost function derived for this approach reduces to that for the information theoretic approach of Sec. 6.4.3.

Methods to cope with multiple sources are detailed in [Scheuing, Yang 2006], [Scheuing, Yang 2007] where the distance structure of peaks in the correlation function and graph-theoretic considerations are exploited. Other approaches project the cross-correlation function $\varphi_{\tilde{y}_\ell \tilde{y}_m}(\tau)$ of all pairs (ℓ, m) onto a common co-ordinate system, e.g., a one-dimensional direction of arrival value [Matsuo et al. 2005], or a two-dimensional grid [Brutti et al. 2005], or the surface of a hemisphere (the accumulative correlation of [Birchfield 2004]) and so on, creating, in essence, a histogram of likelihood values over candidate source locations. The source location then corresponds to the most likely candidate position.

6.5 Direct Localization Approaches

For the indirect approaches, source localization is the result of a two-step approach. First, an estimate of the TDOA is obtained (using an array of two or more microphones) and then, based on the knowledge of the geometry of the array and the time delay estimates, the source position is estimated. Direct approaches, on the other hand, perform TDOA estimation and source localization in one step. Most direct algorithms scan a set of candidate source locations (the so-called *search space*) and then pick the most likely position as an estimate of the source location. This approach makes it easier to incorporate multiple microphones in the optimization criterion. As in the case of the indirect approaches, the algorithms belonging to the direct class may be formulated in the time or the frequency domain.

We discuss localization algorithms – initially designed for narrowband sources – in the frequency domain. The extension to the wideband case could be of the straightforward *incoherent* kind, where the narrowband location estimates at the center frequency of each subband are averaged over all the subbands [Krim, Viberg 1996], [Wax, Kailath 1984b], or they could be of the *coherent* sort, where the data in all the subbands are collectively used when scanning the candidate locations [Krim, Viberg 1996], [Wang, Kaveh 1985], [Hung, Kaveh 1988], [Krolik 1991], [Yoon et al. 2006]. We will also develop their link to the GCC framework (when posed in each subband in the frequency domain, as in Table 6.1). Unless mentioned otherwise, we shall consider a single source located at $\mathbf{r}_s = (x_s, y_s, z_s)^T$, in a three-dimensional Cartesian co-ordinate system. Further, we shall neglect the effect of reverberation.

6.5.1 Steered Response Power Beamforming

The steered response power (SRP) beamforming approach [DiBiase et al. 2001], [Omologo, Svaizer 1994] searches for the candidate source position that maximizes the output power of a filter-and-sum beamformer steered in that direction. While the optimization criterion is of the broadband type, it is again instructive to expand it into a frequency domain formulation. The output power spectral density of the filter-and-sum beamformer may be written as, see, e.g., [Vary, Martin 2006, Chap. 12],

$$\Phi_{\hat{s}\hat{s}}(\mathbf{r}, \Omega) = \mathrm{E}\left\{|\mathbf{H}(\mathbf{r}, \Omega)^H \mathbf{Y}(\Omega)|^2\right\} = \mathbf{H}^H(\mathbf{r}, \Omega)\boldsymbol{\Phi}_{\mathbf{yy}}(\Omega)\mathbf{H}(\mathbf{r}, \Omega) \tag{6.47}$$

$$= \mathbf{H}^H(\mathbf{r}, \Omega) \begin{pmatrix} \Phi_{y_1 y_1}(\Omega) & \cdots & \Phi_{y_1 y_M}(\Omega) \\ \vdots & \ddots & \vdots \\ \Phi_{y_M y_1}(\Omega) & \cdots & \Phi_{y_M y_M}(\Omega) \end{pmatrix} \mathbf{H}(\mathbf{r}, \Omega),$$

where the beam is directed towards $\mathbf{r}$ and $\mathbf{H}(\mathbf{r}, \Omega) = (H_1(\mathbf{r}, \Omega), \ldots, H_M(\mathbf{r}, \Omega))^T$ is the corresponding vector of beamforming filter frequency responses. $\Phi_{y_\ell y_m} = \mathrm{E}\left\{Y_\ell(\Omega) Y_m^*(\Omega)\right\}$ denotes the cross-power spectral density of channels ℓ and m. Then, the source location $\hat{\mathbf{r}}_s$ is found as

$$\hat{\mathbf{r}}_s = \operatorname*{argmax}_{\mathbf{r}} \frac{1}{2\pi} \int_{-\pi}^{\pi} \Phi_{\hat{s}\hat{s}}(\mathbf{r}, \Omega) \, \mathrm{d}\Omega. \tag{6.48}$$

Expanding (6.47) as

$$\Phi_{\hat{s}\hat{s}}(\mathbf{r}, \Omega) = \mathbf{H}^H(\mathbf{r}, \Omega)\boldsymbol{\Phi}_{\mathbf{yy}}(\Omega)\mathbf{H}(\mathbf{r}, \Omega)$$

$$= \sum_{\ell,m} H_\ell^*(\mathbf{r}, \Omega) H_m(\mathbf{r}, \Omega) \Phi_{y_\ell y_m}(\Omega)$$

$$= \sum_{m} |H_m(\mathbf{r}, \Omega)|^2 \Phi_{y_m y_m}(\Omega) + \sum_{\substack{\ell,m \\ \ell \neq m}} H_\ell^*(\mathbf{r}, \Omega) H_m(\mathbf{r}, \Omega) \Phi_{y_\ell y_m}(\Omega) \tag{6.49}$$

$$= \sum_{m} \Phi_{y_m y_m}(\Omega) + \sum_{\substack{\ell,m \\ \ell \neq m}} e^{\jmath \Omega f_S T_\ell(\mathbf{r}) - \jmath \Omega f_S T_m(\mathbf{r})} \Phi_{y_\ell y_m}(\Omega),$$

where for the last step the delay-and-sum beamformer with

$$\mathbf{H}(\mathbf{r}, \Omega) = (e^{-\jmath \Omega f_S T_1(\mathbf{r})}, e^{-\jmath \Omega f_S T_2(\mathbf{r})}, \ldots, e^{-\jmath \Omega f_S T_M(\mathbf{r})})^T \tag{6.50}$$

was assumed. For the delay-and-sum beamformer it may now be seen that the first term is independent of the source location and the second term sums over the cross-power spectral density of all $M(M-1)/2$ microphone pairs.

Similar to the GCC in the case of two channels, the cross-power spectral densities may be weighted according to the criteria outlined in Table 6.1. Thus, we obtain, for instance, the SRP-PHAT approach

$$\hat{\mathbf{r}}_s = \operatorname*{argmax}_{\mathbf{r}} \frac{1}{2\pi} \int_{-\pi}^{\pi} \sum_{\substack{\ell,m \\ \ell \neq m}} e^{\jmath \Omega f_S T_\ell(\mathbf{r}) - \jmath \Omega f_S T_m(\mathbf{r})} \frac{\Phi_{y_\ell y_m}(\Omega)}{|\Phi_{y_\ell y_m}(\Omega)|} \, d\Omega, \qquad (6.51)$$

which may be seen as extensions of the GCC to the M microphone case.

Compared with GCC, the SRP method provides additional degrees of freedom that allows us to smooth over microphone pairs instead over frequency. In fact, the cost function (6.47) may be evaluated for each frequency separately. Thereby, the method can be easily extended to multiple sources with disjoint frequency spectra.

6.5.2 Minimum Mean Square (MMSE) Approach

This approach was developed in [Liu et al. 2000] and extended in [Madhu et al. 2006] and is based on the model of (6.14). The idea behind this approach is to search for appropriate phase compensation factors $e^{\jmath \Omega f_S T_m(\mathbf{r})}$ for each channel m such that the mean-squared error between the phase compensated signals of all pairs is minimized. Note that from the localization point of view, the phase compensation factors are parametrized by the candidate source positions. This may be expressed as

$$\hat{\mathbf{r}}_s = \operatorname*{argmin}_{\mathbf{r}} \sum_{\substack{\ell,m \\ \ell \neq m}} \mathrm{E} \left\{ \left| Y_m e^{\jmath \Omega f_S T_m(\mathbf{r})} - Y_\ell e^{\jmath \Omega f_S T_\ell(\mathbf{r})} \right|^2 \right\} \qquad (6.52)$$

$$= \operatorname*{argmin}_{\mathbf{r}} \sum_m \Phi_{y_m y_m} - \sum_{\substack{\ell,m \\ \ell \neq m}} e^{\jmath \Omega f_S (T_\ell(\mathbf{r}) - T_m(\mathbf{r}))} \Phi_{y_\ell y_m}.$$

It may be seen that the first term is again independent of any phase compensation factors and may thus be neglected, leading to the following simplified cost function

$$\mathcal{J}(\mathbf{r}, \Omega) = - \sum_{\substack{\ell,m \\ \ell \neq m}} e^{\jmath \Omega f_S (T_\ell(\mathbf{r}) - T_m(\mathbf{r}))} \Phi_{y_\ell y_m}. \qquad (6.53)$$

Thus, the MMSE approach is fully *equivalent* to the SRP approach and also falls under the umbrella of the GCC. Similar to PHAT, it is possible to weight the cross-power spectral density using various other criteria. One – rather heuristic – weighting, which gives good results is suggested in [Madhu et al. 2006]:

$$\mathcal{J}(\mathbf{r}, \Omega) = - \sum_{\substack{\ell,m \\ \ell \neq m}} e^{\jmath \Omega f_S (T_\ell(\mathbf{r}) - T_m(\mathbf{r}))} \Phi_{y_\ell y_m} |\Gamma_{\ell m}(\Omega)|^2, \qquad (6.54)$$

where $\Gamma_{\ell m}(\Omega)$ indicates the coherence between channels ℓ and m at the frequency Ω.

6.5.3 Practical aspects

In practice, source location estimates are computed on finite time records of the observed signals using the discrete Fourier transform (DFT) on segmented, windowed frames of the discrete time signals $y_m(k)$. In this case, the expectation is dropped in favor of an instantaneous estimate for each frame λ or the expectation is computed by a first-order, temporal recursive smoothing. For the former case, the cost function may be written in each frequency bin μ as

$$\mathcal{J}(\mathbf{r}, \mu, \lambda) = \mathbf{H}^H(\mathbf{r}, \mu)\mathbf{Y}(\mu, \lambda)\mathbf{Y}^H(\mu, \lambda)\mathbf{H}(\mathbf{r}, \mu) \tag{6.55}$$

$$= \sum_{\ell,m} H_\ell^*(\mathbf{r}, \mu)H_m(\mathbf{r}, \mu)Y_\ell(\mu, \lambda)Y_m^*(\mu, \lambda)$$

$$= \sum_m |H_m(\mathbf{r}, \mu)|^2 |Y_m(\mu, \lambda)|^2 + \sum_{\substack{\ell,m \\ \ell \neq m}} H_\ell^*(\mathbf{r}, \mu)H_m(\mathbf{r}, \mu)Y_\ell(\mu, \lambda)Y_m^*(\mu, \lambda)$$

from which point on, the procedure to localize the source is exactly the same as outlined in Sec. 6.5.1 and 6.5.2, namely

$$\hat{\mathbf{r}}_s(\mu, \lambda) = \underset{\mathbf{r}}{\operatorname{argmax}} \sum_{\substack{\ell,m \\ \ell \neq m}} H_\ell^*(\mathbf{r}, \mu)H_m(\mathbf{r}, \mu)Y_\ell(\mu, \lambda)Y_m^*(\mu, \lambda). \tag{6.56}$$

Note that the cost function may also be weighted as in (6.51) as

$$\hat{\mathbf{r}}_s(\mu, \lambda) = \underset{\mathbf{r}}{\operatorname{argmax}} \sum_{\substack{\ell,m \\ \ell \neq m}} H_\ell^*(\mathbf{r}, \mu)H_m(\mathbf{r}, \mu)\frac{Y_\ell(\mu, \lambda)Y_m^*(\mu, \lambda)}{|Y_\ell(\mu, \lambda)||Y_m(\mu, \lambda)|} \tag{6.57}$$

or in a manner similar to that in (6.54).

When the signals to be localized are broadband, the computed cost function for each bin μ as in (6.57) could, additionally, be averaged across all frequencies as in [DiBiase et al. 2001], [Omologo, Svaizer 1994] yielding an estimate for the source location $\mathbf{r}_s(\lambda)$ per frame λ as

$$\mathcal{J}(\mathbf{r}, \lambda) = \sum_\mu \sum_{\substack{\ell,m \\ \ell \neq m}} H_\ell^*(\mathbf{r}, \mu)H_m(\mathbf{r}, \mu)\frac{Y_\ell(\mu, \lambda)Y_m^*(\mu, \lambda)}{|Y_\ell(\mu, \lambda))||Y_m(\mu, \lambda)|} \tag{6.58}$$

$$\hat{\mathbf{r}}_s(\lambda) = \underset{\mathbf{r}}{\operatorname{argmax}} \, \mathcal{J}(\mathbf{r}, \lambda).$$

6.5.4 Subspace Based Approaches

The MUltiple SIgnal Classification or the MUSIC algorithm proposed in [Schmidt 1981] is a subspace based approach. It works on the farfield model, but may be extended to the nearfield case, too. It was originally proposed as a solution to the problem of localization of Q narrowband, uncorrelated sources, with $Q < M$. This approach can be extended to the wideband scenario in an incoherent [Wax, Kailath 1984b] or a coherent [Wang, Kaveh 1985], [Yoon et al. 2006] manner. In our discussion, we shall restrict ourselves to the narrowband formulation.

We consider again the frequency domain model as in (6.8),

$$\mathbf{Y} = \mathbf{A}\mathbf{S} + \mathbf{V} \,, \tag{6.59}$$

where we drop the frequency variable for convenience. Computing the spectral covariance matrix $\boldsymbol{\Phi_{yy}}$, we obtain

$$\boldsymbol{\Phi_{yy}} = \mathrm{E}\left\{\mathbf{Y}\mathbf{Y}^H\right\} \tag{6.60}$$
$$= \mathbf{A}\mathrm{E}\left\{\mathbf{S}\mathbf{S}^H\right\}\mathbf{A}^H + \mathrm{E}\left\{\mathbf{V}\mathbf{V}^H\right\}$$
$$= \mathbf{A}\boldsymbol{\Phi_{ss}}\mathbf{A}^H + \boldsymbol{\Phi_{vv}} \tag{6.61}$$
$$= \mathbf{A}\boldsymbol{\Phi_{ss}}\mathbf{A}^H + \Phi_{vv}\mathbf{I} \,, \tag{6.62}$$

where the source signals and the noise are assumed to be independent. The last step follows when the noise is spatially uncorrelated and with the same variance at each microphone. The covariance matrix may be decomposed using the eigenvalue decomposition, yielding

$$\boldsymbol{\Phi_{yy}} = \mathbf{U}\left(\mathbf{D} + \Phi_{vv}\mathbf{I}\right)\mathbf{U}^H. \tag{6.63}$$

For $Q < M$ sources, the diagonal matrix $\mathbf{D}$ is singular and contains the Q dominant eigenvalues, corresponding to the spectral power of the sources. Therefore, we may arrange the eigenvalues ρ_{qq} of $\mathbf{D}$ according to decreasing order of magnitude,

$$\rho_{11} > \rho_{22} > \ldots > \rho_{QQ} > \rho_{Q+1Q+1} = \ldots = \rho_{MM} = 0 \,. \tag{6.64}$$

Correspondingly, the first Q eigenvectors $\mathbf{u}_q$ span the so-called signal-plus-noise subspace, whereas the $M - Q$ eigenvectors $\mathbf{u}_q$, $Q < q \le M$ span the noise-only subspace.

From (6.60) and (6.63) it may be seen that the $M - Q$ eigenvectors of the noise-only subspace define the null space of $\mathbf{A}$. Consequently, if we define a spatial spectrum $S_{\mathrm{MUSIC}}(\mathbf{r})$ over all candidate source locations $\mathbf{r}$ as

$$S_{\mathrm{MUSIC}}(\mathbf{r}) = \frac{1}{\mathbf{H}^H(\mathbf{r})\mathbf{U}_v\mathbf{U}_v^H\mathbf{H}(\mathbf{r})} \,, \tag{6.65}$$

where $\mathbf{H}(\mathbf{r}) = \left(e^{\jmath\,\Omega f_S T_1(\mathbf{r})}, \ldots, e^{\jmath\,\Omega f_S T_M(\mathbf{r})}\right)^T$ is a steering vector towards candidate source location $\mathbf{r}$ and $\mathbf{U}_v$ is the $M \times (M - Q)$ matrix containing the eigenvectors corresponding to the noise-only subspace, the locations $\mathbf{r}$ corresponding to the Q peaks of the spectrum are the sought source positions

$$\hat{\mathbf{r}}_{s_q} = \underset{\mathbf{r}}{\mathrm{argmax}}\ S_{\mathrm{MUSIC}}(\mathbf{r}) \,. \tag{6.66}$$

Alternatively, we may obtain the source propagation vectors as

$$\hat{\mathbf{r}}_{s_q} = \operatorname*{argmax}_{\mathbf{r}} \mathbf{H}^H(\mathbf{r})\mathbf{U}_s\mathbf{U}_s^H\mathbf{H}(\mathbf{r}),$$ (6.67)

where $\mathbf{U}_s$ is the $M \times Q$ matrix containing the eigenvectors corresponding to Q dominant eigenvalues, spanning the signal-plus-noise subspace.

Single-Frame MUSIC

The traditional MUSIC approach is batch based: it requires the estimation of the spectral covariance matrix to determine the number of dominant eigenvalues and corresponding eigenvectors. A simple modification of this approach leads to, what we term, the *single-frame* MUSIC approach, which bears a close relation to the SRP algorithm discussed previously.

The idea behind single-frame MUSIC is as follows. The matrix

$$\hat{\mathbf{\Phi}}_{\mathbf{yy}} = \mathbf{YY}^H$$ (6.68)

is of rank one. Thus, an eigenvalue decomposition of this matrix yields one dominant eigenvalue ρ_{11} with its corresponding eigenvector $\mathbf{u}_1$. The single-frame MUSIC spectrum $S_{\text{MUSIC}}(\mathbf{r})$ from (6.65) is then computed, where the matrix $\mathbf{U}_v\mathbf{U}_v^H$ is obtained as

$$\mathbf{U}_v\mathbf{U}_v^H = \mathbf{I} - \mathbf{u}_1\mathbf{u}_1^H.$$ (6.69)

The maxima of $S_{\text{MUSIC}}(\mathbf{r})$ then indicate the propagation vectors.

We shall now discuss the relation of the single-frame MUSIC approach to the SRP in (6.49) for a single source. In this case, $\mathbf{A}(\mathbf{r}_s)$ simplifies to a column vector and (6.68) may be rewritten as

$$\hat{\mathbf{\Phi}}_{\mathbf{yy}} = \left(\mathbf{A}(\mathbf{r}_s)S + \mathbf{V}\right)\left(\mathbf{A}(\mathbf{r}_s)S + \mathbf{V}\right)^H.$$ (6.70)

It is easily verified that the dominant eigenvector is

$$\mathbf{u}_1 = \frac{\mathbf{A}(\mathbf{r}_s)S + \mathbf{V}}{\|\mathbf{A}(\mathbf{r}_s)S + \mathbf{V}\|}$$
$$= \frac{\mathbf{Y}}{\|\mathbf{Y}\|},$$ (6.71)

with the corresponding eigenvalue of $\rho_{11} = \|\mathbf{Y}\|^2$. Maximizing the MUSIC spectrum of (6.65) as in (6.66), we have

$$
\begin{aligned}
\hat{\mathbf{r}}_s &= \operatorname*{argmax}_{\mathbf{r}} \frac{1}{\mathbf{H}(\mathbf{r})^H \mathbf{U}_v \mathbf{U}_v^H \mathbf{H}(\mathbf{r})} \qquad\qquad (6.72)\\
&= \operatorname*{argmax}_{\mathbf{r}} \frac{1}{\mathbf{H}(\mathbf{r})^H \left(\mathbf{I} - \mathbf{u}_1 \mathbf{u}_1^H \right) \mathbf{H}(\mathbf{r})}\\
&= \operatorname*{argmin}_{\mathbf{r}} \frac{1}{\mathbf{H}(\mathbf{r})^H \mathbf{u}_1 \mathbf{u}_1^H \mathbf{H}(\mathbf{r})}\\
&= \operatorname*{argmax}_{\mathbf{r}} \mathbf{H}(\mathbf{r})^H \mathbf{u}_1 \mathbf{u}_1^H \mathbf{H}(\mathbf{r}) ,
\end{aligned}
$$

which, combined with (6.71), is closely related to the SRP cost function (6.57).

6.5.5 Maximum Likelihood Estimation (MLE)

The subspace and beamformer based approaches presented in the previous sections are computationally attractive. However, when multiple partially or fully coherent sources are present, the performance of these estimators is suboptimal. An alternative is to exploit the underlying data model more completely. This leads to the development of the so-called *parametric* methods in the frequency domain, of which maximum likelihood (ML) estimators form an important class. Partial or complete signal coherence does not pose conceptual problems for the MLE approaches [Krim, Viberg 1996], [Jaffer 1988], [Böhme 1986]. Moreover, estimates obtained using MLE approaches can be shown to be asymptotically consistent and attaining the Cramér–Rao lower bound.

ML approaches require models of the probability density function of the signals under consideration. In this chapter we consider the *deterministic* ML approach, where the source signals are modeled as deterministic and unknown and the noise is assumed to be stationary and Gaussian distributed. The *stochastic* ML approach, where both the source signals and the noise are assumed to stationary and Gaussian distributed, with the source signals being independent of the noise, will not be treated here. We consider again the signal model in the DFT domain where μ and λ denote the frequency bin index and the frame index, respectively,

$$
\mathbf{Y}(\mu, \lambda) = \mathbf{A}(\mu)\mathbf{S}(\mu, \lambda) + \mathbf{V}(\mu, \lambda) . \qquad\qquad (6.73)
$$

The noise is assumed to be spatially white. If this is not the case, the received signals at the microphones need to be prewhitened (for whitening approaches see, e.g., [Eldar, Oppenheim 2003]). In what follows, we drop the frequency bin index μ.

If we assume that the source signals are deterministic and unknown and that the noise is Gaussian distributed and spatially white, we have

$$
\mathrm{E}\left\{ \mathbf{V}(\lambda)\mathbf{V}^H(\lambda) \right\} = \Phi_{vv}\mathbf{I} \qquad\qquad (6.74)
$$

$$
p(\mathbf{Y}(\lambda)|\mathbf{A}, \mathbf{S}(\lambda)) = \frac{1}{(\pi\Phi_{vv})^M} \exp\left(-\frac{\|\mathbf{Y}(\lambda) - \mathbf{A}\mathbf{S}(\lambda)\|^2}{\Phi_{vv}} \right) , \qquad\qquad (6.75)
$$

where $p(\cdot \mid \cdot)$ represents the conditional probability density function and $\|\cdot\|$ represents the Euclidean norm. Under the assumption that the measurement at each time frame is independent, we obtain, for Λ time records

$$p(\mathbf{Y}_1^\Lambda | \mathbf{A}, \mathbf{S}_1^\Lambda) = \prod_{\lambda=1}^{\Lambda} p(\mathbf{Y}(\lambda)|\mathbf{A}, \mathbf{S}(\lambda)), \tag{6.76}$$

where $\mathbf{Y}_1^\Lambda = (\mathbf{Y}(1), \ldots, \mathbf{Y}(\Lambda))$ and $\mathbf{S}_1^\Lambda = (\mathbf{S}(1), \ldots, \mathbf{S}(\Lambda))$ are the sensor and source signal matrices, respectively.

The aim of the deterministic ML approach is then to find the optimal parameter vector

$$\boldsymbol{\theta} = (\mathbf{r}_{s_1}^T, \mathbf{r}_{s_2}^T, \ldots, \mathbf{r}_{s_Q}^T, \mathbf{S}^T(1), \ldots, \mathbf{S}^T(\Lambda), \Phi_{vv})^T$$

such that we maximize the *likelihood* function $p(\mathbf{Y}_1^\Lambda|\boldsymbol{\theta})$.

For mathematical tractability, often the log-likelihood function defined as $\mathcal{L}(\boldsymbol{\theta}) = \log_e \left(p(\mathbf{Y}_1^\Lambda|\boldsymbol{\theta})\right)$ is used. Owing to the monotonicity of the $\log_e$ function, the $\boldsymbol{\theta}$ that maximizes the log-likelihood function will also maximize the likelihood function. Thus

$$\mathcal{L}(\boldsymbol{\theta}) = \sum_{\lambda=1}^{\Lambda} \log_e \left(p(\mathbf{Y}(\lambda)|\boldsymbol{\theta})\right) \tag{6.77}$$

$$= -M\Lambda \log_e(\pi\Phi_{vv}) - \frac{1}{\Phi_{vv}} \sum_{\lambda=1}^{\Lambda} \|\mathbf{Y}(\lambda) - \mathbf{A}\mathbf{S}(\lambda)\|^2 \tag{6.78}$$

and

$$\hat{\boldsymbol{\theta}} = \underset{\boldsymbol{\theta}}{\operatorname{argmax}} \ \mathcal{L}(\boldsymbol{\theta}) \tag{6.79}$$

$$= \underset{\boldsymbol{\theta}}{\operatorname{argmin}} \ \left(M\Lambda \log_e(\pi\Phi_{vv}) + \frac{1}{\Phi_{vv}} \sum_{\lambda=1}^{\Lambda} \|\mathbf{Y}(\lambda) - \mathbf{A}\mathbf{S}(\lambda)\|^2\right). \tag{6.80}$$

Minimizing (6.78) with respect to $\boldsymbol{\theta}$ we obtain the well known solutions [Wax 1992], [Chen et al. 2002]

$$\hat{\mathbf{A}}_{\mathrm{ML}} = \underset{\mathbf{H}}{\operatorname{argmin}} \ \sum_\lambda \mathbf{Y}^H(\lambda)\mathbf{P}_{\mathbf{H}}^\perp \mathbf{Y}(\lambda)$$

$$= \underset{\mathbf{H}}{\operatorname{argmax}} \ \sum_\lambda \mathbf{Y}^H(\lambda)\mathbf{P}_{\mathbf{H}} \mathbf{Y}(\lambda) \tag{6.81}$$

$$\hat{\Phi}_{vv} = \frac{1}{M\Lambda} \sum_\lambda \mathbf{Y}^H(\lambda)\mathbf{P}_{\mathbf{H}}^\perp \mathbf{Y}(\lambda) \tag{6.82}$$

$$\hat{\mathbf{S}}(\lambda) = \hat{\mathbf{A}}_{\mathrm{ML}}^\dagger \mathbf{Y}(\lambda), \tag{6.83}$$

where $\mathbf{P_H} = \mathbf{H}\left(\mathbf{H}^H\mathbf{H}\right)^{-1}\mathbf{H}^H$ and $\mathbf{P_H^\perp} = \mathbf{I} - \mathbf{P_H}$ are the projection matrices onto the columnspace and onto the null-space of $\mathbf{H}$, and $\hat{\mathbf{A}}_{\mathrm{ML}}^\dagger = (\hat{\mathbf{A}}_{\mathrm{ML}}^H\hat{\mathbf{A}}_{\mathrm{ML}})^{-1}\hat{\mathbf{A}}_{\mathrm{ML}}^H$ is the *Moore–Penrose pseudoinverse* [Strang 1988] of $\hat{\mathbf{A}}_{\mathrm{ML}}$. Note that $\mathbf{H}$ is the steering vector matrix that is parameterized by the hypothesized source locations. The optimal $\mathbf{H}$ should then provide a good estimate of $\mathbf{A}$ and of the source locations.

For the case of a single source S_1, $\mathbf{A} = \mathbf{A}(\mathbf{r}_s)$ and the optimal source location is obtained as the position that maximizes (from (6.81))

$$\hat{\mathbf{r}}_s = \operatorname*{argmax}_{\mathbf{r}} \sum_\lambda \frac{|\mathbf{H}^H(\mathbf{r})\mathbf{Y}(\lambda)|^2}{\|\mathbf{H}(\mathbf{r})\|^2} \tag{6.84}$$

$$= \operatorname*{argmax}_{\mathbf{r}} \sum_{\substack{\ell,m \\ \ell \neq m}} \hat{\Phi}_{y_\ell y_m} H_\ell^*(\mathbf{r})H_m(\mathbf{r}),$$

where $\hat{\Phi}_{y_\ell y_m} = \frac{1}{\Lambda}\sum_n Y_\ell(\lambda)Y_m^*(\lambda)$ is the estimate of the corresponding power spectal density and $\mathbf{H}(\mathbf{r})$ is the steering vector along $\mathbf{r}$. Recognize that (6.84) is simply the normalized beamformer of (6.49).

6.6 Evaluation of Localization Algorithms

As may be gleaned from the discussions above, most localization approaches utilize only the second-order statistics of the microphone signals and are closely related. This section presents the behavior of representative algorithms – for both indirect and direct methods – in reverberant and noisy situations. While the purpose of this section is not to perform an exhaustive comparison of the various methods, the advantages and disadvantages of the algorithms will be mentioned where appropriate. The algorithms considered are:

indirect methods: GCC-PHAT, CC, LMS, AED
direct methods: SRP-PHAT, MUSIC.

The localization experiments were carried out in a room simulated using the image method [Allen, Berkley 1979], [Habets 2006]. The room dimensions were $3\,\mathrm{m} \times 5\,\mathrm{m} \times 4\,\mathrm{m}$. A microphone array with $M = 5$ microphones was used. The microphones were placed linearly at distances of $3\,\mathrm{cm}$, $8\,\mathrm{cm}$, $15\,\mathrm{cm}$, and $25\,\mathrm{cm}$, respectively, from the first microphone and the array center was located at $\mathbf{r} = (1.5, 2.5, 1.0)^T\mathrm{m}$. The source was placed at $\mathbf{r}_s = (1.0, 3.366, 1.0)^T\mathrm{m}$, see Fig. 6.5.

Further, the simulations were carried out for three different reverberation times $\mathrm{T}_{60}/\mathrm{s} \in \{0.12, 0.3, 0.5\}$ and, for each case, under three different signal-to-noise

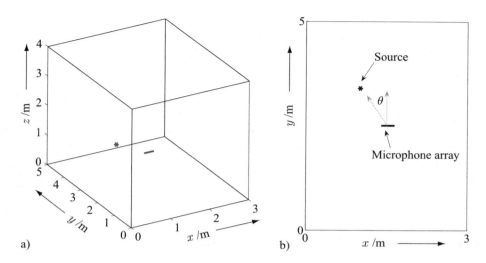

Figure 6.5: Overview of the simulation setup
a) 3-D image of the simulated room
b) Top-view of the simulated room ($\theta = \pi/6$)

ratios SNR/dB $\in \{-5, 5, 15\}$. The sampling frequency was $f_S = 8000\,\text{Hz}$. For the indirect methods, the two outermost microphones were selected, with a resultant inter-microphone distance of $d_{\max} = 25\,\text{cm}$.

For the GCC-PHAT and the simple cross-correlation approach, the required power spectral densities were estimated in the discrete Fourier domain by first-order recursive, temporal smoothing with a smoothing constant η as

$$\Phi_{y_\ell y_m}(\mu, \lambda) = \eta \Phi_{y_\ell y_m}(\mu, \lambda - 1) + (1 - \eta)\, Y_\ell(\mu, \lambda) Y_m^*(\mu, \lambda) \quad \ell, m \in \{1, 2\}. \quad (6.85)$$

The generalized cross-correlation function is then obtained for each frame λ by the inverse discrete Fourier transform of $\Phi_{y_\ell y_m}(\mu, \lambda)$.

For the direct methods considered, the search grid was defined in two ways: over the two-dimensional grid defined in the x–y plane and over a one-dimensional grid computed over the azimuth with respect to the midpoint of the array. For each case, in the SRP approach, the cost function was computed per frame as detailed in Sec. 6.5.3. For the MUSIC approach, the spectrum evaluated over the complete speech signal (5 s duration) was used to build the statistics upon which the estimate of the noise-only subspace $\mathbf{U}_v(\mu)$ was computed, for each frequency bin μ. An estimate for the source location was then obtained over either the 2-D or the 1-D grid in each bin.

The parameters for the various methods are summarized in Table 6.2.

Table 6.2: Algorithm parameters used in the simulations

Algorithm	Filter length (ms)	Window type/ length (ms)	DFT length (ms)	Frame shift (ms)	ς	η
CC / GCC-PHAT	N/A	Hamming/64	128	32	N/A	0.90
LMS	128	Rectangular/128	N/A	0.125	0.005	N/A
AED	128	Rectangular/64	N/A	0.125	0.005	N/A
SRP-PHAT	N/A	Hamming/128	128	64	N/A	N/A
MUSIC	N/A	Hamming/128	128	64	N/A	N/A

6.6.1 Performance of the Indirect Methods

With the chosen positions for the array and the source, the time difference of the direct path between the two outermost microphones corresponds to about three sampling intervals at 8000 Hz. Figure 6.6 depicts the histograms of estimated time delays of arrival between the microphones for the different simulation conditions. The histogram data is accumulated over 120 estimates. The delay axis is limited to the range of $[-5, 5]$ sampling intervals, as the maximum possible delay for the array configuration was about 5.8 sampling intervals. In all plots, the dotted line indicates the true time difference between the direct paths, which was obtained from the room impulse response for each microphone. It may be seen that the performance of the algorithms increases with an increase in the SNR – which is to be expected. Under low reverberation conditions, the estimate of the time delay is almost perfect at high SNRs. However, as the reverberation increases, a spread may be observed about the true value. Apparently, the AED approach, with two simultaneously adaptable filters (one for each microphone) converges better to the true delay difference as compared with the LMS approach as the reverberation increases.

The simple cross-correlator (CC) has the worst performance among the four. This could be explained as follows: speech signals possess maximum energy at the lower frequencies. The CC algorithm applies no explicit weighting to the frequency bins and, thus, implicitly weights each frequency by the energy of the received signal in that frequency bin. Therefore, low frequencies are emphasized and higher frequency contributions are damped in this method. Conversely, time delay information is less accurate at low frequencies, more so in the presence of noise. Additionally, in reverberant conditions, it is the reflections at the higher frequencies that are damped to a greater extent than the lower frequencies. As GCC-PHAT, GCC-Roth, etc., on the other hand, remove this emphasis on the lower frequencies, they lead to an improvement in performance. This is noticeable especially in reverberant environments [Gustafsson et al. 2003] where the higher frequencies – which are less affected by reverberation – contribute more to the delay estimate.

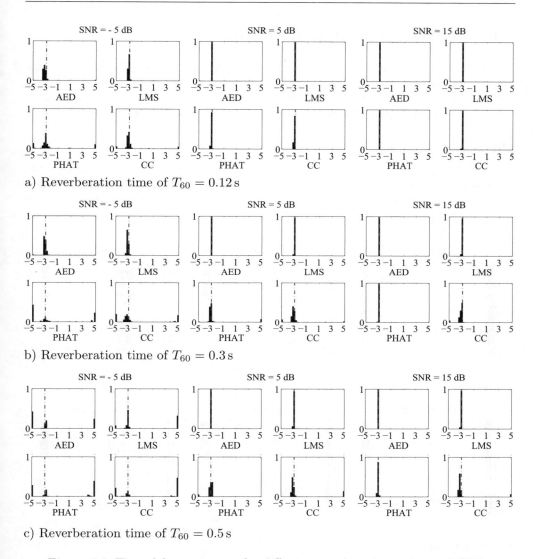

a) Reverberation time of $T_{60} = 0.12\,$s

b) Reverberation time of $T_{60} = 0.3\,$s

c) Reverberation time of $T_{60} = 0.5\,$s

Figure 6.6: Time delay estimates for different reverberation times and SNRs

Another factor affecting the performance of these algorithms is the smoothing factor η for the (GCC-PHAT/CC) and the step-size ς for the LMS/AED approaches. We find that increasing η improves the performance of the GCC-PHAT/CC approaches bringing them – especially GCC-PHAT – close to the performance of the LMS/AED approaches. Similarly, a lower step-size ς improves the robustness of the LMS/AED approaches against noise, but the convergence is slower. In general, the step-size ς and the smoothing parameter η could be made adaptive: one could use a larger value in high SNR environments and a lower one when the noise level increases.

6.6.2 Performance of the Direct Methods

This section deals with the behavior of the SRP-PHAT and the MUSIC algorithms under the same conditions as for the indirect methods except for the number of microphones. The direct approaches use all five microphones of the array. The performance of the SRP-PHAT algorithm will be discussed first, followed by the MUSIC approach.

SRP Approach

Figure 6.7 depicts the cost function $\mathcal{J}(\mathbf{r}, \lambda)$ computed according to (6.58) over the one-dimensional search grid along the azimuth. The x-axis indicates the candidate azimuth angles (measured with respect to the array axis), the y-axis indicates the time frame (in seconds) and the intensity of a point at any co-ordinates is a measure

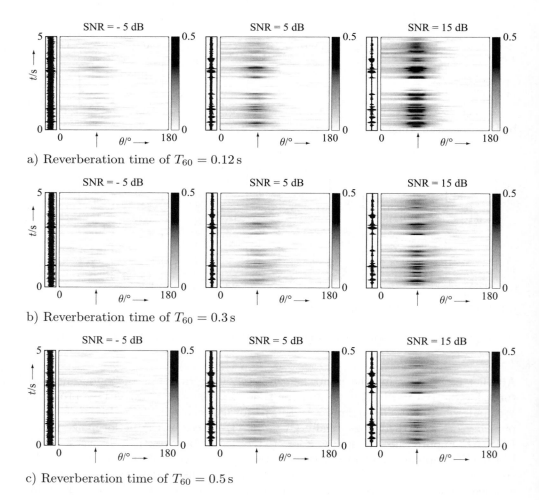

a) Reverberation time of $T_{60} = 0.12\,\mathrm{s}$

b) Reverberation time of $T_{60} = 0.3\,\mathrm{s}$

c) Reverberation time of $T_{60} = 0.5\,\mathrm{s}$

Figure 6.7: $\mathcal{J}(\theta, \lambda)$ estimates for different reverberation times and SNRs

of the cost function value for that time-frame and that candidate location. For the given source-array constellation, the azimuth angle is $\pi/3$ (or $60°$). This forms the 'ground-truth' for comparison and is indicated by an $\uparrow$ along the x-axis in all the plots. The time-domain signal at the first microphone is also plotted parallel to the y-axis.

Three trends are clearly perceivable in all the plots: first, the range of the cost function values increases with the SNR – this is to be expected because, as the noise level decreases, the cost function depends increasingly upon the incident signal and yields a high value only at the true azimuth as perfect phase alignment is obtained among all microphone pairs. When the SNR is low, the SRP cost function gets smeared due to chance phase alignments at different azimuths, between different microphone pairs, reducing the range of values. Secondly, the true azimuth is obtainable even at low SNRs due to the larger spatial diversity available, as compared with the TDE approaches considered in the previous section. Thirdly, notice the broadening of the cost function peaks as reverberation increases – this is to be expected as, due to the multipath propagation, partial phase alignments are possible along a wider range of search locations. However, in terms of localizing the source, the performance does not seem to significantly deteriorate with increasing reverberation – as long as the direct path is dominant.

As an illustration of the capabilities of the SRP algorithm, the cost function is evaluated over a two dimensional grid along the x–y plane and is given in Fig. 6.8 for the time frame λ. The true position of the source is indicated by $+$. The grey bar at the bottom of each plot represents the microphone array.

Physical considerations dictate that, for a linear array with a relatively small aperture, it is difficult to obtain both the range and the azimuth of the source when the source is in the farfield. This can be seen in the plots. Note that there is a broad range of co-ordinates with similar cost function values, lying along a half-hyperbola centered about the array axis. This is sometimes termed the 'cone of confusion' – sources lying anywhere on this cone would generate the same phase deviations at the microphones in the absence of reverberation and noise and it is difficult to pinpoint the location of the source on this cone without additional information, e.g., from a second array mounted perpendicular to the first. The trends with respect to SNR and reverberation observed in the previous case may be perceived here, too.

MUSIC Approach

As mentioned before, the MUSIC approach first estimates the noise-only subspace in each frequency bin from the power spectral density matrix. The latter is obtained, in practice, from a temporal averaging of the signal spectrum. This averaging might be either recursive (in which case, MUSIC may be used to yield a source location estimate in each frame λ) or block based (in which case the location estimates are computed on a batch basis). For the MUSIC approach, the first step lies in determining the number of signals present in the system. This is done either by comparing the values of the eigenvectors or by the application of information-theoretic criteria as described in [Wax, Kailath 1984a]. Once this is done, the identification of the

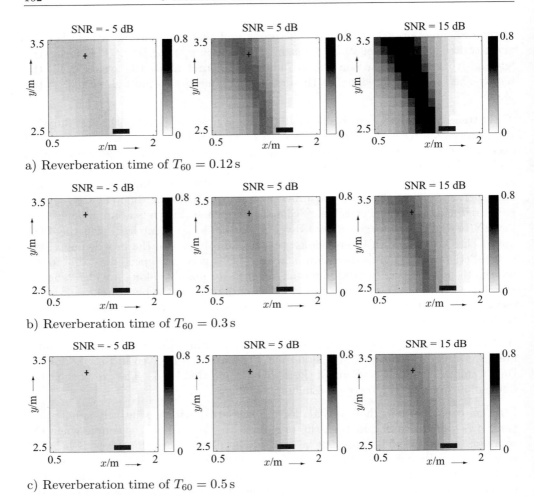

a) Reverberation time of $T_{60} = 0.12$ s

b) Reverberation time of $T_{60} = 0.3$ s

c) Reverberation time of $T_{60} = 0.5$ s

Figure 6.8: $\mathcal{J}(\mathbf{r}, \lambda)$ estimates for different reverberation times and SNRs

noise-only subspace is performed as described in Sec. 6.5.4. For the simulations, the $\mathbf{\Phi_{xx}}(\mu)$ matrix was obtained as a temporal average over the spectrum of the complete speech signal. Figure 6.9 showcases the performance of the MUSIC algorithm over the 1-D search grid.

The plots indicate the MUSIC spectrum values (normalized to a maximum of 1) for each frequency (plotted along the y-axis). Note the lack of directivity at frequencies below 200 Hz. In these bands, it is difficult to obtain an estimate of the source position due to the infinitesimal phase difference between the microphone signals. As the frequencies increase, the MUSIC spectrum spread narrows down – corresponding to increasing directivity at higher frequencies. As expected, the performance improves with an increase in the SNR. Further, as the room becomes more reverberant, the MUSIC spectrum begins to spread out across the azimuth – again an effect that is to be expected, due to the correlated multipath propagation.

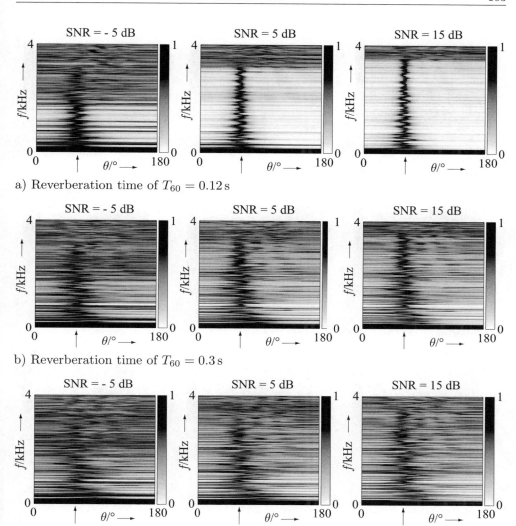

a) Reverberation time of $T_{60} = 0.12\,\text{s}$

b) Reverberation time of $T_{60} = 0.3\,\text{s}$

c) Reverberation time of $T_{60} = 0.5\,\text{s}$

Figure 6.9: MUSIC $\mathcal{J}(\theta, \mu)$ estimates for different reverberation times and SNRs

As in the case of the SRP, a more robust estimate of the source location may be found, for broadband sources, by averaging the normalized MUSIC spectrum along the source bandwidth and then finding the maximum as

$$\mathcal{J}(\theta) = \sum_{\mu} \frac{\mathcal{J}(\theta, \mu)}{\max(\mathcal{J}(\theta, \mu))} \qquad (6.86)$$

$$\hat{\theta}_s = \operatorname*{argmax}_{\theta} \mathcal{J}(\theta) ,$$

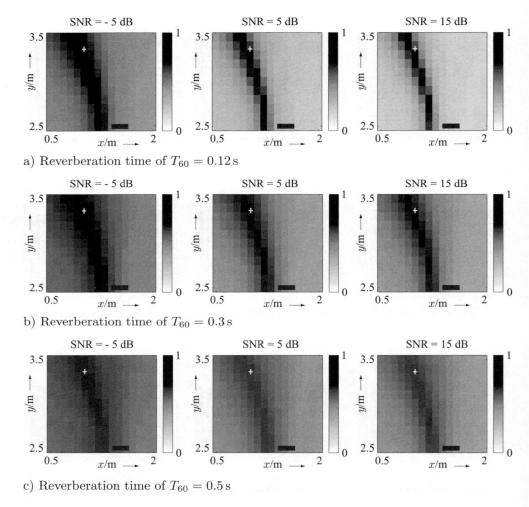

a) Reverberation time of $T_{60} = 0.12\,\text{s}$

b) Reverberation time of $T_{60} = 0.3\,\text{s}$

c) Reverberation time of $T_{60} = 0.5\,\text{s}$

Figure 6.10: MUSIC $\mathcal{J}(\mathbf{r})$ estimates for different reverberation times and SNRs

when evaluating over a one dimensional grid or

$$\mathcal{J}(\mathbf{r}) = \sum_{\mu} \frac{\mathcal{J}(\mathbf{r}, \mu)}{\max\left(\mathcal{J}(\mathbf{r}, \mu)\right)} \tag{6.87}$$

$$\hat{\mathbf{r}}_s = \operatorname*{argmax}_{\mathbf{r}} \mathcal{J}(\mathbf{r}),$$

over the two dimensional grid, leading to the incoherent MUSIC approach. Alternatively, one may choose to weight the estimates in each frequency band before averaging.

The MUSIC spectrum evaluated over a two-dimensional grid is illustrated in Fig. 6.10. Notice, again, the "cone of confusion", and the broadening of this cone with increasing reverberation. As before, the grey bar at the bottom of each plot indicates the microphone array and the + the source position.

6.6.3 The Two-Source Case

A last simulation presented here illustrates the performance of the localization algorithms for a multi-source case. In addition to the first source, we consider another source located at $\mathbf{r}_{s_2} = (2.0, 3.366, 1.0)^T$ m, with a corresponding time delay of -3 samples or an azimuth of $2\pi/3$. The two sources are simultaneously active. The room is simulated with a $T_{60} = 0.12$ s and the SNR is 35 dB.

From Fig. 6.11 it is evident that the indirect methods as discussed here make a single source assumption and thus find it difficult to cope with the multi-speaker scenario.

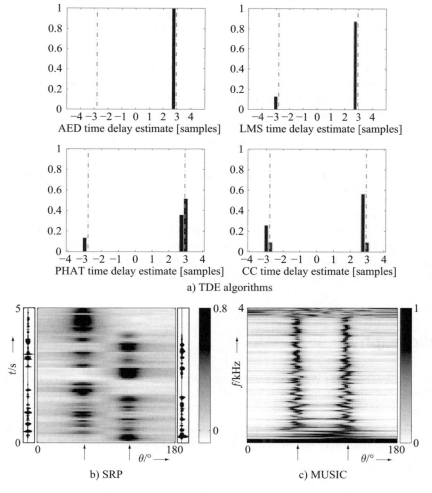

a) TDE algorithms

b) SRP   c) MUSIC

Figure 6.11: Performance in a multi-source scenario at 35 dB SNR

There are, however, methods that locate multiple speaker time delays by examining the secondary peaks of the cross-correlation function, e.g., [Bechler, Kroschel 2003], [Scheuing, Yang 2006], [Scheuing, Yang 2007]. SRP and MUSIC, on the other hand, can exploit the additional freedom at different frequency bins to localize multiple speakers even within a single time-frame.

6.7 Conclusions

This chapter has provided an overview of various, contemporary source localization approaches. These have been classified, for purposes of convenience, into direct approaches and indirect approaches. Algorithms cataloged under indirect approaches first estimate the time delay of arrival (TDOA) between various microphone pairs and then, based on these values and the geometry of the array, estimate the source positions. Direct approaches, on the other hand, pick the most likely set of source positions from a given set of candidate locations.

Further, relations between the various approaches were derived and it was shown that most approaches exploit only the second-order statistics of the observed microphone signals. In general, this second-order dependence comes about as a result of making Gaussian assumptions regarding the signal and noise statistics. As these assumptions might not be realistic for speech, these are, perhaps, not optimal source location estimators. Consequently, future research could focus on incorporating *a priori* knowledge of the signal/noise statistics as done, for instance, in [Aichner et al. 2006]. One could also retrieve the source location information as byproducts of the algorithms on Blind Source Separation.

The performance of representative direct and indirect algorithms were also illustrated for different noise and reverberation scenarios. Generally, the presence of noise and reverberation degrades the algorithm performance. However, while localization is still possible under dominance of the direct path and the corresponding simplified signal model, attempting to model/estimate the source-microphone room impulse does improve the performance.

Bibliography

Aichner, R.; Buchner, H.; Wehr, S.; Kellermann, W. (2006). Robustness of Acoustic Multiple-Source Localization in Adverse Environments, *Proceedings of the 7th German Information Technology Conference on Speech Communication (ITG)*.

Allen, J. B.; Berkley, D. A. (1979). Image Method for Efficiently Simulating Small Room Acoustics, *Journal of the Acoustical Society of America*, vol. 65, no. 4, pp. 943–950.

Bechler, D.; Kroschel, K. (2002). Confidence Scoring of Time Difference of Arrival Estimation for Speaker Localization with Microphone Arrays, *Proceedings of the 13th Conference "Elektronische Sprachsignalverarbeitung" (ESSV)*, Dresden, Germany.

Bechler, D.; Kroschel, K. (2003). Considering the Second Peak in the GCC Function for Multi-Source TDOA Estimation with a Microphone Array, *Proceedings of the International Workshop on Acoustic Echo and Noise Cancellation (IWAENC)*, pp. 315–318.

Bechler, D.; Kroschel, K. (2004). Three Different Reliability Criteria for Time Delay Estimates, *Proceedings of the European Signal Processing Conference (EUSIPCO)*, pp. 1987–1990.

Benesty, J. (2000). Adaptive Eigenvalue Decomposition Algorithm for Passive Source Localization, *Journal of the Acoustical Society of America*, vol. 107, no. 1, pp. 384–391.

Birchfield, S. T. (2004). A Unifying Framework for Acoustic Localization, *Proceedings of the European Signal Processing Conference (EUSIPCO)*.

Böhme, J. F. (1986). Estimation of Spectral Parameters of Correlated Signals in Wavefields, *EURASIP Journal on Applied Signal Processing*, vol. 11, no. 4, pp. 329–337.

Brutti, A.; Omologo, M.; Svaizer, P. (2005). Oriented Global Coherence Field for the Estimation of the Head Orientation in Smart Rooms Equipped with Distributed Microphone Arrays, *Proceedings of the International Conference on Speech Communication and Technology (INTERSPEECH)*, pp. 2337–2340.

Chan, Y.; Ho, K. (1994). A Simple and Efficient Estimator for Hyperbolic Location, *IEEE Transactions on Signal Processing*, vol. 42, no. 8, pp. 1905–1915.

Chan, Y.; Riley, J.; Plant, J. (1981). Modeling of Time Delay and its Application to Estimation of Nonstationary Delays, *IEEE Transactions on Acoustics, Speech and Signal Processing*, vol. 29, no. 3, pp. 577– 581.

Chen, J.; Benesty, J.; Huang, Y. (2003a). Robust Time Delay Estimation Exploiting Redundancy among Multiple Microphones, *IEEE Transactions on Speech and Audio Processing*, vol. 11, no. 6, pp. 549–557.

Chen, J.; Benesty, J.; Huang, Y. (2003b). Time Delay Estimation using Spatial Correlation Techniques, *Proceedings of the International Workshop on Acoustic Echo and Noise Cancellation (IWAENC)*.

Chen, J. C.; Hudson, R. E.; Yao, K. (2002). Maximum Likelihood Source Localization and Unknown Sensor Location Estimation for Wideband Signals in the Near-Field, *IEEE Transactions on Signal Processing*, vol. 50, no. 8, pp. 1843–1854.

Cover, T. M.; Thomas, J. A. (1991). *Elements of Information Theory*, John Wiley & Sons, Inc., New York.

DiBiase, J.; Silverman, H.; Brandstein, M. (2001). Robust Localization in Reverberant Rooms, *in* M. Brandstein; D. Ward (eds.), *Microphone Arrays: Signal Processing Techniques and Applications*, Springer-Verlag, Berlin.

Doclo, S.; Moonen, M. (2003). Robust Adaptive Time Delay Estimation for Speaker Localization in Noisy and Reverberant Acoustic Environments, *EURASIP Journal on Applied Signal Processing*, vol. 11, pp. 1110–1124.

Drews, M. (1995). Time Delay Estimation for Microphone Array Speech Enhancement Systems, *Proceedings of the European Conference on Speech Communication and Technology (EUROSPEECH)*, vol. 3, pp. 2013–2016.

Dvorkind, T. G.; Gannot, S. (2005). Time Difference of Arrival Estimation of Speech Source in a Noisy and Reverberant Environment, *EURASIP Journal on Applied Signal Processing*, vol. 85, pp. 177–204.

Eldar, Y. C.; Oppenheim, A. V. (2003). MMSE Whitening and Subspace Whitening, *IEEE Transactions on Information Theory*, vol. 49, no. 7, pp. 1846–1851.

Etter, D. M.; Stearns, S. D. (1981). Adaptive Estimation of Time Delays in Sampled Data Systems, *IEEE Transactions on Acoustics, Speech and Signal Processing*, vol. 29, no. 3, pp. 582–587.

Gustafsson, T.; Rao, B. D.; Trivedi, M. (2003). Source Localization in Reverberant Environments: Modelling and Statistical Analysis, *IEEE Transactions on Speech and Audio Processing*, vol. 11, no. 6, pp. 791–802.

Habets, E. A. P. (2006). Room Impulse Response Generator, Online resource: http://www.sps.ele.tue.nl/members/E.A.P.Habets/rir-generator/default.asp.

Haykin, S. (1996). *Adaptive Filter Theory*, 3rd edn., Prentice Hall, Englewood Cliffs, New Jersey, USA.

Hertz, D. (1986). Time Delay Estimation by Combining Efficient Algorithms and Generalized Cross-Correlation Methods, *IEEE Transactions on Acoustics, Speech and Signal Processing*, vol. 34, no. 1, pp. 1–7.

Huang, Y.; Benesty, J.; Elko, G. W.; Mersereau, R. M. (2001). Real-time Passive Source Localization: A Practical Linear-Correction Least-Squares Approach, *IEEE Transactions on Speech and Audio Processing*, vol. 9, no. 8, pp. 943–956.

Hung, H.; Kaveh, K. (1988). Focusing Matrices for Coherent Signal-Subspace Processing, *IEEE Transactions on Acoustics, Speech and Signal Processing*, vol. 36, no. 8, pp. 1272–1281.

Jaffer, A. G. (1988). Maximum Likelihood Direction Finding of Stochastic Sources: A Separable Solution, *Proceedings of the IEEE International Conference on Acoustics, Speech and Signal Processing (ICASSP)*, vol. 5, pp. 2893–2896.

Knapp, C. H.; Carter, G. C. (1976). The Generalized Correlation Method for the Estimation of Time Delay, *IEEE Transactions on Acoustics, Speech and Signal Processing*, vol. 24, no. 4, pp. 320–327.

Krim, H.; Viberg, M. (1996). Two Decades of Array Signal Processing Research, *IEEE Signal Processing Magazine*, vol. 13, no. 4, pp. 67–94.

Krolik, J. (1991). Focused Wideband Array Processing for Spatial Spectral Estimation, *in* S. Haykin (ed.), *Advances in Spectrum Analysis and Array Processing, vol. II*, Prentice-Hall.

Liu, C.; Wheeler, B. C.; O'Brien Jr., W. D.; Lansing, C. R.; Bilger, R. C.; Feng, A. (2000). Localization of Multiple Sound Sources with Two Microphones, *Journal of the Acoustical Society of America*, vol. 108, no. 4, pp. 1888–1905.

Madhu, N.; Martin, R. (2005). Robust Speaker Localization Through Adaptive Weighted Pair TDOA (AWEPAT) Estimation, *Proceedings of the International Conference on Speech Communication and Technology (INTERSPEECH)*.

Madhu, N.; Martin, R.; Rehn, H.-W.; Fischer, A. (2006). Brake Squeal Localization, *Proceedings of the Annual Meeting of the German Acoustical Society (DAGA)*.

Matsuo, M.; Hioka, Y.; Hamada, N. (2005). Estimating DOA of Multiple Speech Signals by Improved Histogram Mapping Method, *Proceedings of the International Workshop on Acoustic Echo and Noise Cancellation (IWAENC)*.

Omologo, M.; Svaizer, P. (1994). Acoustic Event Localization using a Crosspower-Spectrum based Technique, *Proceedings of the IEEE International Conference on Acoustics, Speech and Signal Processing (ICASSP)*, vol. II, pp. 273–276.

Oppenheim, A. V.; Schafer, R. W. (1975). *Digital Signal Processing*, Prentice Hall, Englewood Cliffs, New Jersey, USA.

Reed, F.; Feintuch, P.; Bershad, N. (1981). Time Delay Estimation Using the LMS Adaptive Filter-Static Behavior, *IEEE Transactions on Acoustics, Speech and Signal Processing*, vol. 29, no. 3, pp. 561–571.

Scheuing, J.; Yang, B. (2006). Disambiguation of TDOA Estimates in Multi-Path Multi-Source Environments (DATEMM), *Proceedings of the IEEE International Conference on Acoustics, Speech and Signal Processing (ICASSP)*, vol. 4, pp. 837–840.

Scheuing, J.; Yang, B. (2007). Efficient Synthesis of Approximately Consistent Graphs for Acoustic Multi-Source Location, *Proceedings of the IEEE International Conference on Acoustics, Speech and Signal Processing (ICASSP)*, pp. 501–504.

Schmidt, R. O. (1981). *A Signal Subspace Approach to Multiple Emitter Location and Spectral Estimation*, PhD thesis, Stanford University.

So, H. C.; Ching, P. C.; Chan, Y. T. (1994). A New Algorithm for Explicit Adaptation of Time Delay, *IEEE Transactions on Signal Processing*, vol. 42, no. 7, pp. 1816–1820.

Strang, G. (1988). *Linear Algebra and Its Applications*, 3 edn, Thomson / Brooks Cole, USA.

Talantzis, F.; Constantinides, A. G.; Polymenakos, L. C. (2005). Estimation of Direction of Arrival using Information Theory, *IEEE Signal Processing Letters*, vol. 12, no. 8, pp. 561–564.

Vary, P.; Martin, R. (2006). *Digital Speech Transmission: Enhancement, Coding and Error Concealment*, John Wiley & Sons, Ltd., Chichester, England.

Wang, H.; Kaveh, M. (1985). Coherent Signal-Subspace Processing for the Detection and Estimation of Angles of Arrival of Multiple Wideband Sources, *IEEE Transactions on Acoustics, Speech and Signal Processing*, vol. 33, no. 4, pp. 823–831.

Wax, M. (1992). Detection and Localization of Multiple Sources in Noise with Unknown Covariance, *IEEE Transactions on Acoustics, Speech and Signal Processing*, vol. 40, no. 1, pp. 245–249.

Wax, M.; Kailath, T. (1984a). Determining the Number of Signals by Information Theoretic Criteria, *Proceedings of the IEEE International Conference on Acoustics, Speech and Signal Processing (ICASSP)*, vol. 9, pp. 232–235.

Wax, M.; Kailath, T. (1984b). Spatio-Temporal Spectral Analysis by Eigenstructure Methods, *IEEE Transactions on Acoustics, Speech and Signal Processing*, vol. 32, no. 4, pp. 817–827.

Xu, G.; Liu, H.; Tong, L.; Kailath, T. (1995). A Least-Squares Approach to Blind Channel Identification, *IEEE Transactions on Speech and Audio Processing*, vol. 43, no. 12, pp. 2982–2993.

Yoon, Y.; Kaplan, L. M.; McClellan, J. H. (2006). TOPS: New DOA Estimator for Wideband Signals, *IEEE Transactions on Signal Processing*, vol. 54, no. 6, pp. 791–802.

Chapter 7

Multi-Channel System Identification with Perfect Sequences

– Theory and Applications –

Christiane Antweiler

7.1 Introduction

The fundamental problem of multi-channel system identification given the input and the output signals of a *multiple input – single output* (MISO) system arises in many application areas such as speech enhancement, acoustics, or mobile communication. In this chapter a new approach is presented, which is based on the *normalized least-mean-square* (NLMS) algorithm in combination with a special class of excitation signals, the so called *perfect sequences* (PSEQs). It opens up the possibility of uniquely identifying the *true* impulse responses of multiple channels with one measurement in a simple and efficient way for all numbers of channels and all system lengths. Owing to its fast tracking property, this new approach also allows the real-time acquisition of time-variant impulse responses. Furthermore, the method allows an identification of each kind of linear channel, radio or acoustic, and can easily be extended to *multiple input – multiple output* (MIMO) (e.g., wireless) transmission.

Several approaches have addressed the problem of measuring linear and time-invariant (LTI) system responses. In order to characterize a digital linear, time-invariant system, the most direct approach is to apply an impulsive excitation to the system and analyze its response. For many applications, however, we have to deal with system inherent noise (e.g., quantization noise due to A/D conversion) and/or environmental noise, reducing the accuracy of the measurement. To achieve a sufficiently high *signal-to-noise ratio* (SNR) the excitation signal must have high energy uniformly spread over the frequency range of interest. As the maximum amplitude is limited, compared with an impulse-like excitation a higher energy level of the signal is possible, if additionally the energy is spread over time. Thus, all spectrally dense signals, such as binary *maximum-length sequences* (MLS), swept-sine, time-stretched pulses, etc. can be used as excitation signals for the system under test. The desired impulse response can then be recovered by digital signal processing techniques.

One of the most common methods in audio and acoustics is the so called *MLS technique* first proposed by [Schroeder 1979]. The MLS technique is based on the excitation of the unknown linear system by an MLS, i.e., a binary, periodic, pseudo-noise sequence of period length $N = 2^n - 1$, $n \in \mathbb{N}$. It can be generated by shift registers with modulo-feedback taps [Golomb 1982]. MLSs possess an *almost perfect*, i.e., impulse-like, periodic autocorrelation function. The impulse response of the system under test is obtained by circular cross-correlation between the stimulus MLS and its system response, which enables impulse response measurements directly in the time domain. A more detailed analysis of the MLS technique can be found in [Rife, Vanderkooy 1989], [Vanderkooy 1994].

The MLS technique was refined by speeding up the cross-correlation calculation with an algorithm adapted from Hadamard spectroscopy [Nelson, Fredman 1970]. In [Cohn, Lempel 1977] the relationship of MLS to the so called *Hadamard transform* has been shown. It allows the correlation of a maximum length sequence to be computed in a fast algorithm similar to the FFT, called the *fast Hadamard transform* (FHT) or *fast M-sequence transform* (FMT). Since then several authors have proposed different methods for transfer function measurements based on the MLS technique in combination with the FHT (e.g., [Alrutz, Schroeder 1983], [Alrutz 1983], [Borish, Angell 1985], [Borish 1985], [Xiang 1991], [Xiang 1992]).

The major problem of the MLS technique resides in the appearance of distortion artifacts due to non-linearities inherent in the measurement system. These artifacts are more or less uniformly distributed along the deconvolved impulse response. To achieve a higher distortion immunity the *inverse repeated sequence* (IRS) method was introduced by [Ream 1970], [Dunn, Hawksford 1993]. The stimulus sequence (IRS) is generated by alternating the sign of the MLS in each period. The deconvolution process, however, is exactly the same as for the MLS technique.

Two years after Schroeder's proposition, a new method called the *time-stretched pulse* technique for the measurements of impulse responses was proposed by [Aoshima 1981] and further optimized by [Suzuki et al. 1995]. The excitation signal is based on a computer-generated pulse, which is processed with pulse expansion and compression techniques to increase the sound power.

Another well established method is the *time delay spectrometry* (TDS), deriving transfer functions with the help of sweeps, a sinusoid excitation that is swept over the frequency range of interest [Heyser 1967], [Biering, Pedersen 1983]. This basic approach has been further developed by [Poletti 1988b], [Poletti 1988a]. In [Farina 2000] the so-called *logarithmic swept-sine technique* was introduced to overcome the limitations encountered by MLS, IRS and time-stretched pulse methods. Employing a sinusoidal signal with exponentially varied frequency, it is possible to deconvolve simultaneously the linear impulse response of the system, and separate impulse responses for each harmonic distortion order. This method is not limited to linear and time-invariant (LTI) systems, but can also be used to measure strongly non-linear systems as it inherently provides an analysis of the system non-linearities.

Finally, the following methods deserve to be briefly mentioned:

- *stepped sine*, i.e., excitation of the system under test with pure tones in steps of increasing frequency (e.g., [Schoukens et al. 2000]),

- *dual-channel FFT-analysis*, based on a division of the output spectrum of the system by the spectrum of the input signal (e.g., [Herlufsen 1984]), [Mateljan, Ugrinović 2001].

In practice, there is always a certain amount of noise, non-linearity, and time-variance. For the evaluation of the diverse techniques we have to consider that all of them differ in their characteristics such as bandwidth, reproducibility, SNR, handling of nonlinear artifacts, time consumption for the measurement as well as complexity and that the choice of an "optimal" algorithm depends on its application. A comprehensive comparison of the different methods can be studied, e.g., in [Stan et al. 2002], [Mateljan, Ugrinović 2003], and [Müller, Massarani 2001].

In 1994/95 we introduced an alternative approach for the identification of an unknown linear system given the input and the output signal ([Antweiler, Antweiler 1995], [Antweiler, Dörbecker 1994], [Antweiler 1995]). This method can be grouped into the class of cross-correlation based methods. It relies on the well-known *normalized least-mean-square* (NLMS) algorithm ([Widrow, Hoff 1960], [Vary, Martin 2006]) excited by a so called *perfect sequence* (PSEQ) [Ipatov 1979], [Lüke 1992]. PSEQs are special, periodically repeated pseudo-noise signals with a perfect, impulse-like periodic autocorrelation function. Owing to these correlation properties, PSEQs represent the *optimal* excitation signal for the NLMS algorithm. Thus, with a PSEQ excitation the NLMS algorithm is capable of identifying a linear, noiseless system within one period of the sequence. The principle of this NLMS-type identification approach is outlined in Sec. 7.2.

Owing to its simplicity and good properties, we used this technique routinely for a couple of years in diverse applications. However, so far this concept has been used only for single-channel transmission systems. In Sec. 7.3 we will generalize the approach to the multi-channel case in the sense of a *multiple input – single output* (MISO) system. The typical problem of multi-channel system identification as needed, e.g., in stereophonic acoustic echo cancellation [Sondhi et al. 1995], [Benesty et al. 1995] is the non-zero cross-correlation between the excitation signals. As a result, the adaptive filters often don't converge on the *true* system impulse responses or show poor convergence speed. Given only a single measured reference signal, adequate excitation signals are required to identify the *true* impulse response of each channel. As part of this work we will present the idea of how this class of *optimal* excitation signals can be constructed from PSEQs. The theoretical results are verified via simulations.

The technology of multi-channel system identification has a wide range of possible applications in mobile communications, acoustics, digital speech processing, and in the medical community. Finally, in Sec. 7.4 some of our applications will be presented and discussed.

7.2 System Identification with Perfect Sequences

The discrete time model in Fig. 7.1 depicts a system for the identification of an unknown linear transmission system by means of an adaptive filter. For simplicity, we will assume that all signals are digitized with sampling rate f_S. In the following, we do not differentiate between acoustic or analog signals and their digital counterparts. The unknown transmission path is represented by the impulse response $\mathbf{g} = (g_0, g_1, \ldots, g_{N-1})^T$ of length N. In the first instance, we assume that the unknown system is linear and time-invariant (LTI). Later on, time-variant systems will also be studied. The influence of system-inherent or environmental noise on the adaptation process can be taken into account by adding a non-zero noise signal $n(k)$ to the output $y(k)$ of the system under test.

The adaptation of the digital transversal filter is driven by the *normalized least-mean-square* (NLMS) algorithm ([Widrow, Hoff 1960], [Vary, Martin 2006]), i.e., the weights $\mathbf{h}(k)$ of the adaptive filter are updated via the recursion

$$\mathbf{h}(k+1) = \mathbf{h}(k) + \alpha \, \frac{e(k)\,\mathbf{p}(k)}{||\mathbf{p}(k)||^2} \tag{7.1}$$

with stepsize α and the error signal $e(k)$

$$e(k) = \big(\mathbf{g} - \mathbf{h}(k)\big)^T \mathbf{p}(k) + n(k)\,. \tag{7.2}$$

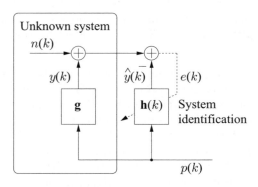

Figure 7.1: Single-channel system identification with PSEQs

The vectors and the squared vector norm are given by

$$\mathbf{h}(k) = \big(h_0(k),\, h_1(k),\, \ldots,\, h_{N-1}(k)\big)^T,$$
$$\mathbf{p}(k) = \big(p(k),\, p(k-1),\, \ldots,\, p(k-N+1)\big)^T,$$
$$\|\mathbf{p}(k)\|^2 = \mathbf{p}^T(k)\,\mathbf{p}(k),$$

where $p(k)$ denotes the excitation signal.

The NLMS algorithm is well known in applications such as acoustic echo control, e.g., [Hänsler 1992], [Antweiler 1995], [Hänsler, Schmidt 2004]. In these applications the adaptation process is normally driven by speech signals, i.e., colored signals, reducing severely the convergence speed. As we will verify below, a white noise excitation provides improved but not optimal convergence speed.

The aim of the identification process, however, is to achieve the best possible match between the adaptive filter with the impulse response $\mathbf{h}(k)$ and the system under test represented by $\mathbf{g}$. The key to our approach is the use of the NLMS algorithm in combination with its *optimal* excitation signal. In the next section, we will study the geometric interpretation of the NLMS algorithm. This interpretation will help us to formulate requirements for the *optimal* excitation signal.

7.2.1 Geometric Interpretation of the NLMS Algorithm

The adaptation process of the NLMS algorithm according to (7.1) can be geometrically interpreted in terms of a vector space representation, e.g., [Claasen, Mecklenbräuker 1981], [Sommen, van Valburg 1989].

For the geometric interpretation we introduce the *distance* vector

$$\mathbf{d}(k) = \mathbf{g} - \mathbf{h}(k), \tag{7.3}$$

defining the misalignment between the impulse responses $\mathbf{g}$ and $\mathbf{h}(k)$. The NLMS algorithm given in (7.1) changes to

$$\mathbf{d}(k+1) = \mathbf{d}(k) - \alpha\,\frac{e(k)\,\mathbf{p}(k)}{\|\mathbf{p}(k)\|^2}. \tag{7.4}$$

Using the distance vector $\mathbf{d}(k)$ and $n(k) \equiv 0$ in (7.2), we obtain for the error signal $e(k) = \mathbf{d}^T(k)\,\mathbf{p}(k)$, leading to

$$\mathbf{d}(k+1) = \mathbf{d}(k) - \alpha\frac{\mathbf{d}^T(k)\,\mathbf{p}(k)}{\|\mathbf{p}(k)\|^2}\,\mathbf{p}(k). \tag{7.5}$$

According to Fig. 7.2, the distance vector $\mathbf{d}(k)$ is decomposed into two components

$$\mathbf{d}(k) = \mathbf{d}^{\perp}(k) + \mathbf{d}^{\|}(k), \tag{7.6}$$

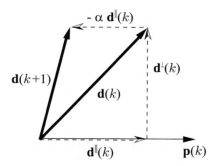

Figure 7.2: Geometric interpretation of the NLMS algorithm
(here: $0 < \alpha < 1$, $N = 2$)

where $\mathbf{d}^{\perp}(k)$ is orthogonal and $\mathbf{d}^{\parallel}(k)$ parallel to vector $\mathbf{p}(k)$. The parallel component $\mathbf{d}^{\parallel}(k)$ can be interpreted as an orthogonal projection of the distance vector $\mathbf{d}(k)$ onto the signal vector $\mathbf{p}(k)$:

$$\mathbf{d}^{\parallel}(k) = \frac{\mathbf{d}^{T}(k)\,\mathbf{p}(k)}{||\mathbf{p}(k)||}\,\frac{\mathbf{p}(k)}{||\mathbf{p}(k)||} = \frac{\mathbf{d}^{T}(k)\,\mathbf{p}(k)}{||\mathbf{p}(k)||^{2}}\,\mathbf{p}(k)\,. \tag{7.7}$$

Combining (7.5) and (7.7) results in

$$\mathbf{d}(k+1) = \mathbf{d}(k) - \alpha\,\mathbf{d}^{\parallel}(k)\,. \tag{7.8}$$

The geometric interpretation of (7.8) is that the update of the distance vector $\mathbf{d}(k+1)$ is achieved by subtracting a part of $\mathbf{d}^{\parallel}(k)$, i.e., the orthogonal projection of distance vector $\mathbf{d}(k)$ onto excitation vector $\mathbf{p}(k)$. Obviously, only the parallel component $\mathbf{d}^{\parallel}(k)$ contributes to the reduction of the length of vector $\mathbf{d}(k)$. Additionally, Fig. 7.2 visualizes that for a reduction the stepsize α has to meet $0 < \alpha < 2$, which represents the stability criterion of the NLMS algorithm. For the noise free condition $(n(k) \equiv 0)$ and a choice of $\alpha = 1$, the parallel component $\mathbf{d}^{\parallel}(k)$ of the distance vector $\mathbf{d}(k)$ can be completely eliminated and the smallest possible length of $|\mathbf{d}(k+1)|$ is obtained.

This interpretation shows that the convergence properties of the adaptation process depend amongst others on the angle between consecutive excitation vectors $\mathbf{p}(k)$ and $\mathbf{p}(k-1)$. This characteristic will now be exploited in the construction of the *optimal* excitation signal.

7.2.2 Optimal Excitation of the NLMS Algorithm

Three main factors determine the convergence performance of the NLMS algorithm: the stepsize α, the filter length N, and the correlation properties of the excitation signal [Widrow, Hoff 1960], [Antweiler 1995]. As indicated by the geometric interpretation of the NLMS algorithm (Sec. 7.2.1), at one time instant k and with $\alpha = 1$

we can completely eliminate the component of distance vector $\mathbf{d}(k)$ in direction of excitation vector $\mathbf{p}(k)$. Consequently, all N components of vector $\mathbf{d}(k)$ can be cancelled, if N consecutive vectors $\mathbf{p}(k)$, $\mathbf{p}(k-1)$, ... , $\mathbf{p}(k-N+1)$ are orthogonal in the N dimensional vector space. In this special case, the adaptive filter with impulse response $\mathbf{h}(k)$ and the system under test $\mathbf{g}$ match exactly after N iterations, i.e., the system is identified.

For a better understanding let us suppose in a first step that $p(k)$ is a white noise process and define the consecutive vectors $\mathbf{p}^*(k)$, $\mathbf{p}^*(k-1)$, ... , $\mathbf{p}^*(k-N+1)$ each of infinite length. These vectors are orthogonal in the infinite vector space. For the adaptation process according to Fig. 7.1, however, we use a vector $\mathbf{p}(k)$ of length N, which represents the projection of $\mathbf{p}^*(k)$ onto the N dimensional vector space. For this projected set of vectors $\mathbf{p}(k)$, $\mathbf{p}(k-1)$, ... , $\mathbf{p}(k-N+1)$ the orthogonality in the N dimensional vector space is not given. Compared with a colored excitation like speech the white noise process represents a quite good – but not optimal – excitation signal of the NLMS driven algorithm.

The key to the *optimal* adaptation of $\mathbf{h}(k)$ bases on the use of N consecutive excitation vectors exactly orthogonal to each other. A special class of pseudo-noise sequences, the so called *perfect sequences* (PSEQs) fulfill this requirement. PSEQs are time discrete, binary, ternary or polyphase sequences of length N. The distinctive attribute, however, is that they show an impulse-like periodic autocorrelation function according to

$$\tilde{\varphi}_{pp}(\lambda) = \sum_{i=0}^{N-1} p(i)\,p(\lambda+i) = \begin{cases} ||\mathbf{p}(\lambda)||^2 & \lambda \bmod N = 0 \\ 0 & \text{otherwise}, \end{cases} \tag{7.9}$$

i.e., $\tilde{\varphi}_{pp}(\lambda)$ vanishes for all out-of-phase values. With this property PSEQs have ideal correlation properties as all N phase-shifted PSEQs are orthogonal in the N dimensional vector space. For each time instant, with the N consecutive phase-shifted vectors $\mathbf{p}(k)$, $\mathbf{p}(k-1)$, ... $\mathbf{p}(k-N+1)$ we obtain a set of orthogonal vectors, i.e., the *optimal* excitation of the NLMS algorithm.

In order to visualize the effect of a PSEQ excitation, the system according to Fig. 7.1 is excited by a spectrally white noise process and a periodically repeated PSEQ. For the validation of the above conclusions, the simulations are performed in a first step under the following constraints:

- the unknown system is linear and time-invariant (LTI)
  (except for a sudden change at one time instant),

- the system is noiseless, i.e., $n(k) \equiv 0$,

- the length of $\mathbf{h}(k)$ is "sufficiently long" with respect to the unknown system $\mathbf{g}$.

Owing to the nature of our approach, the period of the PSEQ has to match the length N of the adaptive filter $\mathbf{h}(k)$. Note, the choice of the adequate PSEQ represents no constraint as PSEQs are available for a large variety of lengths. The optimal stepsize

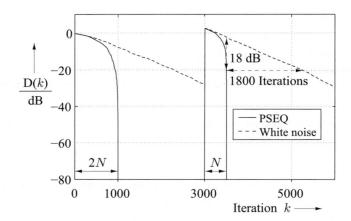

Figure 7.3: System distance for PSEQ and white noise excitation; **g**: time-invariant, except a sudden change at $k = 3000$; $N = 500$, $n(k) \equiv 0$, $\alpha = 1$

for these conditions amounts to $\alpha = 1$. With these assumptions the accuracy of the identification process is limited only by the characteristics of the excitation signal and the available computational precision.

The performance of the system identification is evaluated in terms of the (logarithmic) *system distance*

$$\frac{D(k)}{dB} = 10 \log_{10} \frac{||\mathbf{g} - \mathbf{h}(k)||^2}{||\mathbf{g}||^2} . \tag{7.10}$$

Figure 7.3 shows the results for both stimulus signals. Exploiting PSEQs as stimulus signal for the NLMS takes $2N$ iterations in the initialization phase for an exact identification of the unknown system response **g**. The delay during initialization is caused by the settling time N of the system **g** and by N iterations, which are required to adapt N filter coefficients. In the continuous adaptation process, only N iterations are needed to achieve perfect reconstruction (i.e., $\mathbf{h}(k) = \mathbf{g}$).

The direct comparison with the system distance achieved with a white noise excitation emphasizes how the NLMS benefits from the excitation with deterministic PSEQs. Let us consider, for instance, that we aim at a system distance of $D(k) = -20\,\mathrm{dB}$. At the point where the system distance for a PSEQ excitation meets $-20\,\mathrm{dB}$, the corresponding curve for a white noise stimulation shows only $-1.9\,\mathrm{dB}$. For the white noise excitation the identification process takes 4.6 times longer to reach $-20\,\mathrm{dB}$.

Figure 7.3 thus confirms the expected theoretical results. In practice, however, we have to deal with time-variant systems, physical impulse responses of infinite length, and environmental noise $n(k) \neq 0$. As a result, the generated set of coefficients $\mathbf{h}(k)$ does not normally match exactly the actual impulse response **g**. These aspects relevant for practical applications are presented in the next section.

7.2.3 Influence of Environmental Noise, Stepsize, and Period

The influence of the environmental noise $n(k)$ and the stepsize α on the identification process can be gained from the well-known statistical analysis of the NLMS algorithm. Applying two white noise processes as input signals $n(k)$ and $p(k)$ for the system given in Fig. 7.1 and using the so called *independence assumption* [Haykin 1996], we get for the steady-state system distance

$$\frac{D_\infty}{dB} = 10 \log_{10} \frac{E\{n^2(i)\}}{E\{y^2(i)\}} + 10 \log_{10} \frac{\alpha}{2 - \alpha}. \tag{7.11}$$

It is easy to show that (7.11) holds even if for $p(k)$ a PSEQ is chosen instead of a white noise process, see [Antweiler 1995].

The first term on the right-hand side in (7.11) shows the degradation of the performance in the presence of environmental noise $n(k)$. It states that the achievable steady-state level of identification equals the (logarithmic) power ratio $10 \log_{10}(E\{n^2(i)\}/E\{y^2(i)\})$ of the system response. This dependency can easily be verified via simulation. Figure 7.4-a depicts the performance of the NLMS-type identification algorithm for a PSEQ excitation and various power ratios. Obviously, the achievable steady-state performance D_∞ is limited by the actual power ratio $10 \log_{10}(E\{n^2(i)\}/E\{y^2(i)\})$. As an example Fig. 7.4-a shows additionally the system distance in case of a white noise excitation for a power ratio of -20 dB (dashed line). As expected the steady-state system distance D_∞ is limited accordingly, however, the adaptation process takes longer to reach the -20 dB.

To achieve a higher noise immunity, we make use of a smaller stepsize $0 < \alpha < 1$. This effect is described by the second term of (7.11). The stepsize α determines

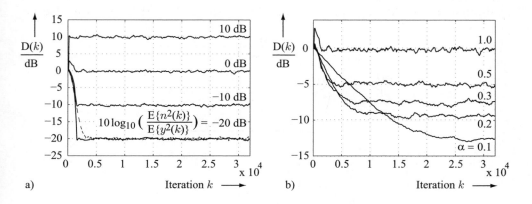

Figure 7.4: System distance for $N = 510$, $n(k)$: white noise, **g**: room impulse response (RIR) and different
a) power ratios $10 \log_{10}(E\{n^2(i)\}/E\{y^2(i)\}) = -20$ dB ... 10 dB, $\alpha = 1$
b) stepsizes $\alpha = 0.1 \ldots 1.0$, $10 \log_{10}(E\{n^2(i)\}/E\{y^2(i)\}) = 0$ dB

the weighting applied to each coefficient update. The choice of a smaller stepsize implies averaging and, thus, reduces the interfering influence of $n(k)$. The effect of the stepsize is visualized in Fig. 7.4-b. The comparison of the curves obtained for different stepsizes α reflects the improvement due to the averaging effect. However, besides the gain of steady-state performance, Fig. 7.4-b also illustrates a decreasing convergence speed.

Note that a similar procedure is applied in the MLS technique to reduce the effects of distortion [Rife, Vanderkooy 1989], [Xiang 1991]. Here, the system output is averaged over a number of MLS periods before the periodic impulse response is computed. In [Antweiler, Dörbecker 1994], [Antweiler 1995] it has been shown that both averaging procedures are equivalent.

Besides the environmental noise and the stepsize, the length of the impulse response $\mathbf{h}(k)$ and the period of the PSEQ also influence the performance of the identification. To distinguish between the length of the impulse response $\mathbf{g}$, the length of the adaptive filter $\mathbf{h}(k)$, and the period of the PSEQ, we introduce the constants N_g, N_h, and N_p, respectively.

Owing to the nature of the approach, the period N_p of the PSEQ has to match the length N_h of the impulse response $\mathbf{h}(k)$, i.e.,

$$N_h = N_p. \tag{7.12}$$

For the choice of a smaller period length ($N_h > N_p$) not all directions of the distance vector $\mathbf{d}(k)$ in the N dimensional vector space can be excited resulting in a limited system distance D_∞. In the other case, with $N_h < N_p$ we obtain similar conditions as with a white noise excitation. Thus, the convergence speed degrades significantly. Figure 7.5 reflects these two effects in terms of the dashed curves. For an actual realization the condition (7.12) can easily be satisfied. As PSEQs are available for

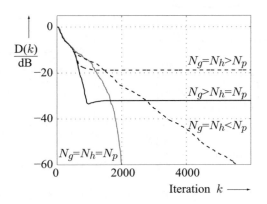

Figure 7.5: System distance for different lengths N_g, N_h, and N_p
$\alpha = 1$, $n(k) \equiv 0$, $\mathbf{g}$: RIR, $N_g = 1022$, N_h and N_p: 510, 1022, 1524

a sufficient number of lengths, the choice of the adequate PSEQ with $N_p = N_h$ represents no constraint. For this reason, we will assume below that $N = N_h = N_p$, if not specified otherwise.

As known from the theory in Sec. 7.2.2, optimal performance is obtained for $N_g = N$ assuming $\alpha = 1$ and $n(k) \equiv 0$ (see also Fig. 7.3 and Fig. 7.5). In most applications, e.g., in acoustics, the systems are of infinite length and the condition $N_g = N$ is difficult to meet. A restriction in the length N inevitably leads to a limitation of the attainable system distance. Figure 7.5 confirms the behavior to be expected from the geometric interpretation. For $\alpha = 1$ and $n(k) \equiv 0$ the identification of the filter coefficients is given by

$$h_{k \bmod N}(k) = \delta(k \bmod N) * \hat{g}_k \,; \quad \text{with } \hat{g}_k = \begin{cases} g_k & k < N_g \\ 0 & k \geq N_g \end{cases}, \ k < 0, \qquad (7.13)$$

where $\delta(k)$ denotes the unit impulse and $*$ the convolution operator. In order to stop the impulse response $\mathbf{h}(k)$ from wrapping back on itself and causing an error known as *time aliasing*, the adaptive filter must be long enough so that the system under test decays to a negligible value.

For the parametrization of our system in a new application we either estimate theoretically the length N_g to be expected or perform initial test measurements with definitely oversized PSEQs. On this basis the length of N is chosen keeping in mind that the choice of N always represents a compromise between convergence speed and time aliasing effects.

7.2.4 Odd-Perfect Sequences

There exist different classes of PSEQs [Lüke 1992]. PSEQs might take two, three, or more amplitudes and differ in their construction method, availability of lengths or energy efficiency

$$\eta = \frac{\sum\limits_{\kappa=0}^{N-1} |p(\kappa)|^2}{N \max\limits_{\kappa} |p(\kappa)|^2} . \qquad (7.14)$$

Hence, they have different effects on the process of system identification. For most applications the sequences should exhibit a high energy efficiency. Therefore, and for ease of implementation, binary sequences are most preferable. Unfortunately, no perfect binary sequence of length $N > 4$ is known. For this reason, we use ternary sequences, most frequently the so called Ipatov sequences [Ipatov 1979] and *odd-perfect* sequences [Lüke, Schotten 1995].

Ipatov sequences are symmetric ternary sequences with a perfect periodic autocorrelation function according to (7.9). A construction is possible for all lengths

$$N = \frac{q^{w \cdot r} + 1}{q^w + 1} \text{ with } q > 2 \text{ prime, } w \in \mathbb{N} \text{ and } r \geq 1, \text{ odd.} \qquad (7.15)$$

Ipatov sequences represent one class of sequences with a generally high energy efficiency.

The orthogonal requirements, however, can also be met with sequences possessing a periodic *odd* autocorrelation function of the form

$$\hat{\varphi}_{pp}(\lambda) = \sum_{i=0}^{N-1} p(i)\, p(\lambda+i) = \begin{cases} \|\mathbf{p}(\lambda)\|^2 & \lambda \bmod 2N = 0 \\ -\|\mathbf{p}(\lambda)\|^2 & \lambda \bmod 2N = N \\ 0 & \text{otherwise}. \end{cases} \tag{7.16}$$

In [Lüke, Schotten 1995] a construction method of such sequences is presented. These so called *odd-perfect* sequences are symmetrical, quasi-binary sequences, which, except for a (leading) zero, only take two amplitudes $p(\kappa) \in \{+a, -a\}$, $\kappa = 1, 2, \ldots N-1$. The odd-perfect sequence of length $N = 6$, e.g., is

$$p(\kappa) = \{0,\, a,\, a,\, a,\, -a,\, a\}. \tag{7.17}$$

For the system identification approach the odd-perfect sequence is periodically applied in an odd-cyclic manner, i.e., the sign is alternated in each period. The example of (7.17) would lead to the periodic excitation signal

$$p(k) = \{0,\, a,\, a,\, a,\, -a,\, a,\, 0,\, -a,\, -a,\, -a,\, +a,\, -a,\, 0,\, a,\, a,\, \ldots\}. \tag{7.18}$$

According to Sec. 7.2.2, also for odd-perfect sequences *optimal* identification of the unknown LTI system is achieved, as the periodic *odd* autocorrelation function $\hat{\varphi}_{pp}(\lambda)$ vanishes, like the periodic autocorrelation function $\tilde{\varphi}_{pp}(\lambda)$ of PSEQs, for all out-of-phase values $\lambda = 1, ..., N-1$.

As the period length must be adapted to the length of the adaptive filter, it is of particular advantage that odd-perfect sequences can be generated for every length

$$N = q^w + 1 \text{ with } q > 2 \text{ prime}, \ w \in \mathbb{N}. \tag{7.19}$$

In Fig. 7.6 the lengths of all possible Ipatov and odd-perfect sequences up to 1000 are depicted. In this range in total 185 odd-perfect sequences can be constructed, whereas only 15 Ipatov sequences exist.

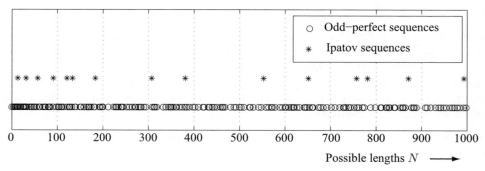

Figure 7.6: Lengths of all possible Ipatov and odd-perfect sequences, $N < 1000$

As odd-perfect sequences contain for all possible lengths only one single zero, the resulting energy efficiency

$$\eta = \frac{N-1}{N} \qquad\qquad (7.20)$$

is even higher than for Ipatov sequences. One example of the impact of the energy efficiency is given in [Antweiler, Antweiler 1995]. Simply the use of an odd-perfect sequence ($N = 122$) instead of an Ipatov sequence ($N = 121$) already results in an improvement of $\Delta \mathrm{D}_\infty = 1.7\,\mathrm{dB}$ ($\alpha = 1$, $n(k)$: white noise).

Owing to the theoretic analogy of perfect and odd-perfect sequences, in this application we will not distinguish between the two classes of sequences. All the following discussions hold for perfect and odd-perfect sequences.

7.2.5 Tracking of Time-Variant Systems

So far the system identification approach with PSEQ excitation was examined under the assumption that the impulse response under study $\mathbf{g}$ is not a function of time. In practice, however, we normally have to deal with time-variant systems $\mathbf{g}(k)$. Reasons for time variations are, e.g., persons in a room, fading in mobile radio channels, or slow heating of a loudspeaker voice coil.

In [van de Kerkhof, Kitzen 1992] the tracking properties of an NLMS-type adaptive filter in the application of modelling an unknown time-variant system were studied. It was shown that the NLMS driven algorithm with a white noise excitation is unable to track time-variant changes fast enough.

In contrast to these investigations, our proposal benefits from the use of the *optimal* excitation signal, i.e., the PSEQ. Owing to the special correlation properties of PSEQs, the NLMS algorithm is capable of identifying an unknown impulse response within N iterations and to keep track of changes in this scale. Furthermore, the most state-of-the-art measuring systems employ typically blockwise oriented methods, e.g., [Alrutz, Schroeder 1983], i.e., a completely new set of filter coefficients is obtained only every N iterations. The NLMS driven approach, however, operates iteratively. In each time instant k a new set of coefficients is available, i.e., $h_i(k) \neq h_i(k+1)$, $\forall i \in \{0, 1, \ldots N-1\}$.

To visualize the tracking properties of the proposed identification algorithm, a synthetic time varying model $\mathbf{g}(k)$ is considered, i.e., $g_i(k) \neq g_i(k+1)$, $\forall i \in \{0, 1, \ldots N_g - 1\}$. In Fig. 7.7 the time fluctuation of one coefficient $g_i(k)$ is depicted. The results of two identification processes are shown: one with a PSEQ excitation and one with a white noise excitation signal. The three curves show that the PSEQ results in a significantly improved tracking.

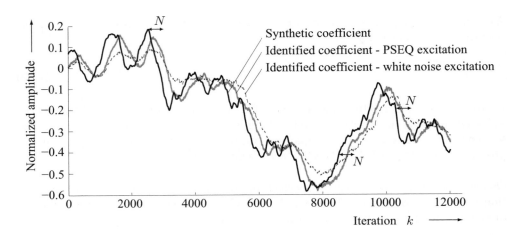

Figure 7.7: Tracking of a time-variant system – display of one exemplary coefficient
($\mathbf{g}(k)$: time-variant, $N_g = N = 500$, $\alpha = 1$, $n(k) \equiv 0$)

As expected the identification with a PSEQ cannot track the synthetic coefficient instantaneously. Obviously, the identified coefficient lags behind to a certain degree ($< N$), but the tendency of the curves does match very well. The comparison with the results obtained for the white noise excitation shows the benefits of the PSEQ.

As a result of its tracking properties, the identification algorithm with PSEQs provides in every single iteration step a close approximation of the actual unknown system. Owing to this capability we use the NLMS-type identification process even to track the changes of slowly varying systems, e.g., to measure time-variant transmission links in medical applications or to reproducibly simulate the fluctuations of RIRs, see Sec. 7.4 for more details.

7.2.6 Complexity

The system identification process is based on the well-known NLMS algorithm. Normally, in each time instant N multiply and add operations are required for filtering and coefficient update, respectively. In order to store the filter states and coefficients, $2N$ storage locations are needed.

However, due to the use of ternary sequences of the form $p(\kappa) \in \{0, +a, -a\}$, $\kappa = 0, 1, \ldots N-1$, the multiply operations for the coefficient update as well as the storage locations for the filter states, i.e., the PSEQ, can be significantly reduced. Thus, in the case of an odd-perfect sequence excitation the NLMS adaptation according to (7.1) simplifies to

$$h_i(k+1) = \begin{cases} h_i(k) & i \bmod N = k \\ h_i(k) \pm \beta \cdot e(k) & \text{otherwise}, \end{cases} \qquad (7.21)$$

where the factor

$$\beta = \frac{\alpha}{(N-1) \cdot a} \tag{7.22}$$

is a constant for all coefficients i and all time-instants k. The multiplication $\beta \cdot e(k)$ has only to be performed once for each time-instant. Therefore, the NLMS-type identification approach is an extremely efficient tracking algorithm.

So far we focused on a system identification algorithm for one single channel. Besides the theoretical background of the approach we discussed its properties and performance. Simulations visualized the various effects. In the following section we will show how to adapt this concept to a multi-channel system in the sense of a *multiple input – single output* (MISO) system.

7.3 Multi-Channel System Identification

As multi-channel system identification is of interest for many applications, in this section we generalize the basic concept of the NLMS-type identification approach to the multi-channel case.

The typical problem of multi-channel system identification is the non-zero cross-correlation between the excitation signals. The adaptive filters either do not converge on the *true* impulse responses or converge extremely slowly. As we have only one single measured reference signal for all observed channels we need an adequate excitation signal to identify all parallel channels simultaneously. In this section we will present the idea of how a class of *optimal* excitation signals can be constructed from PSEQs. First, we will introduce the main idea of the approach and derive an algorithm for the dual-channel case and, in a second step, we will expand it to multi-channel systems.

7.3.1 The Dual-Channel Case

For the dual-channel case we consider the system according to Fig. 7.8. The task is to uniquely identify the *true* impulse responses $\mathbf{g}^{(1)}$ and $\mathbf{g}^{(2)}$ by adapting the digital filters $\mathbf{h}^{(1)}(k)$ and $\mathbf{h}^{(2)}(k)$, given only the error signal $e(k)$.

The dual-channel NLMS algorithm [Benesty et al. 1995] is given by

$$\mathbf{h}^{(1)}(k+1) = \mathbf{h}^{(1)}(k) + \frac{\alpha \, e(k) \, \mathbf{p}^{(1)}(k)}{||\mathbf{p}^{(1)}(k)||^2 + ||\mathbf{p}^{(2)}(k)||^2} \tag{7.23}$$

$$\mathbf{h}^{(2)}(k+1) = \mathbf{h}^{(2)}(k) + \frac{\alpha \, e(k) \, \mathbf{p}^{(2)}(k)}{||\mathbf{p}^{(1)}(k)||^2 + ||\mathbf{p}^{(2)}(k)||^2} \tag{7.24}$$

with all vectors of length N. The two input signals are denoted by $p^{(1)}(k)$ and $p^{(2)}(k)$ and have still to be defined. The indices $^{(1)}$ and $^{(2)}$ denote the signals

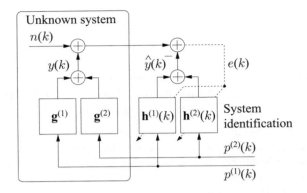

Figure 7.8: Dual-channel system identification with PSEQs

and systems of the first and second channel, respectively. The error signal results in

$$e(k) = \left(\mathbf{g}^{(1)} - \mathbf{h}^{(1)}(k)\right)^T \cdot \mathbf{p}^{(1)}(k) + \left(\mathbf{g}^{(2)} - \mathbf{h}^{(2)}(k)\right)^T \cdot \mathbf{p}^{(2)}(k) + n(k). \quad (7.25)$$

In order to obtain the *optimal* excitation signals $p^{(1)}(k)$ and $p^{(2)}(k)$ for the dual-channel NLMS algorithm, we introduce the PSEQ $\check{p}(k)$ with period length $2N$. For the first channel we periodically apply $\check{p}(k)$ as an excitation signal. For the second channel an N-shifted version of the same periodic excitation signal is applied according to

$$p^{(1)}(k) = \check{p}(k) \quad (7.26)$$
$$p^{(2)}(k) = \check{p}(k - N). \quad (7.27)$$

Thus, the period length $(2N)$ of the PSEQ and the length of the adaptive filters (N) do not match. Below the ' $\check{\ }$ ' sign will indicate that the marked signal or vector is based on a PSEQ of period length $2N$.

The equations for the NLMS recursion according to (7.23) and (7.24) change to

$$\mathbf{h}^{(1)}(k + 1) = \mathbf{h}^{(1)}(k) + \frac{\alpha\, e(k)\, \check{\mathbf{p}}(k)}{||\check{\mathbf{p}}(k)||^2 + ||\check{\mathbf{p}}(k - N)||^2} \quad (7.28)$$

$$\mathbf{h}^{(2)}(k + 1) = \mathbf{h}^{(2)}(k) + \frac{\alpha\, e(k)\, \check{\mathbf{p}}(k - N)}{||\check{\mathbf{p}}(k)||^2 + ||\check{\mathbf{p}}(k - N)||^2}. \quad (7.29)$$

Note that all vectors are still of dimension N. Thus, the excitation vectors $\check{\mathbf{p}}(k)$ and $\check{\mathbf{p}}(k - N)$ contain only half of the underlying PSEQ, i.e.,

$$\check{\mathbf{p}}(k) = \left(\check{p}(k),\, \check{p}(k - 1),\, \ldots,\, \check{p}(k - N + 1)\right)^T \quad (7.30)$$

$$\check{\mathbf{p}}(k - N) = \left(\check{p}(k - N),\, \check{p}(k - N - 1),\, \ldots,\, \check{p}(k - 2N + 1)\right)^T. \quad (7.31)$$

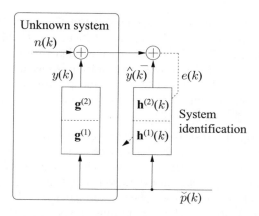

Figure 7.9: Equivalent dual-channel system – serial structure

In order to prove that $p^{(1)}(k)$ and $p^{(2)}(k)$ as defined in (7.26) and (7.27) fulfill the requirements of *optimal* excitation signals for a dual-channel system, we introduce an equivalent model in serial structure. For this reason, we define new combined vectors with

$$\mathbf{g}_{|2N} = \left(\mathbf{g}^{(1)}, \mathbf{g}^{(2)}\right)^{T} \tag{7.32}$$

$$\mathbf{h}(k)_{|2N} = \left(\mathbf{h}^{(1)}(k), \mathbf{h}^{(2)}(k)\right)^{T} \tag{7.33}$$

$$\mathbf{\check{p}}(k)_{|2N} = (\mathbf{\check{p}}(k), \mathbf{\check{p}}(k - N))^{T} \tag{7.34}$$

each of length $2N$. Thus, $\mathbf{\check{p}}(k)_{|2N}$ includes the complete PSEQ of period length $2N$. The system reactions $y(k)$ and $\hat{y}(k)$ can be rewritten as

$$\begin{aligned}
y(k) &= \mathbf{g}^{(1)^{T}} \cdot \mathbf{p}^{(1)}(k) + \mathbf{g}^{(2)^{T}} \cdot \mathbf{p}^{(2)}(k) \\
&= \left(\mathbf{g}^{(1)^{T}}, \mathbf{g}^{(2)^{T}}\right) \cdot \left(\begin{array}{c} \mathbf{p}^{(1)}(k) \\ \mathbf{p}^{(2)}(k) \end{array}\right) \\
&= \mathbf{g}_{|2N}^{T} \cdot \mathbf{\check{p}}(k)_{|2N} \tag{7.35}
\end{aligned}$$

$$\hat{y}(k) = \mathbf{h}^{T}(k)_{|2N} \cdot \mathbf{\check{p}}(k)_{|2N}. \tag{7.36}$$

Exploiting (7.32)–(7.36), we transform the parallel filter structure of Fig. 7.8 into a serial structure according to Fig. 7.9, assuming that the transmission systems $\mathbf{g}^{(1)}$ and $\mathbf{g}^{(2)}$ can be modelled by a direct form FIR filter. In this reorganized system the identification process is defined by

$$\mathbf{h}(k + 1)_{|2N} = \mathbf{h}(k)_{|2N} + \frac{\alpha\, e(k)\, \mathbf{\check{p}}(k)_{|2N}}{||\mathbf{\check{p}}(k)_{|2N}||^{2}} \tag{7.37}$$

$$e(k) = \left(\mathbf{g}_{|2N} - \mathbf{h}(k)_{|2N}\right)^{T} \cdot \mathbf{\check{p}}(k)_{|2N} + n(k). \tag{7.38}$$

Except for the initialization phase, the systems of Fig. 7.8 and Fig. 7.9 are equivalent due to the shift between the two input signals according to (7.26) and (7.27). While in the system given in Fig. 7.8, N iterations are needed to fill all filter states, in the serialized system (Fig. 7.9), $2N$ iterations are required. As we use the serialized system only to prove whether the shifted PSEQ excitation permits optimal performance or not, the difference during the initialization is irrelevant.

With the reorganization of the system we reduced the dual-channel case to the known single-channel problem with all dimensions twice as long. As this problem has been solved in Sec. 7.2, we can also conclude that with a choice of $p^{(1)}(k)$ and $p^{(2)}(k)$ according to (7.26) and (7.27) *optimal* excitation signals for the dual-channel case can easily be generated. Thus, with these special excitation signals the NLMS algorithm is capable of uniquely identifying the *true* impulse responses $\mathbf{g}^{(1)}$ and $\mathbf{g}^{(2)}$ within one period $(2N)$.

7.3.2 Simulation Results

In order to verify the above conclusions, we investigate different excitation strategies for the dual-channel system of Fig. 7.8. Considering the assumptions according to Sec. 7.2 we compare the results for a PSEQ and a white noise excitation. Figure 7.10

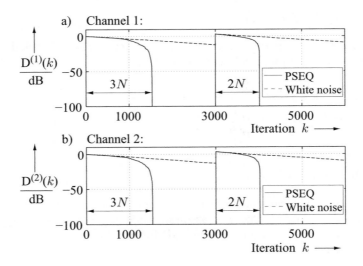

Figure 7.10: System distance of both channels for PSEQ and white noise excitation; $\mathbf{g}^{(1)}, \mathbf{g}^{(2)}$: time-invariant, except sudden change at $k = 3000$; $N_g = N = 511$, $n(k) \equiv 0$, $\alpha = 1$

illustrates the system distances

$$\frac{D^{(1)}(k)}{dB} = 10 \log_{10} \frac{||\mathbf{g}^{(1)} - \mathbf{h}^{(1)}(k)||^2}{||\mathbf{g}^{(1)}||^2} \quad \text{and} \tag{7.39}$$

$$\frac{D^{(2)}(k)}{dB} = 10 \log_{10} \frac{||\mathbf{g}^{(2)} - \mathbf{h}^{(2)}(k)||^2}{||\mathbf{g}^{(2)}||^2}, \tag{7.40}$$

i.e., two objective measures to evaluate the quality of the system identification for each channel. Obviously, the algorithm is capable of perfectly separating and identifying the *true* impulse responses of both channels within computational precision, which is due to the ideal autocorrelation function of $p^{(1)}(k)$ and $p^{(2)}(k)$, and the zero cross-correlation between the two signals. The only deficit in comparison with Fig. 7.3 is the deceleration of convergence speed due to the need for longer PSEQ periods.

The comparison with the results obtained for white noise reflects the benefits of the PSEQs. Note that for white noise the slope of the corresponding system distances in Fig. 7.10-a,b are less steep than in Fig. 7.3, as the short-term cross-correlation between the white noise signals of both channels is not zero.

Finally, we will have a look on the influence of the environmental noise and the averaging effect of a smaller stepsize $0 < \alpha < 1$ for the dual-channel system. Following Sec. 7.2.3, we perform computer simulations for different levels of environmental noise $n(k)$ and different stepsize parameters α. The results are reproduced in Fig. 7.11. Apparently, the curves of the two system distances $D^{(1)}(k)$ and $D^{(2)}(k)$ are close to

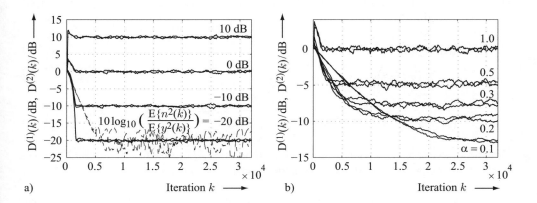

Figure 7.11: System distance $D^{(1)}(k)$ and $D^{(2)}(k)$ for $N_g = N = 511$,
$\quad n(k)$: white noise, $\mathbf{g}$: RIR and different
$\quad$ a) power ratios $10 \log_{10}(E\{n^2(i)\}/E\{y^2(i)\}) = -20 \, dB \dots 10 \, dB$, $\alpha = 1$
$\quad$ b) stepsizes $\alpha = 0.1 \dots 1.0$, $10 \log_{10}(E\{n^2(i)\}/E\{y^2(i)\}) = 0 \, dB$

each other for all conditions. For the computer simulations in Fig. 7.4 and Fig. 7.11 we chose $N = 510$ and 511, i.e., adaptive filters with comparable lengths. Note, however, that the periods of the applied PSEQs differ with $N_p = 510$ and $N_p = 1022$ for the single- and the dual-channel case, respectively. The comparison of the corresponding curves demonstrates the same steady-state performance for the single- and the dual-channel case according to (7.11). However, between the single- and dual-channel case differences might occur during the settling phase that are caused by the need for different PSEQs lengths N_p.

As before in Fig. 7.4 we illustrate in Fig. 7.11-a the steady-state performance for a white noise excitation for a power ratio of $-20\,\mathrm{dB}$ (dashed line). In principle we obtain similar results as in the single-channel case. However, noticeable are the strong fluctuations of $\mathrm{D}^{(1)}(k)$ and $\mathrm{D}^{(2)}(k)$ around the $-20\,\mathrm{dB}$ line, which are caused by the non-perfect short-term cross-correlation between the white noise signals of the two channels.

In the next section, the dual-channel system identification approach is extended to an arbitrary number of channels.

7.3.3 Generalization to the Multi-Channel Case

The generalization to multiple channels is performed along the principles presented in Sec. 7.3.1. In the case of ν channels a PSEQ with period length νN is chosen and submitted to the first channel. All other channels are excited with phase-shifted versions according to

$$p^{(1)}(k) = \check{p}(k) \tag{7.41}$$

$$p^{(2)}(k) = \check{p}(k - N) \tag{7.42}$$

$$\vdots$$

$$p^{(\nu)}(k) = \check{p}(k - (\nu - 1)N). \tag{7.43}$$

This set of excitation signals generated out of one PSEQ represents the *optimal* excitation for the ν channels, as all considerations of Sec. 7.3.1 also hold for $\nu \geq 2$.

So far, we focused on the identification of *multiple input – single output* (MISO) systems. It should be emphasized that the approach allows the identification of various kinds of linear systems such as radio or acoustic channels. Furthermore, the concept can easily be extended to *multiple input – multiple output* (MIMO) systems. Note that the introduction of further microphones only increases the computational complexity, but does not introduce any further fundamental problems.

7.4 Applications

The technology of the identification/measurement of one (or more) linear system(s) by means of an adaptive filter has a wide range of possible applications. Various uses for such measuring systems can be found in room acoustics. In the investigation of existing auditoria, for instance, concert halls, theaters, a system for measuring *room impulse responses* (RIRs) have a practical use [Xiang 1992]. Another application lies in the room-acoustic model technique [Allen, Berkley 1979] for predicting acoustic qualities (e.g., the room reverberation) of a planned acoustic space. Moreover, in present-day audio applications (that is, virtual reality, auralization, spatialization of sounds) the importance of measuring binaural RIRs with a very high signal-to-noise ratio becomes more and more evident [Xiang 1991]. Other possible applications are loudspeaker testing, speech reverberation cancellation, measurement systems for hearing-aid characterization [Schneider, Jamieson 1993], and the realization of *virtual musical instruments* [Farina et al. 1995]. In the context of mobile communications, correlation-based identification algorithms are used for radio channel estimation and fast start-up equalization in synchronous digital communication systems [Milewski 1983], [Molina, Fannin 1993], [Chen et al. 1995].

Owing to its simplicity and generality, the NLMS-based approach with PSEQ excitation has also been used routinely for a couple of years beyond its "classical" application, i.e., the measurement of acoustic transfer functions. As a result of its convergence speed, we use the approach, for instance, to track the fluctuations of time-variant impulse responses. Consecutive impulse responses can be used to simulate reproducibly a real, time-variant transmission link. In [Antweiler, Symanzik 1995] we showed how to simulate time-variant RIRs for the research and design of acoustic echo cancellation algorithms. This concept can now be generalized to, e.g., stereophonic echo cancellation. In a medical application we used the measured time-variant impulse responses to investigate the dynamic behavior of the Eustachian tube function [Antweiler et al. 2006b], [Antweiler et al. 2006a]. With a measurement prototype we can visualize the transmission link between the nose and the ear as a function of time – a new method, which opens up entirely new possibilities in otological diagnostics. These two examples will be discussed in the following sections.

7.4.1 Simulation of Time-Variant RIRs for Stereophonic Echo Control

For applications such as acoustic echo compensation, adaptive noise reduction and acoustic feedback control it is very interesting to reproducibly simulate a real, time-variant RIR and simultaneously allow an objective measurement of the obtained results. For this reason, the principle of the NLMS-based identification with PSEQs has been used for the simulation of time-variant RIRs in the context of acoustic echo cancellation [Antweiler, Symanzik 1995]. This idea can now be extended to a new dual-channel simulation concept for stereophonic echo cancellation, as depicted in Fig. 7.12.

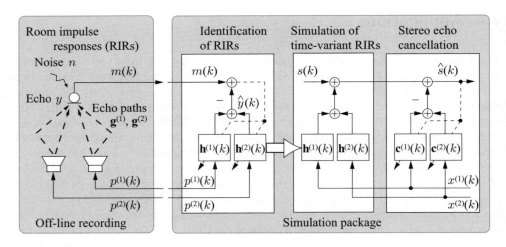

Figure 7.12: Simulation of time-variant RIRs for stereophonic echo cancellation

The complete set-up can be grouped into four blocks.

Off-line recording
The real unknown system, e.g., an office room or a car interior, is excited via loud-speakers with the shifted PSEQs $p^{(1)}(k)$ and $p^{(2)}(k)$. Its reaction $m(k)$ is synchronously recorded and stored on hard disk. These files serve as input signals for the identification process within the simulation package.

Identification of RIRs
The identification algorithm provides in every time instant two sets of coefficients $\mathbf{h}^{(1)}(k)$ and $\mathbf{h}^{(2)}(k)$. As a result of its convergence speed both sets represent a close approximation of the actual acoustic RIRs $\mathbf{g}^{(1)}$ and $\mathbf{g}^{(2)}$. During the run time of the simulation the instantaneous sets of coefficients are transferred to the simulation unit for the time-variant RIRs. The on-line identification can be performed without the necessity of storing large amounts of data: by storing only the reaction of the unknown dual-channel system instead of all sets of filter coefficients for all time-instants k.

Simulation of time-variant RIRs
The convolution of the stereo signals from the far-end $x^{(1)}(k)$ and $x^{(2)}(k)$ with $\mathbf{h}^{(1)}(k)$ and $\mathbf{h}^{(2)}(k)$, respectively, and their addition to the near-end signal $s(k)$ provides a close approximation of an actual room scenario, even taking its time-variance into account.

Stereo echo cancellation
This set-up can now be used for the design and optimization of a stereophonic echo cancellation concept under test. Besides the reproducibility of computer simulations,

objective measures, such as the system distance

$$\frac{\tilde{D}^{(1)}(k)}{dB} = 10 \log_{10} \frac{||\mathbf{h}^{(1)}(k) - \mathbf{c}^{(1)}(k)||^2}{||\mathbf{h}^{(1)}(k)||^2} \quad \text{and} \tag{7.44}$$

$$\frac{\tilde{D}^{(2)}(k)}{dB} = 10 \log_{10} \frac{||\mathbf{h}^{(2)}(k) - \mathbf{c}^{(2)}(k)||^2}{||\mathbf{h}^{(2)}(k)||^2} \tag{7.45}$$

can be evaluated for time-variant conditions in order to enable an objective assessment.

In practice, for time-variant echo paths and $n(k) \neq 0$ the generated sets of coefficients do not exactly match the actual RIRs, i.e., $\mathbf{h}^{(1)}(k) \approx \mathbf{g}^{(1)}$, $\mathbf{h}^{(2)}(k) \approx \mathbf{g}^{(2)}$. However, the crucial point is that the convergence properties of the left and the right block of the simulation set-up given in Fig. 7.12 differ significantly. Owing to the PSEQ excitation the identification process (left block) is capable of efficiently tracking the fluctuations of a time-variant system within much smaller time constants than the echo canceler (right block). In other words, the echo canceler works so slowly that from its point of view $\mathbf{h}^{(1)}(k)$ and $\mathbf{h}^{(2)}(k)$ represent time-variant RIRs that are close to reality.

Figure 7.13 gives a simulation example for different extents of time variance. The variations resulted from a person moving in the acoustic space. For simplicity, the identification was performed only for one channel ($x^{(2)}(k) \equiv 0$), as here the focus lies on the impact of the time-variant RIR. These simulations clearly indicate the degradation of the echo canceler's performance under time-variant conditions.

In order to simulate a full-duplex stereophonic teleconference system another microphone and the corresponding signal processing blocks have to be added. It should be noted that the introduction of a second microphone doubles the complexity but does not introduce any new or unsolved problem.

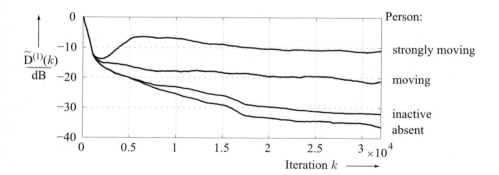

Figure 7.13: System distance for different extents of time variance. © 1995 IEEE
Identification: $N = 2801$, $n(k) \neq 0$, $x^{(2)}(k) \equiv 0$, $\alpha = 1$
Simulation: $N = 553$, $s(k) \equiv 0$, $x^{(2)}(k) \equiv 0$, $\alpha(k)$ adaptive

7.4.2 Acoustic Tube Endoscopy

In [Antweiler et al. 2006b], [Antweiler et al. 2006a] we introduced a new concept for the sonotubometric assessment of the Eustachian tube function – in particular its dynamic behavior.

In a novel real-time acoustic measurement prototype for otological diagnostics the Eustachian tube is treated as a linear transmission system (see Fig. 7.14). We apply a PSEQ in the nasal cavity and record simultaneously the reaction of the nose/ear system using a microphone located in the ear. Its impulse response is obtained by a subsequent NLMS-type system identification.

With digital signal processing algorithms we extract two different features according to Fig. 7.14:

1. The fluctuations of the sound level intensity in the outer ear indicate activity of the Eustachian tube provoked by, e.g., yawning or swallowing. They are mapped with the quadratic norm of the impulse response.

2. Based on techniques known from speech processing such as the acoustic tube model and the Levinson–Durbin algorithm [Vary, Martin 2006], a novel virtual model of the nose/ear transmission link can be built. The dynamic opening and closing process of the tube is visualized by an animation of the virtual tube model over time. By means of this model, a virtual *acoustic tube endoscopy* can be carried out.

As a result, the acoustic measurement system allows real-time monitoring of the Eustachian tube activity under physiological conditions. New insights into the dynamics of the Eustachian tube function can be gained. Future work will aim at a dual-channel measurement prototype to visualize the transmission links between the two nostrils and the ear as a function of time.

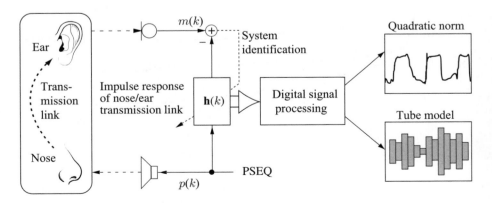

Figure 7.14: Acoustic measurement system for otological diagnostics

7.5 Discussion and Conclusions

High quality videoconferencing systems or multi-media applications increasingly require multi-channel signal processing. Knowledge about the nature of these channels is needed for the research, design, and development of relevant algorithms.

As part of this work we introduced a new concept for multi-channel system identification. The approach relies on the NLMS algorithm excited by a set of *optimal* excitation signals, each signal featuring a perfect autocorrelation and a zero cross-correlation to any other signal out of the set. We introduced a simple technique to generate such sets of *optimal* excitation signals in terms of shifted PSEQs. It is of importance that PSEQs are available for a sufficient variety of lengths. Applying a set of shifted PSEQs to the inputs of a MISO system opens up the possibility of identifying a linear, time-invariant, and noiseless multi-channel system within computational precision. In other words, with one simultaneous measurement of an arbitrary number of channels the approach allows the identification and separation of all unknown systems in parallel. Furthermore, the concept can easily be extended to MIMO systems, as the consideration of further output reference signals, and thus further channels in parallel, would only increase the effort, but does not introduce additional algorithmic problems.

In practice, the advantageous properties of the presented approach permit broadband measurements with an excellent noise tolerance for systems that are reasonably linear and time-invariant. Further notable features are the adjustable sampling frequency, adjustable length of the adaptive filter, ease of implementation, and low computational complexity.

Beyond the possibility to apply the novel approach as a measurement technique for unknown multi-channel systems, it can be used – due to its generality and simplicity – for many other applications. As a result of its convergence speed, we can even use the identification algorithm to track the fluctuations of time-variant impulse responses. Therefore, consecutive sets of coefficients can be used to simulate reproducibly real, time-variant transmission links, e.g., RIRs, for the design of stereophonic echo cancellation algorithms. In a medical application, we use the novel approach to investigate the dynamic behavior of the Eustachian tube function. With a measurement prototype we visualize the transmission link between the nose and the ear as a function of time. Furthermore, the new identification approach opens up the possibility of extending the acoustic tube endoscopy to a dual-channel system.

Bibliography

Allen, J. B.; Berkley, D. A. (1979). Image Method for Efficiently Simulating Small-Room Acoustics, *Journal of the Acoustical Society of America*, vol. 65, no. 4, pp. 943–950.

Alrutz, H. (1983). *Über die Anwendung von Pseudorauschfolgen zur Messung an linearen Übertragungssystemen*, PhD thesis, Georg-August-Universität Göttingen.

Alrutz, H.; Schroeder, M. (1983). A Fast Hadamard Transform Method for the Evaluation of Measurements Using Pseudo-Random Test Signals, *Proceedings of the 11th International Conference on Acoustics (ICA)*, Paris, vol. VI, pp. 235–238.

Antweiler, C. (1995). *Orthogonalisierende Algorithmen für die digitale Kompensation akustischer Echos*, PhD thesis. Aachener Beiträge zu digitalen Nachrichtensystemen, vol. 1, P. Vary (ed.), RWTH Aachen University.

Antweiler, C.; Antweiler, M. (1995). System Identification with Perfect Sequences Based on the NLMS Algorithm, *International Journal of Electronics and Communications (AEÜ)*, vol. 3, pp. 129–134.

Antweiler, C.; Dörbecker, M. (1994). Perfect Sequence Excitation of the NLMS Algorithm and its Application to Acoustic Echo Control, *Annales des Telecommunications*, vol. 49, no. 7–8, pp. 386–397.

Antweiler, C.; Symanzik, H.-G. (1995). Simulation of Time Variant Room Impulse Responses, *Proceedings of the IEEE International Conference on Acoustics, Speech, and Signal Processing (ICASSP)*, Detroit, USA, pp. 3031–3034.

Antweiler, C.; Telle, A.; Vary, P.; Di Martino, E. (2006a). A New Otological Diagnostic System Providing a Virtual Tube Model, *IEEE Biomedical Circuits and Systems Conference (BIOCAS)*, London.

Antweiler, C.; Telle, A.; Vary, P.; Di Martino, E. (2006b). Virtual Time-Variant Model of the Eustachian Tube, *International Symposium on Circuits and Systems (ISCAS)*, Island of Kos, Greece, pp. 5559–5562.

Aoshima, N. (1981). Computer-Generated Pulse Signal Applied for Sound Measurement, *Journal of the Acoustical Society of America*, vol. 69, no. 5, pp. 1484–1488.

Benesty, J.; Amand, F.; Gilloire, A.; Grenier, Y. (1995). Adaptive Filtering Algorithms for Stereophonic Acoustic Echo Cancellation, *Proceedings of the IEEE International Conference on Acoustics, Speech, and Signal Processing (ICASSP)*, Detroit, USA, pp. 3099–3102.

Biering, H.; Pedersen, O. Z. (1983). System Analysis and Time Delay Spectrometry (Part I/Part II), *Brüel & Kjær Technical Review*, vol. 1/2.

Borish, J. (1985). Self-Contained Crosscorrelation Program for Maximum-Length Sequences, *Journal of the Audio Engineering Society*, vol. 33, no. 11, pp. 4–21.

Borish, J.; Angell, J. B. (1985). An Efficient Algorithm for Measuring the Impulse Response Using Pseudorandom Noise, *Journal of the Audio Engineering Society*, vol. 31, pp. 478–488.

Chen, X.; Suzuki, M.; Miki, N.; Nagai, N. (1995). Simultaneous Estimation of Echo Path and Channel Responses Using Full-Duplex Transmitted Training Data Sequences, *IEEE Transactions on Information Theory*, vol. 41, no. 5, pp. 1409–1417.

Claasen, T.; Mecklenbräuker, W. (1981). Comparison of the Convergence of Two Algorithms for Adaptive FIR Digital Filters, *IEEE Transactions on Acoustics, Speech, and Signal Processing (ASSP)*, vol. 3, pp. 670–678.

Cohn, M.; Lempel, A. (1977). On Fast M-Sequence Transforms, *IEEE Transactions on Information Theory*, vol. IT-23, pp. 135–137.

Dunn, C.; Hawksford, M. (1993). Distortion Immunity of MLS-Derived Impulse Response Measurements, *Journal of the Audio Engineering Society*, vol. 41, no. 5, pp. 314–335.

Farina, A. (2000). Simultaneous Measurement of Impulse Response and Distortion with a Swept-Sine Technique, *108 AES Convention*, Paris.

Farina, A.; Langhoff, A.; Tronchin, L. (1995). Realisation of 'Virtual' Musical Instruments: Measurements of the Impulse Response of Violins using MLS Technique, *2nd International Conference on Acoustics and Musical Research (CIARM)*, Ferrara.

Golomb, S. W. (1982). *Shift Register Sequences*, Aegean Park Press, Laguna Hills, CA.

Hänsler, E. (1992). The Hands-Free Telephone Problem – An Annotated Bibliography, *Signal Processing*, vol. 27, pp. 259–271.

Hänsler, E.; Schmidt, G. (2004). *Acoustic Echo and Noise Control: A Practical Approach*, John Wiley & Sons, Ltd., New York.

Haykin, S. (1996). *Adaptive Filter Theory*, 3rd edn, Prentice-Hall, Upper Saddle River, New Jersey 07458.

Herlufsen, H. (1984). Dual Channel FFT Analysis (Part I/Part II), *Brüel & Kjær Technical Review*, vol. 1/2.

Heyser, R. C. (1967). Acoustical Measurements by Time Delay Spectrometry, *Journal of the Audio Engineering Society*, vol. 15, pp. 370–382.

Ipatov, V. (1979). Ternary Sequences with Ideal Periodic Autocorrelation Properties, *Radio Engineering Electronics and Physics*, vol. 24, pp. 75–79.

Lüke, H.-D. (1992). *Korrelationssignale*, Springer-Verlag, Berlin.

Lüke, H.; Schotten, H. (1995). Odd-Perfect, Almost Binary Correlation Sequences, *Transactions on Aerospace and Electronic Systems*, vol. 31, pp. 495–498.

Mateljan, I.; Ugrinović, K. (2001). The Fourier Analyzer for Acoustical Measurement, *Proceedings ELMAR*, Zadar, Croatia.

Mateljan, I.; Ugrinović, K. (2003). The Comparison of Room Impulse Response Measuring Systens, *Proceedings of AAAA Congress*, Portoroz.

Milewski, A. (1983). Periodic Sequences with Optimal Properties for Channel Estimation and Fast Start-Up Equalization, *IBM Journal of Research and Development*, vol. 27, no. 5, pp. 426–431.

Molina, A.; Fannin, P. C. (1993). Application of Mismatched Filter Theory to Bandpass Impulse Response Measurements, *Electronic Letters*, vol. 29, no. 2, pp. 162–163.

Müller, S.; Massarani, P. (2001). Transfer-Function Measurement with Sweeps, *Journal of the Audio Engineering Society*, vol. 49, no. 6, pp. 443–471.

Nelson, E. D.; Fredman, M. L. (1970). Hadamard Spectroscopy, *Journal of the Optical Society of America*, vol. 60, no. 12, pp. 1664–1669.

Poletti, M. A. (1988a). Linearly Swept Frequency Measurements, Time Delay Spectrometry and the Wigner Distribution, *Journal of the Audio Engineering Society*, vol. 36, no. 6, pp. 457–468.

Poletti, M. A. (1988b). The Application of Linearly Swept Frequency Measurement, *Journal of the Acoustical Society of America*, vol. 84, no. 2, pp. 599–610.

Ream, N. (1970). Nonlinear Identification Using Inverse-Repeat m Sequences, *Proc. IEE (London)*, vol. 117, pp. 213–218.

Rife, D.; Vanderkooy, J. (1989). Transfer-Function Measurement with Maximum-Length Sequences, *Journal of the Audio Engineering Society*, vol. 37, June, pp. 419–444.

Schneider, T.; Jamieson, D. G. (1993). A Dual-Channel MLS-Based Test System for Hearing-Aid Characterization, *Journal of the Audio Engineering Society*, vol. 41, no. 7/8, pp. 583–593.

Schoukens, J.; Pintelon, R. M.; Rolain, Y. J. (2000). Broadband Versus Stepped Sine FRF Measurement, *IEEE Transactions on Instrumentation and Measurement*, vol. 49, no. 2, pp. 275–278.

Schroeder, M. R. (1979). Integrated-Impulse Method for Measuring Sound Decay without Using Impulses, *Journal of the Acoustical Society of America*, vol. 66, no. 2, pp. 497–500.

Sommen, P. C. W.; van Valburg, C. J. (1989). Efficient Realisation of Adaptive Filter Using an Orthogonal Projection Method, *Proceedings of the IEEE International Conference on Acoustics, Speech, and Signal Processing (ICASSP)*, pp. 940–943.

Sondhi, M. M.; Morgan, D. R.; Hall, J. L. (1995). Stereophonic Acoustic Echo Cancellation – An Overview of the Fundamental Problem, *IEEE Signal Processing Letters*, vol. 2, no. 8, pp. 148–151.

Stan, G.-B.; Embrechts, J.-J.; Archambeau, D. (2002). Comparison of Different Impulse Response Measurement Techniques, *Journal of the Audio Engineering Society*, vol. 50, no. 4, pp. 249–262.

Suzuki, Y.; Asano, F.; Kim, H. Y.; Sone, T. (1995). An Optimum Computer-Generated Pulse Signal Suitable for the Measurement of Very Long Impulse Responses, *Journal of the Acoustical Society of America*, vol. 97, no. 2, February, pp. 1119–1123.

van de Kerkhof, L. M.; Kitzen, W. J. W. (1992). Tracking of a Time-Varying Acoustic Impulse Response by an Adaptive Filter, *IEEE Transactions on Signal Processing*, vol. 40, no. 6, pp. 1285–1294.

Vanderkooy, J. (1994). Aspects of MLS Measuring Systems, *Journal of the Audio Engineering Society*, vol. 42, no. 4, pp. 219–231.

Vary, P.; Martin, R. (2006). *Digital Speech Transmission*, John Wiley & Sons, Ltd., Chichester.

Widrow, B.; Hoff, M. E. (1960). Adaptive Switching Circuits, *Institute of Radio Engineers, Western Electric Show and Convention, Convention Record*, vol. 4, pp. 96–104.

Xiang, N. (1991). *A Mobile Universal Measuring System for the Binaural Room-Acoustic Modelling Technique*, PhD thesis, Schriftenreihe der Bundesanstalt für Arbeitsschutz, vol. FB611, Ruhr-Universität Bochum.

Xiang, N. (1992). Using M-Sequences for Determining the Impulse Response of LTI-Systems, *Signal Processing*, vol. 28, pp. 139–152.

III

Speech Coding for Heterogeneous Networks

Chapter 8

Embedded Speech Coding: From G.711 to G.729.1

Bernd Geiser, Stéphane Ragot, Hervé Taddei

8.1 Introduction

Speech coding aims at representing speech signals in a format that is suitable for digital communication. Traditionally, the emphasis has been on compression efficiency, i.e., to minimize bit rate subject to some quality requirements. However, in practice, the design of real-world speech coders is mainly governed by application needs and constraints. This chapter considers a special case of speech coding called "*embedded speech coding*". The underlying concept is illustrated by an example in Fig. 8.1. The encoder generates a bitstream that has a three-layer structure with one *core layer* and two *enhancement layers* stacked on top of each other. It is assumed that this structure is *hierarchical* in the sense that a given layer can only be decoded if underlying layers have been received as well. During transmission, a rate adaptation unit allows one to adaptively reduce the number of bitstream layers according, for example, to network conditions or receiver capability. Decoding a downsized bitstream can be viewed as using nested or *embedded* decoding algorithms. If only the core layer is received, the decoder outputs a decoded signal with a *basic* quality. As soon as enhancement layers are received, the decoder produces a signal of *enhanced quality*. Similarly, the encoding algorithm can be viewed as a core encoder nested in enhanced encoders. The key feature of embedded speech coding is *scalability* [Hiwasaki et al. 2004]. Indeed, enhancement layers can bring any kind of functionality, such as audio quality improvement (also called signal-to-noise ratio (SNR) scalability), acoustic bandwidth extension or mono to stereo extension, in addition to the core functionality. The number of layers and the respective bit rate increments between the layers define the so-called coding *granularity*.

Advances in Digital Speech Transmission   Edited by R. Martin, U. Heute and C. Antweiler
© 2008 John Wiley & Sons, Ltd

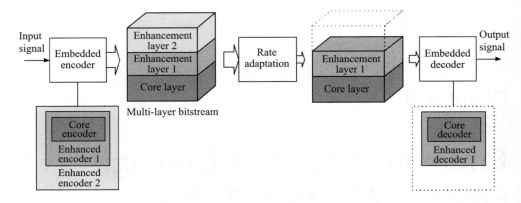

Figure 8.1: Principle of embedded speech coding

There are two main motivations for developing embedded speech coders. First, embedded coding is a possible solution to cope with the heterogeneity and variability in communication systems. Indeed, in real-world applications, links having different capacities and terminals with various capabilities may coexist, and their characteristics may not be known in advance at the transmitter side or may vary over time. For example, users may be connected to a service through a mix of mobile/wireless links (e.g., GSM/GPRS, HSPA, Wi-Fi, Bluetooth) and fixed links (e.g., dial-up modems, DSL, optical fiber access). The scalability of embedded codecs allows one to adapt certain coding attributes, in particular the instantaneous bit rate, in a flexible and efficient way by simple bitstream truncation, i.e., without re-encoding, using an asynchronous bit rate adaptation mechanism that is transparent to the encoder.

Secondly, nowadays a variety of speech coders is deployed in specific networks and applications. An example is the interconnection of circuit switched and Voice over IP (VoIP) networks, which often implement incompatible speech coders, for instance 3GPP AMR in GSM, 3GPP2 EVRC-A/B in cdma2000, ITU-T G.729 and G.711 in VoIP, and G.711 in PSTN. Hence, format conversion (or transcoding) in the respective gateways is inevitable to ensure interoperability. In this context, an "embedded extension" of a widely used core coder (e.g., G.729 or G.711) is a very attractive solution to deploy a new "enhanced coder" while minimizing the required transcoding overhead and ensuring interoperability and backwards compatibility with existing infrastructure and terminals. Note that next generation networks (NGN) will presumably be entirely based on packet-switched techniques, with the possibility for terminals to negotiate which coder to use. The need for transcoding is then virtually eliminated and the bitstream scalability can be exploited to adapt the quality and type of service (e.g., narrowband/wideband, mono/stereo) according to the user settings and other characteristics.

The objective of this chapter is to give a comprehensive overview of embedded speech coding, i.e., from theory to practice, with a focus on conversational applications.

Note that the terminology of embedded coding may be slightly confusing as the related literature alternatively refers to this concept as *embedded, hierarchical, scalable, progressive, multi-resolution, successively refinable, or bit-droppable.* Hereafter, the terms "embedded" and "hierarchical" will be primarily used. In the following, first, the underlying theory and fundamental coding tools are summarized (Sec. 8.2). Then, relevant work on the design and development of embedded speech coders is reviewed and popular methods are outlined (Sec. 8.3). In addition, the most important embedded speech coding *standards* are analyzed (Sec. 8.4). Among others, the ITU-T G.729.1 Voice over IP codec is addressed in particular. To give a more concrete and comprehensive treatment of the subject, the chapter concludes with a discussion on *network related* aspects of embedded speech coding (Sec. 8.5).

8.2 Theory and Tools of Embedded Speech Coding

The purpose of embedded coding techniques is to facilitate the *decoding* of *partially* received messages. For speech transmission applications, this corresponds to a speech reconstruction based on a partially received bitstream. The decoder in Fig. 8.1 is, for instance, able to produce a meaningful output signal if only the "core layer" of the hierarchically structured bitstream has been received.

The multi-layer bitstream structure from Fig. 8.1 suggests that the embedded decoder may operate only at the three *pre-specified* bit rates, but this constraint is actually not essential for partial decoding. As a matter of fact, virtually *every* source decoder can be modified to feature embedded decoding capabilities. It is, for instance, possible to *estimate* missing source parameters based on the received bits. A suitable tool for such an estimation is the computation of a *conditional expectation*, e.g., [Vary, Martin 2006, Chap. 5]. However, the conventional "monolithic" encoding concepts usually lead to a clearly sub-optimum performance of the respective *embedded* decoder. Therefore, this section introduces *encoding concepts* that have been particularly designed to facilitate embedded decoding. Consequently, decoded reproductions that are based on partially received bitstreams can be expected to exhibit an improved quality.

8.2.1 Basic Principles

Early speech coding methods such as PCM and ADPCM rely solely on the encoding of the *speech waveform*. By contrast, today's most efficient speech coders extract certain *parameters* that give a relevant and compact description of the input signal. These parameters are then *quantized* [Vary, Martin 2006, Chap. 7] and transmitted to the corresponding decoder. An important example are *linear predictive speech coders* which decompose the input speech signal into a spectral *envelope* and a *residual signal*, cf. [Vary, Martin 2006, Chap. 6]. The residual signal may be further parameterized, for example, by its *pitch period*.

Based on the notion of "quantized parameters", an embedded coding property can in principle be achieved in two different ways:

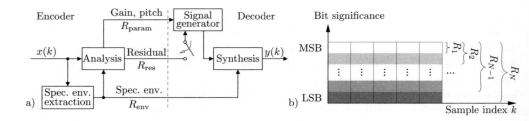

Figure 8.2: Examples for embedded speech coding using a) the "parameter dropping" approach and b) "hierarchical quantization" techniques
a) Only spectral envelope and gain/pitch information received ($R_1 = R_{env} + R_{param}$): *Vocoder* operation; Coded residual signal is received in addition ($R_2 = R_1 + R_{res}$): *Hybrid speech coder*
b) Principle of *bit plane coding* for PCM

1. The decoder may receive only a *part* of the quantized parameters. Hence, it can produce an output signal of intermediate quality. This approach can be termed "parameter dropping".

2. On the other hand, a certain parameter or parameter vector may be quantized in a *hierarchical* fashion. This means that the quantized representation of the parameter can be reconstructed with *different resolutions* depending on the amount of bits received. Such a property is achieved with so-called *hierarchical (vector-)quantization techniques* that — by design — facilitate embedded decoding.

An example of embedded coding by "parameter dropping" is given in Fig. 8.2-a. The decoder may either receive only the spectral envelope and gain/pitch information (rate R_1) or, *in addition*, the coded residual signal (rate $R_2 = R_1 + R_{res}$). In the former case, the decoder acts as a simple *vocoder*, while in the latter case, the availability of a coded residual signal turns the scheme into a typical *hybrid* speech coder [Vary, Martin 2006, Chap. 8]. More elaborate methods for embedded speech coding using "parameter dropping" are introduced in Secs. 8.3 and 8.4.

The second method for embedded coding, the "hierarchical quantization", is in fact also applicable to waveform coders. Figure 8.2-b illustrates a very simple example: the so-called "bit plane coding" of PCM samples. The full PCM resolution is achieved by quantizing the individual speech samples with, for instance, 16 bits (rate R_N). Lower bit rates ($R_{N-1}, R_{N-2}, \ldots, R_1$) can be obtained by successively omitting the *least significant bits* (LSBs). This scheme incurs a loss in terms of signal-to-quantization-noise ratio of approximately 6 dB per omitted bit plane. In fact, there are numerous other advanced possibilities for hierarchical quantization. But before introducing such methods, the information theoretic perspectives and the resulting performance limits with respect to practical implementations will be pointed out.

8.2.2 Approximation Theory

When designing a hierarchical quantizer, one may ask whether or under what circumstances is it possible to realize hierarchical quantization *without loss of "optimality"*. An "optimal" hierarchical quantizer offers exactly the same rate-distortion performance [Berger 1971] at each of its intermediate bit rates as the best monolithic quantizer for the same rate. In information theory, this problem is usually referred to as "successive refinement" [Equitz, Cover 1991] although it has already been formally studied by [Koshelev 1980] as "hierarchical coding".

It is known that optimal embedded coding, or successive refinement, can only be achieved for signal sources that obey certain statistics. Consider the simple example of a uniformly distributed signal versus a signal with Gaussian probability density function (PDF). The decision levels x_i and the corresponding reconstruction values $\hat{x}_i$ of the optimal *scalar* quantizer with w bit per sample satisfy the Lloyd–Max conditions, e.g., [Vary, Martin 2006, Chap. 7]. Table 8.1 lists the respective values of x_i for both sources with $w \in \{1, 2, 3\}$. As a matter of fact, the optimal scalar quantizer for the *uniformly distributed* source is already hierarchically structured, because its decision levels for a given w are contained within the set of optimal decision levels for all $w' > w$. In contrast, such a relationship can obviously not be established for the Gaussian source[1]. The observations for the scalar case can be generalized to higher dimensions d. A source can be hierarchically quantized in an optimal fashion if the $d - 1$ dimensional quantization cell boundaries of the optimal quantizer with w bit per vector form a subset of the cell boundaries for the optimal $w' > w$ bit quantizer. These conditions actually define *tree-structured* quantizers, which are described in more detail in Sec. 8.2.3.

This rather intuitive interpretation of optimal embedded coding for finite vector dimensions corresponds to the formal conditions for *successive refinement* that have been established by [Equitz, Cover 1991] and [Rimoldi 1994] based on the theory of [El Gamal, Cover 1982]. It is shown for high vector dimensions that the "coarsely" quantized description $\hat{X}_1$, the "refined" description $\hat{X}_2$, and the actual source X need to form a Markov chain $X \to \hat{X}_2 \to \hat{X}_1$ if both descriptions $\hat{X}_1$ and $\hat{X}_2$ are required to be optimal in a rate-distortion sense. The Markov property is formally defined by a specific factorization of the joint PDF $p(\hat{x}_1, \hat{x}_2, x)$:

$$p(\hat{x}_1, \hat{x}_2, x) = P(\hat{x}_1|\hat{x}_2) \cdot P(\hat{x}_2|x) \cdot p(x) \;\Leftrightarrow\; P(\hat{x}_1|\hat{x}_2, x) = P(\hat{x}_1|\hat{x}_2) \,. \tag{8.1}$$

In fact, this means that $\hat{X}_1$ needs to be entirely determined by $\hat{X}_2$, i.e., no further knowledge about the source X is required in order to produce $\hat{X}_1$. It is also concluded that a coding scheme to realize successive refinement exhibits a *tree-structure*. In the following, the most important practical approaches for tree-structured quantization are introduced (see Sec. 8.2.3). But first, several theoretical constraints that apply to *any* practical implementation will be addressed.

---

[1]A modified design algorithm for scalar quantizers that enforces this property has been proposed by [Tzou 1986]. Naturally, optimality at all intermediate bit rates is abandoned in this case, i.e., the Lloyd–Max conditions may no longer be satisfied for all w.

Table 8.1: Optimal quantizer decision levels for the uniform and the Gaussian source
using scalar quantization with 1, 2, and 3 bit

w [bit/sample]	Uniform PDF $p_x(u) = 0.5$ for $|u| \leq 1$ and 0 else	Gaussian PDF $p_x(u) = \mathcal{N}(0,1)$
1	0	0
2	$0, \pm 0.5$	$0, \pm 0.98$
3	$0, \pm 0.25, \pm 0.5, \pm 0.75$	$0, \pm 0.5, \pm 1.05, \pm 1.75$

Feasibility of Successive Refinement Coding

As shown, rigorous restrictions apply both to signal sources and to the respective
coders in order to achieve true successive refinement. In general, when compared
with the theoretical rate-distortion bound for monolithic quantizers, practical implementations of hierarchical quantizers either exhibit a certain *quality loss* (increased
distortion) for a given bit rate or, if the quality is to be maintained, they require
a somewhat higher bit rate ("bit rate *penalty*"). Figure 8.3-a addresses both cases.
An ideal embedded source coder should ensure that the quality of the decoded signal
at every intermediate bit rate is equal (or at least close) to the quality of a monolithic coder that has been particularly designed for this specific bit rate. Figure 8.3-b
illustrates an exemplary comparison of a three-stage hierarchical quantizer with its
(asymptotically optimum) monolithic counterparts.

An important question is, what penalty arises if the source does not follow the Markov
condition from (8.1). For large vector dimensions the answer has been given by
[Lastras, Berger 2001] where "near-successive refinement" is found to be achievable for
virtually all sources. Hence, the minimum achievable bit rate penalty in hierarchical
coding is upper-bounded. For a *two stage* coder, it is in particular no more than
either 0.5 or 1 bit/sample, depending on which quantizer stage shall be optimum.
For the mean square error there is an even tighter bound as shown by [Feng, Effros
2003].

Another limitation for practical implementations is found in the coding/decoding
scheme for successive refinement: The generic tree-structured codebooks demanded
by [Equitz, Cover 1991] are sometimes difficult to handle and require a lot of memory.
Actually, the operation of the decoder that is associated with the nth layer of the
quantization tree does, in general, depend on the input of *all* previous tree layers. The
complexity can be greatly reduced if the individual decoding operations are chosen
to be *independent* from all preceding stages. A very popular scheme is the *addition*
of a "refinement" signal to a "base layer" signal. This approach is often called "Multi-
Stage" coding (see Sec. 8.2.3). [Tuncel, Rose 2003] and [Feng, Effros 2003] found that,
using such an additive reconstruction scheme, successive refinement is achievable in
many practically relevant cases.

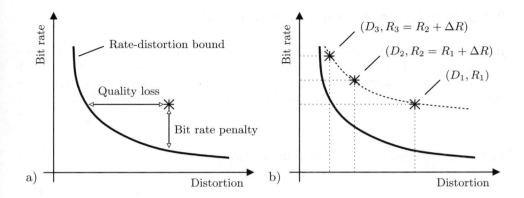

Figure 8.3: a) Illustration of "Quality loss" and "Bit rate penalty"
b) Example for a 3-stage coder with constant bit rate increments ΔR

All previously discussed analyses have been conducted in the spirit of rate-distortion theory (i.e., in the limit of large vector dimensions) and thus exhibit a limited significance for practical implementations that typically employ small dimensions. A *finite* dimension will in general lead to an additional bit rate penalty or, alternatively, to an increased distortion at each coder stage. This has been studied by [Voronov, Feder 2000] and [Yang, Zhang 2004] where the redundancy of successive refinement codes in finite dimensions d is investigated. In fact, there is an additional *accumulative* distortion with each coder stage that scales with $\log(d)/2d$.

It can be concluded from the previous discussion that the inevitable bit rate penalty or performance loss for a practical hierarchical quantizer is determined by several factors as follows.

1. Source properties — Many relevant sources and their respective optimum quantizers do not follow the Markov property from (8.1). This leads to a certain (bounded) performance loss.

2. Memory and complexity constraints — Optimal tree-structured codebooks are not feasible in certain applications.

3. Delay and complexity constraints — Actual implementations only allow for a *finite vector dimension*. This induces an *accumulative loss* for *each stage* of the hierarchical coder.

To finally complement the survey of approximation theory, it shall be noted that there are investigations of hierarchical quantization of a source X for the case that a second *correlated* source is available, e.g., [Viswanathan, Berger 2000], [Steinberg, Merhav 2004], and [Tian, Diggavi 2006]. Potential applications for such methods are embedded *multi-channel* extensions to existing coders.

8.2.3 Hierarchical Vector Quantization Methods

It can be seen from the previous section that the "optimum" hierarchical quantizer exhibits a *tree structure*. Other realizations, such as *multi-stage quantization*, often turn out to be variants of the tree-structured approach. This section introduces and discusses several hierarchical quantization methods that are commonly encountered.

a) Tree-Structured Vector Quantization (TSVQ)

A simple example for a tree-structured vector quantizer with a rate of up to 3 bits per vector and bit-level granularity is depicted in Fig. 8.4. The quantization of the vector $\mathbf{x}$ is realized according to a *binary* tree structure. (In general, a granularity of n bits per stage requires a 2^n-*ary* quantization tree.) For the case of "zero" bit rate (no bits are transmitted), the reproduction vector is simply the centroid $\hat{\mathbf{x}}_{0,0}$ of the probability distribution of $\mathbf{x}$. To obtain the first transmitted bit, $\mathbf{x}$ is compared with the codevectors that are associated with the nodes on the first tree level, i.e., $\hat{\mathbf{x}}_{1,0}$ and $\hat{\mathbf{x}}_{1,1}$. The vector that has the smallest distance to $\mathbf{x}$, i.e., the one yielding the minimum distortion, is chosen. The outcome may be $\hat{\mathbf{x}}_{1,0}$ for instance. Then, in a second step, $\mathbf{x}$ is compared with the vectors that are associated with the respective *child* nodes on the second tree level. Again, the closer one is chosen, e.g., $\hat{\mathbf{x}}_{2,1}$. Having arrived at a leaf of the quantization tree, the *quantizer index i* is obtained by following all vertices that define the path to this leaf. For the vector $\hat{\mathbf{x}}_{3,2}$ this would be the bit

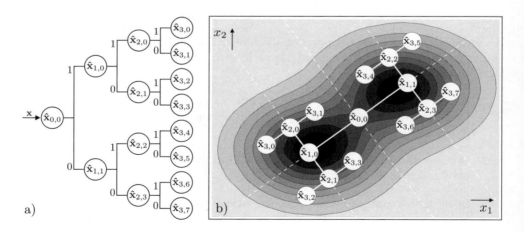

Figure 8.4: Example for Tree-Structured Vector Quantization (TSVQ)
a) Quantization tree for up to 3 bit/vector with bit-level granularity
b) Exemplary PDF of a 2-dimensional random variable $\mathbf{x} = (x_1, x_2)^T$ and graphical representation of the respective quantization tree — the resulting Voronoi partitioning is shown with dashed lines

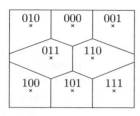 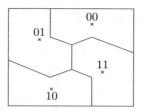 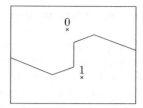

Figure 8.5: Successive reduction of a 3 bit VQ according to [Riskin et al. 1994]. The embedded index assignments with 3, 2, and 1 bit are shown for each centroid. © 1994 IEEE

pattern $i = (101)_2$. The obtained quantizer index i offers embedded decoding. Higher tree layers, i.e., trailing bits of i, may be removed to achieve lower bit rates while sacrificing as little performance as possible. This scheme obviously corresponds to the "bit plane coding" from Fig. 8.2-b. Removing the LSB of $i = (101)_2$ leads to the coarser approximation $\hat{x}_{2,1}$ from the second tree level. It is apparent that TSVQ requires a *dedicated reproduction codebook* for each tree layer. Hence, the storage requirements are rather demanding. However, the *encoding complexity* of TSVQ only grows linearly with the number of bits. In contrast, a fixed rate full search VQ has exponential complexity. In fact, TSVQ is, in addition to being a *hierarchical* quantizer, also attractive for its low computational complexity.

The *design* of tree-structured quantizers is usually carried out in a *greedy* fashion. For bit-level granularity all training vectors **x** that are associated with the current tree node are split into *two* groups (1 bit). This is commonly achieved with well known VQ design algorithms, e.g., [Linde et al. 1980]. Then, both groups of vectors are split further until the desired tree depth (i.e., bit rate) is reached. However, this procedure does not necessarily yield the best TSVQ possible. In fact, only the *first* stage is guaranteed to be equivalent to a single stage VQ of the same bit rate. In contrast to the greedy method, [Riskin et al. 1994] design embedded tree-structured *index assignments* for *given* codebooks. Here, the encoding is performed using the leaf-layer codebook only,[2] which leads to exponential encoding complexity. The lower tree layers are then formed by successively *joining* the Voronoi cells of the fine quantizer. This method, which is illustrated in Fig. 8.5, obviously defines a quantization tree, but optimality is now achieved for the *highest* bit rate. In fact, there are numerous other (usually weighted) "optimality" criteria for TSVQ design as introduced and discussed by [Effros, Dugatkin 2004]. Further theoretical analysis of TSVQ and its performance has been conducted by [Neuhoff, Lee 1991]. Also, some generalized schemes are presented in [Chou et al. 1989] where the quantization tree is optimally *pruned* to allow variable rate coding.

---

[2]In fact, the quantization process could also proceed as usual, i.e., beginning with the *first* stage. However, the quantizers associated with lower tree layers exhibit irregular cell boundaries and can no longer be realized as "nearest neighbor" quantizers (see Fig. 8.5).

b) Multi-Stage Vector Quantization (MSVQ)

The MSVQ principle was originally proposed by [Juang, Gray 1982] as a computationally very efficient VQ method. A simple example with two quantization stages is depicted in Fig. 8.6-a. In particular, MSVQ is based on a requantization of the previously obtained quantization error. The *coarse* (first stage) approximation is given by $\hat{\mathbf{x}}_1 = \mathcal{Q}_1(\mathbf{x})$ while the *fine* (second stage) approximation is computed as $\hat{\mathbf{x}}_2 = \mathcal{Q}_2(\mathbf{x} - \hat{\mathbf{x}}_1) + \hat{\mathbf{x}}_1$. Generalization to more than two stages is straightforward. An upper bound for the performance penalty of this basic MSVQ scheme has been found by [Erdmann, Vary 2004]. Yet, as indicated in Sec. 8.2.2, MSVQ is a realization of the specific *additive* successive refinement problem from [Tuncel, Rose 2003]. There, true successive refinement has been shown to be achievable for many relevant sources, i.e., the elimination of the penalty should in principle be possible. An important idea in this context has already been proposed by [Lee et al. 1991]. The performance of plain MSVQ can be improved by applying specific *orthogonal transformations* to the error vector $\mathbf{x} - \hat{\mathbf{x}}_1$ before the quantization with $\mathcal{Q}_2$ is carried out. The respective system is shown in Fig. 8.6-b. With a given transformation matrix $\mathbf{A}_i$, the quantized vector $\hat{\mathbf{x}}_2$ of the *second* MSVQ stage is then computed as follows:

$$\hat{\mathbf{x}}_2 = \mathbf{A}_i^{-1} \cdot \mathcal{Q}_2(\mathbf{A}_i \cdot (\mathbf{x} - \hat{\mathbf{x}}_1)) + \hat{\mathbf{x}}_1. \tag{8.2}$$

In general, an individual matrix $\mathbf{A}_i$ has to be designed for *each* quantization cell of the first stage quantizer $\mathcal{Q}_1$. This technique, labeled *cell-conditioning*, actually "equalizes" the probability density of the residual error, thus enabling a more effective second stage quantization. A simple example for a two-dimensional Gaussian source and a 1 bit quantizer $\mathcal{Q}_1$ is given in Fig. 8.6-b. In this example, the transformation $\mathbf{A}_0$ is a rotation by an angle of π while $\mathbf{A}_1$ is the identity matrix. In practice, even a simple *scaling* factor a_i instead of the matrix $\mathbf{A}_i$ can yield good results. [Lee et al. 1991] further show that, using (8.2), MSVQ can even reach the performance of single stage VQ under asymptotic conditions.

As an alternative to cell-conditioned MSVQ according to (8.2), the possibility of *successively orthogonalized* MSVQ stages has been used, e.g., [Moreau, Dymarski 1992]. Here, the contribution of the second stage quantizer is (adaptively) orthogonalized to the previous reproduction vector: $(\hat{\mathbf{x}}_2 - \hat{\mathbf{x}}_1) \perp \hat{\mathbf{x}}_1$. The underlying assumption is that the orthogonalized components can be *individually* optimized while retaining a global "near-optimality". The scheme is of particular interest if the individual quantization stages are realized as *gain-shape* quantizers [Vary, Martin 2006, Chap. 7]. In this case, it is ensured that the gain factor for a certain component does not impact all other orthogonalized components. This idea is further discussed in Sec. 8.3.2.

Another aspect of MSVQ ensues when investigating its application to *correlated* sources [Erdmann, Vary 2004]. It turns out that *most* of the source's correlation is already exploited by the *first* MSVQ stage. Hence, subsequent quantization stages are often far less efficient than the first one. This observation provides a plausible explanation for the fact that practical MSVQ schemes usually do not have more than three stages.

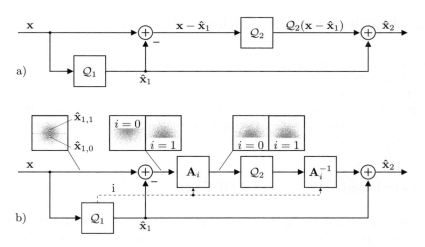

Figure 8.6: a) Multi-Stage Vector Quantization (MSVQ) with two stages
b) *Cell-Conditioned* MSVQ with a 1 bit quantizer Q_1 (index $i \in \{0, 1\}$). The probability density of the quantization error $(\mathbf{x} - \hat{\mathbf{x}}_1)$ depends on the chosen quantizer index i. The orthogonal transformations $\mathbf{A}_i$ are designed to equalize the error densities as much as possible

c) Transform-Domain Vector Quantization with Progressive Decoding

Instead of applying *successive* orthogonalizations in each coder stage as described above for MSVQ, such an operation can also be carried out *in advance*. Popular examples for the decomposition of the input signal into orthogonal components are *spectral transformations* like the DFT or DCT, sub-band decompositions through generic digital filterbanks, and *wavelet* decompositions [Mallat 1999]. A hierarchical quantization is then naturally obtained by computing a suitable *bit allocation* and by applying *individual* (vector) quantizers to the transformed signal components. Hierarchical decoding usually proceeds in the order of ascending importance, for instance from low to high frequencies or from most to least energetic components. This idea has been introduced as *Pyramid Vector Quantization* for embedded image coding [Burt, Adelson 1983]. The approach can, however, not be directly transferred to the speech and audio coding domain since a time-varying acoustical bandwidth is usually not considered acceptable. As a solution, missing frequency components can be filled with an intermediate, possibly artificial, signal (see Sec. 8.3.2 and Sec. 8.4.4). An elaborate discussion of a specific scheme with an octave band QMF-tree decomposition is provided in [Erdmann 2005]. It is concluded that an important advantage over MSVQ is the rather explicit control over the *granularity* of the hierarchical bitstream. Actually such schemes are very attractive if a fairly fine bitstream granularity is required while a larger vector dimension (increased delay) is allowed. In such a configuration, a comparatively large number of hierarchical layers (corresponding to the frequency subbands) can be realized.

8.3 Embedded Speech Coding Methods

This section provides an overview of research conducted in the area of embedded speech coding. Interestingly, embedded variants have been investigated and proposed for almost all relevant speech coding models. This observation emphasizes the evident interest in embedded coding and stresses its practical relevance. Some proposals have led to new coding standards; the most prominent *standardized* embedded codecs are described in Sec. 8.4.

8.3.1 Embedded DPCM and ADPCM

Differential Pulse Code Modulation (DPCM) and Adaptive DPCM (ADPCM) are sampled-based waveform coding techniques that are discussed in [Jayant, Noll 1984, Sec. 6], [Vary, Martin 2006, Sec. 8.3]. The idea of developing *embedded* DPCM and ADPCM schemes was first proposed in [Ching 1973]. Figure 8.7 depicts the DPCM structure with a *hierarchical quantizer* (Sec. 8.2.3) that is applied to the prediction error. The prediction has to be computed based on a signal available to both the encoder and decoder. This requirement can be fulfilled with only the signal reconstructed from the *core* layer in error-free condition. Hence, in embedded DPCM, the enhancement bits are stripped before the prediction is computed. A similar approach is described in [Wassel et al. 1988] for the special case of embedded delta modulation. This technique avoids desynchronization between encoder and decoder as well as noise accumulation in the decoder when the local and distant decoders do not use the same quantized error. The performance penalty of embedded DPCM (compared with the non-embedded version) has been analyzed in [Goodman 1980] and was found to be less than 1 dB in terms of signal-to-noise ratio if the data rate of the core layer is at least 2 bits per sample.

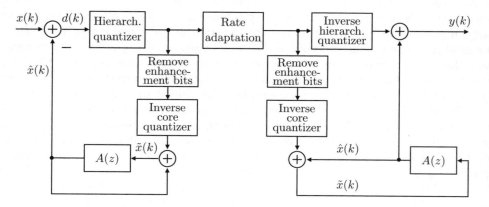

Figure 8.7: Embedded DPCM according to [Goodman 1980]. The (adaptive) linear predictor is represented by its transfer function $A(z)$

An alternative realization of embedded DPCM explicitly encodes the DPCM reconstruction noise by plain PCM to generate enhancement layers [Jayant 1983], [Zhang, Lockhart 1991], [Zhang, Lockhart 1995]. With some encoding delay, it is even possible to use an adaptive bit allocation; no side information is needed to inform the decoder if this bit allocation is based on the past decoded signal.

8.3.2 Embedded CELP

Code excited linear prediction (CELP) [Schroeder, Atal 1985], [Kleijn, Paliwal 1995, Chap. 3] is regarded as the most bit rate efficient solution for high-quality speech coding. Owing to the popularity of this model, many studies have been devoted to *embedded* CELP coding. In the following, the notational conventions of [Vary, Martin 2006, Chap. 8] are adopted. Furthermore, the CELP excitation is viewed as the sum of scaled adaptive *and* fixed codevectors.[3]

a) Embedded Multi-Stage CELP

CELP coding, which is based on codebooks of excitation sequences, is essentially a form of gain-shape vector quantization (VQ) with a time-varying distortion measure. This technique has, in general, large computation requirements, and a great deal of research has been devoted to finding codebook structures that allow for efficient search. In particular, in *multi-stage* CELP [Davidson, Gersho 1988], the fixed part of the excitation sequence is defined as a linear combination of weighted contributions from M different fixed codebooks:

$$\sum_{m=0}^{M-1} g_{\mathrm{f}}^{[m]} \cdot c_{i_m}^{[m]}(\lambda) \quad \text{for} \quad \lambda \in \{0, \dots, L-1\}, \tag{8.3}$$

where $g_{\mathrm{f}}^{[m]}$ is the gain applied to the fixed codevector $c_{i_m}^{[m]}(\lambda)$ of index i_m in the mth fixed codebook. Therefore, the excitation sequence of length L is constructed by a form of closed-loop multi-stage vector quantization (MSVQ, Sec. 8.2.3-b). An important variant of multi-stage CELP coding is vector sum excited linear prediction (VSELP) coding introduced in [Gerson, Jasiuk 1990]. Such techniques are designed to reach a good performance/complexity trade-off at a given *fixed* bit rate, which explains that the codebook gains $g_{\mathrm{f}}^{[m]}$ are typically jointly optimized and quantized.

In multi-stage CELP, the problem of selecting optimal codevectors $c_{i_m}^{[m]}(\lambda)$ and gains $g_{\mathrm{f}}^{[m]}$ in a least-squares sense can be solved by orthogonalizing each codebook contribution with previous ones [Dymarski et al. 1990]; the modified Gram–Schmidt algorithm, Cholesky decomposition, or the Householder transform may be used for this purpose. In particular, the orthogonalization of adaptive and fixed codebooks is described in [Taniguchi et al. 1990] and [Johnson, Taniguchi 1990].

---

[3] An equivalent model defines the CELP excitation as a scaled fixed codevector filtered through a cascade of long-term and short-term predictive filters.

The first realization of an *embedded* multi-stage CELP codec has been proposed in [De Iacovo, Sereno 1991]. This narrowband coder operates at 6.4, 8 and 9.6 kbit/s using $M = 3$ fixed codebooks. If only the adaptive and *first* innovation codebooks are considered, this approach is identical to a classical CELP coder. With one or two additional fixed codebook contributions, the speech quality is gradually improved. The fundamental differences compared with [Davidson, Gersho 1988] are as follows.

- Fixed codebook gains are quantized *separately* to allow for parameter dropping.

- The adaptive codebook is updated using only the first fixed codebook contribution, which modifies long-term prediction (LTP) in a way similar to the idea of *bit stripping* prior to ADPCM prediction (see Sec. 8.3.1).

- The so-called *ringing* (or zero-impulse response) of the perceptually weighted LPC synthesis filter is adjusted with *only* the information available at the lowest (core) bit rate.

These constraints result in a performance degradation at higher bit rates compared with non-embedded multi-stage CELP coding. This degradation is evaluated in [De Iacovo, Sereno 1991] to 0.5 and 1 dB at 8 and 9.6 kbit/s, respectively, in terms of segmental signal-to-noise ratio. Figure 8.8 shows an example of sequential multi-stage CELP search similar to [De Iacovo, Sereno 1991] with one adaptive codebook

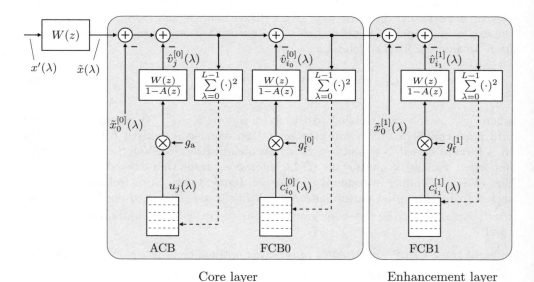

Figure 8.8: Embedded multi-stage CELP encoder with one enhancement layer (two fixed codebooks). $A(z)$ is the transfer function of the linear predictor; $W(z)$ is the transfer function of the perceptual weighting filter. A common choice is $W(z) = (1 - A(z/\gamma_1)) / (1 - A(z/\gamma_2))$ with $0.4 \leq \gamma_2 \leq \gamma_1 \leq 1$

(ACB) and only $M = 2$ fixed codebooks (FCBs). The CELP excitation is constructed as follows:

$$u'(\lambda) = g_a \cdot u_j(\lambda) + \sum_{m=0}^{M-1} g_f^{[m]} \cdot c_{i_m}^{[m]}(\lambda) \quad \text{for} \quad \lambda \in \{0, \ldots, L-1\}, \tag{8.4}$$

where $u_j(\lambda)$ is the adaptive codevector associated with the pitch index j, g_a is the adaptive codebook gain and the latter term is identical to (8.3). Typically, adaptive and fixed codebooks are searched sequentially. In this case, the criteria that are minimized for the adaptive and mth fixed codebook search are:

$$E_a(j, g_a) = \sum_{\lambda=0}^{L-1} |v(\lambda) - \hat{v}_j(\lambda)|^2 \tag{8.5}$$

and

$$E_{f_m}(i_m, g_f^{[m]}) = \sum_{\lambda=0}^{L-1} |v^{[m]}(\lambda) - \hat{v}_{i_m}^{[m]}(\lambda)|^2. \tag{8.6}$$

Note that, in Fig. 8.8, the signal $\tilde{x}_0^{[m]}(\lambda)$ corresponds to the ringing of $W(z)/(1-A(z))$ for $m = 0$ and a ringing adjustment term for $m = 1$.

Several later investigations have shown that codebook orthogonalization (Sec. 8.2.3-b) is well suited for embedded multi-stage coding; one important property of orthogonalized codebooks is that separate codebook gain quantization is equivalent to (optimal) joint quantization [Johnson, Taniguchi 1990]. In [Le Guyader et al. 1992], the orthogonalization of [Dymarski et al. 1990] and [Moreau, Dymarski 1992] for CELP is adapted and it is shown that it is possible to orthogonalize while keeping *fast search algorithms* and producing embedded codes. Furthermore, it is worth noting that the coder described in [Le Guyader et al. 1992] is the first realization of a *wideband* embedded multi-stage CELP coder. In [Le Guyader et al. 1995], more general and faster algorithms have been developed.

Interestingly, embedded multi-stage variants of other types of analysis-by-synthesis coders, namely multi-pulse LPC (MPLPC) and regular pulse excitation (RPE), have been proposed, too — see, for instance, [Singhal, Atal 1989], [Nomura et al. 1998] and [Zhang, Lockhart 1997].

b) Pyramid CELP

The so-called "Pyramid CELP" concept as introduced by [Erdmann, Vary 2002] and [Erdmann 2005] can be viewed as a further specialization of multi-stage CELP with orthogonalized excitation contributions. Here, the core layer of the coder comprises, similar to [De Iacovo, Sereno 1991], an adaptive codebook and a first fixed codebook. Thereafter, a number of subsequent coding stages refine the CELP excitation. In contrast to [Le Guyader et al. 1995], the respective individual contributions are orthogonal in the sense that they represent distinct *frequency subbands* of the excitation signal. In fact, the successive addition of higher frequency bands is called *pyramid coding*, a term which has been adopted from the image coding domain [Burt, Adelson 1983]. Yet, this method results in a varying bandwidth that is not acceptable for *audio* signals. Thus, at intermediate bit rates, missing excitation components can be filled with an artificial signal in order to provide a constant acoustic bandwidth. Specifically, [Erdmann, Vary 2002] use an *octave band decomposition* of the excitation signal. Missing spectral components of the excitation signal are generated by a *spectral folding* of the respective lowpass version. The advantage of this specific realization is that both the octave band decomposition and the spectral folding can be seamlessly integrated with the very popular and efficient ACELP codebooks as, e.g., implemented in [ITU-T Rec. G.729 1996], see also [Erdmann 2005].

c) Subband CELP

The embedded subband CELP method in [Kataoka et al. 1997] is a wideband coding scheme that splits the input speech signal into low band (0–4 kHz) and high band (4–8 kHz) signals by means of a QMF filterbank [Vary, Martin 2006, Chap. 4]. The subband signals are then encoded separately with specifically designed CELP coders. Thereby, the low band coder can also be compatible with a widely deployed standard (e.g., G.729). In the high band, a modified "pitch-less" CELP coder may be used due to the reduced tonality of speech signals for frequencies above 4 kHz.

8.3.3 Embedded Extensions of CELP Coders

Within present speech communication networks, increased quality demands *and* the practical need for *interoperability* with legacy equipment can only be satisfied simultaneously by deploying "add-on" coders on top of existing solutions. These "add-ons" are, in general, designed to introduce new functionality such as, for example, an extended acoustic bandwidth or an improved quality for music signals. The respective techniques can, nevertheless, also be applicable to "from scratch" designs of embedded speech codecs. In fact, already the Subband CELP scheme from Sec. 8.3.2-c can be interpreted as an embedded extension of a given coder. A further selection of such methods is presented in this section.

a) CELP Enhanced by Bandwidth Extension

Being highly bit rate efficient, techniques for an "artificial extension" of the acoustic bandwidth of speech signals [Larsen, Aarts 2004], [Vary, Martin 2006, Chap. 10] have attracted considerable attention in the past. A comprehensive overview of the respective methods and use cases is presented in Chap. 9. In general, they can be roughly categorized as "bandwidth extension (BWE) *without* side information" and "BWE *with* side information", where the latter is closely related to *parametric* speech coding. In the context of an embedded speech coder, especially the application of BWE schemes to conversational CELP codecs is of interest.

The initial motivation for the integration of BWE techniques into *wideband* CELP coding (50–7000 Hz) was the increased coding efficiency and also the decreased complexity [Paulus, Schnitzler 1996], [Schnitzler 1998]. It was found sufficient to encode frequencies above 6 kHz using a parametric model instead of the CELP method, which is used to encode the lower frequencies. This split band concept has been adopted in the AMR-WB codec [3GPP TS 26.190 2001], [Bessette et al. 2002], where frequencies between 6.4 and 7 kHz are artificially regenerated *without* the use of any side information. Only for the highest codec mode, i.e., at 23.85 kbit/s, the gain of the 6.4–7 kHz band is transmitted. In addition to the increased coding efficiency and decreased complexity, a further major advantage of using BWE methods in wideband speech coding is the possibility of enhancing widely deployed *narrowband* (300–3400 Hz) coding standards while preserving interoperability with legacy equipment. In this case, the narrowband coder constitutes the *core layer* in an embedded coding framework, whereas the coded BWE side information forms an *enhancement layer* and is in general used to suitably shape an artificial high band "excitation" signal. There are numerous realizations of this specific setup in the literature such as [McCree 2000], [McCree et al. 2001], [Taori et al. 2000], [Valin, Lefebvre 2000], and [Krishnan et al. 2007] with additional bit rates between 0.5 and 2.3 kbit/s.[4] In particular, the BWE method described in [Jax et al. 2006a] and [Geiser et al. 2007] has been standardized in the embedded ITU-T G.729.1 codec, which is described in Sec. 8.4.4. Furthermore, there are also extended parametric coding models that can perform a BWE of general audio or music signals for frequencies above 8 kHz, e.g., [Dietz et al. 2002], [3GPP TS 26.404 2004]. Of course, compared with the speech-specific solutions, this benefit usually comes at the cost of an increased delay and possibly a slightly "buzzy" sound character for speech signals.

In the literature, there are also proposals that perform a parametric bandwidth extension in the *frequency domain*. For example, [Oshikiri et al. 2007] implements a BWE in the *modified discrete cosine transform* (MDCT) domain with a very low additional bit rate (800 bit/s). This concept has also been applied to extend speech signals beyond the usual wideband range to so-called *super-wideband* speech with frequencies up to 15 kHz [Oshikiri et al. 2002], [Oshikiri et al. 2004]. Alternatively, also a full-fledged *transform coding* of high frequencies may be used to extend the signal bandwidth. A scalable narrowband/wideband coder following this concept has been proposed by [Jung et al. 2004]. Here, the input wideband signal is split in two bands by a QMF filterbank, the low band is coded by multi-stage CELP with a ITU-T G.723.1 [ITU-T Rec. G.723.1 1996] core coder and the high band is encoded in the MDCT domain. A similar approach is taken by [Hiwasaki et al. 2006] based on ITU-T G.711 as the low band coder. In addition to the bandwidth extension application, the transform coding concept can also be employed to encode the residual error of CELP coders. This issue is addressed in the following.

---

[4]Meanwhile, even lower bit rates have been reported to deliver adequate quality, cf. Chap. 9 for the related algorithmic details.

b) CELP Enhanced by Transform Coding

The link between embedded CELP coding and transform coding can be found in
[Lozach 1993] where the so-called CELP target signal is progressively modelled by
orthogonal vectors resulting from an adaptive transform. However this approach still
suffers from CELP limitations: The encoding of *non-speech* signals (e.g., music) usually leads to insufficient quality. [Ramprashad 1999] has clearly described the problems
encountered by LPC-based codecs and the *advantages* of using embedded coding with
a different non-LPC based paradigm to enhance an LPC-based core codec. For generic
audio signals it is usually better to use codecs based on transform coding concepts.
Yet, such techniques are not adapted to the application of speech coding at low bit
rates. The combination of a CELP codec with transform coding in an embedded coding framework promises good speech quality at low bit rates (around 8 to 16 kbit/s)
as well as good music quality at higher bit rates.

The most direct approach is to encode the difference between the original signal and
the CELP decoded signal in the transform domain. Often, the difference signal is
coded jointly with high frequencies for scalability in bit rate *and* bandwidth. This
approach is pursued by [Taddei et al. 1999], where the core coder is ITU-T G.729 enhanced by a second CELP layer based on G.729 Annex E. The transform coding part
is based on the *modified discrete cosine transform* (MDCT), with masking threshold
estimation and bit allocation according to the noise-to-mask ratio per frequency subband. This proposal was modified and improved by [Kövesi et al. 2004], using ITU-T
G.723.1 as a core coder. A similar standardized solution is detailed in Sec. 8.4.3-c.
Moreover, instead of the MDCT transform, the use of a gammatone filterbank was
proposed in [Kim et al. 2002], with G.729 Annex E as a core coder.

Still, CELP and transform coding usually rely on different optimization criteria. One
way to harmonize these two models is to apply a linear-predictive perceptual weighting
filter to the CELP error signal prior to transform coding, and then to encode transform
coefficients with respect to the mean-square error (MSE) criterion. This predictive
transform approach is proposed for instance by [Ragot et al. 2006]. A similar realization is further described in Sec. 8.4.4. Note that an embedded predictive transform
coder is already described in [Ramprashad 2000], whereby embedded transform coding
is directly applied after perceptual weighting filtering.

8.3.4 Embedded Parameter Quantization

The embedded coding concept can not only be applied to a speech codec as a whole
but also to individual *parameters* thereof. An example of embedded coding on the
parameter level is given by the bandwidth scalable LPC coding of [Nomura et al.
1998], [Koishida et al. 2000], and [Ehara et al. 2007]. Here, wideband speech signals
(sampled at 16 kHz) are coded using an embedded codec that is based on a narrowband core coder operating at 8 kHz. The narrowband spectral envelope parameters
(LPC coefficients) are quantized in the core coder while the respective *wideband* parameters are transmitted in an enhancement layer. Thereby, the decoded narrowband

parameters are extended in dimension to differentially code the wideband parameters. Embedded spectral envelope coding is also used in [Aguilar et al. 2000] as part of a general embedded subband sinusoidal coder. Also note that embedded coding of spectral envelopes is related to "bandwidth extension with side information" as described in Sec. 8.3.3-a.

8.4 Standardized Embedded Speech Coders

In this section, several embedded speech codecs are summarized that have been standardized within the "International Telecommunication Union - Telecommunication Standardization Sector" (ITU-T) and the "Moving Pictures Expert Group" of the "International Organization for Standardization" (ISO/MPEG).

8.4.1 ITU-T G.711 PCM Codec

ITU-T G.711 describes a companded (non-uniform) quantization method for speech signals at 64 kbit/s [Jayant, Noll 1984, Sec. 5.3], [Vary, Martin 2006, Chap. 7]. It encodes linear PCM signals (16 bits, sampled at 8 kHz) to 8-bit codewords according to the A-law or μ-law compression characteristic.

In the standard [ITU-T Rec. G.711 1972], the logarithmic compression characteristics are actually approximated by piecewise linear functions. Thus, the 8 bits of a G.711 codeword specify (in that order): the sign, 3 bits for the segment number of the approximated compression characteristic, and 4 bits for the position on the selected segment. Clearly, the resulting bits are *ordered* by their importance (most to least significant bit) such that a *bit plane coding* scheme can be applied. This principle has already been sketched for linear PCM in the introductory example from Fig. 8.2-b in Sec. 8.2.1. When keeping the number of bits being "stolen" rather low, e.g., 1 or 2 least significant bits (LSBs) per sample, the introduction of objectionable artifacts can mostly be avoided.[5] Thus, employing bit plane coding and assuming an appropriate bit reordering before packetization, G.711 can actually be seen as an embedded speech codec. In fact, there are international standards that make use of G.711 at bit rates of 56 kbit/s [ITU-T Rec. H.320 2004] and even 48 kbit/s [ITU-T Rec. H.242 1999].

Apart from embedded coding, a further application of G.711 "bit stealing" is *signaling*, where one or two LSBs (possibly only for every nth sample) are replaced by signaling data. An interesting example is the ETSI standard for "Tandem Free Operation" (TFO) [ETSI GSM 08.62 2000], which replaces the LSBs of G.711 with the bitstream of another speech codec. The intention here is to establish a virtually transparent digital channel between two mobile phones while the core network is not aware of the modified G.711 stream.

---

[5]Yet, for the A-law characteristic, which effectively resembles a *midrise quantizer* [Vary, Martin 2006, Fig. 7.3], inferior performance in coding silence (speech pauses) can be expected.

8.4.2 ITU-T G.727 and G.722 ADPCM Codecs

ITU-T defined several ADPCM recommendations. First, G.721 was adopted for 32 kbit/s ADPCM encoding of signals sampled at 8 kHz. This recommendation has been later extended to 24 and 40 kbit/s bit rates in G.723. Finally, these two standards were superseded by G.726, which is a multi-rate narrowband ADPCM codec at 16, 24, 32, and 40 kbit/s. G.726 is the standard codec used in DECT wireless systems [ETSI EN 300 175-8 2005].

Later on, an *embedded* Adaptive Differential Pulse Code Modulation version of G.726 was adopted under the name ITU-T G.727 [ITU-T Rec. G.727 1990] with the same bit rates. The underlying principle was introduced in Sec. 8.3.1. From the technical viewpoint, on the encoder side a prediction error signal is obtained by subtracting the predicted input signal from the input signal itself based on the output of the inverse quantizer fed by the core bits. Then, the prediction error is quantized by an embedded 4-, 8-, 16-, 32-level quantizer. Depending on the chosen bit rate, it is possible to define the number of bits used for the core quantization. This number varies between 2 and 4. As a result, the number of enhancement bits varies between 0 and 3 for a codec with a total bit rate between 16 and 40 kbit/s. More details of this specific standard can be found in [Sherif et al. 1993]. A use case for ATM networks is described in [Kondo, Ohno 1994].

An embedded *wideband* ADPCM codec is defined in [ITU-T Rec. G.722 1988] for encoding wideband signals (50–7000 Hz) at 48, 56, and 64 kbit/s. It is based on the SB/ADPCM (Sub-Band Adaptive Differential Pulse Code Modulation) model [Maitre 1988], [Mermelstein 1988]. In G.722, a QMF filterbank provides one sample for the low band (0–4 kHz) and one sample for the high band (4–8 kHz), given two wideband input samples. The obtained critically downsampled signals are then separately encoded. The low band is encoded using an embedded ADPCM scheme as in G.727 with 6 bits per sample. Owing to the embedded property of the quantizer, 1 or 2 bits can be stolen. The high band is quantized with 2 bits per sample. Hence, the bit rate is 32, 40, or 48 kbit/s for the low band and 16 kbit/s for the high band. For wideband output, the total core bit rate is 48 kbit/s. Two low band enhancement layers of 8 kbit/s can be added. Though, decoding for example the 32 kbit/s of information from the low band only, G.722 can in fact also be seen as a *bandwidth scalable* embedded coder. An application of bit stealing with G.722 is speech-data multiplexing [Mermelstein 1988]. G.722 is also the mandatory codec for wideband speech coding in New Generation DECT systems [ETSI TS 102 527-1 2007].

8.4.3 MPEG-4 Scalable Speech Coding

The audio coding tools of the MPEG-4 standard [ISO/IEC 14496-3 2005] comprise several solutions for embedded speech coding. The respective codecs are described in the following.

a) MPEG-4 Scalable CELP

MPEG standardized an embedded speech coder [Nomura et al. 1998], the MPEG-4 CELP, which operates in narrowband or in wideband mode (8 or 16 kHz sampling frequency). The CELP excitation sequence is either constructed via "multi-pulse" excitation (MPE, cf. [Vary, Martin 2006, Sec. 8.5.4.1]) or "regular pulse" excitation (RPE, cf. [Vary, Martin 2006, Sec. 8.5.4.2]) with various possibilities for the total bit rate. A further flexibility of this coder is the support of frame lengths between 10 and 40 ms, depending on the chosen bit rate. An overview of the available configurations is given in Table 8.2.

The *embedded* coding feature, i.e., the "bitstream scalability" of the MPEG-4 CELP codec is restricted to the MPE mode for wide- and narrowband signals. Two kinds of enhancement layers can be added to the narrowband or wideband CELP core, which either enhance the quality of the signal or, for a narrowband core, extend the bandwidth to wideband.

The *quality enhancement layers* enrich the CELP excitation by encoding additional MPE contributions. Thereby, the pulse positions are adaptively controlled so that none of them coincides with a position used in the core encoder. Furthermore, the enhancement contributions are not considered for the adaptive codebook of the core CELP. Up to three of these enhancement layers (with 2 kbit/s for narrowband enhancement and 4 kbit/s for wideband enhancement) can be added to each base configuration of the MPEG-4 CELP.

The bit rate of the *bandwidth enhancement* layer is 9–15 kbit/s depending on the chosen narrowband coder. The transmitted information comprises, again, additional pulses that complete the (upsampled) MPE excitation and also a refined pitch delay. In addition, the narrowband spectral envelope needs to be extended to a wideband representation. Therefore, a predictive scheme (see Sec. 8.3.4) is used where the obtained prediction error is encoded and used in the decoder to reconstruct the quantized wideband spectral envelope (line spectrum pairs, LSPs).

Table 8.2: Basic configurations of the MPEG-4 CELP codec

		Narrowband (8 kHz sampling)	Wideband (16 kHz sampling)
MPE mode	bit rate	3.85–12.2 kbit/s	10.9–23.8 kbit/s
	frame length	40, 30, 20, or 10 ms	20 or 10 ms
RPE mode	bit rate	—	14.4–22.533 kbit/s
	frame length		15 or 10 ms

b) MPEG-4 HVXC Coding

The MPEG-4 HVXC narrowband coder [Nishiguchi et al. 1999] operates at very low bit rates of 2 and 4 kbit/s. Thereby, the 4 kbit/s version is obtained by an optional 2 kbit/s enhancement layer, hence, HVXC constitutes an embedded codec. Technically, such very low bit rates can be achieved by making the operation of the coder dependent on a binary voiced/unvoiced decision. In particular, first, conventional LPC analysis is performed. Then, the LPC residual in *unvoiced* frames is coded by "vector excitation coding" (VXC) which is basically a form of CELP coding with a fixed codebook only. The residual in *voiced* frames is treated by pitch estimation and vector quantization of the harmonic magnitude structure. The 2 kbit/s enhancement layer of HVXC comprises refinements of the spectral envelope (LSPs), of the harmonic magnitudes (for voiced frames), and of the VXC excitation (for unvoiced frames). All of these refinement contributions are obtained by multi-stage vector quantization (Sec. 8.2.3-b).

c) MPEG-4 Combined Scalable CELP and AAC

This variant of the MPEG-4 standard extends the MPEG embedded CELP codec with a combination of scalable CELP and Advanced Audio Coding (AAC) tools [Grill 1997]. The latter is basically a state-of-the-art transform coder for general audio or music signals. Such a constellation, i.e., a conversational CELP coder that is combined with a transform coder, is quite advantageous as shown in Sec. 8.3.3-b. The signal flow of the encoder is illustrated in Fig. 8.9. Here, the CELP core codec operates at a lower sampling rate (8 kHz) than the AAC enhancement coder and provides a locally decoded output that is upsampled to the original sampling rate (up to 48 kHz). The perceptual AAC coder encodes the residual signal, which is calculated by subtracting the MDCT spectrum of the core layer signal from the MDCT spectrum of the original signal. As a result, there are three types of spectrum coefficients available: the low band core encoded coefficients $\hat{X}_{\mathrm{LB}}(\mu)$, the low band difference coefficients $\Delta X_{\mathrm{LB}}(\mu)$, and the low band original coefficients $X_{\mathrm{LB}}(\mu)$. In the high band, only the original coefficients $X_{\mathrm{HB}}(\mu)$ have to be considered. In some cases, when the core codec in the low band does not encode the low band signal very well, the difference signal $\Delta X_{\mathrm{LB}}(\mu)$ turns out to be even more difficult to encode than the original signal $\hat{X}_{\mathrm{LB}}(\mu)$. In these cases, a bank of switches (inside the "Frequency Selective Switch" module) can be set such that the original spectrum is directly encoded instead of the difference spectrum. Finally, the high band frequencies (of the original signal spectrum) are joined with this combined spectrum and passed to the quantization and entropy coding module of the AAC coder.

Within this structure, besides the MPEG-4 CELP (Sec.8.4.3-a), several other codecs have been tested [Grill 1997] as core codecs such as [ITU-T Rec. G.723.1 1996] or [ITU-T Rec. G.729 1996].

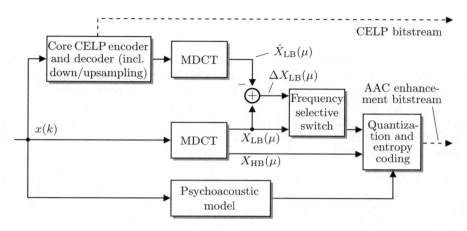

Figure 8.9: MPEG-4 combined scalable CELP and AAC — encoder

8.4.4 Embedded Wideband Coding for VoIP: ITU-T G.729.1

The G.729.1 standard is an 8–32 kbit/s embedded speech and audio coder providing bitstream interoperability with G.729 [ITU-T Rec. G.729 1996], [ITU-T Rec. G.729 Annex A 1996], and [ITU-T Rec. G.729 Annex B 1996]. This coder has been designed to provide better quality and more flexibility than the existing ITU-T G.729 speech coding standard, which is widely used in wireline speech communication, e.g., Voice over IP (VoIP). In fact, G.729.1 offers several kinds of flexibility:

- scalability in bit rate, bandwidth, and complexity,

- support of both 8 and 16 kHz input/output sampling frequency,

- backwards compatibility with G.729/G.729B bitstream format, and

- an option for reduced algorithmic delay at certain bit rates.

Here, "G.729.1" in general refers to the main body of the standard, which is implemented in *fixed-point* arithmetic. In addition, G.729.1 Annex A defines the associated payload format for transmission in IP based networks with the Real-time Transport Protocol (RTP), as well as signaling parameters in H.323 communication sessions [Hersent et al. 2005]. Finally, G.729.1 Annex B defines an alternative implementation using *floating-point* arithmetic. A more detailed algorithmic description of this codec can be found in [ITU-T Rec. G.729.1 2006] and [Ragot et al. 2007].

a) Design Philosophy

G.729.1 has been developed in the context of large scale VoIP deployment on high speed Internet access (e.g., xDSL). Compared with legacy circuit switched telephony (such as PSTN), VoIP represents a major technology shift that requires new service infrastructures and new terminals. VoIP gives the opportunity to improve audio quality greatly by migrating quickly to wideband speech (50–7000 Hz) instead of narrowband (300–3400 Hz). Indeed, the wideband feature is a strong differentiating factor to discriminate VoIP from PSTN.

The standardization of a G.729-based embedded coder for wideband VoIP has been motivated mainly by the fact that G.729 (with its Annexes A and B) is widely used in VoIP networks and terminals, especially in corporate infrastructure. An embedded extension of G.729(AB) can then allow one to migrate services smoothly from narrowband to wideband while keeping interoperability with existing VoIP infrastructure and terminals using G.729(A/AB).

Based on market needs and foreseen applications, the following constraints and requirements have been defined for G.729.1:

- bitstream scalability with an 8 kbit/s core coder which is interoperable with ITU-T G.729 and its Annexes A and B,

- bit rates from 8 to 32 kbit/s, with narrowband output (at least 300–3400 Hz) at 8 and 12 kbit/s and wideband output (50–7000 Hz) from 14 to 32 kbit/s,

- fine bit rate granularity above 14 kbit/s for maximal flexibility in rate adaptation; byte-level granularity was seen as a desired feature, eventually the bit rate steps have been set to 2 kbit/s,

- improved narrowband quality at 12 kbit/s compared with ITU-T G.729 as a reference,

- good clean and noisy speech quality at 24 kbit/s, and

- good music quality at 32 kbit/s.

b) Encoder and Decoder

The G.729.1 encoder and decoder are illustrated in Figs. 8.10 and 8.11. The coder operates on 20 ms frames. By default, both input and output signals are sampled at 16 kHz. The encoder normally operates at the maximal bit rate of 32 kbit/s, however the RTP payload format of G.729.1 allows one to specify a lower encoding bit rate through the Maximum Bit Rate Supported (MBS) field [IETF RFC 4749 2006]. After encoding, the instantaneous bit rate can be adapted in the range 8, 12, 14, 16,..., 32 kbit/s by truncating the bitstream on a 20 ms frame basis.

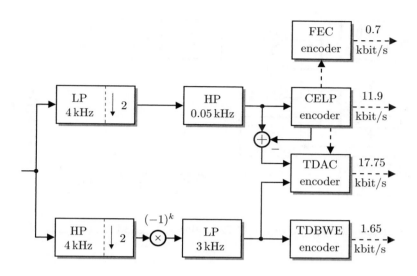

Figure 8.10: Block diagram of the G.729.1 encoder. © 2007 IEEE

At the encoder, the wideband input signal is decomposed into two subbands using a quadrature mirror filterbank (QMF). The *low band* (0–4 kHz) is pre-processed by a 50 Hz high pass (HP) filter and encoded by a cascaded CELP coder [Massaloux et al. 2007] (cf. Sec. 8.3.2). The waveform indices of the CELP enhancement layer are searched within an orthogonalization process that allows for a fast ACELP codebook search [Lee et al. 2003]. This process is actually a special case of embedded ACELP/VSELP coding as described by [Le Guyader et al. 1995]. The embedded CELP coder is based on a structure similar to Fig. 8.8 (see Sec. 8.3.2-a). The main difference to Fig. 8.8 is that, in G.729.1, filtered fixed codevectors $\hat{v}_{i_m}^{[m]}$, $\lambda \in \{0, \ldots, L-1\}$ are orthogonalized to the filtered adaptive codevector $\hat{v}_j(\lambda)$, $\lambda \in \{0, \ldots, L-1\}$. To be specific, it can be shown that the orthogonalized filtered fixed codebooks are defined as:

$$\hat{v}_{i_m,\perp}^{[m]}(\lambda) = \hat{v}_{i_m}^{[m]}(\lambda) - \frac{\displaystyle\sum_{\lambda=0}^{L-1} \hat{v}_{i_m}^{[m]}(\lambda) \cdot \hat{v}_j(\lambda)}{\displaystyle\sum_{\lambda=0}^{L-1} \hat{v}_j(\lambda)^2} \cdot \hat{v}_j(\lambda) \quad \text{for} \quad \lambda \in \{0, \ldots, L-1\}. \quad (8.7)$$

Then, the CELP criterion that is minimized for the fixed codebook search is given by [Le Guyader et al. 1995, Eqs. (23) and (18)]. The orthogonalization process implies only a slight modification in the autocorrelation term of the "standard" CELP criterion, see [Massaloux et al. 2007, Sec. 3.1] for details.

The *high band* components of the G.729.1 input signal (4–8 kHz) are pre-processed by spectral folding and a 3 kHz low pass (LP) filter. They are encoded by a parametric

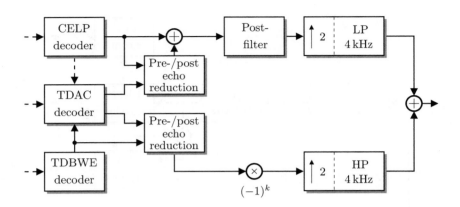

Figure 8.11: Block diagram of the G.729.1 decoder (good frame only). © 2007 IEEE

method, which is called "time-domain bandwidth extension" (TDBWE) [Geiser et al. 2007] (cf. Sec. 8.3.3-a). Then, the residual error in the low band and the (original) high band signal are jointly encoded by the so called "time-domain aliasing cancellation" (TDAC) encoder, which is a transform coder based on the modified discrete cosine transform (MDCT) [Ragot et al. 2006]. To improve the resilience and recovery of the decoder in case of frame erasures, parameters that are useful for frame erasure concealment (FEC) — consisting of signal class (voiced, unvoiced, onset, or voiced/unvoiced transition), phase and energy information — are transmitted by the FEC encoder based on available low band information [Vaillancourt et al. 2007].

The decoder of G.729.1 operates in an embedded manner depending on the received bit rate. At 8 and 12 kbit/s the cascaded CELP decoder reconstructs a low band signal (50 – 4000 Hz), which is then post-filtered in a way similar to G.729; the result is upsampled to 16 kHz using the QMF synthesis filterbank. At 14 kbit/s, the TDBWE decoder reconstructs a high band signal that is combined with the 12 kbit/s synthesis in order to extend the output bandwidth to 50–7000 Hz. From 16 to 32 kbit/s, the TDAC stage decodes both the low band difference and the high band signals, which are then post-processed to reduce pre-/post-echo artifacts that stem from the transform coding module. The resulting low band signal is added to the CELP output, while the resulting high band synthesis is used instead of the TDBWE output.

The G.729.1 decoder operates at narrowband bit rates (8 and 12 kbit/s) and wideband bit rates (14 kbit/s and above). Without any appropriate method, a fast switching between these two sets of bit rates would result in audible artifacts. To avoid this, the decoder includes a cross-fading inside the low band postfilter as well as a slow fade-in (1 second) of the high band in case of narrowband to wideband transition.

Additionally, in the case of lost frames, a frame erasure concealment is performed at the decoder that makes use of the transmitted concealment/recovery parameters and exploits the extra frame delay at the decoder. Efficient concealment and recovery techniques are used including glottal pulse resynchronization, energy control, and artificial onset reconstruction [Vaillancourt et al. 2007].

c) Hierarchical Bitstream Structure

To emphasize the embedded nature of G.729.1, its hierarchical bitstream structure is illustrated in Fig. 8.12.

The respective bitstream layers correspond to the encoder and decoder modules that have been introduced in the previous section. In total, the bitstream comprises 12 hierarchical layers:

- the *core layer* (Layer 1) is interoperable with ITU-T G.729;

- Layer 2 (at 12 kbit/s) is a *narrowband enhancement layer* consisting of cascaded CELP parameters and signal class information (voiced, unvoiced, onset, or voiced/unvoiced transition);

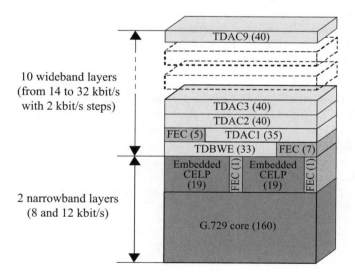

Figure 8.12: G.729.1 bitstream structure of a given 20 ms frame (numbers in parentheses specify the number of bits within the respective part of the bitstream)

- Layer 3 (at 14 kbit/s) is a *wideband extension layer*, comprising time-domain bandwidth extension (TDBWE) parameters and phase information for FEC;

- Layers 4 to 12 (above 14 kbit/s) are *wideband enhancement layers*, comprising energy information and transform coding parameters, which are referred to as time-domain aliasing cancellation (TDAC) parameters.

d) Delay, Quality, and Complexity

An evaluation of the G.729.1 codec in terms of delay, obtained speech and music quality as well as algorithmic complexity is presented here based on the ITU-T characterization.

The maximum algorithmic delay of G.729.1 is 48.9375 ms. The contributions to this delay are 40 ms due to the MDCT windowing used in the TDAC coder (current frame plus one frame of lookahead), 5 ms for the G.729 lookahead, and 3.9375 ms for the QMF analysis and synthesis. However, this delay can be reduced depending on the selected encoder and decoder modes. For instance, for input and output signals sampled at 8 kHz and for the low delay decoder mode, the algorithmic delay of G.729.1 is reduced to 25 ms.

The characterization test results of ITU-T G.729.1 [ITU-T TD258 2006] are summarized in Fig. 8.13-a, -b, and Fig. 8.14 for narrowband clean speech, wideband clean speech and wideband music signals, respectively. The obtained speech and audio quality is compared with selected reference coders of the same acoustical bandwidth. These results are expressed in terms of "mean opinion scores" (MOS) [ITU-T Rec. P.800 1996] with a 95% confidence interval. In the experiments, a residual uncertainty of about ±0.1 MOS was obtained.

According to these results, G.729.1 is better than G.729 Annex A [ITU-T Rec. G.729 Annex A 1996] at 8 kbit/s and equivalent to G.729 Annex E for clean speech at 12 kbit/s. At 14 kbit/s, G.729.1 has a quality similar to AMR-WB [3GPP TS 26.190 2001] at 12.65 kbit/s for clean speech. At 24 and 32 kbit/s, it is similar to AMR-WB at 23.85 kbit/s for clean speech. Furthermore, music quality at 32 kbit/s is good for a conversational coder (close to G.722 [ITU-T Rec. G.722 1988] at 64 kbit/s). In addition, G.729.1 is more robust against frame errors (i.e., packet losses) than the reference coders.

The computational complexity of G.729.1 is scalable with the bit rate as shown in Fig. 8.15. At 32 kbit/s, the total encoder and decoder complexity is around 35.8 weighted million operations per second (WMOPS). Memory requirements (RAM and ROM) are tabulated in [ITU-T Rec. G.729.1 2006].

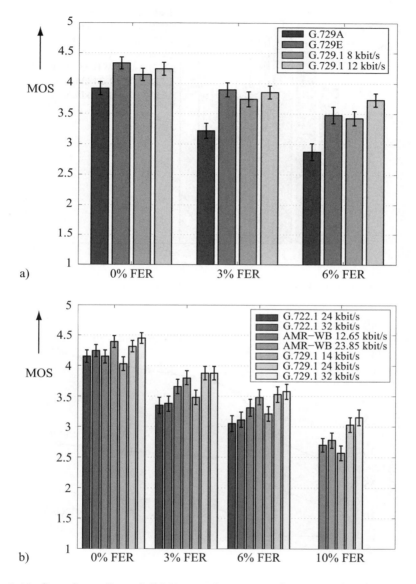

Figure 8.13: Speech quality of G.729.1 and various reference codecs. The quality is evaluated for different frame error rates (FER) in terms of mean opinion score (MOS) with signals at −26 dBov nominal level.
a) Narrowband clean speech quality
b) Wideband clean speech quality

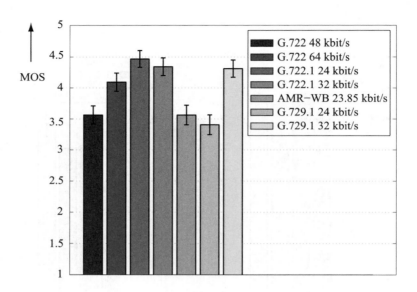

Figure 8.14: Wideband music quality of G.729.1 in error-free condition

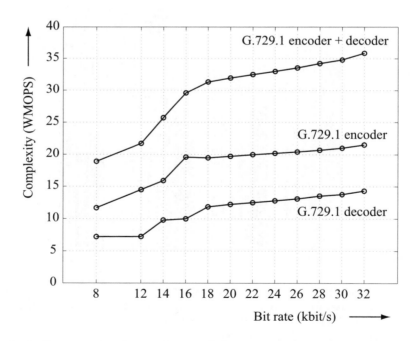

Figure 8.15: Computational complexity of G.729.1. The computational complexity is measured in weighted million operations per second (WMOPS) for 16 kHz-sampled input and output signals

e) Application Example: Wideband Telephony with G.729.1

Figure 8.16 shows an example on how an embedded speech coder, here G.729.1, can be used in a heterogeneous VoIP environment for telephony services. Various terminals are used (e.g., Wi-Fi phones, analog phones and soft clients). Depending on hardware capabilities as well as on the capacity of the respective transmission links, a call agent — managing the quality of service — can decide on the transmitted bit rates. The encoder does not need to be made aware of the decision. The call agent will just "tailor" the bitstream to convey the highest quality that is achievable within the given capacity and complexity constraints.

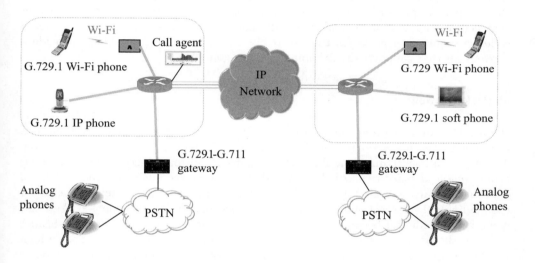

Figure 8.16: Enterprise wideband VoIP based on ITU-T G.729.1

With G.729.1, being an embedded extension of G.729, narrowband terminals based on G.729 are still supported. Furthermore, the interconnection with PSTN involves transcoding between G.729.1 and G.711 in a dedicated gateway. In the example application, G.729.1 at 12 kbit/s can be used instead of G.729/G.729.1 at 8 kbit/s to improve the VoIP/PSTN interconnection quality.

G.729.1 is endorsed as a Korean standard [TTAE.IT-G.729.1 2006] and is the broadband convergence network (BcN) [6] quality reference wideband coder for telephony services in Korea. G.729.1 is also an optional codec for wideband speech coding in New Generation DECT systems [ETSI TS 102 527-1 2007].

---

[6]Broadband convergence Network (BcN) is the name for Next Generation Network (NGN) in Korea.

8.5 Network Aspects of Embedded Speech Coding

So far, the theory, the design as well as standardized solutions for embedded speech coding have been reviewed. In order to complete the discussion of embedded speech coding, this section addresses the *practical* issues that are encountered when network entities utilize the embedded bitstream property of the respective codecs.

8.5.1 Implementation and Utilization of Scalability

Embedded speech codecs offer very attractive features. Yet, the actual utilization of their respective properties in the network is not necessarily trivial. In fact, a packet switched communication network that will take full advantage of an embedded speech codec is required to *packetize* the coded bits into a suitable *format*, and provide a *mechanism* that supports the dropping of bitstream layers.

a) Packetization

In speech communication over packet switched networks, the so-called *payload*, i.e., the coded bits, are arranged in *data packets* with associated *header* information. Thereby, one packet may contain a single or even multiple speech frames. The use of an *embedded codec* requires the payload to be organized such that the individual codec layers can be easily removed. A simple possibility is to generate one packet for each individual layer as the dropping of whole packets is easily implemented. Yet, this can be very inefficient because of a large overhead that is due to the size of the header (e.g., for the IPv4 or IPv6 protocols). A way to reduce the overhead is to pack a high number of speech frames into one packet at the cost of delay and a higher sensitivity to packet loss. For *conversational* applications, the number of speech frames in an IP packet has to be kept low to avoid a high end-to-end delay, and particularly for embedded bitstreams, it is preferable to gather all embedded layers in the same data packet. Typically, *one* speech frame per packet is the preferred number if the frame size is about 20–30 ms. Possibly, a rearrangement of the coded bits is necessary such that the codec layers appear as continuous blocks within the packet. This is, for example, the case for the hierarchically quantized prediction error of the ITU-T G.727 codec (see Sec. 8.4.2). The respective method is specified in [ITU-T Rec. G.764 1990], which defines a protocol for the transport of packetized speech and a way to handle the scalable bitstream of G.727, where the payload header indicates the number of blocks that can be dropped from the packet.

b) Transport Issues

Apart from the bitstream organization, the respective *transport protocols* in packet switched networks need to explicitly support the embedded coding features. First, as shown in the previous section, a precise description of the bitstream format is mandatory. Secondly, the protocols must provide a mechanism to drop certain blocks from a single data packet. In this case, it is also necessary to modify the packet and/or payload header accordingly. For instance, the indicated size and structure of the payload needs to be updated and, possibly, a checksum must be recomputed.

For the application of (embedded) speech transmission it is beneficial to implement a packet prioritization to ensure a certain "Quality of Service" (QoS). In fact, a speech payload can be considered more important than pure data for which a retransmission is in principle possible. As an example, [Kaye, Zhang 1994] focused on the transmission of voice and data over so-called Frame-Relay networks and studied means to apply congestion control mechanisms. First, the speech packets are labeled (classified). As a result, different mechanisms can be applied to the different packet classes. For the particular case of *embedded* bitstreams, a second classification can distinguish bits of high importance (core layer bits) from the bits of low importance (enhancement layer bits) within the payload. Then, a rate adaptation unit, located anywhere in the network, can drop packets or blocks that are labeled as being of lower importance.

In fact, data labeling is already available within IPv4 or IPv6 (Internet Protocol versions 4 and 6) through the so-called "Differentiated Services" (DiffServ) [IETF RFC 2474 1998]. Here, any involved network node conforming to this standard can decide to get rid of bitstream layers of low importance according to the traffic conditions in the network. To do so, the IP packets have to be organized such that the different bitstream layers are packetized in different IP packets. The DiffServ labelling protocol applies to a full IP packet and not to parts of it. As IP packets have a considerable protocol overhead (40–60 bytes) and as a speech codec payload is usually small (about 80 bytes for a frame of G.729.1 at 32 kbit/s), it is more efficient to pack many speech codec frames together. As a result, this method is better suited to audio *streaming* applications. In particular, the transmission of scalable *MPEG-4 audio/video streams* (without using the respective MPEG-4 method) is possible with [IETF RFC 3016 2000]. First, the audio/video data is multiplexed by "Low-overhead MPEG-4 Audio Transport Multiplex" (LATM) and then each audio/video multiplexed layer is packetized into different "Real Time Protocol" (RTP) packets, allowing for the different layers to be treated differently at the IP level, for example, via DiffServ.

c) Rate Adaptation Strategies

In addition to the packetization and protocol issues discussed above, a communication network that works with embedded bitstreams needs to implement so-called "Rate Adaptation Units" that perform the actual removal of embedded bitstream layers. Thereby, an adequate adaptation *strategy* is required in order to decide which bitstream needs to be reduced. The respective decision is usually based on a few basic characteristics, as below.

- The *importance* of the layers within a packet (codec features) — For instance, a classification of the embedded bitstream into high and lower importance layers enables the rate adaptation units to exploit this "tag".

- Network *traffic conditions* — For example in wireless networks, a permanent monitoring can be performed and a network-aware wireless gateway (Wi-Fi access point) can adapt the instantaneous bit rate according to the current network capacity [Mathieu et al. 2005].

- Terminal capabilities — Low cost devices may, e.g., only be capable of decoding the core layer of the bitstream. Hence, all enhancement layers can be stripped.

- User/Customer preferences — "Premium" users could be preferred over "normal" users and thus receive a higher bit rate and hence, a higher quality.

In MPEG, a way to adapt the content of the scalable codec bitstreams to the end terminal, to the access network capabilities as well as the user preferences was specified in the MPEG-21 standard [ISO/IEC 21000-07 2004] through the "Digital Item Adaptation" (DIA) technologies. A new language based on "XML Schema", called "Bitstream Syntax Description Language" (BSDL) describes the syntax of a particular coding format. These "schemas" can be used by a generic processor to automatically parse a bitstream and generate its description, and vice-versa. The BSDL concept requires a resource adaptation engine to be aware of the codec specific schema in order to parse the BSD and generate the corresponding (adapted) bitstreams. As this is not always desired, a second concept was defined in which a "generic Bitstream Syntax" (gBS) which enables the "codec-agnostic" description of the bitstream by describing syntactical bitstream units in a hierarchical fashion. It also provides semantic handles to sections of the bitstream that facilitate semantic-based manipulation or removal of bitstream portions (e.g., to reduce the bit rate of an embedded codec). The (g)BSD of a stream needs to be known for the adaptation process and can be either transmitted as metadata with the codec bitstream or made available off-line. More information on BSD and bitstream adaptation with an application to video transmission is available in [Panis et al. 2003]. Naturally, MPEG-21 can also be applied to streaming of *audio content* as, e.g., described by [Feiten et al. 2005].

d) Application Example: Embedded Speech Coding for Conferencing

The implementation and utilization of scalability can be illustrated with the example of multi-party audio conferencing. Signaling (i.e., how to initiate, modify and terminate sessions) is not discussed here, however it may be implemented, for instance, using the Session Initiation Protocol (SIP) [IETF RFC 3261 2002].

Figure 8.17 shows an example of *centralized architecture* comprising a central unit, known as the *conference bridge*, and end points, which are terminals. All participants in the conference are interconnected through the bridge, which may either mix, multiplex or replicate media streams, depending on the type of bridge. Media streams are typically transported using the RTP/UDP/IP protocol stack [IETF RFC 3550 2003]. For the scenario under consideration, the audio codec is typically negotiated between the end points and bridge through the Session Description Protocol

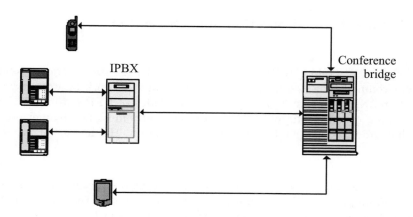

Figure 8.17: Centralized audio conferencing with bit rate adaptation in the bridge. Two IP terminals access the conference through an IP branch exchange (IPBX)

(SDP) [IETF RFC 4566 2006]. The bridge centralizes all information about negotiated audio codecs and other system characteristics. There are two typical problems in the conferencing system shown in Fig. 8.17:

1. how to handle heterogeneous capabilities, i.e., allow conferences between participants with various transmission or computation capacities, without forcing all participants to a default low bit rate mode;

2. how to adapt to variation of link capacities (including congestion) to keep continuous data stream and avoid packet losses.

These problems may be addressed using embedded coding for up- and down-streams, and implementing bit rate adaptation (i.e., bitstream truncation) in the conference bridge [Deléam et al. 2005]. In particular the adaptation decision can be based on static link capacities given by SDP, or dynamic link capacities estimated by the Real Time Control Protocol (RTCP). Other factors, such as the importance (or priority) of each embedded layer or terminal, may also be included in the decision. The main advantage of embedded coding is to be able to solve the problems mentioned above through simple bitstream manipulation. Still, if the bridge is acting as a mixer, decoding, mixing and re-encoding is needed. Though, in this case, embedded coding can still be beneficial using the concept of *partial mixing* [Hiwasaki et al. 2006].

If audio streams are encoded by ITU-T G.729.1, several important issues for the transmission of such embedded bitstreams are addressed in the the RTP payload format specification [IETF RFC 4749 2006]. This standard explicitly allows one to adapt bit rate, possibly in a dynamic way during a session, taking into account service requirements and network constraints.

8.5.2 Unequal Error Protection and Encryption

Unequal error protection (UEP) is often implemented for speech transmission in a mobile environment, for example, in the GSM cellular network between a base station (BS) and the mobile station (MS). The basic idea of UEP is to apply a robust channel coding over the bits of high importance and almost no channel protection to bits that are less sensitive to bit errors. As a result, the impact of bit errors is minimized.

UEP is easily applicable to scalable codecs as their layered structure already indicates which bits are important. For example, it is clear that most of the bits from the core codec should be well protected. The easiest solution would be to protect each bit-stream layer with a dedicated channel code. But still, some bits from the enhancement layers can also be considered as important as the core. In order to ensure an optimal quality, one must conduct a *bit error sensitivity analysis* to evaluate the relative importance of the bits. This was, for example, applied to the Multi-Mode Transform Predictive Coding scheme in [Taddei et al. 2002]. Here, according to the bit error sensitivity, bits were ranked and grouped in different classes with their associated channel protections. The sensitivity values have mainly been obtained through informal listening tests and segmental SNR measurements. As a further example, [Bernard et al. 2002] applied UEP to the embedded ITU-T G.727 codec (Sec. 8.4.2) using rate-compatible punctured convolutional codes (RCPC) and rate-compatible punctured trellis codes (RCPT) for channel coding.

When transmitting Voice over IP, it is possible to use IP security (IPsec) in order to encrypt the voice data. IPsec can be run either in transport mode or in tunnel mode. Transport mode is used to help protect end-to-end communications. In this mode, the IP payload is encrypted and the original headers are left intact. Tunnel mode is most commonly used to encrypt site-to-site traffic and traffic between networks. When IPsec tunnel mode is used, a new IP packet encapsulates the entire original IP that is then protected by one of the IPsec protocol formats. Then, it becomes difficult to profit from the scalability feature of the codec bitstream in the network as either the payload (no access to the data) or the IP headers are encrypted (no possibility to detect the type of content).

Nevertheless, there have been some studies on how to encrypt the payload that can be applied to have secured communications without using IPsec. A first idea is to encrypt the core bitstream and to leave the other layers untouched [Gibson et al. 2004]. Yet, this method is not really designed for safe communications as unencrypted higher codec layers potentially suffice to regenerate a coarse (intelligible) speech signal. Another solution, [Hofbauer et al. 2006], proposed an encryption scheme that preserves the bitstream scalability in the encrypted domain. In this case, enhancement layers can be simply dropped by any equipment in the network that is aware of the structure of such an encrypted stream.

8.6 Conclusions and Perspectives

Embedded speech coding is in essence reflected by its layered bitstream format that renders the decoding of partially received data possible. This feature, which is often called *bitstream scalability*, can be exploited in many ways to cope efficiently with the heterogeneity of network capacity and terminal capability, or to adapt to time-varying conditions such as network congestion.

Recently, embedded speech coding techniques have become the subject of intense research, and the need for embedded speech coding has been clearly recognized by the industry, resulting in new standardization activities. Indeed, bitstream scalability facilitates the deployment of new codecs that are built as embedded extensions of widely deployed codecs such as [ITU-T Rec. G.729 1996] or [ITU-T Rec. G.711 1972]. For this reason, ITU-T Study Group 16 (SG16) has standardized G.729.1 [Ragot et al. 2007] which is an 8–32 kbit/s embedded coder providing bitstream interoperability with G.729 and its Annexes A and B. Also, a wideband extension of G.711 called G.711WB is being developed within ITU-T (cf. [Hiwasaki et al. 2006]), targeting audio conferencing applications on optical fiber access and interoperability with the public switched telephone network (PSTN). In parallel, ITU-T SG16 is developing a coder called "Embedded Variable Bit Rate" (EV-VBR) [Gibbs 2006], which has attributes similar to G.729.1. EV-VBR is a narrowband/wideband coder operating at 8, 12, 16, 24, and 32 kbit/s with a 20 ms frame length. However, its core coder has no bitstream interoperability constraint. The emphasis of EV-VBR is on high quality and robustness against packet losses with relaxed complexity constraints, and its foreseen applications are 4th generation (4G) mobile communications. Besides, super-wideband (50–14000 Hz) and stereo extensions of both G.729.1 and EV-VBR are under study in ITU-T which will extend the operating bit rates of both coders to 8–64 kbit/s.

The bitstream of an embedded coder is normally hierarchical in the sense that enhancement layers are meaningful only if the underlying layers are also received. Then, bit rate adaptation boils down to simple bitstream truncation. There are some cases where the constraint of layer hierarchy can be partially relaxed to allow more flexible bitstream configurations [Hiwasaki et al. 2006]. For instance in G.711WB, the bitstream at the maximum rate (96 kbit/s) consists of a 64 kbit/s narrowband G.711 core layer, one 16 kbit/s low band (50–4000 Hz) enhancement layer, and one 16 kbit/s high band (4000–7000 Hz) enhancement layer. At 80 kbit/s, the bitstream can be formed by adding either the low band enhancement or the high band enhancement to the core layer, which violates the strict layer hierarchy but provides more flexibility. A similar idea is pursued in the super-wideband and stereo extension of EV-VBR, wherein super-wideband and stereo capabilities are associated with different enhancement layers and it is allowed to drop intermediate layers to obtain more flexible modes: super-wideband mono, wideband stereo, or super-wideband stereo. Finally, as shown, embedded coding techniques are often used to enhance the quality of present transmission systems. The approach of adding "enhancement layers" to the standardized bitstream ensures interoperability with legacy equipment meaning that the additional bits can be discarded if necessary. However, there may be situations that require a true *backwards compatibility*. This means that any modification of the bitstream format is prohibited. Consequently, conventional embedded coding techniques can not be applied anymore as additional bitstream layers are disallowed. A solution to provide an enhanced quality despite such strong restrictions is the use of *data hiding* techniques or *steganography*. A respective transmission system "hides" the enhancement bits in the core layer signal by inserting a so-called "watermark" signal. For example in real-time speech transmission, the data hiding mechanism can be efficiently *integrated* with the speech encoder [Geiser, Vary 2007] and the hidden data can be used for *bandwidth extension* purposes. This idea is discussed in more detail in the following chapter.

Interestingly, the basic tools of embedded coding as introduced in Sec. 8.2 are quite generic and fundamental, and the applications of embedded coding are by far not limited to real-time speech transmission. In fact, embedded audio, image and video coding have been studied extensively. Examples of embedded audio coders are MPEG-4 "Bit Sliced Arithmetic Coding" (BSAC) [Park et al. 1997], which is based on the idea of bit plane coding, and the MPEG Surround standard [Herre et al. 2007]. Examples of embedded image and video coders are JPEG2000 [Taubman 2000] and the joint ITU/MPEG-4 standard for *Scalable Video Coding* (SVC) [ITU-T Rec. H.264 Amd. 3 2007]. In particular, embedded coding is also very interesting for multimedia streaming or storage applications since bitstream scalability provides an efficient way to support a variety of bit rates and device capabilities without the need for multiple re-encodings.

Bibliography

3GPP TS 26.190 (2001). *AMR Wideband Speech Codec; Transcoding Functions*, 3rd Generation Partnership Project (3GPP).

3GPP TS 26.404 (2004). *Enhanced aacPlus General Audio Codec; Encoder Specification; Spectral Band Replication (SBR) Part*, 3rd Generation Partnership Project (3GPP).

Aguilar, G.; Chen, J.-H.; Dunn, R.; McAulay, R.; Sun, X.; Wang, W.; Zopf, R. (2000). An Embedded Sinusoidal Transform Codec with Measured Phases and Sampling Rate Scalability, *Proceedings of the IEEE International Conference on Acoustics, Speech, and Signal Processing (ICASSP)*, Istanbul, Turkey, pp. 1141–1144.

Berger, T. (1971). *Rate Distortion Theory*, Prentice-Hall, Inc., Englewood Cliffs, NJ, USA.

Bernard, A.; Liu, X.; Wesel, R. D.; Alwan, A. (2002). Speech Transmission Using Rate-Compatible Trellis Codes and Embedded Source Coding, *IEEE Transactions on Communications*, vol. 50, no. 2, pp. 309–320.

Bessette, B.; Lefebvre, R.; Jelínek, M.; Rotola-Pukkila, J.; Mikkola, H.; Järvinen, K. (2002). The Adaptive Multirate Wideband Speech Codec (AMR-WB), *IEEE Transactions on Speech and Audio Processing*, vol. 10, no. 8, pp. 620–636.

Burt, P.; Adelson, E. H. (1983). The Laplacian Pyramid as a Compact Image Code, *IEEE Transactions on Communications*, vol. 31, April, pp. 532–540.

Ching, Y.-C. (1973). Differential Pulse Code Communications System Having Dual Quantization Schemes, U.S. Patent 3 781 685.

Chou, P. A.; Lookabaugh, T.; Gray, R. M. (1989). Optimal Pruning with Applications to Tree-Structured Source Coding and Modeling, *IEEE Transactions on Information Theory*, vol. 35, no. 2, pp. 299–315.

Davidson, A.; Gersho, A. (1988). Multiple-Stage Vector Excitation Coding of Speech Waveforms, *Proceedings of the IEEE International Conference on Acoustics, Speech, and Signal Processing (ICASSP)*, Tokyo, Japan, pp. 163–166.

De Iacovo, R. D.; Sereno, D. (1991). Embedded CELP Coding for Variable Bit-Rate Between 6.4 and 9.6 kbit/s, *Proceedings of the IEEE International Conference on Acoustics, Speech, and Signal Processing (ICASSP)*, Toronto, ON, Canada, pp. 681–684.

Deléam, D.; Sollaud, A.; Chmitt, J.-C.; Boissonade, P.; Calliger, O. (2005). Videoconference System, U.S. Patent 2005088514.

Dietz, M.; Liljeryd, L.; Kjörling, K.; Kunz, O. (2002). Spectral Band Replication: A Novel Approach in Audio Coding, *Proceedings of 112th Convention of the AES*, Munich, Germany. Preprint 5553.

Dymarski, P.; Moreau, N.; Vigier, A. (1990). Optimal and Sub-Optimal Algorithms for Selecting the Excitation in Linear Predictive Coders, *Proceedings of the IEEE International Conference on Acoustics, Speech, and Signal Processing (ICASSP)*, Albuquerque, NM, USA, pp. 485–488.

Effros, M.; Dugatkin, D. (2004). Multiresolution Vector Quantization, *IEEE Transactions on Information Theory*, vol. 50, no. 12, pp. 3130–3145.

Ehara, H.; Morii, T.; Yoshida, K. (2007). Predictive Vector Quantization of Wideband LSF Using Narrowband LSF for Bandwidth Scalable Coders, *Speech Communication*, vol. 49, no. 6, pp. 490–500.

El Gamal, A.; Cover, T. M. (1982). Achievable Rates for Multiple Descriptions, *IEEE Transactions on Information Theory*, vol. 28, no. 6, pp. 851–857.

Equitz, W. H. R.; Cover, T. M. (1991). Successive Refinement of Information, *IEEE Transactions on Information Theory*, vol. 37, no. 2, pp. 269–275.

Erdmann, C. (2005). *Hierarchical Vector Quantization: Theory and Application to Speech Coding*, PhD thesis. Aachener Beiträge zu digitalen Nachrichtensystemen, vol. 19, P. Vary (ed.), RWTH Aachen University.

Erdmann, C.; Vary, P. (2002). Embedded Speech Coding Based on Pyramid CELP, *IEEE Workshop on Speech Coding (SCW)*, pp. 29–31.

Erdmann, C.; Vary, P. (2004). Performance of Multistage Vector Quantization in Hierarchical Coding, *European Transactions on Telecommunications*, vol. 15, no. 4, pp. 363–372.

ETSI EN 300 175-8 (2005). *Digital Enhanced Cordless Telecommunications (DECT); Common Interface (CI); Part 8: Speech Coding and Transmission*, European Telecommunications Standards Institute (ETSI).

ETSI GSM 08.62 (2000). *Digital Cellular Telecommunication System (Phase 2+); Inband Tandem Free Operation (TFO) of Speech Codecs; Service Description; Stage 3*, European Telecommunications Standards Institute (ETSI).

ETSI TS 102 527-1 (2007). *Digital Enhanced Cordless Telecommunications (DECT); New Generation DECT; Part 1: Wideband Speech*, European Telecommunications Standards Institute (ETSI).

Feiten, B.; Wolf, I.; Oh, E.; Seo, J.; Kim, H.-K. (2005). Audio Adaptation According to Usage Environment and Perceptual Quality Metrics, *IEEE Transactions on Multimedia*, vol. 7, no. 3, pp. 446–453.

Feng, H.; Effros, M. (2003). Improved Bounds for the Rate Loss of Multiresolution Source Codes, *IEEE Transactions on Information Theory*, vol. 49, no. 4, pp. 809–821.

Geiser, B.; Jax, P.; Vary, P.; Taddei, H.; Schandl, S.; Gartner, M.; Guillaumé, C.; Ragot, S. (2007). Bandwidth Extension for Hierarchical Speech and Audio Coding in ITU-T Rec. G.729.1, *IEEE Transactions on Audio, Speech, and Language Processing*, vol. 15, no. 8, pp. 2496–2509.

Geiser, B.; Vary, P. (2007). Backwards Compatible Wideband Telephony in Mobile Networks: CELP Watermarking and Bandwidth Extension, *Proceedings of the IEEE International Conference on Acoustics, Speech, and Signal Processing (ICASSP)*, Honolulu, HI, USA, pp. 533–536.

Gerson, I.; Jasiuk, M. (1990). Vector Sum Excited Linear Prediction (VSELP) Speech Coding at 8 kbps, *Proceedings of the IEEE International Conference on Acoustics, Speech, and Signal Processing (ICASSP)*, Albuquerque, NM, USA, pp. 461–464.

Gibbs, J. A. (2006). The ITU-T Embedded Variable Bit Rate Speech Coder – Requirements and Standardization, *Proceedings of 44th Annual Allerton Conference*, Allerton House, IL, USA.

Gibson, J.; Servetti, A.; Dong, H.; Gersho, A.; Lookabaugh, T.; De Martin, J. (2004). Selective Encryption and Scalable Speech Coding for Voice Communications over Multi-Hop Wireless Links, *IEEE Military Communications Conference (MILCOM)*, Monterey, CA, USA, pp. 792–798.

Goodman, D. J. (1980). Embedded DPCM for Variable Bit Rate Transmission, *IEEE Transactions on Communications*, vol. 28, no. 7, pp. 1040–1046.

Grill, B. (1997). A Bit Rate Scalable Perceptual Coder for MPEG-4 Audio, *Proceedings of 103rd Convention of the AES*, New York, NY, USA. Preprint 4620.

Herre, J.; Kjörling, K.; Breebaart, J.; Faller, C.; Disch, S.; Purnhagen, H.; Koppens, J.; Hilpert, J.; Rödén, J.; Oomen, W.; Linzmeier, K.; Chong, K. (2007). MPEG Surround – The ISO/MPEG Standard for Efficient and Compatible Multi-Channel Audio Coding, *Proceedings of 122nd Convention of the AES*, Vienna, Austria. Preprint 7084.

Hersent, O.; Petit, J.-P.; Gurle, D. (2005). *IP Telephony : Deploying Voice-over-IP Protocols*, John Wiley & Sons, Ltd, Chichester, UK.

Hiwasaki, Y.; Mori, T.; Ohmuro, H.; Ikedo, J.; Tokumoto, D.; Kataoka, A. (2004). Scalable Speech Coding Technology for High-Quality Ubiquitous Communications, *NTT Technical Review*, vol. 2, no. 3, pp. 53–58.

Hiwasaki, Y.; Ohmuro, H.; Mori, T.; Kurihara, S.; Kataoka, A. (2006). A G.711 Embedded Wideband Speech Coding for VoIP Conferences, *IEICE Transactions on Information and Systems*, vol. E89-D, no. 9, pp. 2542–2552.

Hofbauer, H.; Stütz, T.; Uhl, A. (2006). Selective Encryption for Hierarchical MPEG, *Proceedings of the 10th IFIP International CMS 2006 Conference*, pp. 151–160.

IETF RFC 2474 (1998). *Definition of the Differentiated Services Field (DS Field) in the IPv4 and IPv6 Headers)*, Internet Engineering Task Force (IETF).

IETF RFC 3016 (2000). *RTP Payload Format for MPEG-4 Audio/Visual Streams*, Internet Engineering Task Force (IETF).

IETF RFC 3261 (2002). *SIP: Session Initiation Protocol*, Internet Engineering Task Force (IETF).

IETF RFC 3550 (2003). *RTP: A Transport Protocol for Real-Time Applications*, Internet Engineering Task Force (IETF).

IETF RFC 4566 (2006). *SDP: Session Description Protocol*, Internet Engineering Task Force (IETF).

IETF RFC 4749 (2006). *RTP Payload Format for G.729.1*, Internet Engineering Task Force (IETF).

ISO/IEC 14496-3 (2005). *Information technology – Coding of Audio-Visual Objects – Part 3: Audio*, ISO/IEC JTC1/SC29/WG11 MPEG.

ISO/IEC 21000-07 (2004). *Information Technology – Multimedia Framework (MPEG-21) – Part 7: Digital Item Adaptation*, ISO/IEC JTC1/SC29/WG11 MPEG.

ITU-T Rec. G.711 (1972). *Pulse Code Modulation (PCM) of Voice Frequencies*, International Telecommunication Union (ITU).

ITU-T Rec. G.722 (1988). *7 kHz Audio-Coding within 64 kbit/s*, International Telecommunication Union (ITU).

ITU-T Rec. G.723.1 (1996). *Dual Rate Speech Coder for Multimedia Communications Transmitting at 5.3 and 6.3 kbit/s*, International Telecommunication Union (ITU).

ITU-T Rec. G.727 (1990). *5-, 4-, 3- and 2-bits Sample Embedded Adaptive Differential Pulse Code Modulation (ADPCM)*, International Telecommunication Union (ITU).

ITU-T Rec. G.729 (1996). *Coding of Speech at 8 kbit/s Using Conjugate Structure Algebraic-Code-Excited Linear-Prediction (CS-ACELP)*, International Telecommunication Union (ITU).

ITU-T Rec. G.729 Annex A (1996). *Reduced Complexity 8 kbit/s CS-ACELP Speech Codec*, International Telecommunication Union (ITU).

ITU-T Rec. G.729 Annex B (1996). *A Silence Compression Scheme for G.729 Optimized for Terminals Conforming to Recommendation V.70*, International Telecommunication Union (ITU).

ITU-T Rec. G.729.1 (2006). *An 8-32 kbit/s Scalable Wideband Coder Bitstream Interoperable with G.729*, International Telecommunication Union (ITU).

ITU-T Rec. G.764 (1990). *Voice Packetization - Packetized Voice Protocols*, International Telecommunication Union (ITU).

ITU-T Rec. H.242 (1999). *System for Establishing Communication between Audiovisual Terminals Using Digital Channels up to 2 Mbit/s*, International Telecommunication Union (ITU).

ITU-T Rec. H.264 Amd. 3 (2007). *Advanced Video Coding for Generic Audiovisual Services: Scalable Video Coding*, International Telecommunication Union (ITU).

ITU-T Rec. H.320 (2004). *Narrow-Band Visual Telephone Systems and Terminal Equipment*, International Telecommunication Union (ITU).

ITU-T Rec. P.800 (1996). *Methods for Subjective Determination of Transmission Quality*, International Telecommunication Union (ITU).

ITU-T TD258 (2006). Executive Summary of G729.1 Characterisation Step 2 – Experiments 1, 2 & 3, TD-258-GEN/16 Attachment 2 (Source: Q7/12 Rapporteurs).

Jax, P.; Geiser, B.; Schandl, S.; Taddei, H.; Vary, P. (2006). An Embedded Scalable Wideband Codec Based on the GSM EFR Codec, *Proceedings of the IEEE International Conference on Acoustics, Speech, and Signal Processing (ICASSP)*, Toulouse, France, pp. 5–8.

Jayant, N. (1983). Variable rate ADPCM Based on Explicit Noise Coding, *Proceedings of the IEEE Global Telecommunications Conference (GLOBECOM)*, San Diego, CA, USA, pp. 657–677.

Jayant, N.; Noll, P. (1984). *Digital Coding of Waveforms: Principles and Applications to Speech and Video*, Prentice Hall, Englewood Cliffs, NJ, USA.

Johnson, M.; Taniguchi, T. (1990). Pitch-Orthogonal Code-Excited LPC, *Proceedings of the IEEE Global Telecommunications Conference (GLOBECOM)*, Dallas, TX, USA, pp. 542–546.

Juang, B.-H.; Gray, A. H. (1982). Multiple Stage Vector Quantization for Speech Coding, *Proceedings of the IEEE International Conference on Acoustics, Speech, and Signal Processing (ICASSP)*, Paris, France, pp. 597–600.

Jung, S.-K.; Kini, K.-T.; Kang, H.-G. (2004). A Bit-Rate/Bandwidth Scalable Speech Coder Based on ITU-T G.723.1 Standard, *Proceedings of the IEEE International Conference on Acoustics, Speech, and Signal Processing (ICASSP)*, Montreal, QC, Canada, pp. 285–288.

Kataoka, A.; Kurihara, S.; Sasaki, S.; Hayashi, S. (1997). A 16-kbit/s Wideband Speech Codec Scalable with G.729, *Proceedings of the European Conference on Speech Communication and Technology (EUROSPEECH)*, Rhodes, Greece, pp. 1491–1494.

Kaye, A. R.; Zhang, S. (1994). Congestion Control in Integrated Voice-Data Frame Relay Networks and the Case for Embedded Coding, *Proceedings of the IEEE Global Telecommunications Conference (GLOBECOM)*, San Francisco, CA, USA, pp. 1565–1570.

Kim, K.-T.; Jung, S.-K.; Clark, Y.-C.; Youn, D. (2002). A New Bandwidth Scalable Wideband Speech/Audio Coder, *Proceedings of the IEEE International Conference on Acoustics, Speech, and Signal Processing (ICASSP)*, Orlando, FL, USA, pp. 657–660.

Kleijn, W. B.; Paliwal, K. K. (eds.) (1995). *Speech Coding and Synthesis*, Elsevier Science Inc., New York, NY, USA.

Koishida, K.; Cuperman, V.; Gersho, A. (2000). A 16 kbit/s Bandwidth Scalable Audio Coder Based on the G.729 Standard, *Proceedings of the IEEE International Conference on Acoustics, Speech, and Signal Processing (ICASSP)*, Istanbul, Turkey, pp. 1149–1152.

Kondo, K.; Ohno, M. (1994). Packet Speech Transmission on ATM Networks Using a Variable Rate Embedded ADPCM Coding Scheme, *IEEE Transactions on Communications*, vol. 42, February, pp. 243–247.

Koshelev, V. N. (1980). Hierarchical Coding of Discrete Sources, *Problemy Peredachi Informatsii*, vol. 16, no. 3, pp. 31–49.

Krishnan, V.; Rajendran, V.; Kandhadai, A.; Manjunath, S. (2007). EVRC-Wideband: The New 3GPP2 Wideband Vocoder Standard, *Proceedings of the IEEE International Conference on Acoustics, Speech, and Signal Processing (ICASSP)*, Honolulu, HI, USA, pp. 333–336.

Kövesi, B.; Massaloux, D.; Sollaud, A. (2004). A Scalable Speech and Audio Coding Scheme with Continuous Bitrate Flexibility, *Proceedings of the IEEE International Conference on Acoustics, Speech, and Signal Processing (ICASSP)*, Montreal, QC, Canada, pp. 273–276.

Larsen, E.; Aarts, R. M. (eds.) (2004). *Audio Bandwidth Extension*, John Wiley & Sons, Ltd, New York, NY, USA.

Lastras, L.; Berger, T. (2001). All Sources are Nearly Successively Refinable, *IEEE Transactions on Information Theory*, vol. 47, no. 3, pp. 918–926.

Le Guyader, A.; Lamblin, C.; Boursicaut, E. (1995). Embedded Algebraic CELP/VSELP Coders for Wideband Speech Coding, *Speech Communication*, vol. 16, no. 4, pp. 319–328.

Le Guyader, A.; Lozach, B.; Moreau, N. (1992). Embedded Algebraic CELP Coders for Wideband Speech Coding, *Proceedings of European Signal Processing Conference (EUSIPCO), Signal Processing VI: Theory and Applications*, Brussels, Belgium, pp. 527–530.

Lee, D. H.; Neuhoff, D. L.; Paliwal, K. K. (1991). Cell-Conditioned Multistage Vector Quantization, *Proceedings of the IEEE International Conference on Acoustics, Speech, and Signal Processing (ICASSP)*, Toronto, ON, Canada, pp. 653–656.

Lee, E. D.; Lee, M. S.; Kim, D. Y. (2003). Global Pulse Replacement Method for Fixed Codebook Search of ACELP Speech Codec, *Proceedings of IASTED Communications, Internet, and Information Technology Conference (CIIT)*, Scottsdale, AZ, USA.

Linde, Y.; Buzo, A.; Gray, R. M. (1980). An Algorithm for Vector Quantizer Design, *IEEE Transactions on Communications*, vol. 28, no. 1, pp. 84–95.

Lozach, B. (1993). System for Predictive Coding/Decoding of a Digital Speech Signal by Embedded-Code Adaptive Transform, U.S. Patent 5583963.

Maitre, X. (1988). 7 kHz Audio Coding within 64 kbit/s, *IEEE Journal on Selected Areas on Communications*, vol. 2, no. 6, pp. 283–298.

Mallat, S. (1999). *A Wavelet Tour of Signal Processing*, (2nd edn), Academic Press, New York, NY, USA.

Massaloux, D.; Trilling, R.; Lamblin, C.; Ragot, S.; Ehara, E.; Lee, M.; Kim, D.; Bessette, B. (2007). An 8–12 kbit/s Embedded CELP Coder Interoperable with ITU-T G.729 Coder: First Stage of the New G.729.1 Standard, *Proceedings of the IEEE International Conference on Acoustics, Speech, and Signal Processing (ICASSP)*, Honolulu, HI, USA, pp. 1105–1108.

Mathieu, B.; Carlinet, Y.; Massaloux, D.; Kövesi, B.; Deléam, D. (2005). A Network Aware Truncating Module for Scalable Streams Saving Bandwidth for Overused Networks, *Proceedings of 2nd International Workshop on Mobility Aware Technologies and Applications (MATA)*, Montreal, QC, Canada, pp. 12–21.

McCree, A. (2000). A 14 kb/s Wideband Speech Coder with a Parametric Highband Model,, *Proceedings of the IEEE International Conference on Acoustics, Speech, and Signal Processing (ICASSP)*, Istanbul, Turkey, pp. 1153–1156.

McCree, A.; Unno, T.; Anandakumar, A.; Bernard, A.; Paksoy, E. (2001). An Embedded Adaptive Multi-Rate Wideband Speech Coder, *Proceedings of the IEEE International Conference on Acoustics, Speech, and Signal Processing (ICASSP)*, Salt Lake City, UT, USA, pp. 761–764.

Mermelstein, P. (1988). G.722: A New CCITT Coding Standard for Digital Transmission of Wideband Audio Signals, *IEEE Communications Magazine*, vol. 26, no. 1, pp. 8–15.

Moreau, N.; Dymarski, P. (1992). Successive Orthogonalizations in the Multistage CELP Coder, *Proceedings of the IEEE International Conference on Acoustics, Speech, and Signal Processing (ICASSP)*, San Francisco, CA, USA, pp. 61–64.

Neuhoff, D. L.; Lee, D. H. (1991). On the Performance of Tree-Structured Vector Quantization, *Proceedings of the IEEE International Conference on Acoustics, Speech, and Signal Processing (ICASSP)*, Toronto, ON, Canada, pp. 2277–2280.

Nishiguchi, M.; Inoue, A.; Maeda, Y.; Matsumoto, J. (1999). Parametric Speech Coding-HVXC at 2.0–4.0 kbps, *IEEE Workshop on Speech Coding (SCW)*, Porvoo, Finland, pp. 84–86.

Nomura, T.; Iwadare, M.; Serizawa, M.; Ozawa, K. (1998). A Bitrate and Bandwidth Scalable CELP Coder, *Proceedings of the IEEE International Conference on Acoustics, Speech, and Signal Processing (ICASSP)*, Seattle, WA, USA, pp. 341–344.

Oshikiri, M.; Ehara, H.; Morii, T.; Yamanashi, T.; Satoh, K.; Yoshida, K. (2007). An 8–32 kbit/s Scalable Wideband Coder Extended with MDCT-Based Bandwidth Extension on Top of a 6.8 kbit/s Narrowband CELP Coder, *Proceedings of the European Conference on Speech Communication and Technology (INTERSPEECH)*, Antwerp, Belgium.

Oshikiri, M.; Ehara, H.; Yoshida, K. (2002). A Scalable Coder Designed for 10-kHz Bandwidth Speech, *IEEE Workshop on Speech Coding (SCW)*, Tsukuba, Ibaraki, Japan, pp. 111–113.

Oshikiri, M.; Ehara, H.; Yoshida, K. (2004). Efficient Spectrum Coding for Super-Wideband Speech and its Application to 7/10/15 kHz Bandwidth Scalable Coders, *Proceedings of the IEEE International Conference on Acoustics, Speech, and Signal Processing (ICASSP)*, Montreal, QC, Canada, pp. 481–484.

Panis, G.; Hutter, A.; Heuer, J.; Hellwagner, H.; Kosch, H.; Timmerer, C.; Devillers, S.; Amielh, M. (2003). Bitstream Syntax Description: A Tool for Multimedia Resource Adaptation within MPEG-21, *Signal Processing: Image Communications*, vol. 18, no. 8, pp. 721–747.

Park, S.; Y.B. Kim, S. K.; Seo, Y. (1997). Multi-Layer Bit-Sliced Bit-Rate Scalable Audio Coding, *Proceedings of 103rd Convention of the AES*, New York, NY, USA. Preprint 4520.

Paulus, J. W.; Schnitzler, J. (1996). 16 kbit/s Wideband Speech Coding Based on Unequal Subbands, *Proceedings of the IEEE International Conference on Acoustics, Speech, and Signal Processing (ICASSP)*, Atlanta, GA, USA, pp. 255–258.

Ragot, S.; Kövesi, B.; Trilling, R.; Virette, D.; Duc, N.; Massaloux, D.; Proust, S.; Geiser, B.; Gartner, M.; Schandl, S.; Taddei, H.; Gao, Y.; Shlomot, E.; Ehara, H.; Yoshida, K.; Vaillancourt, T.; Salami, R.; Lee, M. S.; Kim, D. Y. (2007). ITU-T G.729.1: An 8-32 kbit/s Scalable Coder Interoperable with G.729 for Wideband Telephony and Voice over IP, *Proceedings of the IEEE International Conference on Acoustics, Speech, and Signal Processing (ICASSP)*, Honolulu, HI, USA, pp. 529–532.

Ragot, S.; Kövesi, B.; Virette, D.; Trilling, R.; Massaloux, D. (2006). A 8-32 kbit/s Scalable Wideband Speech and Audio Coding Candidate for ITU-T G.729EV Standardization, *Proceedings of the IEEE International Conference on Acoustics, Speech, and Signal Processing (ICASSP)*, Toulouse, France, pp. 1–4.

Ramprashad, S. (1999). Embedded Coding Using a Mixed Speech and Audio Coding Paradigm, *International Journal of Speech Technology, Special Issue "Speech Coding"*, vol. 2, no. 4, pp. 359–372.

Ramprashad, S. (2000). High Quality Embedded Wideband Speech Coding Using an Inherently Layered Coding Paradigm, *Proceedings of the IEEE International Conference on Acoustics, Speech, and Signal Processing (ICASSP)*, Istanbul, Turkey, pp. 1145–1148.

Rimoldi, B. (1994). Successive Refinement of Information: Characterization of the Achievable Rates, *IEEE Transactions on Information Theory*, vol. 40, no. 1, January, pp. 253–259.

Riskin, E. A.; Ladner, R.; Wand, R.-Y.; Atlas, L. E. (1994). Index Assignment for Progressive Transmission of Full-Search Vector Quantization, *IEEE Transactions on Image Processing*, vol. 3, no. 3, pp. 307–312.

Schnitzler, J. (1998). A 13.0 kbit/s Wideband Speech Codec Based on SB-ACELP, *Proceedings of the IEEE International Conference on Acoustics, Speech, and Signal Processing (ICASSP)*, Seattle, WA, USA, pp. 157–160.

Schroeder, M. R.; Atal, B. S. (1985). Code-Excited Linear Prediction(CELP): High-quality Speech at Very Low Bit Rates, *Proceedings of the IEEE International Conference on Acoustics, Speech, and Signal Processing (ICASSP)*, Tampa, FL, USA, pp. 937–940.

Sherif, M.; Bowker, D.; Bertocci, G.; Orford, B.; A., M. G. (1993). Overview and Performance of CCITT/ANSI Embedded ADPCM Algorithms, *IEEE Transactions on Communications*, vol. 41, no. 2, pp. 391–399.

Singhal, S.; Atal, B. S. (1989). Amplitude Optmization and Pitch Prediction in Multipulse Coders, *IEEE Transactions on Acoustics, Speech and Signal Processing*, vol. 37, no. 3, pp. 317–327.

Steinberg, Y.; Merhav, N. (2004). On Successive Refinement for the Wyner–Ziv Problem, *IEEE Transactions on Information Theory*, vol. 50, no. 8, pp. 1636–1654.

Taddei, H.; Massaloux, D.; Le Guyader, A. (1999). A Scalable Three Bitrate (8, 14.2, and 24 kbit/s) Audio Coder, *Proceedings of 107th Convention of the AES*, New York, NY, USA. Preprint 5034.

Taddei, H.; Ramprashad, S.; Sundberg, C.-E.; Lou, H. (2002). Mode Adaptive Unequal Error Protection for Transform Predictive Speech and Audio Coders, *Proceedings of the IEEE International Conference on Acoustics, Speech, and Signal Processing (ICASSP)*, Orlando, FL, USA, pp. 165–168.

Taniguchi, S.; Johnson, M.; Ohta, Y. (1990). Multi-Vector Pitch-Orthogonal LPC: Quality Speech with Low Complexity at Rates between 4 and 8 kbps, *Proceedings of International Conference on Spoken Language Processing (ICSLP)*, Kobe, Japan, pp. 113–116.

Taori, R.; Sluijter, R. J.; Gerrits, A. J. (2000). Hi-BIN: An Alternative Approach to Wideband Speech Coding, *Proceedings of the IEEE International Conference on Acoustics, Speech, and Signal Processing (ICASSP)*, Istanbul, Turkey, pp. 1157–1160.

Taubman, D. (2000). High Performance Scalable Image Compression with EBCOT, *IEEE Transactions on Image Processing*, vol. 9, no. 7, pp. 1158 – 1170.

Tian, C.; Diggavi, S. N. (2006). Multistage Successive Refinement for Wyner–Ziv Source Coding with Degraded Side Informations, *Proceedings of the IEEE International Symposium on Information Theory (ISIT)*, Seattle, WA, USA, pp. 1594–1598.

TTAE.IT-G.729.1 (2006). *G.729 Based Embedded Variable Bit-Rate Coder: An 8–32 kbit/s Scalable Wideband Coder Bitstream Interoperable with G.729*, Telecommunications Technology Association of Korea (TTA).

Tuncel, E.; Rose, K. (2003). Additive Successive Refinement, *IEEE Transactions on Information Theory*, vol. 49, no. 8, pp. 1983–1991.

Tzou, K.-H. (1986). Embedded Max Quantization, *Proceedings of the IEEE International Conference on Acoustics, Speech, and Signal Processing (ICASSP)*, Tokyo, Japan, pp. 505–508.

Vaillancourt, T.; Jelinek, M.; Salami, R.; Lefebvre, R. (2007). Efficient Frame Erasure Concealment in Predictive Speech Codecs Using Glottal Pulse Resynchronisation, *Proceedings of the IEEE International Conference on Acoustics, Speech, and Signal Processing (ICASSP)*, Honolulu, HI, USA, pp. 1113–1116.

Valin, J.-M.; Lefebvre, R. (2000). Bandwidth Extension of Narrowband Speech for Low Bit-Rate Wideband Coding, *IEEE Workshop on Speech Coding (SCW)*, Delavan, WI, USA, pp. 130–132.

Vary, P.; Martin, R. (2006). *Digital Speech Transmission: Enhancement, Coding and Error Concealment*, John Wiley & Sons, Ltd, Chichester, UK.

Viswanathan, H.; Berger, T. (2000). Sequential Coding of Correlated Sources, *IEEE Transactions on Information Theory*, vol. 46, no. 1, pp. 236–246.

Voronov, G.; Feder, M. (2000). The Redundancy of Successive Refinement Codes and Codes with Side Information, *Proceedings of the IEEE International Symposium on Information Theory (ISIT)*, Sorrento, Italy, p. 126.

Wassel, I.; Goodman, D.; Steele, R. (1988). Embedded Delta Modulation, *IEEE Transactions on Acoustics, Speech and Signal Processing*, vol. 36, no. 8, pp. 1236–1243.

Yang, J.; Zhang, Z. (2004). Redundancy-Complexity Tradeoff of Multi-Resolution Coding for Successively Refinable Sources, *Proceedings of the IEEE International Symposium on Information Theory (ISIT)*, Chicago, IL, USA, p. 93.

Zhang, S.; Lockhart, G. (1995). An Efficient Embedded ADPCM Coder, *Proceedings of Fifth IEE Conference on Telecommunications*, Brighton, UK, pp. 210–214.

Zhang, S.; Lockhart, G. (1997). Embedded RPE Based on Multistage Coding, *IEEE Transactions on Speech and Audio Processing*, vol. 5, no. 4, pp. 367–371.

Zhang, Z.; Lockhart, G. (1991). Design and Performance of Robust Embedded ADPCM Coder, *Electronics Letters*, vol. 27, no. 8, pp. 1786–1788.

Chapter 9

Backwards Compatible Wideband Telephony

Peter Jax

9.1 Introduction

The limited frequency range of about 300 Hz to 3.4 kHz of today's *narrowband* (NB) telephone networks leads to restricted audio quality compared with *wideband* (WB) telephony (50 Hz to 7 kHz). Wideband speech codecs have been standardized and are ready to be used, providing significant improvements in terms of speech intelligibility and naturalness. However, the conversion from NB to WB telephony requires investments by operators and customers. In the transition period NB and WB terminals will coexist for a long time, and compatibility of operation is a mandatory requirement. Therefore, each WB terminal has to be equipped with an NB codec to allow interoperability with any far-end NB terminal. The WB mode can only be used if the far-end terminal, the network, and the near-end terminal all have the improved WB capabilities.

In this chapter we will focus on techniques for extending the capabilities of existing NB voice communication systems to provide WB speech quality. In particular, we concentrate on embedded, hierarchical codecs that make use of *bandwidth extension* (BWE) techniques. Several such parametric BWE schemes have been published recently; they show that the step from NB quality to acceptable WB quality requires only a low to moderate additional data rate on top of the NB bit stream.

Such backwards compatible WB speech transmission techniques target at cost-efficient implementation of the encoder and decoder algorithms based on existing systems. Furthermore, network operators have to modify their infrastructure only moderately

---

Advances in Digital Speech Transmission   Edited by R. Martin, U. Heute and C. Antweiler
© 2008 John Wiley & Sons, Ltd

or, for some approaches, not at all. It is possible to deploy WB speech services step-by-step, starting with new terminals. It is inherently guaranteed that the new terminals will interoperate seamlessly with existing NB terminals.

We will start this chapter by discussing the major application scenarios in Sec. 9.2. We introduce the principles of state-of-the-art stand-alone BWE approaches in Sec. 9.3, and in Sec. 9.4 further describe how BWE techniques with transmission of BWE information can be used as part of a WB speech codec. In Sec. 9.5 a very recent proposal of embedding the stream of BWE information in the NB speech signal will be discussed. Finally, some advanced encoding schemes for the BWE information will be described in Sec. 9.6.

Major parts of Sec. 9.2 and 9.3 have been previously published in [Jax, Vary 2006].[1]

9.2 From Narrowband Telephony to Wideband Telephony

As a matter of fact, the limited quality of narrowband telephone speech is widely accepted. However, in certain situations we clearly become aware of the impacts of the bandwidth limitation. For example, the limited intelligibility of syllables becomes apparent when we try to understand unfamiliar words or names on the phone. In these cases, we often need a spelling alphabet, especially to distinguish certain unvoiced or plosive phones, such as /s/ and /f/ or /p/ and /t/. Another drawback is that many speaker-specific characteristics are not retained transparently in the narrowband speech signal. Therefore, it is sometimes difficult to distinguish similar-sounding speakers on the phone.

The bandwidth of wideband transmission is comparable to that of AM radio transmission, and it allows for excellent speech intelligibility and very good speech quality. An example of a speech signal with significant frequency content beyond 3.4 kHz is given in Fig. 9.1, which shows a series of short-term spectrograms of natural speech with an indication of the limited frequency bands covered by the narrowband and wideband versions. A closer look at Fig. 9.1 reveals that narrowband speech may lack significant parts of the spectrum, especially for unvoiced sounds. Even the difference between wideband speech and original speech is still considerable.

The introduction of wideband transmission in a telephone network requires at least new terminals with better electro-acoustic front ends, improved analog-to-digital converters, and new speech codecs. In addition, signaling procedures are needed for

---

[1] Portions reprinted, with permission, from P. Jax, P. Vary, "Bandwidth Extension of Speech Signals: A Catalyst for the Introduction of Wideband Speech Coding?", IEEE Communications Magazine, 44(5), May 2006. ©2006 IEEE.

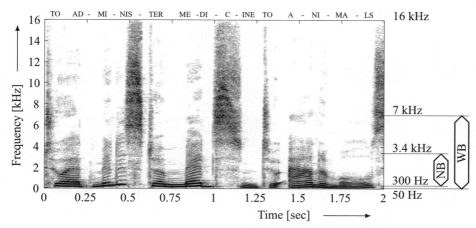

Figure 9.1: Example short-term spectrogram of the sentence "to administer medicine to animals" spoken by a female voice. Dark regions indicate a strong short-term power spectrum

detection and activation of the wideband capability. In cellular radio networks, expensive modifications are necessary, since error protection (speech codec specific channel coding) is implemented in the base stations and not in the centralized switching centers.

Several wideband speech codecs have been standardized in the past. In 1985, a first wideband speech codec (G.722) was specified by CCITT (now ITU-T) for ISDN and teleconferencing with bit rates of 64, 56 and 48 kbit/s. It is mainly applied in the context of radio broadcast stations by external reporters using special terminals and ISDN connections from outside to the studio. In 1999, a second wideband codec (G.722.1) was introduced by ITU-T that produces almost comparable speech quality at reduced bit rates of 32 and 24 kbit/s. Most recently, the *adaptive multi-rate wideband* (AMR-WB) speech codec was specified by ETSI and 3GPP for CDMA cellular networks such as UMTS. The AMR-WB codec has also been adopted for fixed network applications by ITU-T (G.722.2). By the AMR-WB standard a family of wideband codecs with data rate modes between 6.6 and 23.85 kbit/s is defined together with control mechanisms to adapt the codec mode to channel conditions. A further extension, the AMR-WB+ codec, supports general audio in mono/stereo with frequency bandwidths from 7 to more than 16 kHz and bit rates of between 6.6 and 32 kbit/s. For an overview of recent work in wideband speech standards see [de Campos Neto, Järvinen 2006] and the related technical papers in the same issue of the journal.

Even if cellular phones are replaced by new models much more often than fixed line telephones, there will be a long transitional period with narrowband and wideband terminals in mixed use in both cellular and fixed networks. During this transition period different technical solutions may be employed, as illustrated in Fig. 9.2. All of these solutions produce WB speech at the near-end terminal.

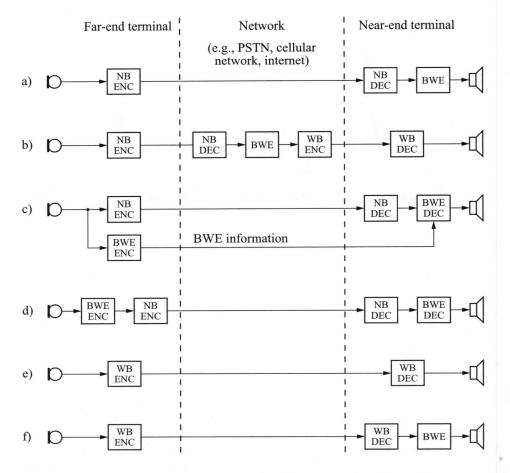

Figure 9.2: Steps from narrowband to wideband telephony (and beyond). © 2006 IEEE
    a) Narrowband transmission and bandwidth extension in the receiver
    b) Narrowband transmitter and bandwidth extension in the network
    c) Transmission of parallel BWE information for bandwidth extension
    d) Embedding of BWE information into narrowband signal
    e) Speech transmission using true wideband coding
    f) Wideband transmission plus bandwidth extension for "super-
       wideband" speech quality

There may be a narrowband terminal at the far end and narrowband transmission over
the network, while the electro-acoustic front end of the near-end terminal has already
got wideband capabilities, see Fig. 9.2-a. Owing to the increased audio bandwidth of
the near-end terminal (sampling rate 16 kHz), BWE can be applied to enhance the
received speech signal. This produces more natural sounding speech, and the user
can benefit from the improved wideband capabilities of the terminal. This approach
does not require any modification of the sending terminal and the network. The

implementation of BWE is particularly attractive for manufacturers with respect to the competition on the terminal market. For reasons of compatibility, the narrowband encoder has to be used in the WB terminal for the reverse direction.

Alternatively, the BWE processing can be placed within the core network, as illustrated in Fig. 9.2-b. With this setup, the network operator can offer connections with improved quality at any time to any customer using a wideband terminal, i.e., even if the far-end terminal provides only NB capabilities. During call setup the network can detect mixed connections between NB and WB terminals. Then, it can route the connection via a transcoding unit located inside the core network. The transcoding unit consists of a NB decoder, bandwidth extension, and WB encoder. The near-end terminal does not have to implement any BWE algorithms itself.

Many other setups are imaginable for which BWE in the network is reasonable, especially if a heterogeneous mixture of NB and WB terminals is involved. Examples include multi-party conference bridges, or mechanisms to prevent temporary switching from WB to NB, e.g., in the case of inter-cell handovers in cellular networks.

A third solution is shown in Fig. 9.2-c, which provides a significantly improved quality in comparison to the approaches of Fig. 9.2-a,b. At the far end some BWE information is determined and communicated to the near-end terminal in parallel with the narrowband speech signal. The BWE information allows decoding of the wideband speech signal on top of the already decoded narrowband speech. Accordingly, in certain cases, this approach can be interpreted as a variant of layered or embedded speech coding. In layered speech coding the bitstream consists of several layers built on each other. At the receiver the base layer of the bitstream is sufficient to decode an acceptable speech signal. With each additional layer that is received, the speech quality is improved successively.

A promising new approach is to embed the side information into the NB speech signal as a digital watermark message before encoding, see Fig. 9.2-d. The proper watermarking method makes this BWE system inherently backwards-compatible without the need for any signaling procedure: if the watermarked speech signal is presented to a human listener by a conventional NB receiver, he or she will not perceive any difference to the encoded original NB speech. If both sides support the BWE side information transmission, the receiver can produce wideband speech with very good quality, almost comparable to that of true wideband codecs. If, on the other hand, the BWE-receiver does not detect the embedded watermark in the NB speech, a stand-alone BWE approach (Fig. 9.2-a) can still be activated.

Finally, the true wideband connection, as shown in Fig. 9.2-e, requires modifications to the transmitter, possibly the network, and the receiver by introducing new encoders and decoders. This solution can obviously provide the best speech quality.

Even if wideband coding (50 Hz to 7 kHz) has already been implemented in the network, wideband extension beyond 7 kHz can be applied in addition to produce a *super-wideband* speech signal, e.g., with frequency components up to 15 kHz. This situation is depicted in Fig. 9.2-f. It is obvious from Fig. 9.1 that the subjective speech quality can be further improved compared with the transmitted wideband speech. Naturally, all of the schemes described above to take the step from NB to WB quality can be applied again to realize efficient transmission of super-wideband speech based on a WB codec.

9.3 Stand-Alone Bandwidth Extension

To assess the prospects and limitations of BWE techniques it is necessary to understand the underlying principles. From Nyquist's theorem it is evident that it would be virtually impossible to perform non-trivial bandwidth extension for *arbitrary* signals directly and solely in the signal domain. Frequency components beyond half of the sampling frequency cannot be directly recovered. If a mathematical model of the signal generation process could be assumed, on the other hand, BWE becomes feasible indirectly via the parameters of this model. Knowing that both the NB signal and the WB signal are governed by the same source model, we can estimate the source parameters from the NB signal, and then use these estimates to produce a corresponding WB speech signal.

Here, we restrict our view to speech signals. Therefore, we can make use of the well-known source-filter model of speech production. The human speech production process can be divided into two parts. A periodic, noise-like, or mixed excitation signal is produced by the vocal chords (glottis), or by constrictions of the vocal tract. Then, the sound is shaped by the acoustic resonances of the vocal tract cavities.

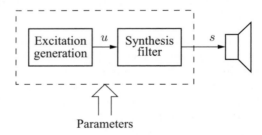

Figure 9.3: Signal processing model of the speech production process. © 2006 IEEE

The modeling is shown in Fig. 9.3. In analogy to the human physiology, the mathematical source–filter model of speech production consists of two parts: a signal generator producing a spectrally flat[2] *excitation signal* u, and a synthesis filter shaping the *spectral envelope* of the speech signal s. This source–filter model has been used extensively in many areas of speech signal processing, e.g., for speech synthesis, coding, recognition, and enhancement.

---

[2]Strictly speaking, the glottis signal is not spectrally flat due to the shape of the glottis pulses. However, the shape of the glottis pulse can be modeled by a glottis filter with a spectrally flat excitation u. In practice the glottis filter is merged into the synthesis filter.

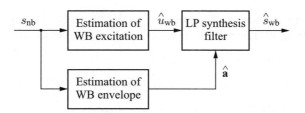

Figure 9.4: BWE with separate extension of spectral envelope and excitation signal.
© 2006 IEEE

Almost all state-of-the-art approaches to bandwidth extension of speech signals are built on this simple source–filter model. Following the two-stage structure of the model, the bandwidth extension is performed separately for the excitation signal u and for the spectral envelope of the speech signal [Carl, Heute 1994], [Cheng et al. 1994]. These two constituent parts of the speech signal can be assumed to be mutually independent to a certain extent, such that more or less separate optimization of the two parts of the algorithm is possible. In Fig. 9.4 a generic block diagram of this concept is shown.

9.3.1 Estimation of the Wideband Spectral Envelope

The bandwidth extension algorithm starts with the estimation of the spectral envelope of the wideband speech signal; see the lower signal path in Fig. 9.4. This block is shown in more detail in Fig. 9.5. In most adaptive BWE algorithms, statistical estimation methods are used that are to some extent similar to approaches from pattern recognition or speech recognition. The estimation scheme is based on a vector $\mathbf{b}_{nb}$ of features that is extracted from each frame of the narrowband input signal s_{nb}. Typically, this feature vector is composed of information on the spectral envelope of the narrowband speech signal (for example, LSF or reflection coefficients, [Carl, Heute 1994]) plus, in addition, features reflecting voiced/unvoiced attributes of the speech (for example short-term power, zero crossing rate, etc. [Jax, Vary 2000]).

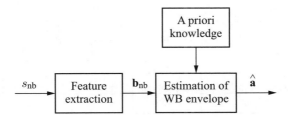

Figure 9.5: Estimation of the spectral envelope in stand-alone BWE. © 2006 IEEE

Many different schemes have been proposed in the literature for estimating the wideband spectral envelope. The most important basic techniques include:

- codebook mapping [Carl, Heute 1994],

- linear or piece-wise linear mapping [Nakatoh et al. 1997], and

- Bayesian estimation based on *Gaussian mixture models* (GMMs) [Park, Kim 2000] or *hidden Markov models* (HMMs) [Jax, Vary 2000].

Within any estimation scheme, *a priori* knowledge on the joint behavior of the observation (feature vector) and the estimated quantity is needed. This *a priori* knowledge is contained in a statistical model, whose form depends on the employed estimation method. For example, in the case of codebook mapping the statistical model is composed of two vector quantizer codebooks for the LP or LSF coefficients, both for the narrowband and wideband speech. The statistical model has to be acquired and stored during an *off-line training phase* using a database of representative wideband speech signals.

The result of the estimation block is the wideband spectral envelope of the speech frame, represented by the filter coefficient vector $\hat{\mathbf{a}}$ of the *linear predictor* (LP) synthesis filter.

9.3.2 Extension of the Excitation Signal

The next step in the BWE system consists of substituting the missing frequency components in the excitation signal. Owing to the assumed spectral flatness of the excitation signal u, and because of the fact that the human ear is quite insensitive to variations of the spectral fine structure at high frequencies, the extension can be realized very efficiently.

The basic functional principle of most algorithms can be described by Fig. 9.6. After interpolation of the sampling rate from 8 kHz to 16 kHz, the narrowband excitation $\hat{u}_{nb}$ is estimated by applying the interpolated signal $\tilde{s}_{nb}$ to the wideband LP analysis filter $1 - \hat{A}(z)$. The actual extension is performed in the blocks labeled HFR (*high frequency re-synthesis*, beyond 3.4 kHz) and LFR (*low frequency re-synthesis*,

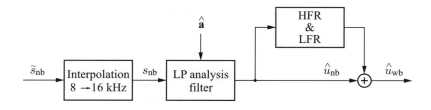

Figure 9.6: Extension of the excitation signal. © 2006 IEEE

below 300 Hz). The techniques typically used for extension of the excitation signal are (see, e.g., [Fuemmeler et al. 2001], [Jax 2004], [Chan, Hui 1996] for more details)

- mirroring, shifting or frequency scaling of the NB spectral components,
- generation of harmonics by non-linear distortion and filtering, or
- synthetic generation of the new frequency components.

The extended frequency components are added to the estimated narrowband excitation. The output signal $\hat{u}_{wb}$ is the desired estimate of the wideband excitation signal. Listening tests have shown that the extension of the excitation signal has much less influence on the quality of the enhanced speech than the estimation of the WB spectral envelope. Many of the listed techniques produce output signals with similar quality.

9.3.3 Performance and State-of-the-Art

Stand-alone bandwidth extension algorithms for speech have reached a stable baseline quality: the artificial wideband output of a BWE system is in general preferred to narrowband telephone speech, even for a speaker- and language-independent setup. Results of many *informal* listening tests have been reported in research papers (for example, [Fuemmeler et al. 2001], [Chan, Hui 1996]) that consistently indicate this preference. To our knowledge, *formal* listening tests of stand-alone BWE algorithms have not been performed to date.

The best results have been obtained for systems trained for a specific language, or, even better, for an individual speaker. Nevertheless, in any case the quality of the enhanced speech does not reach the quality of the original wideband speech. This observation is supported by theoretical investigations on the amount of mutual information between the set of features of the NB speech that is typically used for BWE and the envelope parameters of the highband signal components [Nilsson et al. 2000], [Nilsson et al. 2002], [Jax, Vary 2002]. The mutual information is only about 1–2.5 bit/frame (depending on the set of NB features), which has been shown to be lower than that required for high-quality wideband speech representation [Jax, Vary 2002].

For more details on stand-alone bandwidth extension of telephone speech signals the reader is referred to [Jax 2004], [Vary, Martin 2006] and to the references therein.

9.4 Embedded Wideband Coding Using Bandwidth Extension Techniques

As explained above, most of the BWE algorithms proposed in literature are based on the source–filter model of speech production. The extension of the source signal (excitation) and of the frequency response of the synthesis filter (spectral envelope) can be treated separately. The former is less challenging because the ear is fairly insensitive with respect to coarse quantization or approximation of the excitation signal. It is much more important to find a good approximation of the spectral envelope.

Therefore, BWE can be implemented with great success if information on the complete (wideband) spectral envelope is transmitted, while the extension of the excitation is performed at the receiver without additional information. In fact, some very special and effective variants of bandwidth extension techniques have been used as an integral part of various speech codecs for many years. A very prominent early example in this respect is the GSM full-rate codec (cf. [Vary, Martin 2006]).

More recently, BWE techniques have been applied in the context of wideband speech coding, e.g., in the 3GPP/ETSI AMR-WB standard. In this codec, *code excited linear predictive* (CELP) coding is applied to the speech components up to 6.4 kHz, and artificial bandwidth extension is used to synthesize a supplementary signal for the narrow frequency range from 6.4 to 7 kHz. The extension is supported by transmitting side information, which controls the spectral envelope and the level of noise excitation in the extension band. A more flexible version of this approach is used in the AMR-WB+ codec, which produces spectral components up to 16 kHz.

In this section, we will focus on *embedded* wideband speech codecs, i.e., we assume that an existing narrowband speech codec is used for transmission of the low frequency range from 0 Hz up to 4 kHz at most. In this setup, BWE techniques are used to synthesize the high signal components between maximally 4 kHz up to a cutoff frequency of about 7 kHz. Compared with the aforementioned AMR-WB codec this task is more challenging because the frequency gap to be filled with BWE techniques is broader.

Besides the parametric techniques described below, other approaches have been taken to realize embedded WB speech coding on top of existing NB speech codecs, for example, analysis-by-synthesis coding of the highband signal components [Kataoka et al. 1997], [Nomura et al. 1998], [Koishida et al. 2000].

9.4.1 Transmission of BWE Information

The generic signal processing concept of a typical approach to this challenge is shown in Figs. 9.7 and 9.8. Figure 9.7 shows the encoder. First, the wideband input speech signal s_{wb} (sample rate 16 kHz) is split into two sub-band signals using a pair of low-pass and high-pass filters. The lowband signal s_{nb} is decimated to a sample rate of 8 kHz and fed into the encoder of the embedded narrowband codec standard. The narrowband encoder produces the embedded bit stream, labeled "NB info" in the Fig. 9.7. This information is self-contained and can be decoded by any existing standard-conform narrowband decoder.

The highband signal s_{hb} is input to a BWE encoder module. This block analyzes the spectral envelope and time envelope structure of the highband signal and determines a

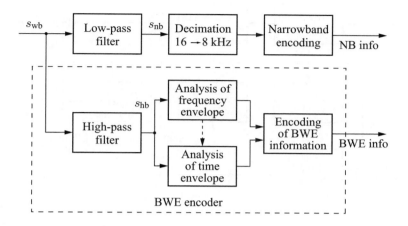

Figure 9.7: Principle of embedded wideband encoding. © 2006 IEEE

set of corresponding parameters. We will denote these parameters by the term *BWE parameters* in the sequel. As indicated by the dashed arrow in Fig. 9.7, information on the spectral envelope may be used for extracting characteristics of the time envelope. In particular, often the time envelope of a prediction residual (where the predictor is based on the highband spectral envelope information) is transmitted rather than the time envelope of the original highband signal s_{hb}.

The BWE parameters are quantized and the obtained set of quantizer indices forms the extension bit stream, labeled "BWE info" in the figure. For quantizing the BWE parameters, features of the narrowband speech signal can optionally be used as side information in order to increase the quantizer's efficiency, cf. Sec. 9.6.1. Typically, such side information can be extracted from the available NB codec parameters, such as LP coefficients, gain factors, etc.

Figure 9.8 shows the corresponding decoder of a wideband embedded speech codec. The embedded bit stream of the narrowband codec is used to decode the narrowband speech signal s_{nb} at a sample rate of 8 kHz. To allow for the later addition of the highband signal components, this signal is interpolated to the target sample rate of 16 kHz.

Generation of the highband signal components is shown in the lower signal processing branch of Fig. 9.8. The BWE parameters are decoded from the hierarchical bit stream and applied in a three-step procedure to synthesize the highband speech components s_{hb}. First, an excitation signal is generated. In the simplest case, this excitation can be a plain noise signal, but more sophisticated schemes have been proposed that make use of decoded parameters from the narrowband decoder. In a second step, gain factors are applied in order to shape the time envelope of the excitation. Finally, a time-variant filter is applied to form the spectral envelope of the synthetic highband signal.

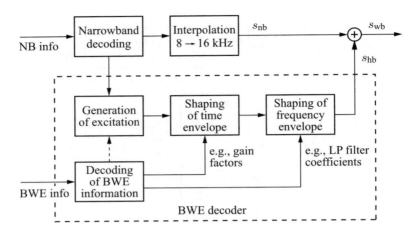

Figure 9.8: Principle of embedded wideband decoding

Typically, the BWE decoder closely follows the well-established *linear prediction coding* (LPC) paradigm, motivated by the (simplified) source-filter model of speech production, cf. Sec. 9.3. Then, the excitation is a mixture of noise and pulse contributions. The time envelope of the excitation signal is formed with a sub-frame resolution. Finally, shaping of the frequency envelope is performed by an all-pole, LP synthesis filter.

9.4.2 Examples of Embedded Wideband Speech Codecs

Maybe the first proposal of a wideband embedded speech codec was made in 1983 by Patrick [Patrick 1983]. This proposal was based on the observation that in stand-alone bandwidth extension some specific fricatives, especially /s/ sounds, are difficult to distinguish by analyzing the narrowband speech signal. Hence, Patrick proposed transmitting one bit per frame in addition to the narrowband signal in order to signal /s/ sounds to the BWE algorithm at the receiver side.

A decade later, in 1993, McElroy et al. published an experimental wideband speech codec with an embedded proprietary narrowband CELP codec [McElroy et al. 1993]. For the highband signal the quantized coefficients of a second-order linear prediction filter and one gain factor per frame are transmitted at a data rate of only 640 bit/s.

By 2000, wideband embedded speech coding with parametric transmission of the highband signal components had gained much interest in the research community, and several approaches were developed independently [Taori et al. 2000], [McCree 2000], [Valin, Lefebvre 2000], [Epps 2000], [Aguilar et al. 2000]. In all of these proposals, a linear prediction synthesis filter is applied for shaping the spectral envelope, and gain factors are transmitted for each signal frame or for sub-frames. The gross data rates for the BWE information are in the range 500 bit/s up to 3.2 kbit/s.

In [Taori et al. 2000] and [McCree 2000] embedding of a standardized narrowband speech codec was investigated for the first time. In both publications strong emphasis is put on concise reconstruction of the time envelope of the synthetic highband signal. To achieve this target, McCree proposed modulating white noise, the raw excitation signal, with the time envelope of the 3–4 kHz sub-band components of the decoded narrowband speech signal [McCree 2000]. This method produces a pitch-dependent time envelope contour in the highband signal, and it consistently reduces the noisy sound characteristic that is produced when plain white noise is used as the excitation signal. McCree's codec is based on the G.729 Annex E narrowband codec (data rate 11.8 kbit/s) and achieves with 2.2 kbit/s of additional BWE information (gross data rate of 14 kbit/s) a subjective speech quality that is between that of G.722 at 48 and 56 kbit/s.

In contrast, Taori et al. applied explicit transmission of the highband time envelope with high temporal resolution [Taori et al. 2000]. Gain factors for the excitation signal are determined every 1–2.5 ms. Thus, similar to McCree's approach described above, the pitch-dependent time envelope contour can be reproduced by the BWE system. Combining gain factors with a time resolution of 1 ms and spectral envelope information (LP coefficients) with a time resolution of 10 ms, the BWE information results in a data rate of 3.8 kbit/s. On top of the GSM *enhanced full-rate* (EFR) codec (data rate 12.2 kbit/s), this approach uses a gross data rate of 16 kbit/s. It achieves a subjective quality that is equivalent to G.722 at 48 kbit/s.

McCree et al. later modified their approach to include the GSM *adaptive multi-rate* (AMR) codec as the embedded narrowband codec [McCree et al. 2001]. The proposed wideband codec comprises several data rates in the range 8.05 kbit/s to 31.8 kbit/s. For the low data rates (8.05–15.95 kbit/s) the BWE information uses only 1.35 kbit/s and for the higher data rates 2.3 kbit/s are applied. In addition to the increased flexibility in terms of data rate, in comparison with [McCree 2000], the BWE algorithm has been specifically tailored in order to improve the quality of the wideband codec in background noise conditions.

Another wideband embedded speech coding approach was proposed in [Jax et al. 2006a], [Geiser et al. 2007]. Compared with previous approaches, this method differs mainly in the spectral envelope shaping technique: instead of the typical auto-regressive LP synthesis filter, a linear-phase FIR filter-bank equalizer is applied here. The adaption of the filter coefficients is performed *by the decoder* every 10 ms. This adaptation is based on a comparison between a target spectral envelope, transmitted by the BWE encoder, and the measured spectral envelope of the excitation signal. The excitation signal is synthesized according to excitation parameters (like codebook gains and adaptive codebook lag) taken from the NB speech decoder. In addition, the target time envelope of the highband signal components is transmitted with a high time resolution of 1.25 ms. This BWE algorithm recently was standardized by ITU-T as part of the hierarchical G.729.1 speech and audio codec [ITU G.729.1 2006].

Figure 9.9 shows exemplary results of an informal subjective listening test from [Jax et al. 2006b]. Here, the bandwidth extension scheme is based on an embedded GSM EFR speech codec. Three different data rates, 300, 600 and 1500 bit/s, were used

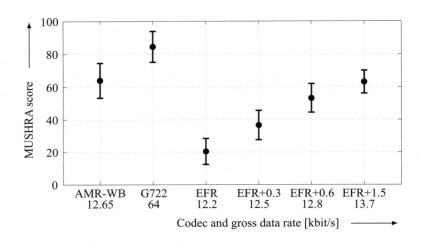

Figure 9.9: Results of a MUSHRA-style subjective listening tests for GSM EFR codec plus BWE information [Jax et al. 2006b]. © 2006 IEEE

for transmitting the BWE information. The subjective speech quality was compared with the GSM EFR base layer and with two reference codecs, the AMR-WB codec at 12.65 kbit/s and the G.722 codec at 64 kbit/s.

9.4.3 Audio Coding

Somewhat related approaches to wideband speech transmission with BWE techniques have been introduced in the context of MPEG general audio coding as *spectral band replication* (SBR). Basic differences to the techniques applied for speech signals are that SBR for audio signals cannot rely on a signal model, and that the extension starts with an audio signal that already has quite a high cutoff frequency, e.g., of 8 kHz. The psycho-acoustic characteristics of the human ear can be exploited, especially the reduced resolution at higher frequencies. SBR has successfully been used to enhance the low-rate coding efficiency of the MPEG-1/2 layer 3 codec (MP3, the extended version has been named MP3pro) and MPEG-2/4 *Advanced Audio Coding* (AAC, extended versions AACplus and HE-AAC). Further information can be found in [Dietz et al. 2002], [Wolters et al. 2003].

9.5 Combination of Bandwidth Extension with Watermarking

A significant step towards a truly backwards-compatible wideband voice communication system may be taken by combining informed bandwidth extension, as introduced in the previous section, with digital watermarking technology. This promising combination has emerged only very recently in literature [Ding 2004], [Chen, Leung 2005], [Geiser et al. 2005], [Sagi, Malah 2007].

The appeal of this approach lies in the fact that transmission of the BWE information, including all signaling between encoder and decoder, is performed "hidden" within the narrowband speech signal. The BWE information becomes almost intrinsically tied to the narrowband speech signal. The voice transmission system need not be aware of the hidden communication channel, and, therefore, the approach is capable of transporting wideband speech via many existing legacy narrowband voice transmission systems. A prerequisite for this target is that the BWE information survives modification of the watermarked narrowband signal as, e.g., produced by tandem coding, transmission errors, DA/AD conversion, or additive noise.

9.5.1 Digital Watermarking of Speech Signals

Digital watermarking technology provides a "hidden" communication channel that is inherently linked to a certain host signal. The watermarking encoder (embedder) modifies specific signal components of the host signal in such a manner that the signal modification is imperceptible to a human listener. The watermarking decoder (detector), however, can reconstruct the hidden message by analyzing the signal mixture.

Digital watermarking is still a young technology. The first work on digital watermarking of multimedia content was published at the beginning of the 1990s. About ten years ago in the mid-1990s the first practical systems were proposed, especially for image and video watermarking. Today, watermarking systems have been designed and deployed for a large variety of applications, including forensics, copyright protection, audience metering among many others [Hartung, Kutter 1999]. In the sequel, we will concentrate on the basic principles of watermarking for speech signals, e.g., [Cheng, Sorensen 2001], [Sagi, Malah 2004].

Designing a watermarking system is a unique discipline because the hidden communication has to be mixed with an existing, much stronger host signal. In general, a trade-off between at least three imperatives has to be found.

1. *Inaudibility* – Typically, the difference between the raw host signal and the signal mixture of host-plus-watermark shall not be perceptible by a human listener. In some specific applications, a certain audibility may be acceptable.

2. *Robustness* – In general, the watermark message shall remain detectable even if the host plus watermark signal is modified. Potential signal modifications include manipulation by a malicious "attacker" (not to be expected for transmission of BWE information) or any degradation of the host plus watermark signal, e.g., by channel distortions or transmission errors.

3. *Data rate* – Depending on the application, a certain fixed or minimum data rate shall be communicated via the watermarking channel.

In addition to these watermarking-specific constraints, additional implementation aspects may play a role, e.g., computational complexity, algorithmic latency, etc. Corresponding to the wide area of watermarking applications there is not a single optimal tradeoff, but an individual tradeoff has to be found for each particular application.

Designing the watermarking approach for a specific application involves the basic modulation scheme, but it may include further essential aspects of a digital communication system such as synchronization, forward error correction, channel estimation, error concealment, etc.

The watermarking process can be described mathematically by addition of a watermark signal $w(k)$ to the host signal $s(k)$

$$\tilde{s}(k) = s(k) + w(k; m, \mathbf{s}(k)), \tag{9.1}$$

where m is the message to be embedded, and $\mathbf{s}(k)$ denotes the set (frame) of samples adjacent to the kth sample $s(k)$. The dependency of the watermark signal $w(k)$ from the host signal $\mathbf{s}(k)$ is necessary for two reasons: first, knowledge of $\mathbf{s}(k)$ is required to render the hidden signal $w(k)$ inaudible, and, second, knowledge of the host signal can be utilized in order to limit or even completely remove interferences of the watermark with the host signal in the watermark detector [Costa 1983], [Chen, Wornell 2001].

The algorithms to determine the watermark signal $w(k)$ in general can be split into two blocks. A perceptual analysis is performed first in order to determine the masking properties of the current segment of the host signal. This analysis can explicitly make use of a psycho-acoustic masking model [Sagi, Malah 2004]. Alternatively, a more heuristic approach can be taken, e.g., using linear prediction based noise shaping techniques [Geiser et al. 2005]. The result of the perceptual analysis controls the modulation of the watermark signal. The additive signal $w(k)$ can be shaped so that its frequency components lie beneath the masking threshold defined by the host signal $s(k)$.

A variety of basic modulation techniques for audio and speech watermarking have been devised. The most important methods are spread spectrum and dither modulation techniques.

Spread spectrum modulation [Cheng, Sorensen 2001] has the advantage that the decoding can be done by applying a simple correlation detector. The detection results are very consistent over wide ranges of channel conditions. However, in the simple correlation detection the watermark sequence interferes with the much stronger host signal. Therefore, the maximum achievable data rate is limited according to the local statistical properties of the host signal.

In contrast, dither modulation takes interferences with the host signal into account already in the embedding process [Chen, Wornell 2001]. The maximum achievable data rate is not constrained by the host signal but only by distortions and attacks of the watermarked signal. The vector version of dither modulation combines the capacity of dither modulation with the consistency and reliability of spread spectrum modulation [Fischer et al. 2004], [Geiser et al. 2005].

9.5.2 Transmission of BWE Information via Watermarking

The system architecture for embedded wideband speech transmission using water-marking techniques is shown in Fig. 9.10. In the upper branch of the encoder, Fig. 9.10-a, the wideband input signal is low-pass filtered and decimated to a sample rate of 8 kHz. The resulting speech signal s_{nb} will be the host signal for the digital watermark. In parallel, the wideband voice signal is fed into a BWE encoder, which produces a continuous stream of BWE information, cf. Sec. 9.4.

The BWE information constitutes the message that is input to the watermarking embedder in order to be transmitted via the "hidden" watermark channel to the decoder. The watermarking embedder uses knowledge of the narrowband host signal in order to optimize modulation and frequency shaping of the watermark signal $w(k)$. The watermarked signal $\tilde{s}_{nb}(k) = s_{nb}(k) + w(k)$ is finally input to the narrowband encoder.

Using a well-designed watermarking scheme, the transmitted signal $\tilde{s}_{nb}$ should be subjectively indistinguishable from the original host signal s_{nb}. This is one key to providing the best possible backwards-compatibility, because a narrowband speech decoder will produce subjectively identical results as if it were connected with any legacy speech encoder.

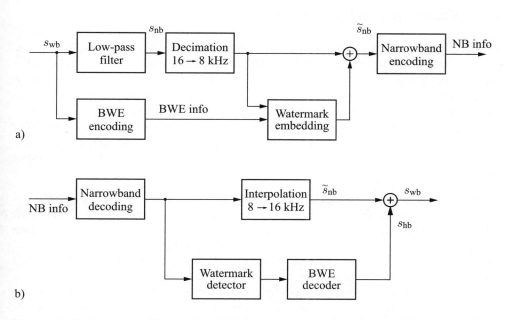

Figure 9.10: Principle of embedded wideband speech transmission with watermarking
a) Encoder
b) Decoder

Because the narrowband speech encoder is positioned behind the watermark embedder, the non-linear coding distortions introduced by the voice codec already contribute to the channel distortions that will be seen by the watermark detector at the receiver site of the communication link.

The receiver, shown in Fig. 9.10-b, first has to decode the narrowband voice signal. The narrowband speech components are interpolated to a sample rate of 16 kHz. In parallel, the watermarked narrowband signal is fed into the watermark detector that recovers the BWE information. The BWE algorithm decodes the BWE information and produces the highband signal components s_{hh}, which are finally added to s_{nb} to produce the full wideband output signal s_{wb} of the decoder. The same BWE decoding principles as described in Sec. 9.4 can be applied.

Again, the decoder can easily be designed to be fully backwards-compatible to plain NB communication links. If no watermark is being detected, i.e., if no BWE encoder has been used at the transmitter or the watermark information has been destroyed by modification of the watermarked signal, the bandwidth extension can either be completely deactivated, or a stand-alone BWE algorithm can be applied, see Sec. 9.3. In Sec. 9.6 we will describe two algorithms that *inherently* switch between stand-alone BWE and utilization of the watermark information, if available.

9.5.3 Challenges and Status

The above described concept produces high demands for both the watermarking system and the BWE algorithm. The BWE information is produced frame-by-frame and requires a transmission data rate of at least 100–300 bit/s for consistent results, the more the better. It is a strong challenge for a watermarking system to provide this capacity continuously and with low latency, while guaranteeing that the watermark is virtually inaudible. Furthermore, typical transmission channels for voice communication, with their low-rate speech codecs and radio transmission, may introduce strong signal distortion that, in turn, requires a very robust watermarking scheme.

On the other hand, the concept introduced in this section also requires sophisticated BWE techniques that produce consistent wideband speech signals with the little available capacity of the hidden watermarking channel. As an example, the maximum data rate has been reported by [Sagi, Malah 2007] to be 600 bit/s with a bit error rate of about $1 \ldots 5 \cdot 10^{-4}$ in typical analog telephony AWGN channel models (SNR 35 dB) and with μ-law quantization. For channels including state-of-the-art low bit rate speech codecs the data rate achievable by watermark communication can be expected to be even lower. Besides the limited data rate, the BWE decoder must be capable of addressing varying channel conditions as to be expected if the watermarked signal is significantly degraded. It is still an open research topic how to make best use of this very limited and time-varying capacity of the watermarking channel. Some first approaches to this problem will be described in Sec. 9.6.

Nevertheless, several experimental BWE plus watermarking systems have already been proposed recently [Ding 2004], [Chen, Leung 2005], [Geiser et al. 2005], [Sagi, Malah 2007], [Geiser, Vary 2007]. In these papers it was reported that the resulting subjective quality of the decoded wideband voice signals is consistently better than that of their narrowband pendants. Furthermore, it was found in [Geiser et al. 2005], [Chen, Leung 2005] that a BWE plus watermarking system consistently outperforms a stand-alone BWE algorithm.

Before BWE plus watermarking systems can be deployed in practice, some challenges remain to be addressed. In particular, a complete system approach has to be specified and tested, including algorithms for synchronization, signaling, etc. Maybe the schemes that are nearest to practical applications are those that embed the BWE information directly into the bit stream information of a narrowband speech codec [Chétry, Davies 2006], [Geiser, Vary 2007]. However, also for these approaches important questions that arise in practice remain to be answered, e.g., to what extent does such bit stream watermarking survive transcoding in the transmission network.

9.6 Advanced Transmission of Highband Parameters

Besides the signal processing approaches described in the previous sections, specific quantization and source (de)coding methods have been developed that make use of the inherent hierarchical structure of embedded wideband speech codecs. Using these technologies the mutual information between highband parameters and parameters of the narrowband speech signal can be exploited in order to improve the fidelity of BWE information that is transmitted with a certain data rate.

According to independent investigations, the amount of mutual information to be exploited is in the order of about 1–2.5 bit/frame, corresponding to about 50–250 bit/second at a typical frame rate of 50 frames/second [Nilsson et al. 2000], [Nilsson et al. 2002], [Jax, Vary 2002]. At first sight these values appear rather low, but they have to be seen in relation to the typical data rates of only 300–3000 bit/s that have been applied for signaling of BWE information in literature (see Sec. 9.4 and 9.5). In comparison, the exploitable mutual information is not negligible and may allow for significant performance improvements, particularly for low-rate applications such as watermark embedding of BWE information.

Two variants to exploit side information from the narrowband signal in order to improve transmission of BWE information are described in this section. The first variant, Sec. 9.6.1, explicitly uses the side information in both the encoder and decoder in order to improve the performance of the BWE information quantizer. The second variant, Sec. 9.6.2, exploits side information extracted from the narrowband speech in order to apply sophisticated error concealment of the BWE information at the receiver only.

9.6.1 Coding with Side Information

The term *side information* is used here to denote any information that is extracted from the narrowband speech signal, or from parameters of the narrowband codec, which can be utilized to improve the performance of the quantization of the BWE information. The principle is illustrated in Fig. 9.11. The same side information $\mathbf{b}_{\text{nb}}$ is used in the transmitter and receiver in order to control encoding and decoding of the BWE parameters.

This may lead to problems if the narrowband signal is subject to distortion in the transmission system. Then, the BWE decoder uses different side information from the BWE encoder which may produce substantial distortion of the decoded BWE parameters as well. Therefore, the coding with side information approach is restricted to scenarios in which no or negligible distortion of the narrowband parameters is to be expected.

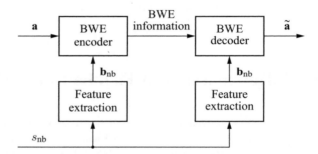

Figure 9.11: Coding with side information in order to improve the performance of transmission of the BWE information

Conditional Quantization

The basic idea of conditional quantization is to modify the quantization of the BWE information $\mathbf{a}$ depending on the observed side information $\mathbf{b}_{\text{nb}}$. Normally, i.e., without using side information, quantization of the BWE parameters $\mathbf{a}$ would be performed by searching a codebook $\mathcal{C}_{\mathbf{a}}$ for the candidate representative $\hat{\mathbf{a}}_i$ that minimizes a distortion criterion $d(\mathbf{a}, \hat{\mathbf{a}}_i)$

$$Q(\mathbf{a}) = \arg \min_{\hat{\mathbf{a}}_i \in \mathcal{C}_{\mathbf{a}}} d(\mathbf{a}, \hat{\mathbf{a}}_i). \tag{9.2}$$

The index i of the found representative codebook entry is transmitted to the BWE decoder, which can reconstruct the corresponding representative $\hat{\mathbf{a}}_i$. Normally, the quantizer codebook $\mathcal{C}_{\mathbf{a}}$ is globally optimized to minimize the average distortion of the quantizer

$$\mathcal{C}_{\mathbf{a}} = \arg \min_{\mathcal{C}} \mathrm{E}_{p(\mathbf{a})} \big\{ d(\mathbf{a}, Q(\mathbf{a})) \big\}. \tag{9.3}$$

For this purpose, for example, the well-known LBG training algorithm can be used in combination with a sufficiently large data base of training vectors.

Now, in order to utilize the side information $\mathbf{b}_{nb}$ to improve the quantizer performance further, an *individual* codebook $\mathcal{C}_{\mathbf{a}|\mathbf{b}_{nb}}$ can be used that is optimized for the conditional *probability density function* (PDF) $p(\mathbf{a}|\mathbf{b}_{nb})$ instead of the unconditional PDF $p(\mathbf{a})$ applied above

$$\mathcal{C}_{\mathbf{a}|\mathbf{b}_{nb}} = \arg\min_{\mathcal{C}} \mathrm{E}_{p(\mathbf{a}|\mathbf{b}_{nb})} \big\{ d(\mathbf{a}, Q(\mathbf{a}))|\mathbf{b}_{nb} \big\} . \tag{9.4}$$

Without resource limitations this would mean training an individual quantizer codebook $\mathcal{C}_{\mathbf{a}|\mathbf{b}_{nb}}$ for each value of the side information $\mathbf{b}_{nb}$. Naturally, this is not possible in practice due to constraints on memory consumption and a lack of training material.

For practical systems, it has been proposed that one clusters the side information data $\mathbf{b}_{nb}$ into a limited number of classes, e.g., up to 16, and stores one conditional codebook $\mathcal{C}_{\mathbf{a}|\mathbf{b}_{nb}}$ for each class. Then, for each signal segment both encoder and decoder classify the side information data to select the quantizer codebook to be used for quantizing the BWE information $\mathbf{a}$ [Epps 2000], [Epps, Holmes 2001], [Agiomyrgiannakis, Stylianou 2004], [Agiomyrgiannakis, Stylianou 2007].

Quantization of a Prediction Residual

An alternative approach to coding with side information is based on prediction techniques, for example, [Nomura et al. 1998]. The side information $\mathbf{b}_{nb}$ is used to calculate a prediction $\mathrm{E}\{\mathbf{a}|\mathbf{b}_{nb}\}$ of the measured BWE parameters $\mathbf{a}$, which is subtracted from the true parameters to form the prediction residual $\mathbf{e}$

$$\mathbf{e} = \mathbf{a} - \mathrm{E}\{\mathbf{a}|\mathbf{b}_{nb}\} . \tag{9.5}$$

The prediction residual is quantized, encoded and transmitted to the BWE receiver. The decoder has to determine the same prediction as the encoder to reconstruct the quantized BWE parameter set $\hat{\mathbf{a}}$

$$\hat{\mathbf{a}} = \mathrm{E}\{\mathbf{a}|\mathbf{b}_{nb}\} + Q(\mathbf{e}) . \tag{9.6}$$

If side information $\mathbf{b}_{nb}$ and BWE information $\mathbf{a}$ share mutual information $I(\mathbf{a}; \mathbf{b}_{nb}) > 0$, then the prediction residual will have a lower differential entropy $h(\mathbf{e})$ than the original BWE information. In general, $h(\mathbf{a}) - I(\mathbf{a}; \mathbf{b}_{nb}) \leq h(\mathbf{e}) \leq h(\mathbf{a})$. Therefore, as compared with direct quantization of $\mathbf{a}$, less data rate is required to encode and transmit a BWE parameter set with a specific fidelity.

The functionality of the prediction in (9.5) and (9.6) is more or less identical to that of the parameter estimation applied in stand-alone bandwidth extension, see Sec. 9.3.1. Similar approaches can be taken. This property allows an elegant fall-back strategy in case the transmission of the encoded prediction residual fails, e.g., in the case of frame erasures for packet-oriented networks or if a watermarking channel is severely degraded. Then it is sufficient to replace the decoded residual $Q(\mathbf{e})$ in (9.6) with zero to obtain a conventional stand-alone BWE system.

9.6.2 Error Concealment with Side Information

One of the major applications for embedded speech coding is transmission of the individual bit streams over channels with unequal error behavior. An example is packet-oriented transmission, e.g., voice over IP, where the narrowband base layer bit stream may be better protected against packet loss than the BWE information, e.g., by forward error correction or quality of service mechanisms. Another application may be radio transmission with unequal error protection, i.e., with stronger forward error protection for the narrowband base layer and less error protection for the enhancement layer(s).

Erroneous decoding or even loss of the BWE information might lead to short-term switchings between narrowband decoding and wideband decoding. Such alternate narrowband and wideband characteristics of the decoded speech signal is very annoying to a listener and, thus, the subjective quality of the codec may drop significantly, even below that of the embedded narrowband codec. In order to provide graceful degradation characteristics in such scenarios, it is mandatory to perform error concealment at the receiver. In this section we will present a suitable error concealment technique that is tailored to the specific application of embedded wideband speech coding with parametric BWE information [Geiser et al. 2005].

In the sequel, we will use the notation from [Vary, Martin 2006, Chap. 10.3.2]. The system setup is depicted in Fig. 9.12. The lowest branch results in the original set of BWE parameters $\mathbf{a}$, which can be directly calculated from the original wideband speech signal s_{wb}.

The upper branch illustrates how these BWE parameters can be estimated from features $\mathbf{b}_{\text{nb}}$ of the narrowband speech signal, as applied in stand-alone bandwidth extension, see Sec. 9.3.

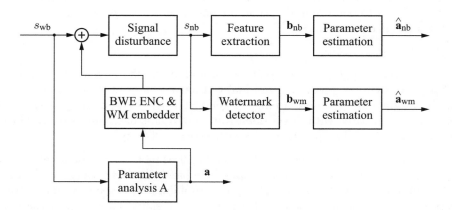

Figure 9.12: System model of parallel implicit and explicit transmission of BWE information

The middle branch in Fig. 9.12 shows explicit transmission of the BWE information via a watermarking channel, cf. Sec. 9.5. Here, the BWE decoder has to estimate the parameters from the output $\mathbf{b}_{\mathrm{wm}}$ of the watermarking detector. We assume that the watermark detector produces "soft information" such that it is possible to derive the likelihood values $p(\mathbf{b}_{\mathrm{wm}}|\mathbf{a})$ required for error concealment [Fingscheidt, Vary 2001].

The BWE information is effectively transmitted via two channels: explicitly as part of the hierarchical bit stream, and implicitly through the characteristics of the narrowband speech signal. This can be interpreted as a kind of diversity transmission. In the error concealment scheme, both parallel information paths will be taken into account. Nevertheless, for comprehensibility we will start below with the definition of typical statistical estimation rules for a single information path.

Estimation from One Information Path

As an example, we will describe statistical estimation of the BWE parameters $\mathbf{a}$ from the set of narrowband features $\mathbf{b}_{\mathrm{nb}}$. This case corresponds to stand-alone BWE, cf. Sec. 9.3.1. The derivations for the estimator in the parallel branch in Fig. 9.12 are the same, but $\mathbf{b}_{\mathrm{nb}}$ has to be exchanged with the channel output $\mathbf{b}_{\mathrm{wm}}$ from the watermark detector.

A very common estimation criterion targets at *minimization of the mean square error* (MMSE) $\mathrm{E}\{(\mathbf{a}-\mathbf{a})^{\mathrm{T}}(\mathbf{a}-\mathbf{a})\}$. The optimal estimate is [Vary, Martin 2006]

$$\hat{\mathbf{a}}_{\mathrm{mmse}} = \int_{\mathcal{R}^n} \mathbf{a}\, p(\mathbf{a}|\mathbf{b}_{\mathrm{nb}})\, \mathrm{d}\mathbf{a}. \tag{9.7}$$

Alternatively, *maximum a posteriori* (MAP) detection can be applied in order to find that coefficient vector $\mathbf{a}$ that has the largest conditional probability given the observation vector $\mathbf{b}_{\mathrm{nb}}$

$$\hat{\mathbf{a}}_{\mathrm{map}} = \arg\max_{\mathbf{a}} p(\mathbf{a}|\mathbf{b}_{\mathrm{nb}}). \tag{9.8}$$

The common key ingredient of these two estimators is the *a posteriori* PDF $p(\mathbf{a}|\mathbf{b}_{\mathrm{nb}})$ of the desired coefficient vector $\mathbf{a}$, given the observation vector $\mathbf{b}_{\mathrm{nb}}$. Since the *a posteriori* PDF is not readily available in practice, it has to be split into measurable components using Bayes' rule

$$p(\mathbf{a}|\mathbf{b}_{\mathrm{nb}}) = \frac{p(\mathbf{b}_{\mathrm{nb}}|\mathbf{a})\, p(\mathbf{a})}{p(\mathbf{b}_{\mathrm{nb}})} \tag{9.9}$$

$$= \frac{p(\mathbf{b}_{\mathrm{nb}}|\mathbf{a})\, p(\mathbf{a})}{\int_{\mathcal{R}^n} p(\mathbf{b}_{\mathrm{nb}}|\mathbf{a})\, p(\mathbf{a})\, \mathrm{d}\mathbf{a}}. \tag{9.10}$$

Both the observation probability $p(\mathbf{b}_{\mathrm{nb}}|\mathbf{a})$ and the probability density $p(\mathbf{a})$ can be modeled, with the model parameters trained off-line and stored for later use by the estimator. Accordingly, it is possible with (9.10) to compute the *a posteriori* PDF from the observation vector $\mathbf{b}_{\mathrm{nb}}$. Subsequently, any *a posteriori* based estimation rule can be applied, e.g., one of the variants from (9.7) and (9.8).

Error Concealment Using Two Information Paths

The example estimators derived above shall now be modified for two parallel input observations. The target is to define a single estimation rule that makes best use of *both* available information paths. For the application of embedded wideband speech coding with BWE techniques, we have explicit signaling of the BWE information in the watermark message m leading to the channel output $\mathbf{b}_{\mathrm{wm}}$, and implicit information in the narrowband speech signal via the observation vector $\mathbf{b}_{\mathrm{nb}}$. Therefore, we have to replace the *a posteriori* PDF $p(\mathbf{a}|\mathbf{b}_{\mathrm{nb}})$ in the example estimation rules (9.7) and (9.8) by the *a posteriori* PDF $p(\mathbf{a}|\mathbf{b}_{\mathrm{wm}}, \mathbf{b}_{\mathrm{nb}})$ that is conditioned on both available information paths.

Before applying Bayes' rule, we introduce the assumption that the two observations $\mathbf{b}_{\mathrm{wm}}$ and $\mathbf{b}_{\mathrm{nb}}$ are independent under the condition of a fixed value of $\mathbf{a}$. That is, we assume

$$p(\mathbf{b}_{\mathrm{wm}}, \mathbf{b}_{\mathrm{nb}}|\mathbf{a}) \approx p(\mathbf{b}_{\mathrm{wm}}|\mathbf{a}) \cdot p(\mathbf{b}_{\mathrm{nb}}|\mathbf{a}) \,. \tag{9.11}$$

The basis for this assumption is that for a given BWE parameter vector $\mathbf{a}$ the distribution $p(\mathbf{b}_{\mathrm{wm}}|\mathbf{a})$ depends on stochastic channel perturbations affecting explicit transmission of the BWE information, while the observation probability $p(\mathbf{b}_{\mathrm{nb}}|\mathbf{a})$ depends on dependencies between the BWE parameters and the narrowband speech signal.

Applying Bayes' rule for this case of two information paths and using the above assumption leads to the modified *a posteriori* PDF

$$p(\mathbf{a}|\mathbf{b}_{\mathrm{wm}}, \mathbf{b}_{\mathrm{nb}}) = \frac{p(\mathbf{b}_{\mathrm{wm}}, \mathbf{b}_{\mathrm{nb}}|\mathbf{a})\, p(\mathbf{a})}{p(\mathbf{b}_{\mathrm{wm}}, \mathbf{b}_{\mathrm{nb}})} \tag{9.12}$$

$$\approx \frac{p(\mathbf{b}_{\mathrm{wm}}|\mathbf{a})\, p(\mathbf{b}_{\mathrm{nb}}|\mathbf{a})\, p(\mathbf{a})}{p(\mathbf{b}_{\mathrm{wm}}, \mathbf{b}_{\mathrm{nb}})} \tag{9.13}$$

$$\approx \frac{p(\mathbf{b}_{\mathrm{wm}}|\mathbf{a})\, p(\mathbf{b}_{\mathrm{nb}}|\mathbf{a})\, p(\mathbf{a})}{\int_{\mathcal{R}^n} p(\mathbf{b}_{\mathrm{wm}}|\mathbf{a})\, p(\mathbf{b}_{\mathrm{nb}}|\mathbf{a})\, p(\mathbf{a})\, \mathrm{d}\mathbf{a}} \,. \tag{9.14}$$

This formulation of the *a posteriori* PDF allows for statistical estimation of the BWE parameter set $\mathbf{a}$ using both information sources, $\mathbf{b}_{\mathrm{nb}}$ and $\mathbf{b}_{\mathrm{wm}}$. It is, however, necessary to have certain *a priori* knowledge. For watermarking channels the likelihood values $p(\mathbf{b}_{\mathrm{wm}}|\mathbf{a})$ can be derived from an AWGN model of the equivalent watermark transmission path [Geiser et al. 2005]. For this purpose an estimate of the effective distortion of the watermarking channel has to be determined at the receiver. The *a priori* knowledge required to determine $p(\mathbf{a})$ and $p(\mathbf{b}_{\mathrm{nb}}|\mathbf{a})$ has to be trained off-line, as for stand-alone BWE, see above and Sec. 9.3.1.

The signal flow of the joint estimation rule is illustrated in Fig. 9.13. For more details and an application example for this joint estimation rule the reader is referred to [Geiser et al. 2005].

The derivation of the *a posteriori* PDFs and the two example estimators has been shown above for the case of continuous PDFs. The same procedure can be applied

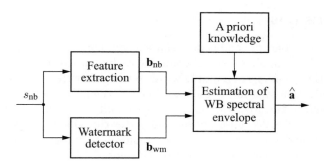

Figure 9.13: Joint estimation of the BWE parameters **a** from both available information paths, the output $\mathbf{b}_{\mathrm{wm}}$ of the watermark channel and the features
$\mathbf{b}_{\mathrm{nb}}$ of the NB speech signal

for discrete probability functions, or for a mixture of discrete and continuous quantities using the mixed form of Bayes' rule. Furthermore, the same methodology
can be applied with more sophisticated source models, e.g., with a hidden Markov
model.

Fall-back Behavior

The *a posteriori* PDF (9.14) has been derived in order to exploit both available information sources. What happens if the information from one of the two paths is degraded? In the extreme case one of the two information paths may get completely lost.
Then, the mutual information of the channel output and the original parameter set de-
grades to zero. That is, if, for example, the watermark channel is lost, $I(\mathbf{a}; \mathbf{b}_{\mathrm{wm}}) = 0$,
and the corresponding likelihood can be approximated as

$$p(\mathbf{b}_{\mathrm{wm}}|\mathbf{a}) \approx p(\mathbf{b}_{\mathrm{wm}})\,. \tag{9.15}$$

Thereby, the *a posteriori* PDF is reduced to

$$p(\mathbf{a}|\mathbf{b}_{\mathrm{wm}}, \mathbf{b}_{\mathrm{nb}}) = \frac{p(\mathbf{b}_{\mathrm{wm}}|\mathbf{a})\,p(\mathbf{b}_{\mathrm{nb}}|\mathbf{a})\,p(\mathbf{a})}{\int\limits_{\mathcal{R}^n} p(\mathbf{b}_{\mathrm{wm}}|\mathbf{a})\,p(\mathbf{b}_{\mathrm{nb}}|\mathbf{a})\,p(\mathbf{a})\,\mathrm{d}\mathbf{a}} \tag{9.16}$$

$$\approx \frac{p(\mathbf{b}_{\mathrm{wm}})}{p(\mathbf{b}_{\mathrm{wm}})} \cdot \frac{p(\mathbf{b}_{\mathrm{nb}}|\mathbf{a})\,p(\mathbf{a})}{\int\limits_{\mathcal{R}^n} p(\mathbf{b}_{\mathrm{nb}}|\mathbf{a})\,p(\mathbf{a})\,\mathrm{d}\mathbf{a}} \tag{9.17}$$

$$= \frac{p(\mathbf{b}_{\mathrm{nb}}|\mathbf{a})\,p(\mathbf{a})}{\int\limits_{\mathcal{R}^n} p(\mathbf{b}_{\mathrm{nb}}|\mathbf{a})\,p(\mathbf{a})\,\mathrm{d}\mathbf{a}}\,. \tag{9.18}$$

The result (9.18) is identical to the *a posteriori* PDF for the case of only one observation source, cf. (9.10). This means that, if one information source gets lost, the
joint estimator will gracefully fall back to exploit only the single remaining observation.

9.7 Conclusions

All of the described concepts and codecs share a common core technology: parametric representation of the highband signal components by a set of BWE parameters. The difference is in the questions of if and how this set of parameters is transmitted via the speech communication system. These questions are the major key to both backwards-compatibility and achievable wideband speech quality. The larger the data rate of explicitly transmitted BWE information, the better the wideband speech quality that can be obtained, and the greater the effort required to integrate the approach with existing narrowband speech communication systems.

Although bandwidth extension is a rather young technology, some BWE techniques have already found their way into speech coding standards. Other, more sophisticated approaches need further research and testing to become mature enough for deployment in practical systems.

Bandwidth extension and "true" wideband speech coding should not be regarded as alternatives. In fact, there are several applications for which the two techniques complement each other. In particular for embedded wideband speech coding, estimation and coding techniques need to go hand in hand in order to provide robust and high-quality speech communication services over adverse transmission channels.

Bibliography

Agiomyrgiannakis, Y.; Stylianou, Y. (2004). Combined Estimation/Coding of Highband Spectral Envelopes for Speech Spectrum Expansion, *Proceedings of IEEE International Conference on Acoustics, Speech, and Signal Processing (ICASSP)*, Montreal, Canada, vol. 1, pp. 469–472.

Agiomyrgiannakis, Y.; Stylianou, Y. (2007). Conditional Vector Quantization for Speech Coding, *IEEE Transactions on Audio, Speech, and Language Processing*, vol. 15, no. 2, pp. 377–386.

Aguilar, G.; Chen, J.-H.; Dunn, R. B.; McAulay, R. J.; Sun, X.; Wang, W.; Zopf, R. (2000). An Embedded Sinusoidal Transform Codec with Measured Phases and Sampling Rate Scalability, *Proceedings of IEEE International Conference on Acoustics, Speech, and Signal Processing (ICASSP)*, Istanbul, Turkey, vol. 2, pp. 1141–1144.

Carl, H.; Heute, U. (1994). Bandwidth Enhancement of Narrow-Band Speech Signals, *Proceedings of European Signal Processing Conference (EUSIPCO)*, Edinburgh, Scotland, vol. 2, pp. 1178–1181.

Chan, C.-F.; Hui, W.-K. (1996). Wideband Re-synthesis of Narrowband CELP Coded Speech Using Multiband Excitation Model, *Proceedings of International Conference on Spoken Language Processing (ICSLP)*, Philadelphia, PA, USA, vol. 1, pp. 322–325.

Chen, B.; Wornell, G. W. (2001). Quantization Index Modulation: A Class of Provably Good Methods for Digital Watermarking and Information Embedding, *IEEE Transactions on Information Theory*, vol. 47, no. 4, pp. 1423–1443.

Chen, S.; Leung, H. (2005). Artificial Bandwidth Extension of Telephony Speech by Data Hiding, *Proceedings of International Symposium on Circuits and Systems (ISCAS)*, vol. 4, pp. 3151–3154.

Cheng, Q.; Sorensen, J. (2001). Spread Spectrum Signaling for Speech Watermarking, *Proceedings of IEEE International Conference on Acoustics, Speech, and Signal Processing (ICASSP)*, Salt Lake City, UT, USA, vol. 3, pp. 1337–1340.

Cheng, Y. M.; O'Shaughnessy, D.; Mermelstein, P. (1994). Statistical Recovery of Wideband Speech from Narrowband Speech, *IEEE Transactions on Speech and Audio Processing*, vol. 2, no. 4, pp. 544–548.

Chétry, N.; Davies, M. (2006). Embedding Side Information Into a Speech Codec Residual, *Proceedings of European Signal Processing Conference (EUSIPCO)*, Florence, Italy.

Costa, M. H. M. (1983). Writing on Dirty Paper, *IEEE Transactions on Information Theory*, vol. IT-29, no. 3, pp. 439–441.

de Campos Neto, S. F.; Järvinen, K. (2006). Guest Editorial: Wideband Speech Coding Standards and Wireless Services, *IEEE Communications Magazine*, vol. 44, no. 5, pp. 56–57.

Dietz, M.; Liljeryd, L.; Kjorling, K.; Kunz, O. (2002). Spectral Band Replication: A Novel Approach in Audio Coding, *AES Convention*, Munich, Germany. Preprint 5553.

Ding, H. (2004). Wideband Audio over Narrowband Low-Resolution Media, *Proceedings of IEEE International Conference on Acoustics, Speech, and Signal Processing (ICASSP)*, Montreal, Canada, vol. 1, pp. 489–492.

Epps, J. (2000). *Wideband Extension of Narrowband Speech for Enhancement and Coding*, PhD thesis, School of Electrical Engineering and Telecommunications, The University of New South Wales.

Epps, J. R.; Holmes, W. H. (2001). A New Very Low Bit Rate Wideband Speech Coder with a Sinusoidal Highband Model, *Proceedings of International Symposium on Circuits and Systems (ISCAS)*, Sydney, Australia, vol. 2, pp. 349–352.

Fingscheidt, T.; Vary, P. (2001). Softbit Speech Decoding: A New Approach to Error Concealment, *IEEE Transactions on Speech and Audio Processing*, vol. 9, no. 3, pp. 240–251.

Fischer, R. F. H.; Tzschoppe, R.; Bäuml, R. (2004). Lattice Costa Schemes Using Subspace Projection for Digital Watermarking, *Proceedings of International ITG Conference on Source and Channel Coding*, Erlangen, Germany, pp. 127–134.

Fuemmeler, J. A.; Hardie, R. C.; Gardner, W. R. (2001). Techniques for the Regeneration of Wideband Speech from Narrowband Speech, *EURASIP Journal on Applied Signal Processing*, vol. 2001, no. 4, pp. 266–274.

Geiser, B.; Jax, P.; Vary, P. (2005). Artificial Bandwidth Extension of Speech Supported by Watermark Transmitted Side Information, *Proceedings of European Conference on Speech Communication and Technology (INTERSPEECH)*, Lisbon, Portugal, pp. 1497–1500.

Geiser, B.; Jax, P.; Vary, P.; Taddei, H.; Schandl, S.; Gartner, M.; Guillaumé, C.; Ragot, S. (2007). Bandwidth Extension for Hierarchical Speech and Audio Coding in ITU-T Rec. G.729.1, *IEEE Transactions on Audio, Speech, and Language Processing*, vol. 15, no. 8, pp. 2496–2509.

Geiser, B.; Vary, P. (2007). Backwards Compatible Wideband Telephony in Mobile Networks: CELP Watermarking and Bandwidth Extension, *Proceedings of IEEE International Conference on Acoustics, Speech, and Signal Processing (ICASSP)*, Honolulu, HW, USA, vol. 4, pp. 533–536.

Hartung, F.; Kutter, M. (1999). Multimedia Watermarking Techniques, *Proceedings of the IEEE*, vol. 87, no. 7, July, pp. 1079–1107.

ITU G.729.1 (2006). G.729 Based Embedded Variable Bitrate Coder: An 8–32 kbit/s Scalable Wideband Coder Bitstream Interoperable with G.729.

Jax, P. (2004). Bandwidth Extension for Speech, *in* E. Larsen; R. M. Aarts (eds.), *Audio Bandwidth Extension*, John Wiley & Sons, Ltd, chapter 6, pp. 171–236.

Jax, P.; Geiser, B.; Schandl, S.; Taddei, H.; Vary, P. (2006a). A Scalable Wideband "Add-On" for the G.729 Speech Codec, *ITG-Fachtagung Sprachkommunikation*, Kiel, Germany.

Jax, P.; Geiser, B.; Schandl, S.; Taddei, H.; Vary, P. (2006b). An Embedded Scalable Wideband Codec Based on the GSM EFR Codec, *Proceedings of IEEE International Conference on Acoustics, Speech, and Signal Processing (ICASSP)*, Toulouse, France, vol. 1, pp. 5–8.

Jax, P.; Vary, P. (2000). Wideband Extension of Telephone Speech Using a Hidden Markov Model, *Proceedings of the IEEE Workshop on Speech Coding*, Delavan, WI, USA, pp. 133–135.

Jax, P.; Vary, P. (2002). An Upper Bound on the Quality of Artificial Bandwidth Extension of Narrowband Speech Signals, *Proceedings of IEEE International Conference on Acoustics, Speech, and Signal Processing (ICASSP)*, Orlando, FL, USA, vol. 1, pp. 237–240.

Jax, P.; Vary, P. (2006). Bandwidth Extension of Speech Signals: A Catalyst for the Introduction of Wideband Speech Coding?, *IEEE Communications Magazine*, vol. 44, no. 5, pp. 106–111.

Kataoka, A.; Kurihara, S.; Sasaki, S.; Hayashi, S. (1997). A 16-kbit/s Wideband Speech Codec Scalable With G.729, *Proceedings of European Conference on Speech Communication and Technology (EUROSPEECH)*, Rhodes, Greece, pp. 1491–1494.

Koishida, K.; Cuperman, V.; Gersho, A. (2000). A 16 kbit/s Bandwidth Scalable Audio Coder Based on the G.729 Standard, *Proceedings of IEEE International Conference on Acoustics, Speech, and Signal Processing (ICASSP)*, Istanbul, Turkey, vol. 2, pp. 1149–1152.

McCree, A. (2000). A 14 kb/s Wideband Speech Coder with a Parametric Highband Model, *Proceedings of IEEE International Conference on Acoustics, Speech, and Signal Processing (ICASSP)*, Istanbul, Turkey, vol. 2, pp. 1153–1156.

McCree, A.; Unno, T.; Anandakumar, A.; Bernard, A.; Paksoy, E. (2001). An Embedded Adaptive Multi-Rate Wideband Speech Coder, *Proceedings of IEEE International Conference on Acoustics, Speech, and Signal Processing (ICASSP)*, Salt Lake City, UT, USA, vol. 2, pp. 761–764.

McElroy, C.; Murray, B.; Fagan, A. D. (1993). Wideband Speech Coding in 7.2 kb/s, *Proceedings of IEEE International Conference on Acoustics, Speech, and Signal Processing (ICASSP)*, Minneapolis, MN, USA, vol. 2, pp. 620–623.

Nakatoh, Y.; Tsushima, M.; Norimatsu, T. (1997). Generation of Broadband Speech from Narrowband Speech using Piecewise Linear Mapping, *Proceedings of European Conference on Speech Communication and Technology (EUROSPEECH)*, Rhodos, Greece, vol. 3, pp. 1643–1646.

Nilsson, M.; Andersen, S. V.; Kleijn, W. B. (2000). On the Mutual Information Between Frequency Bands in Speech, *Proceedings of IEEE International Conference on Acoustics, Speech, and Signal Processing (ICASSP)*, Istanbul, Turkey, vol. 3, pp. 1327–1330.

Nilsson, M.; Gustafsson, H.; Andersen, S. V.; Kleijn, W. B. (2002). Gaussian Mixture Model Based Mutual Information Estimation Between Frequency Bands in Speech, *Proceedings of IEEE International Conference on Acoustics, Speech, and Signal Processing (ICASSP)*, Orlando, FL, USA, vol. 1, pp. 525–528.

Nomura, T.; Iwadare, M.; Serizawa, M.; Ozawa, K. (1998). A Bitrate and Bandwidth Scalable CELP Coder, *Proceedings of IEEE International Conference on Acoustics, Speech, and Signal Processing (ICASSP)*, Seattle, WA, USA, vol. 1, pp. 341–344.

Park, K.-Y.; Kim, H. S. (2000). Narrowband to Wideband Conversion of Speech using GMM-based Transformation, *Proceedings of IEEE International Conference on Acoustics, Speech, and Signal Processing (ICASSP)*, Istanbul, Turkey, vol. 3, pp. 1847–1850.

Patrick, P. J. (1983). *Enhancement of Bandlimited Speech Signals*, PhD thesis, Loughborough University of Technology.

Sagi, A.; Malah, D. (2004). Data Embedding in Speech Signals Using Perceptual Masking, *Proceedings of European Signal Processing Conference (EUSIPCO)*, Vienna, Austria, pp. 1657–1660.

Sagi, A.; Malah, D. (2007). Bandwidth Extension of Telephone Speech Aided by Data Embedding, *EURASIP Journal on Advances in Signal Processing*, vol. 2007, no. 1, p. 37.

Taori, R.; Sluijter, R. J.; Gerrits, A. J. (2000). Hi-BIN: An Alternative Approach to Wideband Speech Coding, *Proceedings of IEEE International Conference on Acoustics, Speech, and Signal Processing (ICASSP)*, Istanbul, Turkey, vol. 2, pp. 1157–1160.

Valin, J.-M.; Lefebvre, R. (2000). Bandwidth Extension of Narrowband Speech for Low Bit-Rate Wideband Coding, *Proceedings of the IEEE Workshop on Speech Coding*, Delavan, WI, USA, pp. 130–132.

Vary, P.; Martin, R. (2006). *Digital Speech Transmission: Enhancement, Coding and Error Concealment*, John Wiley & Sons, Ltd.

Wolters, M.; Kjörling, K.; Homm, D.; Purnhagen, H. (2003). A Closer Look Into MPEG-4 High Efficiency AAC, *AES Convention*, New York, NY, USA. Preprint 5871.

IV

Joint Source-Channel Coding

Chapter 10

Parameter Models and Estimators in Soft Decision Source Decoding

Tim Fingscheidt

10.1 Introduction

The typical goals of transmission system design such as high capacity, but also high signal quality and robustness usually contradict each other. Thanks to Shannon's landmark contribution to information theory [Shannon 1948] one can treat the issues of bit rate reduction (source coding) and error protection (channel coding) separately. This however in principle presumes unlimited computational power and delay. In real-time conversational and streaming services for speech, audio, and video transmission in particular these assumptions are of course not met. To guarantee a limited delay and to achieve a certain level of robustness, frame-based transmission is employed, which limits the ability to remove interframe correlations. In virtually all source coders, this fact is practically observed in terms of residual redundancy in the bit stream.

In the very same publication [Shannon 1948], Shannon mentioned further: *"However, any redundancy in the source will usually help if it is utilized at the receiving point. In particular, if the source already has a certain redundancy and no attempt is made to eliminate it in matching the channel, this redundancy will help combat noise."* In the last twenty years this led to a remarkable interest in *joint source and channel coding.* The common idea of many of the proposed approaches is the exploitation of this residual redundancy in the channel decoder and/or in the source decoder. This is

Advances in Digital Speech Transmission   Edited by R. Martin, U. Heute and C. Antweiler
© 2008 John Wiley & Sons, Ltd

often performed in conjunction with the use of so-called *soft values* through parts of the receiver processing chain equalizer and demodulator, channel decoder, and source decoder.

Previous work on joint source channel coding can roughly be grouped as follows. First, there are approaches that do not touch conventionally designed transmission systems except the receiver. These can be further subdivided into proposals where the conventional system design does not require additional explicit redundancy for channel encoding, e.g., [Gerlach 1993], [Phamdo, Farvardin 1994], [Fingscheidt, Vary 1996], and those where some explicit redundancy is used in the system, e.g., [Hagenauer 1995], [Weiss et al. 1996], [Alajaji et al. 1996], [Fazel, Fuja 2000].

Secondly, there are approaches that allow a redesign of the encoder. Well-known schemes without adding explicit redundancy for channel coding are channel-optimized codebooks and index assignments for vector quantization [Farvardin, Vaishampayan 1987], [Farvardin 1990], [Farvardin, Vaishampayan 1991], [Sayood, Borkenhagen 1991], [Knagenhjelm 1993], [Cuperman et al. 1994], [Skoglund, Hedelin 1999]. Finally, one can think of course as well of a full system optimization with modified source encoding, channel encoding, and receiver-sided functions.

Chapter 10 and Chapter 11 deal with system-compatible improvements to existing transmission systems where in principle no explicit redundancy for channel coding is required (although allowed!). The technique presented will exploit residual redundancy in the decoder and is called *Soft Decision Source Decoding* (SDSD). The later chapters of this part of the book comprise typical channel coding schemes as well and show possibilities for a better total system design in the sense of joint source channel coding.

The mere SDSD technique is presented in this chapter. It can be seen as an advanced *error concealment* scheme [Fingscheidt, Vary 1996], [Fingscheidt, Vary 1997a], [Fingscheidt, Vary 2001]. Besides the modification of the source decoder, SDSD only requires the availability of soft information at the channel decoder's output. As an introduction to this chapter and also the following chapters we will briefly revise the principle of SDSD in Sec. 10.2 – in the fashion in which it was first published [Fingscheidt, Vary 1996]. In Sec. 10.3 we will then introduce some basics to Markovian parameter modelling, a prerequisite to the estimators presented in the subsequent sections. Basic extrapolative estimators for different Markov model orders will be discussed in Sec. 10.4. Sections 10.5 and 10.6 will introduce two closely related types of estimators: The first exploiting statistical dependencies between two different parameters, the second making use of the redundancy that occurs in a repeated parameter transmission. Techniques of interpolative parameter estimation are discussed in Sec. 10.7.

10.2 Overview to Soft Decision Source Decoding

10.2.1 Source Encoding

In the following we are interested in robust transmission of so-called *parameters*. These parameters could be, for example, gray values of pixels in an image or samples of a speech signal. In practice, of course, image and speech signals are subject to source coding before being transmitted. Such source encoders can follow quite different coding philosophies such as waveform coding, parametric coding, or hybrid coding. Any of these coding paradigms generates one or more parameters in a certain period of time or per frame. In the case of hybrid speech coding, examples of these parameters are spectral coefficients, gain factors, and pitch values.

In the following, we will regard a parameter that is typically a vector of scalar parameter components $\tilde{\underline{v}}_k = (\tilde{v}_{1,k}, \ldots, \tilde{v}_{m,k}, \ldots, \tilde{v}_{M,k})$ with M being the vector dimension, and k being the frame index. We shall assume real-valued parameters, i.e., $\tilde{v}_{m,k} \in \mathbb{R}$. In the context of hybrid speech coders, $\tilde{v}_{m,k}$ may represent a gain factor or a spectral coefficient, for example. A typical $M = 2$ case is the vector quantization of an algebraic codebook gain and an adaptive codebook gain (see, e.g., [3GPP-AMR 1999]).

As shown in Fig. 10.1, we assume this parameter is subject to a (vector) quantizer Q yielding $\underline{v}_k \in \mathbb{V}$, with $\mathbb{V}$ being a codebook with a number of 2^w (vectorial) codebook entries $\underline{v}^{(i)}, i = 0, 1, \ldots, 2^w - 1$. A bit mapping scheme (BM) chooses the respective bit pattern $\mathbf{x}_k = (x_k(1), \ldots, x_k(\kappa), \ldots, x_k(w)) = \mathbf{x}^{(i)} \in \mathbb{X}$ that will be transmitted via a so-called *equivalent channel*. The bit index κ is in the range $\kappa = 1, \ldots, w$. $\mathbb{X}$ is called the codebook of bit patterns, and it can be addressed by the index i to yield a specific bit pattern $\mathbf{x}^{(i)}$.

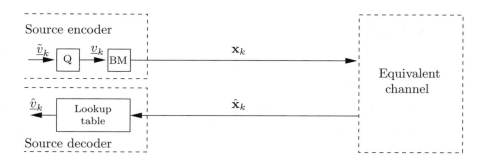

Figure 10.1: Conventional *Hard Decision* (HD) transmission system

10.2.2 Equivalent Channel

The equivalent channel may comprise channel coding, modulation, the physical channel itself, and also synchronization, demodulation, equalization, and channel decoding. In this chapter we are interested only in the output of such an equivalent channel, which classically consists of a received bit pattern $\hat{\mathbf{x}}_k$. The simplest channel we can imagine is a *Binary Symmetric Channel* (BSC) that merely assigns a received bit $\hat{x}_k(\kappa)$ to a transmitted bit $x_k(\kappa)$. The bits transmitted over the BSC are subject to errors with a certain bit error probability $0 \le p_k(\kappa) \le 1$, which may be time-varying. As shown in Fig. 10.2 we expect the equivalent channel not only to yield the received bit pattern $\hat{\mathbf{x}}_k$, but also to provide an estimate of the instantaneous bit error probabilities $\mathbf{p}_k = (p_k(1), \ldots, p_k(w))$.

A more powerful yet simple model of an equivalent channel is given if we assume transmission over a *Binary Phase Shift Keying* (BPSK) modulated *Additive White Gaussian Noise* (AWGN) or *Rayleigh fading* channel with coherent demodulation. Postulating bipolar bits with $x_k(\kappa) \in \{-1, +1\}$, we can express the real-valued demodulator output as

$$z_k(\kappa) = a_k(\kappa) \cdot x_k(\kappa) + n_k(\kappa), \qquad z_k(\kappa) \in \mathbb{R}, \tag{10.1}$$

with $a_k(\kappa) \in \mathbb{R}^+$ being the fading factor ($a_k(\kappa) = 1$ for AWGN) and $n_k(\kappa) \in \mathbb{R}$ being a sample of the additive white Gaussian noise with variance $N_0/2$. The respective output bit of the equivalent channel is then

$$\hat{x}_k(\kappa) = \text{sign}(z_k(\kappa)). \tag{10.2}$$

Utilizing this model of an equivalent channel, however, yields much more information than the received bit alone. If the fading factor $a_k(\kappa)$ is known (at least in average)

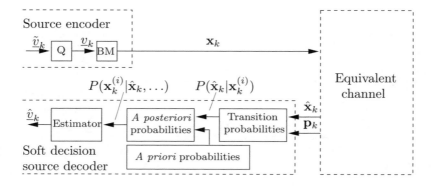

Figure 10.2: SDSD transmission system

and also the noise variance $N_0/2$, this equivalent channel also yields estimates for the instantaneous bit error probability

$$p_k(\kappa) = (1 + \exp|L(x_k(\kappa))|)^{-1} \tag{10.3}$$

with

$$L(x_k(\kappa)) = 4a_k(\kappa) \cdot \frac{E_s}{N_0} \cdot z_k(\kappa) \tag{10.4}$$

being a so-called log-likelihood ratio $L(x_k(\kappa)) \in \mathbb{R}$ of the transmitted bit $x_k(\kappa)$. The BPSK symbol energy will hold $E_s = 1$, the expectation value $E\{a^2\} = 1$.

So far, this simple channel model is very suitable for the evaluation of the SDSD approach. In practice, however, error correction means are employed and are part of the equivalent channel. It is important to note that, in this case, the equivalent channel can also be designed to yield (estimations of) bit error probabilities along with the received bits. We expect the channel decoder then to be a soft-input and soft-output algorithm, like, e.g., the *Soft-Output Viterbi Algorithm* (SOVA) as a sequence estimator [Battail 1987], [Hagenauer, Höher 1989], [Huber, Rüppel 1990]. While the SOVA is a computationally efficient suboptimal algorithm, the channel decoder by Bahl et al. [Bahl et al. 1974] is the optimal symbol-by-symbol estimator. All these algorithms (can be implemented in a way that they) deliver log-likelihood ratios $L(x_k(\kappa)) \in \mathbb{R}$ that can directly be converted into bit error probabilities by (10.3).

10.2.3 Hard Decision and Soft Decision Source Decoding

In Fig. 10.1 a conventional transmission scheme is shown. Any possible error correction capability is completely dedicated to the functions of the equivalent channel. Therefore, no log-likelihood ratios or bit error probabilities are passed to the source decoder. Consequently, decoding of the parameter $\hat{v}_k$ is simply performed by a table lookup. Henceforth, we will call this principle *Hard Decision* (HD) decoding.

In Fig. 10.2 the concept of *Soft Decision* (SD) source decoding is shown: Along with the received bit pattern $\hat{\mathbf{x}}_k$ the vector of instantaneous bit error probabilities $\mathbf{p}_k$ is delivered by the equivalent channel. In a first step, SDSD computes so-called *transition probabilities*

$$P(\hat{\mathbf{x}}_k | \mathbf{x}_k^{(i)}) = \prod_{\kappa=1}^{w} P(\hat{x}_k(\kappa) | x_k^{(i)}(\kappa)), \qquad i = 0, 1, \ldots, 2^w - 1, \tag{10.5}$$

exploiting the bit error probability vector by

$$P(\hat{x}_k(\kappa) | x_k^{(i)}(\kappa)) = \begin{cases} 1 - p_k(\kappa) & \text{if } \hat{x}_k(\kappa) = x_k^{(i)}(\kappa) \\ p_k(\kappa) & \text{if } \hat{x}_k(\kappa) \neq x_k^{(i)}(\kappa). \end{cases} \tag{10.6}$$

The transition probabilities are an important statistical entity for the further processing within the SDSD framework, since they are assumed to describe the equivalent channel fully. It becomes clear at this point that regardless what functionalities the equivalent channel may comprise, it is modeled as memoryless by SDSD. It is important to note that this has been found not to be a problem even in the case of equivalent channels with a lot of memory as is the case if convolutional coders are used [Fingscheidt, Scheufen 1997].

For each received parameter $\hat{\mathbf{x}}_k$ coded by w bits, a number of 2^w transition probabilities $P(\hat{\mathbf{x}}_k|\mathbf{x}_k^{(i)})$ are available now that yield the likelihood of the observation $\hat{\mathbf{x}}_k$ under the assumption that $\mathbf{x}_k^{(i)}$ has been transmitted ($i = 0, 1, \ldots, 2^w - 1$). The exact transmitted bit pattern $\mathbf{x}_k$ is of course not known.

The parameter (vector) is now to be modeled as a Markov process of order N. This model decides to which extent residual redundancy that is present in the bit stream is exploited in the SDSD computation. The parameter models are used as *a priori knowledge*, i.e., as statistical information that is available about the bit pattern $\mathbf{x}_k$ without knowledge of the received sequence of bit patterns $\hat{\mathbf{x}}_1^k = \{\hat{\mathbf{x}}_1, \ldots, \hat{\mathbf{x}}_k\}$. In Sec. 10.3 we will analyze the process of Markov modeling in more detail.

After the transition probabilities are available, the next computational step is the derivation of *a posteriori* probabilities $P(\mathbf{x}_k^{(i)}|\hat{\mathbf{x}}_k, \ldots)$, where "..." stands for further (past) received bit patterns. Here the residual redundancy is exploited. It requires the stored parameter model, i.e., the *a priori* knowledge.

Having available the *a posteriori* probabilities, one can perform the final step in SDSD: The estimation of the parameter vector $\underline{v}_k$ according to a given error criterion. In [Melsa, Cohn 1978] and for SDSD in [Fingscheidt, Vary 1997b] the interested reader finds advice on how to design estimators. In this chapter we assume that the *Minimum Mean Square Error* (MMSE) with its error criterion $\mathrm{E}\{||\hat{\underline{v}}_k - \underline{v}_k||^2\} \rightarrow \min$ is appropriate. Hence the optimal SDSD estimator as we use it in the following is simply

$$\hat{\underline{v}}_k = \sum_{i=0}^{2^w-1} \underline{v}^{(i)} \cdot P(\mathbf{x}_k^{(i)}|\hat{\mathbf{x}}_k, \ldots). \tag{10.7}$$

The presentation of new but simple MMSE estimators in the context of SDSD will be the main goal of Secs. 10.4 through 10.7.

Note that although the MMSE estimator minimizes the mean squared Euclidean distance to the (unknown) true quantized and transmitted parameter vector $\underline{v}_k$, we should evaluate the performance of such an estimator always including the distortion due to quantization. For a scalar parameter $\tilde{v}_{1,k} = \tilde{v}_k$ transmitted over any of the systems in Figs. 10.1 or 10.2, an appropriate measure of quality is therefore the *parameter Signal-to-Noise Ratio*, or briefly *parameter SNR*

$$\text{parameter SNR} = 10 \log_{10} \frac{\mathrm{E}\{\tilde{v}^2\}}{\mathrm{E}\{(\hat{v} - \tilde{v})^2\}} \quad \text{dB} . \tag{10.8}$$

10.3 The Markovian Parameter Model

10.3.1 Description of A Priori Knowledge

When developing new estimators in this chapter we can either work with w bit quantized parameter vectors $\underline{v}_k$ with frame index k, or, alternatively, with their respective bit patterns $\mathbf{x}_k$. Similar to Sec. 10.1 we augment it by the quantization table index i according to $\mathbf{x}_k = \mathbf{x}_k^{(i)}$ whenever needed.

If one sample (parameter vector) $\underline{v}_k$ taken from a discrete-time (vectorial) random process V at time k does not depend on all, but only on the previous sample parameter $\underline{v}_{k-1}$, we say this process has the *Markov property* (see for example, [Papoulis 1965]).

Denoting the probability of occurrence of the quantized parameter vector $\underline{v}_k$ (of the bit pattern $\mathbf{x}_k$, respectively) at time k by $P(\underline{v}_k) = P(\mathbf{x}_k)$, one can characterize a Markov process as follows:

$$P(\mathbf{x}_k|\mathbf{x}_{k-1}, \mathbf{x}_{k-2}, \ldots) = P(\mathbf{x}_k|\mathbf{x}_{k-1}) \ . \tag{10.9}$$

In the following, the transition from $\mathbf{x}_{k-1}$ to $\mathbf{x}_k$ is assumed to be stationary, i.e., the probabilities $P(\mathbf{x}_k|\mathbf{x}_{k-1})$ depend only on the bit patterns $\mathbf{x}_k$ and $\mathbf{x}_{k-1}$, but not on the absolute index of time k. This property is usually called *homogeneity* [Cox, Miller 1965]. The question of whether a certain parameter can be modeled as taken from an homogeneous Markov process is to be decided on a case-by-case basis.

In general we cannot assume that, e.g., a quantized speech codec parameter fulfills (10.9). Therefore, in the following we will use a more generalized definition of a Markov process. Let us define a Markov process of *order* N by [Takàcs 1968]:

$$P(\mathbf{x}_k|\mathbf{x}_{k-1}, \ldots, \mathbf{x}_{k-N}, \mathbf{x}_{k-N-1}, \ldots) = P(\mathbf{x}_k|\mathbf{x}_{k-1}, \ldots, \mathbf{x}_{k-N}) \ . \tag{10.10}$$

The property classically denoted as Markov property is characterized in this notation as Markov property of 1st order. Moreover, the special case of statistically independent parameters is also covered, using the term Markov process of 0th order: $P(\mathbf{x}_k|\mathbf{x}_{k-1}, \ldots) = P(\mathbf{x}_k)$.

The statistical knowledge about a homogeneous generalized Markov process of order N is given by the joint probabilities $P(\mathbf{x}_k, \mathbf{x}_{k-1}, \ldots, \mathbf{x}_{k-N})$. In some cases the conditional probabilities $P(\mathbf{x}_k|\mathbf{x}_{k-1}, \ldots, \mathbf{x}_{k-N})$ are sufficient. In the following, both terms will be called *a priori knowledge* or *a priori probabilities* of the respective parameter.

Assuming homogeneity, the *a priori* probabilities can be measured in the form of the joint probabilities $P(\mathbf{x}_k, \mathbf{x}_{k-1}, \ldots, \mathbf{x}_{k-N})$ by processing a large database through the source encoder and generating statistics of the sequences of parameters (i.e., bit patterns). The conditional *a priori* probabilities $P(\mathbf{x}_k|\mathbf{x}_{k-1},\ldots,\mathbf{x}_{k-N})$ can be computed from the measured joint *a priori* probabilities by

$$P(\mathbf{x}_k \mid \mathbf{x}_{k-1},\ldots,\mathbf{x}_{k-N}) = \frac{P(\mathbf{x}_k, \mathbf{x}_{k-1},\ldots,\mathbf{x}_{k-N})}{\displaystyle\sum_{i=0}^{2^w-1} P(\mathbf{x}_k^{(i)}, \mathbf{x}_{k-1},\ldots,\mathbf{x}_{k-N})} \ . \tag{10.11}$$

The step from the joint *a priori* probabilities to the conditional *a priori* probabilities in (10.11) can be interpreted as a normalization of the joint *a priori* probabilities over their first dimension $\mathbf{x}_k$, since we always require

$$\sum_{i=0}^{2^w-1} P(\mathbf{x}_k^{(i)} \mid \mathbf{x}_{k-1},\ldots,\mathbf{x}_{k-N}) = 1 \ . \tag{10.12}$$

10.3.2 Quantification of Utilizable Residual Redundancy

Once we have *a priori* probabilities of order N available, we would like to know how much residual redundancy (in bits!) is captured by this model order and is therefore *utilizable* for SDSD. For this reason we express the *conditional entropy* in generalized form as [Gallager 1968]

$$H(\mathbf{x}_k \mid \mathbf{x}_{k-1},\ldots,\mathbf{x}_{k-N}) =$$
$$-\sum_{i=0}^{2^w-1}\sum_{j=0}^{2^w-1}\cdots\sum_{l=0}^{2^w-1} P(\mathbf{x}_k^{(i)}, \mathbf{x}_{k-1}^{(j)},\ldots,\mathbf{x}_{k-N}^{(l)}) \cdot \log_2 P(\mathbf{x}_k^{(i)} \mid \mathbf{x}_{k-1}^{(j)},\ldots,\mathbf{x}_{k-N}^{(l)}) \ . \tag{10.13}$$

In analogy to the generalized definition of the Markov property we will call the expression in (10.13) *conditional entropy of order* N. The conditional entropy can be interpreted as the required information to specify $\mathbf{x}_k$, averaged over all permutations of bit patterns $\mathbf{x}_k^{(i)},\ldots,\mathbf{x}_{k-N}^{(l)}$, if $\mathbf{x}_{k-1},\ldots,\mathbf{x}_{k-N}$ are already known.

The simplest case is the conditional entropy of order 0, usually called simply *entropy* (the frame index k can be omitted here):

$$H(\mathbf{x}) = -\sum_{i=0}^{2^w-1} P(\mathbf{x}^{(i)}) \log_2 P(\mathbf{x}^{(i)}) \ . \tag{10.14}$$

The residual redundancy ΔR that can be utilized in a SDSD with Markov modeling of order N can be written as

$$\Delta R = H_0 - H(\mathbf{x}_k \mid \mathbf{x}_{k-1},\ldots,\mathbf{x}_{k-N}), \tag{10.15}$$

since the *a priori* knowledge according to Sec. 10.3.1 is captured by the conditional entropy of (10.13). $H_0 = w$ is called the *perfect information content* of the respective quantized parameter. ΔR with the unit [bit] denotes the potential to reduce the bit rate[1] of the respective parameter from $R = w$ bit down to $R = w - \Delta R$ bit without changing the quantizer. In principle, this could be done by using a Huffman or a Fano code [Hamming 1986], [Fano 1961].

The residual redundancy given in (10.15) that can be utilized for SDSD depends in general on the distribution $P(\mathbf{x})$ of the quantized parameter and also on its temporal statistical dependencies. Both contributions to the utilizable residual redundancy are additive according to

$$\Delta R = \Delta R_d + \Delta R_c .\tag{10.16}$$

The *utilizable, distribution-dependent* residual redundancy ΔR_d is given by the difference of the perfect information content and the entropy (index "*d*" denotes d̲istribution-dependent)

$$\Delta R_d = H_0 - H(\mathbf{x}) .\tag{10.17}$$

The *utilizable, correlation-dependent* residual redundancy ΔR_c equals the difference between the entropy and the respective conditional entropy (index "*c*" stands for c̲orrelation-dependent)

$$\Delta R_c = H(\mathbf{x}) - H(\mathbf{x}_k \mid \mathbf{x}_{k-1}, \ldots, \mathbf{x}_{k-N}) .\tag{10.18}$$

In the literature ΔR_c is often denoted as *mutual information* $\mathcal{I}(\mathbf{x}_k; \mathbf{x}_{k-1}, \ldots, \mathbf{x}_{k-N})$ between $H(\mathbf{x}_k)$ and $H(\mathbf{x}_{k-1}, \ldots, \mathbf{x}_{k-N})$ and is alternatively given as [Gallager 1968]

$$\Delta R_c = \mathcal{I}(\mathbf{x}_k; \mathbf{x}_{k-1}, \ldots, \mathbf{x}_{k-N}) = H(\mathbf{x}_k) - H(\mathbf{x}_k \mid \mathbf{x}_{k-1}, \ldots, \mathbf{x}_{k-N}) =$$

$$\sum_{i=0}^{2^w-1} \sum_{j=0}^{2^w-1} \cdots \sum_{l=0}^{2^w-1} P(\mathbf{x}_k^{(i)}, \mathbf{x}_{k-1}^{(j)}, \ldots, \mathbf{x}_{k-N}^{(l)}) \cdot \log_2 \frac{P(\mathbf{x}_k^{(i)}, \mathbf{x}_{k-1}^{(j)}, \ldots, \mathbf{x}_{k-N}^{(l)})}{P(\mathbf{x}_k^{(i)}) \cdot P(\mathbf{x}_{k-1}^{(j)}, \ldots, \mathbf{x}_{k-N}^{(l)})} .\tag{10.19}$$

10.3.3 Choice of the Model Order

When deciding about the optimal Markov model order N a useful theorem states that the conditional entropy is a monotonically decreasing function with increasing order N [Gallager 1968]. In addition, the conditional entropy is always less than or equal to the perfect information content. This leads to

$$\begin{aligned} H_0 \;\geq\; H(\mathbf{x}) \;&\geq\; H(\mathbf{x}_k \mid \mathbf{x}_{k-1}, \ldots, \mathbf{x}_{k-N}) \\ &\geq\; H(\mathbf{x}_k \mid \mathbf{x}_{k-1}, \ldots, \mathbf{x}_{k-(N+\nu)}) \quad \text{for all} \quad N, \nu = 0, 1, 2, \ldots . \end{aligned}\tag{10.20}$$

---

[1]In the following the term *bit rate* is related to the *number of bits* that are used for the transmission of a single (source codec) parameter vector $\mathbf{x}$. For simplification we use the unit [bit] instead of the correct unit [bit/parameter] for R, w, H_0, H, as well as for all measures of residual redundancy. This also follows the common terminology of rate distortion theory.

The identity $H_0 = H(\mathbf{x})$ holds only in the case of a uniformly distributed quantized parameter with $P(\mathbf{x}) = 2^{-w}$. Given a tolerable loss ϵ of utilizable residual redundancy, using inequality (10.20), the Markov model order N can be estimated as the lowest value N for which we have

$$H(\mathbf{x}_k \mid \mathbf{x}_{k-1}, \ldots, \mathbf{x}_{k-N}) - H(\mathbf{x}_k \mid \mathbf{x}_{k-1}, \ldots, \mathbf{x}_{k-(N+\nu)}) \leq \epsilon \quad \text{for all } \nu \geq 0 \,.$$
$$(10.21)$$

If $\mathbf{x}$ has the exact Nth order Markov property, equality is achieved even for $\epsilon = 0$ for all $\nu \geq 0$.

Storage of the Nth order *a priori* probabilities of a parameter quantized by w bits requires $2^{w \cdot (N+1)}$ words of memory. Particularly for parameters with high bit rate w the trade-off between Markov model order N (and therefore memory) and utilizable residual redundancy ΔR should be kept in mind, and actually often decides in practice on an appropriate choice of N.

10.4 Basic Extrapolative Estimators

10.4.1 Introduction and Simulation Settings

We are interested now in formulations of concrete MMSE parameter vector estimators. The degrees of freedom we have are the Markov model order N, and the question whether estimation should be extrapolative (here in the sense that no additional delay is used) or interpolative (including lookahead frames). If there are correlations between two parameter vectors of the same frame, it is advantageous to estimate them in common. Finally, we can also think of soft decoding of a parameter vector that has been transmitted twice. It is important to note that in all cases (10.7) can be used to perform the actual MMSE estimation. Therefore, in this section we can focus fully on the respective computation of the *a posteriori* probabilities $P(\mathbf{x}_k^{(i)} \mid \ldots)$ which in some cases will be marginal distributions of higher dimensional *a posteriori* probabilities.

In this chapter, we are mainly interested in *simple* estimators, and also the simulations we present refer to parameters drawn from simple statistical processes. Without loss of generality we have simulated only scalar parameters which allows us to easily use (10.8) for evaluation. All other formulae in this chapter can also be used directly for vector quantized parameters.

We assume the parameter $\tilde{v}_k$ to be a parameter of a zero mean and unit variance autoregressive Gaussian process of order $\bar{N} \in \{0, 1, 2\}$, named the AR($\bar{N}$) process. Note that $\bar{N}$ must not necessarily be equal to the Markov model order N as used in the SDSD. The auto-correlation coefficients of the AR($\bar{N}$) process $\rho = \rho(1)$ (and $\rho(2)$ for $\bar{N} = 2$) are given wherever reasonable. The unquantized parameter is subject to scalar Lloyd–Max quantization with 2^w quantization levels coded by w bits. The bit mapping in Figs. 10.1 and 10.2 is done according to the *Natural Binary Code* (NBC) [Jayant,

Noll 1984]. NBC actually corresponds to the binary representation of the codebook index i, if the negative codebook entry with largest amplitude refers to $\mathbf{x}^{(0)}$, and the largest positive codebook entry is represented by $\mathbf{x}^{(2^w-1)}$.

The equivalent channel used in the simulations of this chapter is AWGN and thus we have $a_k(\kappa) = 1$ in (10.1) and (10.3). We assume bit-individual availability of the error probability according to (10.3) with E_s/N_0 being constant and perfectly known. All SDSD simulations are performed with an MMSE estimator according to (10.7).

10.4.2 Estimators

Parameter estimation without any algorithmic delay is possible if the *a posteriori* probabilities $P(\mathbf{x}_k^{(i)}|\ldots)$ are computed only for the current frame k and no future received bit patterns $\hat{\mathbf{x}}_{k+K}$ with $K > 0$ are required. Therefore, the transition probabilities $P(\hat{\mathbf{x}}_k|\mathbf{x}_k^{(i)})$ must be known only for frames $1, \ldots, k$. We are interested in the *a posteriori* probabilities

$$P(\mathbf{x}_k^{(i)}|\ldots) = P(\mathbf{x}_k^{(i)}|\hat{\mathbf{x}}_k, \ldots, \hat{\mathbf{x}}_1) = P(\mathbf{x}_k^{(i)}|\hat{\mathbf{x}}_1^k) \tag{10.22}$$

given the full history of received bit patterns $\hat{\mathbf{x}}_1^k = \{\hat{\mathbf{x}}_1, \ldots, \hat{\mathbf{x}}_k\}$.

Applying the chain rule [Papoulis 1965] we can write (10.22) as

$$P(\mathbf{x}_k^{(i)}|\hat{\mathbf{x}}_1^k) = P(\mathbf{x}_k^{(i)}|\hat{\mathbf{x}}_k, \hat{\mathbf{x}}_1^{k-1}) = \frac{P(\mathbf{x}_k^{(i)}, \hat{\mathbf{x}}_k \mid \hat{\mathbf{x}}_1^{k-1})}{P(\hat{\mathbf{x}}_k \mid \hat{\mathbf{x}}_1^{k-1})} = \frac{1}{C} \cdot P(\mathbf{x}_k^{(i)}, \hat{\mathbf{x}}_k \mid \hat{\mathbf{x}}_1^{k-1}) \tag{10.23}$$

with the probability $C = P(\hat{\mathbf{x}}_k \mid \hat{\mathbf{x}}_1^{k-1})$ being invariant w.r.t. the indices $i = 0, 1, \ldots,$ 2^w-1. In practice, the computation of this constant is done by ensuring that

$$\sum_{i=0}^{2^w-1} P(\mathbf{x}_k^{(i)} \mid \hat{\mathbf{x}}_k, \hat{\mathbf{x}}_1^{k-1}) = 1 . \tag{10.24}$$

As a direct consequence we get[2]

$$C = \sum_{i=0}^{2^w-1} P(\mathbf{x}_k^{(i)}, \hat{\mathbf{x}}_k \mid \hat{\mathbf{x}}_1^{k-1}) . \tag{10.25}$$

Using the memoryless property of the equivalent channel as pointed out in Sec. 10.2.3 the *a posteriori* probabilities we look for can be expressed as

$$P(\mathbf{x}_k^{(i)} \mid \hat{\mathbf{x}}_k, \hat{\mathbf{x}}_1^{k-1}) = \frac{1}{C} \cdot P(\hat{\mathbf{x}}_k|\mathbf{x}_k^{(i)}) \cdot P(\mathbf{x}_k^{(i)} \mid \hat{\mathbf{x}}_1^{k-1}) . \tag{10.26}$$

---

[2]In the following, the constant C is always used in a way that the sum of all probabilities on the left hand side of the equation is normalized to one. The absolute values of C may differ among the following equations.

The transition probabilities $P(\hat{\mathbf{x}}_k \mid \mathbf{x}_k^{(i)})$ are known from the first computational step of SDSD, Eq. (10.5). Only the so-called *prediction probabilities* $P(\mathbf{x}_k^{(i)} \mid \hat{\mathbf{x}}_1^{k-1})$ are still to be computed. They represent the receiver-sided information on the transmitted bit pattern $\mathbf{x}_k$, before the bit pattern $\hat{\mathbf{x}}_k$ has become known.

The computation of the prediction probabilities depends on the modeling of the parameter and on the known history of transition probabilities $P(\hat{\mathbf{x}}_{k-1} \mid \mathbf{x}_{k-1}^{(j)})$, $P(\hat{\mathbf{x}}_{k-2} \mid \mathbf{x}_{k-2}^{(l)})$, ... The remainder of Sec. 10.4.2 is now concerned with solutions of the prediction probability term in (10.26).

Looking for the solution of (10.26), the modeling of the parameter as a Markov process of Nth order requires the interpretation of the prediction probability as marginal distribution

$$P(\mathbf{x}_k^{(i)} \mid \hat{\mathbf{x}}_1^{k-1}) = \sum_{j=0}^{2^w-1} \cdots \sum_{h=0}^{2^w-1} P(\mathbf{x}_k^{(i)}, \mathbf{x}_{k-1}^{(j)}, \ldots, \mathbf{x}_{k-N+1}^{(h)} \mid \hat{\mathbf{x}}_1^{k-1}). \tag{10.27}$$

As shown in [Fingscheidt 1998], the term $P(\mathbf{x}_k^{(i)}, \mathbf{x}_{k-1}^{(j)}, \ldots, \mathbf{x}_{k-N+1}^{(h)} \mid \hat{\mathbf{x}}_1^{k-1})$ can be computed recursively according to

$$
\begin{aligned}
P(&\mathbf{x}_k^{(i)}, \mathbf{x}_{k-1}^{(j)}, \ldots, \mathbf{x}_{k-N+1}^{(h)} \mid \hat{\mathbf{x}}_1^{k-1}) \\
&= \frac{1}{C} \cdot P(\hat{\mathbf{x}}_{k-1} \mid \mathbf{x}_{k-1}^{(j)}) \cdot \ldots \cdot P(\hat{\mathbf{x}}_{k-N+1} \mid \mathbf{x}_{k-N+1}^{(h)}) \\
&\quad \cdot \sum_{l=0}^{2^w-1} P(\mathbf{x}_k^{(i)} \mid \mathbf{x}_{k-1}^{(j)}, \ldots, \mathbf{x}_{k-N}^{(l)}) \quad \cdot \sum_{\cdots} \cdots \\
&\quad \cdot \sum_{m=0}^{2^w-1} P(\mathbf{x}_{k-N+1}^{(h)} \mid \mathbf{x}_{k-N}^{(l)}, \ldots, \mathbf{x}_{k-2N+1}^{(m)}) \\
&\quad \cdot P(\hat{\mathbf{x}}_{k-N} \mid \mathbf{x}_{k-N}^{(l)}) \cdot P(\mathbf{x}_{k-N}^{(l)}, \ldots, \mathbf{x}_{k-2N+1}^{(m)} \mid \hat{\mathbf{x}}_1^{k-N-1}).
\end{aligned}
\tag{10.28}
$$

The prediction probabilities in (10.27) are obviously a one-dimensional marginal distribution of an N-dimensional joint probability term, which can be computed recursively according to (10.28) from the joint probability term computed N frames earlier. The recursion comprises the *a priori* knowledge terms $P(\mathbf{x}_k \mid \mathbf{x}_{k-1}, \ldots, \mathbf{x}_{k-N})$ as well as the transition probabilities $P(\hat{\mathbf{x}}_{k-1} \mid \mathbf{x}_{k-1}^{(j)})$ through $P(\hat{\mathbf{x}}_{k-N} \mid \mathbf{x}_{k-N}^{(l)})$ representing the last N received bit patterns and their reliability. To probe further on higher order Markov models in SDSD, the interested reader is also referred to [Lahouti, Khandani 2004].

With a Markov model of order $N = 0$ the computation of the marginal distribution in (10.27) becomes superfluous. The missing temporal correlations lead to a simplification of the *a priori* knowledge $P(\mathbf{x}_k \mid \mathbf{x}_{k-1}, \ldots, \mathbf{x}_{k-N}) = P(\mathbf{x}_k)$. The recursion and consequently the prediction probability are then simply

$$P(\mathbf{x}_k^{(i)} \mid \hat{\mathbf{x}}_1^{k-1}) = P(\mathbf{x}_k^{(i)}). \qquad \text{(SD/AK0)} \tag{10.29}$$

If the *a posteriori* probabilities are computed according to (10.26) and (10.29), we refer to this approach in the following as "SD/AK0" for *SDSD with a priori knowledge of 0th order*.

In case no *a priori* knowledge at all is available about the parameter, the prediction probabilities result in a uniform distribution according to

$$P(\mathbf{x}_k^{(i)} \mid \hat{\mathbf{x}}_1^{k-1}) = 2^{-w} \qquad \text{(SD/NAK)} \qquad (10.30)$$

and will be named "SD/NAK" for *SDSD, no a priori knowledge*.

Since in the computation of the marginal distribution $N - 1$ nested summations are performed, a Markov model of order $N = 1$ does not yet require the computation of (10.27). The recursion (10.28) however leads to the (forward) prediction probabilities

$$P(\mathbf{x}_k^{(i)} \mid \hat{\mathbf{x}}_1^{k-1}) = \frac{1}{C} \cdot \sum_{j=0}^{2^w-1} P(\mathbf{x}_k^{(i)} \mid \mathbf{x}_{k-1}^{(j)}) \cdot P(\hat{\mathbf{x}}_{k-1} \mid \mathbf{x}_{k-1}^{(j)}) \cdot P(\mathbf{x}_{k-1}^{(j)} \mid \hat{\mathbf{x}}_1^{k-2}). \qquad (10.31)$$

In the special case of a 1st order Markov model (10.26) can also be written directly as recursion

$$P(\mathbf{x}_k^{(i)} \mid \hat{\mathbf{x}}_k, \hat{\mathbf{x}}_1^{k-1}) = \frac{1}{C} \cdot P(\hat{\mathbf{x}}_k \mid \mathbf{x}_k^{(i)}) \cdot \sum_{j=0}^{2^w-1} P(\mathbf{x}_k^{(i)} \mid \mathbf{x}_{k-1}^{(j)}) \cdot P(\mathbf{x}_{k-1}^{(j)} \mid \hat{\mathbf{x}}_{k-1}, \hat{\mathbf{x}}_1^{k-2}) \qquad (10.32)$$

and will consequently be denoted by "SD/AK1". An expression such as (10.32) also occurs in the context of channel decoding with symbol decision [Bahl et al. 1974], called the *forward recursion*. Beyond that, it has also been reported for detection of Markov sources over discrete, memoryless channels [Phamdo, Farvardin 1994].

In the case of a Markov model of order $N = 2$, however, both (10.27) and (10.28) must be evaluated separately. We then get

$$P(\mathbf{x}_k^{(i)} \mid \hat{\mathbf{x}}_1^{k-1}) = \sum_{j=0}^{2^w-1} P(\mathbf{x}_k^{(i)}, \mathbf{x}_{k-1}^{(j)} \mid \hat{\mathbf{x}}_1^{k-1}) \qquad \text{(SD/AK2)} \qquad (10.33)$$

with

$$P(\mathbf{x}_k^{(i)}, \mathbf{x}_{k-1}^{(j)} \mid \hat{\mathbf{x}}_1^{k-1}) = \frac{1}{C} \cdot P(\hat{\mathbf{x}}_{k-1} \mid \mathbf{x}_{k-1}^{(j)}) \cdot \sum_{l=0}^{2^w-1} P(\mathbf{x}_k^{(i)} \mid \mathbf{x}_{k-1}^{(j)}, \mathbf{x}_{k-2}^{(l)})$$

$$\cdot \sum_{m=0}^{2^w-1} P(\mathbf{x}_{k-1}^{(j)} \mid \mathbf{x}_{k-2}^{(l)}, \mathbf{x}_{k-3}^{(m)}) \cdot P(\hat{\mathbf{x}}_{k-2} \mid \mathbf{x}_{k-2}^{(l)}) \cdot P(\mathbf{x}_{k-2}^{(l)}, \mathbf{x}_{k-3}^{(m)} \mid \hat{\mathbf{x}}_1^{k-3}). \qquad (10.34)$$

Storage of the Nth order *a priori* knowledge requires $2^{w(N+1)}$ words of memory. The computational complexity to compute the *a posteriori* probabilities via (10.28), (10.27), and (10.26), is of order $O(2^{2wN})$ (multiply and/or accumulate) operations without normalization. This holds also for $N = 1$ and the use of recursion (10.32).

10.4.3 Simulation Results

Influence of Estimation Order and Bit Rate

At first we want to investigate the influence of the bit rate w and of the Markov model of order N as used in the MMSE estimator. We use a highly correlated AR(1) Gaussian parameter (i.e., $\bar{N} = 1$) with $\rho = 0.9$.

As we can see in Fig. 10.3, HD decoding reveals the worst SNR values for $w = 1$ as well as for $w = 4$. SDSD on the other hand shows slightly improved performance if no *a priori* knowledge is used (SD/NAK). The reason for this improvement can be found in the exploitation of the bit error probabilities from the equivalent channel.

A further performance gain occurs for $w \geq 2$ bit with the (SD/AK0) approach. In the case of 1 bit quantization both quantizer table entries $i = 0, 1$ are equiprobable, therefore (10.30) and (10.29) lead to the same formulation. This is also confirmed by simulations, since the (SD/NAK) and (SD/AK0) curves in Fig. 10.3-a are identical.

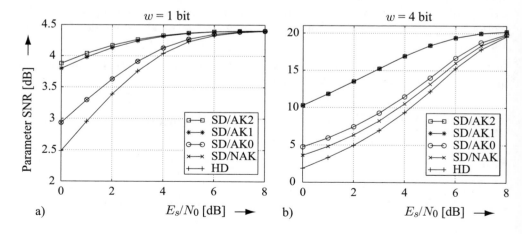

Figure 10.3: Transmission of a w bit quantized Gaussian AR(1) parameter with $\rho = 0.9$

As expected, the largest performance gain is achievable when the temporal correlation of parameters is exploited in the SDSD process. Independently of w we observe a graceful degradation of SNR performance towards low channel qualities. The choice of $N = 2 > \bar{N} = 1$ (i.e., AK2) allows the exploitation of further correlation-dependent residual redundancy, as we can see from the slightly better estimation results of the AK2 curve compared with the AK1 curve in Fig. 10.3-a. For the bit rate $w = 4$ bit in Fig. 10.3-b we observe that an estimation order $N > \bar{N}$ does not lead to measurable additional gains. For high bit rates this is expected from theory, since for $N = \bar{N}$ one can show that the utilizable correlation-dependent residual redundancy ΔR_c approximately equals the total *correlation-dependent residual redundancy* $\Delta R'_c$. The total correlation-dependent residual redundancy for a Gaussian AR(1) parameter is given from rate distortion theory [Berger 1971], [Jayant, Noll 1984] for a not too large correlation ρ (or alternatively: for high bit rate w) by

$$\Delta R'_c = -\frac{1}{2} \log_2(1 - \rho^2) . \tag{10.35}$$

For a correlation $\rho = 0.9$, the bit rate $w = 4$ bit can be considered as high, so (10.35) is applicable and results in $\Delta R'_c = 1.2$ bit of correlation-dependent residual redundancy if scalar quantization is employed.

Estimation of Parameters of Higher Markov Order

Let us now consider a parameter of order $\bar{N} = 2$, i.e., an AR(2) parameter. Figure 10.4 shows simulation results for an AR(2) parameter taken from a bandpass process with $\rho = \rho(1) = 0$ and $\rho(2) = 0.81$. Consecutive parameter values $\tilde{v}_k$, $\tilde{v}_{k+1}$ are uncorrelated, while $\tilde{v}_k$, $\tilde{v}_{k+2}$ show clear statistical dependencies.

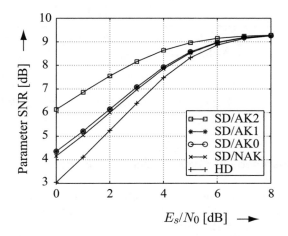

Figure 10.4: Transmission of a $w = 2$ bit quantized Gaussian AR(2) parameter with $\rho(1) = 0.0$, $\rho(2) = 0.81$

As expected, the (SD/AK0) and the (SD/AK1) scheme achieve the same estimation performance, since consecutive parameters are uncorrelated. A further gain by SDSD is only achievable using the (SD/AK2) scheme as given in (10.33).

Influence of Correlation and Bit Rate

The influence of different correlations of a $w = 3$ bit scalar quantized Gaussian AR(1) parameter on the performance of the (SD/AK1) scheme is shown in Fig. 10.5-a. It turns out that, depending on the amount of correlation, good or even very good estimation results are achieved.

Figure 10.5-b shows the performance of the (SD/AK1) scheme for a highly correlated parameter ($\rho = 0.9$) for the bit rates $w = 1, 2, 3, 4, 5$ bit. With increasing bit rate the curves degrade faster for HD as well as for SD decoding. In very bad channels it turns out that hard decoding of high rate quantizers may lead to slightly worse results than hard decoding of lower rate quantizers. This effect is not observed with SDSD: The estimation quality for higher bit rates is always better than the SD performance for lower rates.

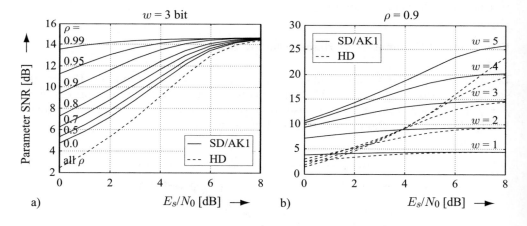

a)   b)

Figure 10.5: Transmission of a w bit quantized Gaussian AR(1) parameter

Influence of the Parameter Probability Density Function

Since not all real-world parameters are well described by a Gaussian *Probability Density Function* (PDF), we also investigate the performance of SDSD for parameters with Laplacian and with uniform PDF. Figure 10.6-a shows the results for uncorrelated parameters, Fig. 10.6-b for correlated parameters. For the generation of AR(1) parameters with Laplacian PDF the reader is referred to [Fingscheidt 1998].

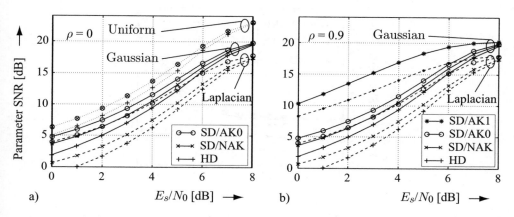

Figure 10.6: Transmission of a $w = 4$ bit quantized AR(1) parameter
  a) $\rho = 0$ (uncorrelated random process)
  b) $\rho = 0.9$

As we know already from the 1 bit quantization of the Gaussian parameter, we can expect that if the quantization indices are equiprobable, the (SD/NAK) and the (SD/AK0) approach lead to identical results. This is true for a parameter with uniform distribution independent of the bit rate w, as can be seen in the simulation in Fig. 10.6-a and the $w = 4$ bit case. Furthermore, we find in Fig. 10.6-a that the SDSD estimation gain of (SD/AK0) is significantly greater for a Laplacian distributed parameter as for a Gaussian distributed parameter. The advantage of the (SD/AK0) approach is therefore greater, the more the respective parameter PDF deviates from the uniform PDF. Lloyd–Max quantization of Laplacian parameters leads to a high degree of utilizable, distribution-dependent residual redundancy $\Delta R_d = w - H(\mathbf{x})$, which directly results in a better performance of SDSD.

Figure 10.6-b also shows the SDSD performances for a Laplacian and a Gaussian distributed parameter for the (SD/AK1) case. With the chosen correlation coefficient $\rho = 0.9$ the influence of the correlation-dependent residual redundancy is dominant. Hence, the curves of both PDFs assume similar shapes, however, fixed to their individual basic quality at error-free channels.

Concerning the SNR of SDSD *with the MMSE estimator* in [Fingscheidt 1998] it has been shown that

$$\text{SNR}_{\text{SD/AK0}} \geq 0 \text{ dB} \tag{10.36}$$

as well as

$$\text{SNR}_{\text{SD/AK0}} \geq \text{SNR}_{\text{HD}} . \tag{10.37}$$

Figure 10.7 presents the simulations of Fig. 10.6-a extended towards very bad channel conditions. An effect that can be seen for all SDSD approaches is that in the case of

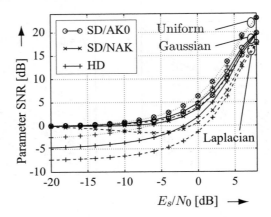

Figure 10.7: Transmission of a $w = 4$ bit quantized, uncorrelated random parameter

very bad transmission the SNR tends towards zero. The reason for this behavior is that SDSD decodes the mean value of the parameter v if no reasonable information is available from the channel. In our simulations however the mean value of all simulated parameters equals zero. Therefore, we have $\hat{v} = 0$ and as a consequence from (10.8) we achieve an SNR of 0 dB. Note that this effect may be highly desirable, e.g., if gain factors of zero mean parameters are transmitted, as in the case of very bad channel conditions SDSD performs *inherent muting*.

The curves in Fig. 10.7 also confirm (10.36), which means that the SNR of the (SD/AK0) scheme with MMSE estimation will never become negative. On the other hand, the (SD/NAK) curve of the Laplacian parameter shows that without the exploitation of *a priori* knowledge a positive SNR cannot be guaranteed.

Finally, the validity of (10.37) is shown by simulation even for the simplest scheme of SDSD, which is (SD/NAK). Here we find that SDSD using the MMSE estimator – independent of the amount of *a priori* knowledge used – will never lead to a weaker SNR than hard decoding does.

10.5 Joint Extrapolative Estimation of Two Different Parameters

10.5.1 Estimators

The concept of SDSD can easily be extended to a number of *different* parameters that reveal correlations among each other. An example in the context of $f_S = 8$ kHz speech coding are the *Line Spectral Frequency* (LSF) parameters where 10 different but highly correlated parameters are transmitted per frame [Lahouti, Khandani 2001], [Fingscheidt et al. 2002]. A more general treatment of this so-called *intra-frame* correlation will be given in Chap. 11.

Here, we want to focus on simple estimators, i.e., we consider only two different parameters, one represented by the bit pattern $\mathbf{x}_k$ with w_x bits, the second one represented by $\mathbf{y}_k$ with w_y bits. Similar to (10.26) the *a posteriori* probabilities for each of these parameters can be computed as the respective marginal distribution of the joint *a posteriori* probability

$$
\begin{aligned}
P(\mathbf{x}_k^{(i)}, \mathbf{y}_k^{(j)} \mid \hat{\mathbf{x}}_1^k, \hat{\mathbf{y}}_1^k) &= \frac{1}{C} \cdot P(\hat{\mathbf{x}}_k, \hat{\mathbf{y}}_k \mid \mathbf{x}_k^{(i)}, \mathbf{y}_k^{(j)}) \cdot P(\mathbf{x}_k^{(i)}, \mathbf{y}_k^{(j)} \mid \hat{\mathbf{x}}_1^{k-1}, \hat{\mathbf{y}}_1^{k-1}) \\
&= \frac{1}{C} \cdot P(\hat{\mathbf{x}}_k \mid \mathbf{x}_k^{(i)}) \cdot P(\hat{\mathbf{y}}_k \mid \mathbf{y}_k^{(j)}) \cdot P(\mathbf{x}_k^{(i)}, \mathbf{y}_k^{(j)} \mid \hat{\mathbf{x}}_1^{k-1}, \hat{\mathbf{y}}_1^{k-1}) \, .
\end{aligned}
$$
(10.38)

The joint prediction probability terms occurring in the formula above can be given in general form in analogy to (10.27) and (10.28). More details can be found in [Fingscheidt 1998].

Modeling both parameters as 0th order Markov processes, the joint prediction probabilities in (10.38) become

$$
P(\mathbf{x}_k^{(i)}, \mathbf{y}_k^{(j)} \mid \hat{\mathbf{x}}_1^{k-1}, \hat{\mathbf{y}}_1^{k-1}) = P(\mathbf{x}_k^{(i)}, \mathbf{y}_k^{(j)}) \quad \text{(SD/JAK0)}
$$
(10.39)

with the joint *a priori* knowledge $P(\mathbf{x}_k^{(i)}, \mathbf{y}_k^{(j)})$. In analogy to (10.32), a 1st order Markov model of both parameters leads to the recursion

$$
\begin{aligned}
P(\mathbf{x}_k^{(i)}, \mathbf{y}_k^{(j)} \mid \hat{\mathbf{x}}_1^k, \hat{\mathbf{y}}_1^k) &= \frac{1}{C} \cdot P(\hat{\mathbf{x}}_k \mid \mathbf{x}_k^{(i)}) \cdot P(\hat{\mathbf{y}}_k \mid \mathbf{y}_k^{(j)}) \quad \text{(SD/JAK1)} \\
&\cdot \sum_{l=0}^{2^{w_x}-1} \sum_{m=0}^{2^{w_y}-1} P(\mathbf{x}_k^{(i)}, \mathbf{y}_k^{(j)} \mid \mathbf{x}_{k-1}^{(l)}, \mathbf{y}_{k-1}^{(m)}) \cdot P(\mathbf{x}_{k-1}^{(l)}, \mathbf{y}_{k-1}^{(m)} \mid \hat{\mathbf{x}}_1^{k-1}, \hat{\mathbf{y}}_1^{k-1})
\end{aligned}
$$
(10.40)

with the joint *a priori* knowledge $P(\mathbf{x}_k^{(i)}, \mathbf{y}_k^{(j)} \mid \mathbf{x}_{k-1}^{(l)}, \mathbf{y}_{k-1}^{(m)})$. Assuming $w_x = w_y = w$, we observe that the memory requirements for the *a priori* knowledge in comparison with a separate (SD/AKN) estimation has been increased from $2 \cdot 2^{w(N+1)}$ to $2^{2w(N+1)}$ words. Therefore a practical application of this scheme will only be possible for low bit rate parameters and low Markov model order N. The computational complexity for the *a posteriori* probabilities without normalization for (SD/JAK0) amounts to 2^{2w+1} operations and for (SD/JAK1) to 2^{4w+1} operations.

10.5.2 Simulation Results

In the following, the joint extrapolation of two different but mutually correlated parameters $\tilde{u}$, $\tilde{v}$ shall be investigated. Parameter $\tilde{u}$ has been taken from a zero-mean unit-variance white Gaussian noise process, while parameter $\tilde{v}$ has been generated in a way that both parameters reveal a (normalized) cross-correlation factor [Papoulis 1965] of $\delta \in \{0.5, 0.9\}$.

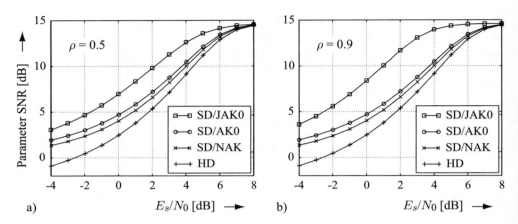

Figure 10.8: Transmission of two Gaussian parameters with cross-correlation coefficient δ. Quantization of both parameters with $w = 3$ bit

The simulations we conducted will give an impression of the performance of the (SD/JAK0) scheme exploiting joint *a priori* knowledge, especially compared with separate SDSD of both parameters by (SD/AK0). Note that $\tilde{u}$ is an uncorrelated parameter, while $\tilde{v}$ exhibits correlation[3] $\rho = \delta$, which, however, is not exploited by any of the simulated schemes. Figure 10.8-a presents the results for a weak cross-correlation $\delta = 0.5$, while Fig. 10.8-b documents the curves for a high cross-correlation $\delta = 0.9$ of both parameters.

It has to be noted that all curves in Fig. 10.8 are valid for each of the two parameters. As known from previous sections, the parameter individual SDSD with 0th order *a priori* knowledge leads to performance gains over HD or SD without *a priori* knowledge. The best results in both subfigures are achieved by joint extrapolation (SD/JAK0) using (10.39). The performance gains are impressive, especially in the case of a high cross-correlation between both parameters. Note that possible *temporal* correlations of the parameters have not even been exploited. If they exist, in addition further significant performance gains can be achieved if they are exploited by SDSD. It is instructive to compare the (SD/JAK0) curve in Fig. 10.8-b with the (SD/AK1) performance for $w = 3$ bit in Fig. 10.5-b. In the first case a cross-correlation δ is exploited, while in the latter case a temporal correlation $\rho = \delta$ is exploited. In this direct comparison the (SD/AK1) scheme performs slightly better (for parameter $\tilde{v}$), which is due to the fact that the actual *a posteriori* probabilities $P(\mathbf{x}_k^{(i)}, \mathbf{y}_k^{(j)} \mid \hat{\mathbf{x}}_k, \hat{\mathbf{y}}_k)$ in (10.38) for (SD/JAK0) are unable to exploit cross-correlations δ in a recursive manner as (10.32) does for the temporal correlation ρ.

---

[3]In this chapter and the subsequent chapters in this part of the book, δ denotes a cross-correlation instead of a unit impulse response.

10.6 Extrapolative Estimation with Repeated Parameter Transmission

10.6.1 Estimators

Now we apply the case of two different correlated parameters as discussed in Sec. 10.5 to a situation where the bit pattern of a single parameter $\mathbf{x}_k$ is always received twice. In this case the formerly two separate parameters are not only mutually correlated, but they are identical. In practical systems we can find this situation in the case where a repeated transmission of the same bit pattern $\mathbf{x}_k$ is performed. Alternatively, the bit pattern $\mathbf{x}_k$ may be transmitted only once but received twice, e.g., with a receiver that allows receive diversity over two different transmission channels.

While in the first case we assume a memoryless channel as usual, in the latter case we additionally have to postulate the statistical independence of both transmission paths. In both cases we then have $\mathbf{x}_k = \mathbf{y}_k$ and $w = w_x = w_y$. On the receiver side, however, we have to deal in general with two different received bit patterns $\hat{\mathbf{x}}_k$ and $\hat{\mathbf{y}}_k$, along with their individual bit error probabilities.

For the derivation of *a posteriori* probabilities for an estimator, both cases can be treated together. Without loss of generality w.r.t. the two application cases we call the scheme "SD/RPT" standing for SDSD with *Repeated Parameter Transmission*. The *a posteriori* probabilities for an extrapolative estimation of a single parameter that has been received twice can be given in analogy to (10.26) as

$$P(\mathbf{x}_k^{(i)} \mid \hat{\mathbf{x}}_1^k, \hat{\mathbf{y}}_1^k) = \frac{1}{C} \cdot P(\hat{\mathbf{x}}_k, \hat{\mathbf{y}}_k \mid \mathbf{x}_k^{(i)}) \cdot P(\mathbf{x}_k^{(i)} \mid \hat{\mathbf{x}}_1^{k-1}, \hat{\mathbf{y}}_1^{k-1})$$

$$= \frac{1}{C} \cdot P(\hat{\mathbf{x}}_k \mid \mathbf{x}_k^{(i)}) \cdot P(\hat{\mathbf{y}}_k \mid \mathbf{y}_k^{(i)}) \cdot P(\mathbf{x}_k^{(i)} \mid \hat{\mathbf{x}}_1^{k-1}, \hat{\mathbf{y}}_1^{k-1}), \quad (10.41)$$

with $\hat{\mathbf{x}}_1^k$ being the sequence of bit patterns that was received first, and $\hat{\mathbf{y}}_1^k$ being the sequence of bit patterns that was received second.

For a 0th order Markov model we get prediction probabilities in analogy to (10.39)

$$P(\mathbf{x}_k^{(i)} \mid \hat{\mathbf{x}}_1^{k-1}, \hat{\mathbf{y}}_1^{k-1}) = P(\mathbf{x}_k^{(i)}), \quad \text{(SD/AK0/RPT)} \quad (10.42)$$

while the *a posteriori* probabilities for a 1st order Markov model are given as a recursion in analogy to (10.40) as

$$P(\mathbf{x}_k^{(i)} \mid \hat{\mathbf{x}}_1^k, \hat{\mathbf{y}}_1^k) = \frac{1}{C} \cdot P(\hat{\mathbf{x}}_k \mid \mathbf{x}_k^{(i)}) \cdot P(\hat{\mathbf{y}}_k \mid \mathbf{y}_k^{(i)}) \quad \text{(SD/AK1/RPT)}$$

$$\cdot \sum_{j=0}^{2^w-1} P(\mathbf{x}_k^{(i)} \mid \mathbf{x}_{k-1}^{(j)}) \cdot P(\mathbf{x}_{k-1}^{(j)} \mid \hat{\mathbf{x}}_1^{k-1}, \hat{\mathbf{y}}_1^{k-1}). \quad (10.43)$$

Since both transmitted bit patterns are equal, instead of the joint *a priori* knowledge (JAKN) we get the simple *a priori* knowledge terms (AKN) of the respective parameter again, as was the case in the discussion of a single transmitted and received parameter in Sec. 10.4.2.

Independently of the Markov model order the bit error probabilities of both received bit patterns $\hat{\mathbf{x}}_0$ and $\hat{\mathbf{y}}_0$ are separately used to compute the two transition probability terms $P(\hat{\mathbf{x}}_k|\mathbf{x}_k^{(i)})$ and $P(\hat{\mathbf{y}}_k|\mathbf{y}_k^{(i)})$. As can be seen in (10.41) both transition probabilities can be multiplied yielding a single transition probability term $P(\hat{\mathbf{x}}_k, \hat{\mathbf{y}}_k \mid \mathbf{x}_k^{(i)})$.

As mentioned before, the assumption of a memoryless equivalent channel in (10.41) allows the repeated parameter transmission to be interpreted alternatively as transmission over two statistically independent diversity channels of the same quality. In transmission systems with diversity reception an important issue is the receiver-located approaches for signal combination [Jakes 1974], [Parsons, Gardiner 1989], [Papen 1996]. Assuming the availability of log-likelihood ratios (L-values) $L(x_k(\kappa))$ and $L(y_k(\kappa))$ to each received bit $\hat{x}_k(\kappa)$ and $\hat{y}_k(\kappa)$, the received bits of both diversity paths can be combined to a new (bipolar) bit $\hat{z}_k(\kappa) = \text{sign}[L(x_k(\kappa))]$ or $\hat{z}_k(\kappa) = \text{sign}[L(y_k(\kappa))]$, depending on which L-value determined the modulus of the combined L-value $L(z_k(\kappa))$, where [Papen 1996]

$$|L(z_k(\kappa))| = \max\left(|L(x_k(\kappa))|, |L(y_k(\kappa))|\right) . \tag{10.44}$$

This signal combination technique is commonly called *Selection Combining* (SC). It always selects the received bit from the channel with the currently lower bit error probability (i.e., the higher L-value magnitude).

In general a lower error probability $p_k(\kappa) = (1 + \exp|L(z_k(\kappa))|)^{-1}$ of the combined bit $\hat{z}_k(\kappa)$ can be achieved by applying the *Maximal Ratio Combining* (MRC) approach, which simply performs a summation of the L-values according to [Kahn 1954], [Papen 1996]

$$L(z_k(\kappa)) = L(x_k(\kappa)) + L(y_k(\kappa)) . \tag{10.45}$$

In the case of two diversity channels and bit-individual combination schemes following (10.44) or (10.45), SC and MRC always yield the same received bit $\hat{z}_k(\kappa)$. However, the MRC bit error probability $p_k(\kappa)$ resulting from $L(z_k(\kappa))$ using (10.3) is in general closer to reality.

Computing the transition probabilities $P(\hat{\mathbf{z}}_k|\mathbf{x}_k^{(i)})$ based on the MRC log-likelihood values $L(z_k(\kappa))$, $\kappa = 1, \ldots, w$, using (10.3), (10.6), and (10.5), one can prove that these are – apart from a constant factor – equal to the product of the parameter transition probabilities in (10.41) [Fingscheidt 1998]:

$$P(\hat{\mathbf{z}}_k|\mathbf{x}_k^{(i)})\Big|_{\text{MRC}} = \frac{1}{C} \cdot P(\hat{\mathbf{x}}_k, \hat{\mathbf{y}}_k \mid \mathbf{x}_k^{(i)}) . \tag{10.46}$$

As a consequence we can state that SDSD with repeated parameter transmission (or, equivalently, with diversity reception) (SD/RPT) performs a combination of the received signals according to the maximal ratio combining principle, before *a priori* knowledge is used for even better decoding of the parameter.

Looking back to Sec. 10.5 the multiplication of the transition probabilities of different parameters with in general different bit rates resulting in a new joint transition probability

$$P(\hat{\mathbf{x}}_k, \hat{\mathbf{y}}_k | \mathbf{x}_k^{(i)}, \mathbf{y}_k^{(j)}) = P(\hat{\mathbf{x}}_k | \mathbf{x}_k^{(i)}) \cdot P(\hat{\mathbf{y}}_k | \mathbf{y}_k^{(j)}) \qquad (10.47)$$

can be interpreted as a combination of signals in the sense of a *generalized* MRC.

10.6.2 Simulation Results

RPT of two parameters each with bit rate R or, equivalently, diversity reception over two channels, is a special case for SDSD. The utilizable residual redundancy amounts to $R + \Delta R_d + \Delta R_c$. Therefore, we expect high gains even in the case of a temporally uncorrelated parameter.

Figure 10.9 shows the performance of the (SD/AK0/RPT) approach with repeated transmission of a $w = 2$ bit quantized Gaussian parameter. The gains of the (SD/AK0/RPT) scheme amount to about 3 dB with respect to E_s/N_0, which is not a surprise, since the total transmitter power has been doubled. As outlined in Sec. 10.6.1, the performance of the (SD/AK0/RPT) curve can be reached as well in a transmission system with 2-channel diversity reception, if both signal paths are statistically independent, if signal combination is performed by MRC following (10.45), and if SDSD is performed using the (SD/AK0) scheme. Note that in this case transmitter power has not been doubled!

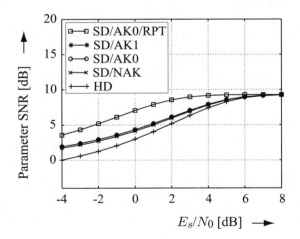

Figure 10.9: Transmission of a $w = 2$ bit quantized Gaussian AR(0) parameter

With these results a comparison with channel coding schemes adding explicit redundancy becomes possible. Instead of a conventional *Forward Error Correction* (FEC) with, e.g., rate $r = 1/2$ convolutional channel coding, one can compare the performance at equal transmitter power and equal SNR for error-free channels to the scheme with repeated parameter transmission and SDSD by joint extrapolation. In comparison with channel coding approaches, the loss of quality in terms of SNR will of course start already at higher E_s/N_0 ratios for (SD/RPT), but it will exhibit a much more graceful degradation than the FEC scheme towards further decreasing channel quality [Fingscheidt 1998]. Instead of repeating the parameter transmission one could think of adding parity bits $\mathbf{y}_k$ as explicit redundancy to the transmitted parameter bit pattern $\mathbf{x}_k$, and to exploit the parity bits in the context of a (SD/JAKN) scheme. Depending on the parity code the *a priori* knowledge then reveals a number of zero entries. The question whether to employ SDSD with any of the discussed alternatives or to use FEC depends on many issues. To probe further, the interested reader is referred to [Fingscheidt et al. 1999], and to Chap. 12, Sec. 12.3, where this approach is further developed to so-called *Source Optimized Channel Codes* (SOCC).

10.7 Interpolative Estimation of a Parameter

10.7.1 Estimators

In this remaining section we will assume that when estimating the parameter $\hat{v}_k$ a number of K future bit patterns $\hat{\mathbf{x}}_{k+1}^{k+K} = \hat{\mathbf{x}}_{k+1}, \ldots, \hat{\mathbf{x}}_{k+K}$ of the respective parameter have already been received and can be used for the estimation. For $K \geq 1$ the decoding scheme becomes interpolative resulting in additional decoding delay. Looking a bit deeper into common hybrid speech coders, one finds that they often operate in frame and subframe structures. With the exception of the last subframe in each frame, in principle such a frame structure allows interpolative decoding of subframe parameters without introducing additional algorithmic delay.[4]

The *a posteriori* probabilities we are interested in can be given in analogy to (10.26) as

$$P(\mathbf{x}_k^{(i)} \mid \hat{\mathbf{x}}_{k+1}^{k+K}, \hat{\mathbf{x}}_k, \hat{\mathbf{x}}_1^{k-1}) = \frac{1}{C} \cdot P(\hat{\mathbf{x}}_k, \mathbf{x}_k^{(i)} \mid \hat{\mathbf{x}}_{k+1}^{k+K}, \hat{\mathbf{x}}_1^{k-1})$$

$$= \frac{1}{C} \cdot P(\hat{\mathbf{x}}_k \mid \mathbf{x}_k^{(i)}) \cdot P(\mathbf{x}_k^{(i)} \mid \hat{\mathbf{x}}_{k+1}^{k+K}, \hat{\mathbf{x}}_1^{k-1}) . \qquad (10.48)$$

The prediction probabilities $P(\mathbf{x}_k^{(i)} \mid \hat{\mathbf{x}}_{k+1}^{k+K}, \hat{\mathbf{x}}_1^{k-1})$ can be divided into *backward* joint prediction probabilities, *forward* joint prediction probabilities, and a residual probability term. A general solution is given in [Bahl et al. 1974], [Fingscheidt 1998].

---

[4]An extrapolative decoding scheme is to be applied only in the last subframe of each frame in order to prevent additional algorithmic decoding delay.

Here, we just want to focus on the case $N = 1$, which is the interpolation under the assumption of a 1st order Markov parameter. At first, we assume that only a single future bit pattern is known to SDSD ($K = 1$). The prediction probabilities are then

$$P(\mathbf{x}_k^{(i)} \mid \hat{\mathbf{x}}_{k+1}, \hat{\mathbf{x}}_1^{k-1}) = \frac{1}{C} \cdot P(\hat{\mathbf{x}}_{k+1} \mid \mathbf{x}_k^{(i)}) \cdot P(\mathbf{x}_k^{(i)} \mid \hat{\mathbf{x}}_1^{k-1}) \tag{10.49}$$

with $P(\mathbf{x}_k^{(i)} \mid \hat{\mathbf{x}}_1^{k-1})$ being the already discussed forward prediction probabilities (see (10.31)) and the backward prediction probabilities being

$$P(\hat{\mathbf{x}}_{k+1} \mid \mathbf{x}_k^{(i)}) = \sum_{h=0}^{2^w-1} P(\hat{\mathbf{x}}_{k+1}, \mathbf{x}_{k+1}^{(h)} \mid \mathbf{x}_k^{(i)})$$

$$= \sum_{h=0}^{2^w-1} P(\hat{\mathbf{x}}_{k+1} \mid \mathbf{x}_{k+1}^{(h)}) \cdot P(\mathbf{x}_{k+1}^{(h)} \mid \mathbf{x}_k^{(i)}) . \tag{10.50}$$

The two multiplicative probability terms in (10.49) represent the contribution of the future and the past to the total prediction probabilities, respectively.

As already discussed, in some practical applications interpolative estimation and extrapolative estimation have to be chosen depending on the time index k (here: subframe parameters). It is therefore advisable to apply the recursion (10.32) to compute the extrapolative *a posteriori* probabilities $P(\mathbf{x}_k^{(i)} \mid \hat{\mathbf{x}}_k, \hat{\mathbf{x}}_1^{k-1})$. In the desired case with $N = K = 1$, they can easily be augmented with the backward prediction probabilities (10.50) to yield the interpolative *a posteriori* probabilities

$$P(\mathbf{x}_k^{(i)} \mid \hat{\mathbf{x}}_{k+1}, \hat{\mathbf{x}}_k, \hat{\mathbf{x}}_1^{k-1}) = \frac{1}{C} \cdot P(\hat{\mathbf{x}}_{k+1} \mid \mathbf{x}_k^{(i)}) \cdot P(\mathbf{x}_k^{(i)} \mid \hat{\mathbf{x}}_k, \hat{\mathbf{x}}_1^{k-1}) \quad \text{(SD/AK1/INT1)} . \tag{10.51}$$

This approach will be referred to as "SD/AK1/INT1", since 1st order *a priori* probabilities are used and one future received bit pattern is exploited for interpolation.

If *two* future received bit patterns are available for SDSD, (10.51) will change to

$$P(\mathbf{x}_k^{(i)} \mid \hat{\mathbf{x}}_{k+2}, \hat{\mathbf{x}}_{k+1}, \hat{\mathbf{x}}_k, \hat{\mathbf{x}}_1^{k-1})$$

$$= \frac{1}{C} \cdot P(\hat{\mathbf{x}}_{k+2}, \hat{\mathbf{x}}_{k+1} \mid \mathbf{x}_k^{(i)}) \cdot P(\mathbf{x}_k^{(i)} \mid \hat{\mathbf{x}}_k, \hat{\mathbf{x}}_1^{k-1}) \quad \text{(SD/AK1/INT2)} \tag{10.52}$$

with the new backward prediction probabilities

$$P(\hat{\mathbf{x}}_{k+2}, \hat{\mathbf{x}}_{k+1} \mid \mathbf{x}_k^{(i)}) = \sum_{g=0}^{2^w-1} P(\hat{\mathbf{x}}_{k+2} \mid \mathbf{x}_{k+2}^{(g)}) \cdot \sum_{h=0}^{2^w-1} P(\mathbf{x}_{k+2}^{(g)} \mid \mathbf{x}_{k+1}^{(h)})$$

$$\cdot P(\hat{\mathbf{x}}_{k+1} \mid \mathbf{x}_{k+1}^{(h)}) \cdot P(\mathbf{x}_{k+1}^{(h)} \mid \mathbf{x}_k^{(i)}) . \tag{10.53}$$

The additional computational load required for (SD/AK1/INTK) as compared with the extrapolation (SD/AK1) with (10.32) is due to the backward prediction probabilities, e.g., as in (10.50) or (10.53). As a consequence the interpolative (SD/AK1/INTK) approach has a total computational complexity of order $O(2^{2w})$ operations for extrapolation plus about $O(2^{w+wK})$ operations for the backward prediction probabilities. In the case of $K = 1$ the complexity turns out to be just twice as high as that with extrapolation.

10.7.2 Simulation Results

As Fig. 10.3-b did for extrapolative SDSD, Fig. 10.10 shows the results for interpolative SDSD for a $w = 4$ bit quantized, highly correlated parameter ($\rho = 0.9$). The high amount of residual redundancy leads to good performance gains of the (SD/AK1) approach when compared with the (SD/AK0) scheme.

If even a future received bit pattern $\hat{\mathbf{x}}_{k+1}$ can be used for SDSD with (10.51), i.e., (SD/AK1/INT1), further significant gains of about 1 dB are found in terms of E_s/N_0. The reason for higher gains through interpolation in the case of a 1st order Markov parameter is that the past and the future bit pattern are not exactly known. The exploitation of even more future received bit patterns leads to further slight performance improvements, as shown with the (SD/AK1/INT2) approach using (10.52). The additional gain increases with decreasing channel quality. In analogy to the Viterbi decoding of convolutional codes [Johannesson, Zigangirov 1999] we can expect that with increasing K the SNR performance will be saturated in the case of $N = \bar{N}$, as soon as K equals a multiple of the order $\bar{N}$ of the Markovian parameter, which is then related to a multiple of the constraint length.

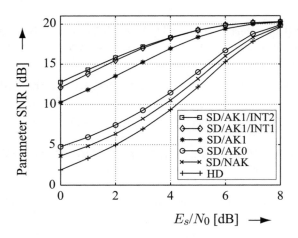

Figure 10.10: Transmission of a $w = 4$ bit quantized Gaussian AR(1) parameter with $\rho = 0.9$

10.8 Discussion and Conclusions

In this chapter we have given a concise summary of the principle of SDSD as the means for bit and frame error concealment. Without introducing changes to the encoder, this technique can be employed at the decoder side to perform robust source decoding, if some reliability information about the received bits is available. In practice, such reliability information stems from the demodulator, or from soft output values of a channel decoder. The most attractive application of SDSD is in the context of mobile communications, e.g., mobile speech telephony with low bit rate parameters. In many communication systems the standards allow manufacturer-dependent implementations of error concealment schemes.

We have discussed the modeling of the transmitted signal or parameters as Markov process of order N, and have introduced extrapolative techniques to estimate them assuming a certain model order. Depending on the correlation, significant gains could be achieved compared with conventional HD decoding. Diversity reception in the SDSD framework was shown to provide further gains.

Then we have investigated the joint extrapolation of two parameters, which turns out to support the decoding process even if no significant temporal correlation is available, but if two parameters in the same frame are correlated instead. The issue of joint extrapolation of an arbitrary number of parameters including the exploitation of temporal correlations will be the main topic of the next chapter.

Finally we have presented the technique of interpolative soft decision source decoding. Whenever a parameter reveals measurable correlations and at least one frame of delay is available for improved decoding, considerable gains can be achieved. This is often the case in real-time streaming services.

Bibliography

3GPP-AMR (1999). Mandatory Speech Codec Speech Processing Functions: AMR Speech Codec; Transcoding Functions (3G TS 26.090), 3GPP; TSG SA.

Alajaji, F.; Phamdo, N.; Fuja, T. (1996). Channel Codes that Exploit the Residual Redundancy in CELP-Encoded Speech, *IEEE Transactions on Speech and Audio Processing*, vol. 4, no. 5, September, pp. 325–336.

Bahl, L. R.; Cocke, J.; Jelinek, F.; Raviv, J. (1974). Optimal Decoding of Linear Codes for Minimizing Symbol Error Rate, *IEEE Transactions on Information Theory*, vol. 20, no. 2, pp. 284–287.

Battail, G. (1987). Pondération des symboles décodés par l'algorithme de Viterbi, *Annales des Télécommunications*, vol. 42, no. 1, pp. 31–38 (in French).

Berger, T. (1971). *Rate Distortion Theory*, Prentice-Hall, Inc., Englewood Cliffs, New Jersey.

Cox, D. R.; Miller, H. D. (1965). *The Theory of Stochastic Processes*, Chapman and Hall, Ltd., London.

Cuperman, V.; Liu, F.-H.; Ho, P. (1994). Robust Vector Quantization for Noisy Channels Using Soft Decision and Sequential Decoding, *European Transactions on Telecommunications*, vol. 5, no. 5, pp. 7–18.

Fano, R. M. (1961). *Transmission of Information: Statistical Theory of Communications*, The MIT Press and John Wiley & Sons, Ltd, New York.

Farvardin, N. (1990). A Study of Vector Quantization for Noisy Channels, *IEEE Transactions on Information Theory*, vol. 36, no. 4, pp. 799–809.

Farvardin, N.; Vaishampayan, V. (1987). Optimal Quantizer Design for Noisy Channels: An Approach to Combined Source-Channel Coding, *IEEE Transactions on Information Theory*, vol. 33, no. 6, pp. 827–838.

Farvardin, N.; Vaishampayan, V. (1991). On the Performance and Complexity of Channel Optimized Vector Quantizers, *IEEE Transactions on Information Theory*, vol. 37, no. 1, pp. 155–160.

Fazel, T.; Fuja, T. (2000). Joint Source-Channel Decoding of Block-Encoded Compressed Speech, *Conference on Information Sciences and Systems*, pp. FA5–1 – FA5–6.

Fingscheidt, T.; Vary, P. (1996). Error Concealment by Softbit Speech Decoding, *ITG-Fachtagung "Sprachkommunikation"*, VDE–Verlag, Frankfurt a.M., Germany, pp. 7–10.

Fingscheidt, T. (1998). *Softbit-Sprachdecodierung in digitalen Mobilfunksystemen*, PhD thesis, Aachener Beiträge zu digitalen Nachrichtensystemen, edited by P. Vary, vol. 9, (ISBN 3-86073-438-5) (in German).

Fingscheidt, T.; Heinen, S.; Vary, P. (1999). Joint Speech Codec Parameter and Channel Decoding of Parameter Individual Block Codes (PIBC), *IEEE Workshop on Speech Coding*, Porvoo, Finland.

Fingscheidt, T.; Hindelang, T.; Cox, R. V.; Seshadri, N. (2002). Joint Source-Channel (De-)Coding for Mobile Communications, *IEEE Transactions on Communications*, vol. 50, no. 2, pp. 200–212.

Fingscheidt, T.; Scheufen, O. (1997). Robust GSM Speech Decoding Using the Channel Decoder's Soft Output, *European Conference on Speech Communication and Technology (EUROSPEECH)*, Rhodos, Greece, pp. 1315–1318.

Fingscheidt, T.; Vary, P. (1997a). Robust Speech Decoding: A Universal Approach to Bit Error Concealment, *IEEE International Conference on Acoustics, Speech, and Signal Processing (ICASSP)*, Munich, Germany, vol. 3, pp. 1667–1670.

Fingscheidt, T.; Vary, P. (1997b). Speech Decoding With Error Concealment Using Residual Source Redundancy, *IEEE Workshop on Speech Coding*, Pocono Manor, Pennsylvania, USA, pp. 91–92.

Fingscheidt, T.; Vary, P. (2001). Softbit Speech Decoding: A New Approach to Error Concealment, *IEEE Transactions on Speech and Audio Processing*, vol. 9, no. 3, pp. 240–251.

Gallager, R. G. (1968). *Information Theory and Reliable Communication*, John Wiley & Sons, Ltd, New York.

Gerlach, C. (1993). A Probabilistic Framework for Optimum Speech Extrapolation in Digital Mobile Radio, *IEEE International Conference on Acoustics, Speech, and Signal Processing (ICASSP)*, Minneapolis, Minnesota, vol. 2, pp. 419–422.

Hagenauer, J. (1995). Source-Controlled Channel Decoding, *IEEE Transactions on Communications*, vol. 43, no. 9, pp. 2449–2457.

Hagenauer, J.; Hoeher, P. (1989). A Viterbi Algorithm with Soft-Decision Outputs and its Applications, *IEEE Global Telecommunications Conference (GLOBECOM)*, Dallas, Texas, pp. 1680–1686.

Hamming, R. W. (1986). *Coding and Information Theory*, Prentice Hall, Englewood Cliffs, N.J.

Huber, J.; Rüppel, A. (1990). Zuverlässigkeitsschätzung für die Ausgangssymbole von Trellis-Decodern, *International Journal of Electronics and Communications (AEÜ)*, vol. 44, no. 1, January, pp. 8–21 (in German).

Jakes, W. C. (ed.) (1974). *Microwave Mobile Communications*, IEEE Press, Piscataway, New Jersey.

Jayant, N. S.; Noll, P. (1984). *Digital Coding of Waveforms*, Prentice-Hall, Inc., Englewood Cliffs, New Jersey.

Johannesson, R.; Zigangirov, K. S. (1999). *Fundamentals of Convolutional Coding*, IEEE Press, Inc., Piscataway, New Jersey.

Kahn, L. R. (1954). Ratio Squarer, *Proceedings of the Institute of Radio Engineers (IRE)*, vol. 42, no. 1, pp. 1704.

Knagenhjelm, P. (1993). How Good is Your Index Assignment?, *IEEE International Conference on Acoustics, Speech, and Signal Processing (ICASSP)*, Minneapolis, Minnesota, vol. 2, pp. 423–426.

Lahouti, F.; Khandani, A. K. (2001). Approximating and Exploiting the Residual Redundancies – Applications to Efficient Reconstruction of Speech over Noisy Channels, *IEEE International Conference on Acoustics, Speech, and Signal Processing (ICASSP)*, Salt Lake City, Utah, vol. 2, pp. 721–724.

Lahouti, F.; Khandani, A. K. (2004). Efficient Source Decoding over Memoryless Noisy Channels Using Higher Order Markov Models, *IEEE Transactions on Information Theory*, vol. 50, no. 9, pp. 2103–2118.

Melsa, J. L.; Cohn, D. L. (1978). *Decision and Estimation Theory*, McGraw-Hill Kogakusha, Tokyo, Japan.

Papen, W. (1996). *Makrodiversität und Signalkombination in zellularen digitalen Mobilfunksystemen*, PhD thesis, Aachener Beiträge zu Digitalen Nachrichtensystemen (ABDN), edited by P. Vary, vol. 4, ISBN 3-86073-433-4, RWTH Aachen University (in German).

Papoulis, A. (1965). *Probability, Random Variables, and Stochastic Processes*, McGraw-Hill, New York.

Parsons, J. D.; Gardiner, J. G. (1989). *Mobile Communication Systems*, Blackie, Glasgow, London.

Phamdo, N.; Farvardin, N. (1994). Optimal Detection of Discrete Markov Sources Over Discrete Memoryless Channels – Applications to Combined Source-Channel Coding, *IEEE Transactions on Information Theory*, vol. 40, no. 1, pp. 186–193.

Sayood, K.; Borkenhagen, J. (1991). Use of Residual Redundancy in the Design of Joint Source/Channel Coders, *IEEE Transactions on Communications*, vol. 39, no. 6, pp. 838–846.

Shannon, C. (1948). A Mathematical Theory of Communication, *Bell Systems Technical Journal*, vol. 27, pp. 379–423.

Skoglund, M.; Hedelin, P. (1999). Hadamard Based Soft-Decoding for Vector Quantization Over Noisy Channels, *IEEE Transactions on Information Theory*, vol. 45, no. 2, pp. 515–532.

Takàcs, L. (1968). *Stochastic Processes*, Methuen & Co. Ltd. and Science Paperbacks, London.

Weiss, C.; Riedel, S.; Hagenauer, J. (1996). Sequential Decoding Using A Priori Information, *IEE Electronics Letters*, vol. 32, no. 13, pp. 1190–1191.

Chapter 11

Optimal MMSE Estimation for Vector Sources with Spatially and Temporally Correlated Elements

Stefan Heinen, Marc Adrat

11.1 Introduction

Originally, *Soft Decision Source Decoding* (SDSD) was proposed for scalar sources [Fingscheidt, Vary 1997], [Fingscheidt 1998], [Fingscheidt, Vary 2001]. As already stated in Chap. 10, modern source codecs usually make use of several parameters that not only exhibit residual temporal correlation but also spatial correlation, i.e., they are mutually correlated. Typical examples are parameters describing the spectral envelope of a source signal. On the one hand, subsequent realizations of a spectral envelope do not change abruptly: This leads to temporal correlation. On the other hand, a realization of the envelope is usually smooth, which means that the amplitudes of neighboring frequency bins are correlated as well.

In principle, extending SDSD from scalar to vector context is straightforward and was shown in Chap. 10 for the example of parameter tuples. The SDSD decoder of Chap. 10 was formulated as a vector MMSE estimator performing the estimation jointly for the elements of the parameter vector and actually handles the vector in the same way as one single composite parameter. However, the drawback of this ad-hoc approach is obvious; the estimation complexity grows exponentially with the number

Advances in Digital Speech Transmission   Edited by R. Martin, U. Heute and C. Antweiler
© 2008 John Wiley & Sons, Ltd

of parameters jointly estimated. A practical implementation based on this approach would be extremely costly if not impossible.

Inspired by the famous algorithm of Bahl et al. [Bahl et al. 1974], we approach the solution to the complexity problem by claiming a Markov property for the vector elements. While SDSD is already doing this in the time direction, we add another Markov property for the spatial (mutual) correlation of the vector elements. Based on this key idea we derive an optimal MMSE estimator in the presence of *Additive White Gaussian Noise* (AWGN), whose complexity grows only linearly with the parameter vector length. As this optimal estimator is still computationally demanding, we also develop an approximation for practical applications.

The present chapter is structured as follows. In Sec. 11.2 we develop a model of a vector source with the mentioned two-fold Markov property and determine its correlation properties. Section 11.3 describes the assumed transmission channel. Based on these prerequisites, the optimal MMSE estimator is derived in Sec. 11.4. The approximation with further reduced complexity but with comparable estimation performance is presented in Sec. 11.5. Both approaches are compared and evaluated in Sec. 11.6. Finally, in order to predict the estimation enhancement for a real system, we emulate the statistical properties of important codec parameters of the *Digital Audio Broadcast* (DAB) [ETSI, Standard ETS 300 401 1997] system as well as of the *Global System for Mobile communication* (GSM) [ETSI TC-SMG 1999] and we measure the resulting parameter SNR gain.

11.2 Source Model

The source model makes use of the fact that the sum of two random variables is statistically dependent of the summands. Temporal and spatial correlation can thus be established by adding up delayed realizations of the same parameter and neighboring parameters.

Figure 11.1 depicts the source model in detail. A parameter at position m and time instant k, denoted by $\tilde{v}_{m,k}$, is emitted from a time discrete, white, zero mean, unit variance scalar prototype source with *Probability Density Function* (PDF) $p_{\tilde{v}}$, which is added to the scaled value of the spatial predecessor $\tilde{v}_{m-1,k}$ (except for $\tilde{v}_{1,k}$, which will be explained later). In this way the spatial correlation is established. Furthermore, the subsequent one tap IIR filter (*Infinite Impulse Response*) gives rise to temporal correlation by adding the scaled temporal predecessor value $\tilde{v}_{m,k-1}$. Variance and correlation properties of the parameters result from the choice of the scaling constants a, b, and c.

The continuous parameter values $\tilde{v}_{m,k}$ are individually quantized by Q_m yielding value-discrete parameters $v_{m,k}$, which take values in the reproduction sets

$$v_{m,k} \in \mathbb{V}_m = \{v_m^{(0)}, \dots, v_m^{(V_m-1)}\} \,, \tag{11.1}$$

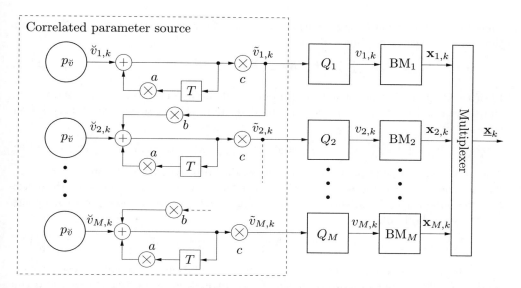

Figure 11.1: Model of a source encoder generating temporally and spatially correlated parameters (Q_m: quantizer, BM_m: index assignment and bit mapping)

with V_m denoting the number of reproduction values of quantizer Q_m. To each reproduction level a unique bit combination

$$\mathbf{x}_{m,k} \in \mathbb{X}_m = \{\mathbf{x}_m^{(0)}, \ldots, \mathbf{x}_m^{(V_m-1)}\} \tag{11.2}$$

is assigned by the bit mapping unit BM_m. The multiplexer composes the bit codes $\mathbf{x}_{m,k}$ to a bit vector $\underline{\mathbf{x}}_k$, where

$$\underline{\mathbf{x}}_k \in \mathbb{X} = \{\underline{\mathbf{x}}^{(0)}, \ldots, \underline{\mathbf{x}}^{(\check{V}-1)}\}, \quad \check{V} = \prod_{m=1}^{M} V_m . \tag{11.3}$$

In order to abbreviate sequences of scalars or vectors, we introduce the notation $v_{1,k}^m \doteq (v_{1,k}, \ldots, v_{m,k})$ and $\underline{v}_1^k \doteq (\underline{v}_1, \ldots, \underline{v}_k)$, respectively. To keep equations short, we further use the notation $P(\mathbf{x}_k) \doteq \Pr\{\mathbf{X}_k = \mathbf{x}_k\}$. Similarly, we omit the index of PDFs, i.e., $p(\tilde{v}_k) \doteq p_{\tilde{v}_k}(\tilde{v}_k)$, unless the more detailed notation is required to avoid ambiguity.

Thanks to the particular layout of the source in Fig. 11.1, the components $v_{m,k}$ exhibit the wanted Markov property

$$p(\tilde{v}_{m,k} \mid \tilde{v}_{1,k}^{m-1}, \tilde{\mathbf{v}}_1^{k-1}) = p(\tilde{v}_{m,k} \mid \tilde{v}_{m-1,k}, \tilde{v}_{m,k-1}) . \tag{11.4}$$

Hence, to predict $\tilde{v}_{m,k}$ the knowledge of the values $\tilde{v}_{m-1,k}$ and $\tilde{v}_{m,k-1}$ is sufficient. Taking further past values into consideration does not enhance a prediction, where "past" here means both preceding in time k and position m.

If the applied quantizer Q_m has sufficient precision the same is valid for the quantized parameters $v_{m,k}$ and simultaneously for the bit combinations $\mathbf{x}_{m,k}$. Representing the source output at time k and position m by the discrete random variable $\mathbf{X}_{m,k}$ and the complete vector at time k by $\underline{\mathbf{X}}_k$, we get

$$P(\mathbf{X}_{m,k} \mid \underline{\mathbf{X}}_1^{k-1}, \mathbf{X}_{1,k}^{m-1}) = P(\{\mathbf{X}_{m,k} \mid \mathbf{X}_{m,k-1}, \mathbf{X}_{m-1,k}) \ . \tag{11.5}$$

In order to quantify the correlation properties of the model parameters generated by the source in Fig. 11.1, it is required to determine the auto- and cross-correlation series

$$\varphi_{\tilde{v}_m \tilde{v}_m}(\lambda) = \mathrm{E}\{\, v_{m,k} \cdot v_{m,k+\lambda}\,\} \quad \text{and} \quad \varphi_{\tilde{v}_m \tilde{v}_{m+1}}(\mu) = \mathrm{E}\{\, v_{m,k} \cdot v_{m+1,k+\mu}\,\} \ . \tag{11.6}$$

The above equations indicate that the correlation properties depend on the index m. The reason is that there is no contribution from a predecessor at $m = 1$. However, this "boundary effect" vanishes with increasing m, which can be shown by proving the existence of the limits

$$\lim_{m \to \infty} \varphi_{\tilde{v}_m \tilde{v}_m}(\lambda) = \rho(\lambda) \quad \text{and} \quad \lim_{m \to \infty} \varphi_{\tilde{v}_m \tilde{v}_{m+1}}(\mu) = \delta(\mu) \ , \tag{11.7}$$

in the following referred to as auto- and cross-correlation coefficients[1], respectively. In [Heinen 2001a] an in-depth analysis of the correlation properties is given and the conditions of convergence are determined. Under consideration of the convergence criterion $c^2 < (1-a)^2/b^2$ we can now easily determine some important correlation properties. For large m the approximation $\varphi_{\tilde{v}_m \tilde{v}_m} \approx \varphi_{\tilde{v}_{m+1} \tilde{v}_{m+1}}$ becomes arbitrarily accurate. For $m \to \infty$ even identity holds. By applying the Wiener–Khintchine theorem this relation can be transformed into the frequency domain. Considering that the prototype sources are white and have unit power $\sigma_{\tilde{v}}^2 = 1$ we obtain

$$\left[1 + b^2\, \Phi_{\tilde{v}\tilde{v}}(\Omega)\right] \cdot \left|\frac{1}{1 - a\,\mathrm{e}^{-j\Omega}}\right|^2 c^2 = \Phi_{\tilde{v}\tilde{v}}(\Omega) \ . \tag{11.8}$$

Solving (11.8) for $\Phi_{\tilde{v}\tilde{v}}(\Omega)$ yields the power spectral density (PSD) for $m \to \infty$. The auto-correlation coefficient for $\lambda = 0$ is given by Parseval's theorem

$$\rho(0) = \frac{1}{2\pi} \int\limits_{-\pi}^{\pi} \Phi_{\tilde{v}\tilde{v}}(\Omega)\mathrm{e}^{-j\,0\cdot\Omega}\,\mathrm{d}\Omega = \frac{c^2}{\sqrt{((1-a)^2 - c^2 b^2)((1+a)^2 - c^2 b^2)}} \ . \tag{11.9}$$

Without loss of generality we define the power of the model parameters $\tilde{v}_m$ to $\rho(0) = 1$. This allows us to express c as a function of a and b

$$c = \frac{1 - a^2}{\sqrt{b^2(1 + a^2) + \sqrt{4b^4 a^2 + (1 - a^2)^2}}} \ . \tag{11.10}$$

In analogy to (11.9) the 1st order auto-correlation coefficient can be determined to

$$\rho(1) = \frac{1}{2\pi} \int\limits_{-\pi}^{\pi} \Phi_{\tilde{v}_m \tilde{v}_m}(\Omega)\,\mathrm{e}^{-j\,1\cdot\Omega}\,\mathrm{d}\Omega = \frac{1 + a^2 - c^2(1 + b^2)}{2a} \ . \tag{11.11}$$

---

[1]Note that in the context of this chapter and related literature the symbol δ specifies cross-correlation coefficients rather than the Dirac pulse.

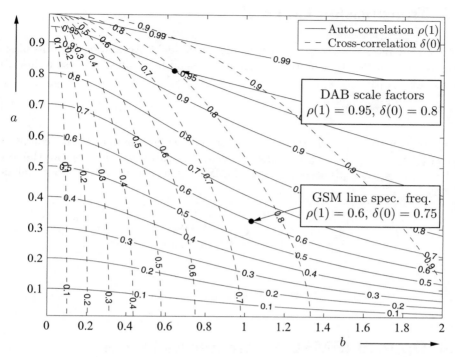

Figure 11.2: Contour plot of auto-correlation $\rho(1)$ and cross-correlation $\delta(0)$ coefficients

Finally, for the cross-correlation coefficient $\delta(0)$ it can be shown [Heinen 2001a] that

$$\delta(0) = \frac{1}{bc}(\rho(0) - a\rho(1) - c^2) = \frac{1 - a^2 + c^2(b^2 - 1)}{2bc} \tag{11.12}$$

holds. The inverse functions allow us to compute a and b for a given tuple $\rho(1), \delta(0)$

$$a = \frac{\rho(1)(1 - \delta^2(0))}{1 - \rho^2(1)\delta^2(0)}, \quad b = \frac{(1 - \rho^2(1))\delta(0)}{\sqrt{(1 - \rho^2(1)\delta^2(0))(1 + \rho^2(1)\delta^2(0) - \rho^2(1) - \delta^2(0))}}. \tag{11.13}$$

With the quantities $\rho(1)$ and $\delta(0)$ we have now two essential correlation properties that allow to classify a real codec and align our model with it. Figure 11.2 shows a contour plot of the relation between a, b and $\rho(1), \delta(0)$, which allows us to determine the coefficient tuple a, b graphically for a given correlation constellation $\rho(1), \delta(0)$. Two prominent examples of source codecs are also mapped into the diagram.

11.3 Transmission Channel

Figure 11.3 depicts the considered transmission model. We assume a transmission, which can be modeled by a stationary memoryless additive noise process $\mathbf{n}$ described by the PDF

$$p(\mathbf{n}) = \prod_m p(n_m) \ . \tag{11.14}$$

Hence, we can uniquely characterize the transmission channel by the conditional PDF $p(\hat{\underline{\mathbf{x}}}_k \,|\, \underline{\mathbf{x}}_k) = p_{\mathbf{n}}(\hat{\underline{\mathbf{x}}}_k - \underline{\mathbf{x}}_k)$. The received *soft decision vectors* (in the sense of Chap. 10) $\hat{\underline{\mathbf{x}}}_k$ are processed by the MMSE estimator to get the optimal estimates $\hat{\underline{v}}_k$.

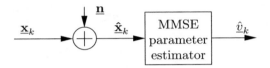

Figure 11.3: Transmission model

11.4 Optimal MMSE Parameter Estimator

The task of the parameter estimator in Fig. 11.3 is to determine MMSE optimal estimates $\hat{\underline{v}}_k$ from the noisy observation vector $\hat{\underline{\mathbf{x}}}_k$ [Heinen 2001a], [Heinen 2001b]. Since the source is not white an optimal estimator has to take into account the entire history $\hat{\underline{v}}_k = \hat{\underline{v}}_k(\hat{\underline{\mathbf{x}}}_1, \ldots, \hat{\underline{\mathbf{x}}}_k)$ to obtain an estimate[2] $\hat{\underline{v}}_k$. The mean square error is given by the expectation of $\|\underline{v}_k - \hat{\underline{v}}_k\|^2$ over all involved random variables. With the total number of bits $W = \sum_{m=1}^{M} w_m$ per transmission $\underline{\mathbf{x}}_k$ the expectation can be written as

$$\mathrm{E}\big\{\, \|\underline{v}_k - \hat{\underline{v}}_k\|^2 \,\big\} = \int\limits_{\hat{\underline{\mathbf{x}}}_1 \in \mathbb{R}^W} \cdots \int\limits_{\hat{\underline{\mathbf{x}}}_k \in \mathbb{R}^W} \int\limits_{\tilde{\underline{v}}_k \in \mathbb{R}^M} \|\tilde{\underline{v}}_k - \hat{\underline{v}}_k\|^2 \, p(\tilde{\underline{v}}_k, \hat{\underline{\mathbf{x}}}_1^k) \, \mathrm{d}\tilde{\underline{v}}_k \, \mathrm{d}\hat{\underline{\mathbf{x}}}_k \cdots \mathrm{d}\hat{\underline{\mathbf{x}}}_1 \ . \tag{11.15}$$

From estimation theory it is known [Melsa, Cohn 1978], [Cover, Thomas 1991] that minimizing the mean square error (11.15) leads to a conditional expectation. Applying this to a single parameter $v_{m,k}$, we obtain

$$\hat{v}_{m,k} = \frac{1}{p(\hat{\underline{\mathbf{x}}}_1^k)} \int\limits_{\tilde{v}_{m,k}=-\infty}^{\infty} \tilde{v}_{m,k} \, p(\tilde{v}_{m,k}, \hat{\underline{\mathbf{x}}}_1^k) \, \mathrm{d}\tilde{v}_{m,k} \ . \tag{11.16}$$

---

[2]Actually, it also depends on future observation vectors. However, to exploit these dependencies, additional delay would be necessary, which is usually prohibitive in many practical applications. For this reason we do not consider dependencies on future observations here, yet the extension of the presented algorithm to this case is straightforward. Note that in Chap. 10 the exploitation of past and future observations was discussed for a scalar parameter.

Without loss of generality we assume that the quantizer reproduction values $v_m^{(i)}$ are centroids of their respective cells. Then, the optimal estimator is given by

$$\hat{v}_{m,k} = \frac{1}{p(\hat{\underline{\mathbf{x}}}_1^k)} \sum_{i=0}^{2^{w_m}-1} v_m^{(i)}\, p(\mathbf{x}_m^{(i)}, \hat{\underline{\mathbf{x}}}_1^k) \,, \tag{11.17}$$

which is already also known from Chap. 10. Note that due to the deterministic bit mapping there is a one-to-one correspondence between $v_m^{(i)}$ and $\mathbf{x}_m^{(i)}$. The challenging part of (11.17) is the marginal PDF $p(\mathbf{x}_m^{(i)}, \hat{\underline{\mathbf{x}}}_1^k)$, which has to be computed for each realization $v_m^{(i)}$. Therefore, in the following an algorithm for an efficient computation of this PDF will be derived.

The formal solution

$$p(\mathbf{x}_m^{(i)}, \hat{\underline{\mathbf{x}}}_1^k) = \sum_{\ell_1=0}^{\check{V}-1} \cdots \sum_{\ell_{k-1}=0}^{\check{V}-1} \sum_{\ell_{k,1}=0}^{V_{\ell_k,1}-1} \cdots \sum_{\ell_{k,m-1}=0}^{V_{\ell_k,m-1}-1} \sum_{\ell_{k,m+1}=0}^{V_{\ell_k,m+1}-1} \cdots \sum_{\ell_{k,M}=0}^{V_{\ell_k,M}-1}$$
$$p(\underline{\mathbf{x}}^{(\ell_1)}, \ldots, \underline{\mathbf{x}}^{(\ell_{k-1})}, \mathbf{x}^{(\ell_{k,1})}, \ldots, \mathbf{x}^{(\ell_{k,m-1})}, \mathbf{x}_m^{(i)}, \mathbf{x}_m^{(\ell_{k,m+1})}, \ldots, \mathbf{x}_m^{(\ell_{k,M})}, \hat{\underline{\mathbf{x}}}_1^k)$$
$$\tag{11.18}$$

is by far too complex for direct evaluation. Therefore, our aim is to develop a recursion formula that reduces the number of required operations significantly. To enhance readability of equations, instead of using explicit limits as in (11.18) we apply the notation $\sum_{\forall \mathbf{x}_1} p(\underline{\mathbf{x}}_1, \ldots, \underline{\mathbf{x}}_k) = p(\underline{\mathbf{x}}_2, \ldots, \underline{\mathbf{x}}_k)$ to express a marginal distribution.

In a first step we split the joint PDF $p(\underline{\mathbf{x}}_1^k, \hat{\underline{\mathbf{x}}}_1^k)$ at time k and position m in order to exploit the Markov properties of the source

$$p(\underline{\mathbf{x}}_1^k, \hat{\underline{\mathbf{x}}}_1^k) = p(\mathbf{x}_{m+1,k}^M, \hat{\mathbf{x}}_{m+1,k}^M \mid \mathbf{x}_{1,k}^m, \underline{\mathbf{x}}_1^{k-1}, \hat{\mathbf{x}}_{1,k}^m, \hat{\underline{\mathbf{x}}}_1^{k-1})$$
$$\cdot\, p(\mathbf{x}_{m,k}, \hat{\mathbf{x}}_{m,k} \mid \mathbf{x}_{1,k}^{m-1}, \underline{\mathbf{x}}_1^{k-1}, \hat{\mathbf{x}}_{1,k}^{m-1}, \hat{\underline{\mathbf{x}}}_1^{k-1})$$
$$\cdot\, p(\mathbf{x}_{1,k}^{m-1}, \hat{\mathbf{x}}_{1,k}^{m-1} \mid \underline{\mathbf{x}}_1^{k-1}, \hat{\underline{\mathbf{x}}}_1^{k-1}) \cdot p(\underline{\mathbf{x}}_1^{k-1}, \hat{\underline{\mathbf{x}}}_1^{k-1}) \,. \tag{11.19}$$

Making use of the assumption of a memoryless channel and the Markov properties of the source allows us to simplify (11.19) significantly. Figure 11.4 illustrates how the source properties can be exploited. Each of the three diagrams a), b) and c) represents one of the conditional PDFs in (11.19). Terms left of the conditional symbol "|" are marked with grey boxes, while terms right of the conditional symbol are framed by a black line. Owing to the spatial and temporal Markov property of the source it turns out that only terms corresponding to the filled bullets are relevant. Therefore, all other terms can be omitted, which leads to

$$p(\underline{\mathbf{x}}_1^k, \hat{\underline{\mathbf{x}}}_1^k) = p(\mathbf{x}_{m+1,k}^M, \hat{\mathbf{x}}_{m+1,k}^M \mid \mathbf{x}_{m+1,k-1}^M, \mathbf{x}_{m,k})$$
$$\cdot\, p(\mathbf{x}_{m,k}, \hat{\mathbf{x}}_{m,k} \mid \mathbf{x}_{m,k-1}, \mathbf{x}_{m-1,k})$$
$$\cdot\, p(\mathbf{x}_{1,k}^{m-1}, \hat{\mathbf{x}}_{1,k}^{m-1} \mid \mathbf{x}_{1,k-1}^{m-1}) \cdot p(\underline{\mathbf{x}}_1^{k-1}, \hat{\underline{\mathbf{x}}}_1^{k-1}) \,. \tag{11.20}$$

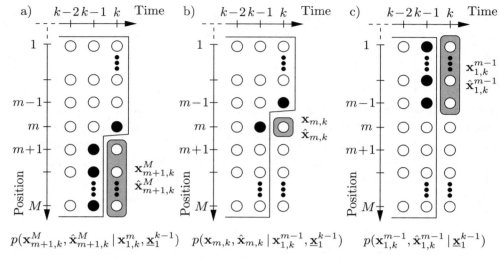

$$p(\mathbf{x}_{m+1,k}^M, \hat{\mathbf{x}}_{m+1,k}^M \,|\, \mathbf{x}_{1,k}^m, \underline{\mathbf{x}}_1^{k-1}) \qquad p(\mathbf{x}_{m,k}, \hat{\mathbf{x}}_{m,k} \,|\, \mathbf{x}_{1,k}^{m-1}, \underline{\mathbf{x}}_1^{k-1}) \qquad p(\mathbf{x}_{1,k}^{m-1}, \hat{\mathbf{x}}_{1,k}^{m-1} \,|\, \underline{\mathbf{x}}_1^{k-1})$$

Figure 11.4: Graphical representation of the relevant terms in (11.19)

The second factor can be factorized to

$$p(\mathbf{x}_{m,k}, \hat{\mathbf{x}}_{m,k} \,|\, \mathbf{x}_{m,k-1}, \mathbf{x}_{m-1,k}) = p(\hat{\mathbf{x}}_{m,k} \,|\, \mathbf{x}_{m,k})\, P(\mathbf{x}_{m,k} \,|\, \mathbf{x}_{m,k-1}, \mathbf{x}_{m-1,k}) \quad (11.21)$$

and it describes the channel and source properties.

In the next step we consider the last factor of (11.20), which has the same form as on the left hand side of (11.20), but with a decremented time index $k-1$, i.e.,

$$p(\underline{\mathbf{x}}_{k-1}, \hat{\underline{\mathbf{x}}}_1^{k-1}) = \sum_{\forall \underline{\mathbf{x}}_1} \cdots \sum_{\forall \underline{\mathbf{x}}_{k-2}} p(\underline{\mathbf{x}}_1, \ldots, \underline{\mathbf{x}}_{k-1}, \hat{\underline{\mathbf{x}}}_1^{k-1}) , \qquad (11.22)$$

and can be interpreted as the basis for recursion step k. Substituting (11.22) into (11.18) we see that the sums up to time instant $k-2$ have already been evaluated by the time recursion. Provided that this time recursion exists, we can therefore now focus on $p(\underline{\mathbf{x}}_{k-1}, \hat{\underline{\mathbf{x}}}_1^{k-1})$. By splitting this term in a similar way as above but now with respect to the position m we get

$$p(\underline{\mathbf{x}}_{k-1}, \hat{\underline{\mathbf{x}}}_1^{k-1}) = P(\mathbf{x}_{m+1,k-1}^M \,|\, \mathbf{x}_{m,k-1}, \hat{\underline{\mathbf{x}}}_1^{k-1})$$
$$\cdot P(\mathbf{x}_{m,k-1} \,|\, \mathbf{x}_{m-1,k-1}, \hat{\underline{\mathbf{x}}}_1^{k-1})\, p(\mathbf{x}_{1,k-1}^{m-1}, \hat{\underline{\mathbf{x}}}_1^{k-1}) . \quad (11.23)$$

The second factor can be expressed by

$$P(\mathbf{x}_{m,k-1} \,|\, \mathbf{x}_{m-1,k-1}, \hat{\underline{\mathbf{x}}}_1^{k-1}) = \frac{p(\mathbf{x}_{m,k-1}, \mathbf{x}_{m-1,k-1}, \hat{\underline{\mathbf{x}}}_1^{k-1})}{p(\mathbf{x}_{m-1,k-1}, \hat{\underline{\mathbf{x}}}_1^{k-1})} \qquad (11.24)$$

where the denominator is given by the marginal distribution of a single parameter and is therefore a direct product of the recursion step $k-1$. The numerator is a joint marginal distribution of two spatially adjacent parameters, which can easily be computed as a side-product, as we will see a bit later.

Before substituting (11.20)–(11.23) into (11.18), we define the following abbreviations to simplify notation

$$
\alpha_{m-1}(\mathbf{x}_{m-1,k}, \mathbf{x}_{m-1,k-1}) \doteq \sum_{\forall \mathbf{x}_{1,k}^{m-2}} \sum_{\forall \mathbf{x}_{1,k-1}^{m-2}} p(\mathbf{x}_{1,k}^{m-1}, \hat{\mathbf{x}}_{1,k}^{m-1} \mid \mathbf{x}_{1,k-1}^{m-1}) \, p(\mathbf{x}_{1,k-1}^{m-1}, \hat{\underline{\mathbf{x}}}_1^{k-1})
$$

$$(11.25)$$

$$
\beta_m(\mathbf{x}_{m,k}, \mathbf{x}_{m,k-1}) \doteq \sum_{\forall \mathbf{x}_{m+1,k}^{M}} \sum_{\forall \mathbf{x}_{m+1,k-1}^{M}} p(\mathbf{x}_{m+1,k}^{M}, \hat{\mathbf{x}}_{m+1,k}^{M} \mid \mathbf{x}_{m+1,k-1}^{M}, \mathbf{x}_{m,k})
$$
$$
\cdot P(\mathbf{x}_{m+1,k-1}^{M} \mid \mathbf{x}_{m,k-1}, \hat{\underline{\mathbf{x}}}_1^{k-1}) . \quad (11.26)
$$

This allows us to rewrite our starting point, the marginal distribution (11.18) in a compact way as

$$
p(\mathbf{x}_m^{(i)}, \hat{\underline{\mathbf{x}}}_1^{k}) = \sum_{\forall \mathbf{x}_{m,k-1}} \alpha_m(\mathbf{x}_m^{(i)}, \mathbf{x}_{m,k-1}) \cdot \beta_m(\mathbf{x}_m^{(i)}, \mathbf{x}_{m,k-1}) . \quad (11.27)
$$

Matching (11.20) and (11.23) with the definitions of α and β shows that multiplying α_{m-1} by

$$
\gamma_m(\mathbf{x}_{m,k}, \mathbf{x}_{m-1,k}, \mathbf{x}_{m,k-1}, \mathbf{x}_{m-1,k-1}) \doteq
$$
$$
p(\mathbf{x}_{m,k}, \hat{\mathbf{x}}_{m,k} \mid \mathbf{x}_{m,k-1}, \mathbf{x}_{m-1,k}) \, P(\mathbf{x}_{m,k-1} \mid \mathbf{x}_{m-1,k-1}, \hat{\underline{\mathbf{x}}}_1^{k-1}) \quad (11.28)
$$

and summing up over all $x_{m-1,k}$ and $x_{m-1,k-1}$ yields the next recursion step α_m. Similarly, a recursion step in the opposite direction can be identified for β_{m-1}. Thus, we obtain the recursion formulas

$$
\alpha_{m+1}(\mathbf{x}_{m+1,k}, \mathbf{x}_{m+1,k-1}) = \sum_{\forall \mathbf{x}_{m,k}} \sum_{\forall \mathbf{x}_{m,k-1}} \gamma_{m+1}(\mathbf{x}_{m+1,k}, \mathbf{x}_{m,k}, \mathbf{x}_{m+1,k-1}, \mathbf{x}_{m,k-1})
$$
$$
\cdot \alpha_m(\mathbf{x}_{m,k}, \mathbf{x}_{m,k-1}) \quad (11.29)
$$

$$
\beta_{m-1}(\mathbf{x}_{m-1,k}, \mathbf{x}_{m-1,k-1}) = \sum_{\forall \mathbf{x}_{m,k}} \sum_{\forall \mathbf{x}_{m,k-1}} \gamma_m(\mathbf{x}_{m,k}, \mathbf{x}_{m-1,k}, \mathbf{x}_{m,k-1}, \mathbf{x}_{m-1,k-1})
$$
$$
\cdot \beta_m(\mathbf{x}_{m,k}, \mathbf{x}_{m,k-1}) . \quad (11.30)
$$

The analysis of the boundary conditions for $m = 1$ and $m = M$ shows that the recursions have to be initialized by

$$
\alpha_1(\mathbf{x}_{1,k}, \mathbf{x}_{1,k-1}) = p(\mathbf{x}_{1,k-1}, \hat{\underline{\mathbf{x}}}_1^{k-1}) \cdot p(\hat{\mathbf{x}}_{1,k} \mid \mathbf{x}_{1,k}) \, P(\mathbf{x}_{1,k} \mid \mathbf{x}_{1,k-1}) \quad (11.31)
$$

$$
\beta_M(\mathbf{x}_{M,k}, \mathbf{x}_{M,k-1}) = 1 . \quad (11.32)
$$

Finally, to establish a complete recursion we have to provide the marginal distribution in the numerator of (11.24), which is given by

$$p(\mathbf{x}_{m,k-1}, \mathbf{x}_{m-1,k-1}, \hat{\underline{\mathbf{x}}}_1^{k-1}) = \sum_{\forall \mathbf{x}_{m,k-1}} \sum_{\forall \mathbf{x}_{m-1,k-1}} \alpha_{m-1}(\mathbf{x}_{m-1,k}, \mathbf{x}_{m-1,k-1})$$

$$\cdot \beta_m(\mathbf{x}_{m,k}, \mathbf{x}_{m,k-1}) \gamma_m(\mathbf{x}_{m,k}, \mathbf{x}_{m-1,k}, \mathbf{x}_{m,k-1}, \mathbf{x}_{m-1,k-1}) . \quad (11.33)$$

To summarize, we give the entire algorithm at a glance in Table 11.1 [Heinen 2001b].

Table 11.1: MMSE-optimal estimation algorithm for vector sources with spatially and temporally correlated elements

---

1) Compute $P(\mathbf{x}_{m,1} \,|\, \hat{\underline{\mathbf{x}}}_1)$ and $P(\mathbf{x}_{m,1}, \mathbf{x}_{m-1,1} \,|\, \hat{\underline{\mathbf{x}}}_1)$ (forward/backward recursion).
2) $k \leftarrow k + 1$.
3) Initialize α_1 and β_M according to (11.31) and (11.32).
4) Forward/backward recursion according to (11.29), (11.30).
5) Compute $p(\mathbf{x}_m^{(i)}, \hat{\underline{\mathbf{x}}}_1^k)$ according to (11.27).
6) Determine $P(\mathbf{x}_{m,k-1} \,|\, \mathbf{x}_{m-1,k-1}, \hat{\underline{\mathbf{x}}}_1^{k-1})$ by (11.33) and (11.24).
7) Compute estimates according to (11.17).
8) Return to 2) and process the next received vector $\hat{\underline{\mathbf{x}}}_k$.

11.5 Near-Optimal MMSE Parameter Estimator

The optimal MMSE estimation rule introduced in the preceding section exploits the spatial and temporal Markov property jointly via the conditional probability mass function $P(\mathbf{x}_{m,k}|\mathbf{x}_{m,k-1}, \mathbf{x}_{m-1,k})$ (see, e.g., (11.21)). Even if the efficient *forward–backward* algorithm (11.29), (11.30) is applied, the computation of the marginal distribution $p(\mathbf{x}_m^{(i)}, \hat{\underline{\mathbf{x}}}_1^k)$ according to (11.27) is still computationally demanding. Further complexity savings become possible by exploiting the auto- and cross-correlation separately via independent probability distributions $P(\mathbf{x}_{m,k}|\mathbf{x}_{m,k-1})$ (temporal Markov property) and $P(\mathbf{x}_{m,k}|\mathbf{x}_{m-1,k})$ (spatial Markov property).

A couple of solutions for such near-optimal MMSE parameter estimators have been proposed in the literature, e.g., [Adrat et al. 2000], [Lahouti, Khandani 2001], [Kliewer, Görtz 2001], [Hindelang 2001], [Fingscheidt et al. 2002], [Adrat 2003], [Adrat et al. 2004]. While most of these approaches exhibit comparable complexity, they reveal different performances. The most capable approach, which was first introduced in [Hindelang 2001], [Fingscheidt et al. 2002] and afterwards analyzed in detail in [Adrat 2003], [Adrat et al. 2004], will be reviewed next.

Again, the formal solution (11.18) for the computation of the marginal distribution $p(\mathbf{x}_m^{(i)}, \hat{\underline{\mathbf{x}}}_1^K)$ serves as starting point. Notice, in the following considerations a look-ahead of K future frames $\hat{\underline{\mathbf{x}}}_{k+1}, \ldots, \hat{\underline{\mathbf{x}}}_{k+K}$ is assumed to be acceptable. If the resulting additional delay of K frames is not tolerable in a real-world application (see also Footnote 2 on page 316 as well as Sec. 10.7), it can be avoided by setting $K = 0$.

In contrast to (11.19), the factorization of the overall joint PDF $p(\underline{\mathbf{x}}_1^K, \hat{\underline{\mathbf{x}}}_1^K)$ is realized on complete frames $\underline{\mathbf{x}}_k$,

$$p(\underline{\mathbf{x}}_1^K, \hat{\underline{\mathbf{x}}}_1^K) = p(\underline{\mathbf{x}}_{k+1}^K, \hat{\underline{\mathbf{x}}}_{k+1}^K \mid \underline{\mathbf{x}}_k, \hat{\underline{\mathbf{x}}}_k, \underline{\mathbf{x}}_1^{k-1}, \hat{\underline{\mathbf{x}}}_1^{k-1})$$
$$\cdot p(\underline{\mathbf{x}}_k, \hat{\underline{\mathbf{x}}}_k \mid \underline{\mathbf{x}}_1^{k-1}, \hat{\underline{\mathbf{x}}}_1^{k-1}) \cdot p(\underline{\mathbf{x}}_1^{k-1}, \hat{\underline{\mathbf{x}}}_1^{k-1}) . \quad (11.34)$$

Similarly to (11.20), assuming a memoryless transmission channel and a 1st order Markov property on frames $\underline{\mathbf{x}}_k$ permits us to simplify (11.34) to

$$p(\underline{\mathbf{x}}_1^K, \hat{\underline{\mathbf{x}}}_1^K) = p(\underline{\mathbf{x}}_{k+1}^K, \hat{\underline{\mathbf{x}}}_{k+1}^K \mid \underline{\mathbf{x}}_k) \cdot p(\underline{\mathbf{x}}_k, \hat{\underline{\mathbf{x}}}_k \mid \underline{\mathbf{x}}_{k-1}) \cdot p(\underline{\mathbf{x}}_1^{k-1}, \hat{\underline{\mathbf{x}}}_1^{k-1}) . \quad (11.35)$$

Substituting (11.35) into (11.18) allows us to reorganize the nested summations of the marginal distribution as

$$p(\mathbf{x}_{m,k}, \hat{\underline{\mathbf{x}}}_1^K) = \sum_{\forall \mathbf{x}_{1,k}^{m-1}} \sum_{\forall \mathbf{x}_{m+1,k}^M} \left\{ \left(\sum_{\forall \underline{\mathbf{x}}_{k+1}^K} p(\underline{\mathbf{x}}_{k+1}^K, \hat{\underline{\mathbf{x}}}_{k+1}^K \mid \underline{\mathbf{x}}_k) \right) \right.$$
$$\left. \cdot \sum_{\forall \underline{\mathbf{x}}_{k-1}} \left[p(\underline{\mathbf{x}}_k, \hat{\underline{\mathbf{x}}}_k \mid \underline{\mathbf{x}}_{k-1}) \cdot \left(\sum_{\forall \underline{\mathbf{x}}_1^{k-2}} p(\underline{\mathbf{x}}_1^{k-1}, \hat{\underline{\mathbf{x}}}_1^{k-1}) \right) \right] \right\} . \quad (11.36)$$

Both summations in the parentheses denote *forward*, respectively *backward* recursions on frames $\underline{\mathbf{x}}_k$, which can efficiently be calculated by

$$\alpha_{k-1}(\underline{\mathbf{x}}_{k-1}) \doteq \sum_{\forall \underline{\mathbf{x}}_1^{k-2}} p(\underline{\mathbf{x}}_1^{k-1}, \hat{\underline{\mathbf{x}}}_1^{k-1})$$
$$= \sum_{\forall \underline{\mathbf{x}}_{k-2}} p(\underline{\mathbf{x}}_{k-1}, \hat{\underline{\mathbf{x}}}_{k-1} \mid \underline{\mathbf{x}}_{k-2}) \cdot \alpha_{k-2}(\underline{\mathbf{x}}_{k-2}) \quad (11.37)$$

and

$$\beta_k(\underline{\mathbf{x}}_k) \doteq \sum_{\forall \underline{\mathbf{x}}_{k+1}^K} p(\underline{\mathbf{x}}_{k+1}^K, \hat{\underline{\mathbf{x}}}_{k+1}^K \mid \underline{\mathbf{x}}_k)$$
$$= \sum_{\forall \underline{\mathbf{x}}_{k+1}} p(\underline{\mathbf{x}}_{k+1}, \hat{\underline{\mathbf{x}}}_{k+1} \mid \underline{\mathbf{x}}_k) \cdot \beta_{k+1}(\underline{\mathbf{x}}_{k+1}) . \quad (11.38)$$

The only unknown term in (11.36), (11.37), and (11.38) is the innovation $p(\underline{\mathbf{x}}_k, \hat{\underline{\mathbf{x}}}_k \mid \underline{\mathbf{x}}_{k-1})$. It can be factorized according to the contributions of the preceding $\mathbf{x}_{1,k}^{m-1}$, the present $\mathbf{x}_{m,k}$ and the succeeding bit patterns $\mathbf{x}_{m+1,k}^M$ in the spatial direction m

$$\gamma_k(\underline{\mathbf{x}}_k, \underline{\mathbf{x}}_{k-1}) \doteq p(\underline{\mathbf{x}}_k, \hat{\underline{\mathbf{x}}}_k \mid \underline{\mathbf{x}}_{k-1})$$
$$= p(\mathbf{x}_{m+1,k}^M, \hat{\mathbf{x}}_{m+1,k}^M \mid \mathbf{x}_{1,k}^m, \hat{\mathbf{x}}_{1,k}^m, \underline{\mathbf{x}}_{k-1}) \quad (11.39)$$
$$\cdot p(\mathbf{x}_{1,k}^{m-1}, \hat{\mathbf{x}}_{1,k}^{m-1} \mid \mathbf{x}_{m,k}, \hat{\mathbf{x}}_{m,k}, \underline{\mathbf{x}}_{k-1}) \cdot p(\mathbf{x}_{m,k}, \hat{\mathbf{x}}_{m,k} \mid \underline{\mathbf{x}}_{k-1}) .$$

Taking a 1st order Markov property in the spatial dimension into account as well as the clear separation of the temporal and spatial dependencies $P(\mathbf{x}_{m,k}|\mathbf{x}_{m,k-1})$ (temporal Markov property) and $P(\mathbf{x}_{m,k}|\mathbf{x}_{m-1,k})$ (spatial Markov property) results in

$$
\gamma_k(\underline{\mathbf{x}}_k, \underline{\mathbf{x}}_{k-1}) = \overbrace{p(\mathbf{x}_{m,k}, \hat{\mathbf{x}}_{m,k} \mid \mathbf{x}_{m,k-1})}^{\text{temporal dependencies of } \mathbf{x}_{m,k}}
$$
$$
\underbrace{\cdot\, p(\mathbf{x}_{m+1,k}^M, \hat{\mathbf{x}}_{m+1,k}^M \mid \mathbf{x}_{m,k}) \cdot p(\mathbf{x}_{1,k}^{m-1}, \hat{\mathbf{x}}_{1,k}^{m-1} \mid \mathbf{x}_{m,k})}_{\text{spatial dependencies of } \mathbf{x}_{m,k}} . \qquad (11.40)
$$

The term in the first line of (11.40) considers temporal dependencies of bit patterns $\mathbf{x}_{m,k}$ and the terms in the second line the respective spatial dependencies. Inserting (11.37), (11.38), and (11.40) into (11.36) and rearranging the summations yields

$$
p(\mathbf{x}_{m,k}, \hat{\underline{\mathbf{x}}}_1^K) = \sum_{\forall \mathbf{x}_{m,k-1}} \left[p(\mathbf{x}_{m,k}, \hat{\mathbf{x}}_{m,k} \mid \mathbf{x}_{m,k-1}) \cdot \sum_{\forall \mathbf{x}_{1,k-1}^{m-1}} \sum_{\forall \mathbf{x}_{m+1,k-1}^M} \alpha_{k-1}(\underline{\mathbf{x}}_{k-1}) \right]
$$
$$
\cdot \sum_{\forall \mathbf{x}_{1,k}^{m-1}} \sum_{\forall \mathbf{x}_{m+1,k}^M} \left[\beta_k(\underline{\mathbf{x}}_k) \cdot p(\mathbf{x}_{m+1,k}^M, \hat{\mathbf{x}}_{m+1,k}^M \mid \mathbf{x}_{m,k}) \cdot p(\mathbf{x}_{1,k}^{m-1}, \hat{\mathbf{x}}_{1,k}^{m-1} \mid \mathbf{x}_{m,k}) \right] .
$$
$$
(11.41)
$$

Some parts of the nested structure of (11.36) have been resolved. The first line of (11.41) exhibits information about the impact of past frames $\underline{\mathbf{x}}_1^{k-1}$ on $\mathbf{x}_{m,k}$. The second line comprises knowledge of some possibly given future frames $\underline{\mathbf{x}}_{k+1}^K$ via the *backward* recursion $\beta_k(\underline{\mathbf{x}}_k)$ as well as of bit patterns in adjacent positions $\mathbf{x}_{1,k}^{m-1}$, $\mathbf{x}_{m+1,k}^M$.

In order to enhance the readability and comprehensibility we define

$$
\alpha_k^{[\mathrm{TIM}]}(\mathbf{x}_{m,k}) \doteq \sum_{\forall \mathbf{x}_{1,k}^{m-1}} \sum_{\forall \mathbf{x}_{m+1,k}^M} \alpha_k(\underline{\mathbf{x}}_k)
$$
$$
= \sum_{\forall \mathbf{x}_{m+1,k}^M} p(\mathbf{x}_{m+1,k}^M, \hat{\mathbf{x}}_{m+1,k}^M \mid \mathbf{x}_{m,k}) \cdot \sum_{\forall \mathbf{x}_{1,k}^{m-1}} p(\mathbf{x}_{1,k}^{m-1}, \hat{\mathbf{x}}_{1,k}^{m-1} \mid \mathbf{x}_{m,k})
$$
$$
\cdot \sum_{\forall \mathbf{x}_{m,k-1}} \left[p(\mathbf{x}_{m,k}, \hat{\mathbf{x}}_{m,k} \mid \mathbf{x}_{m,k-1}) \cdot \alpha_{k-1}^{[\mathrm{TIM}]}(\mathbf{x}_{m,k-1}) \right] \qquad (11.42)
$$
$$
= \alpha_m^{[\mathrm{POS}]}(\mathbf{x}_{m,k}) \cdot \beta_m^{[\mathrm{POS}]}(\mathbf{x}_{m,k})
$$
$$
\cdot \sum_{\forall \mathbf{x}_{m,k-1}} p(\mathbf{x}_{m,k}, \hat{\mathbf{x}}_{m,k} \mid \mathbf{x}_{m,k-1}) \cdot \alpha_{k-1}^{[\mathrm{TIM}]}(\mathbf{x}_{m,k-1})
$$

and

$$
\beta_k^{[\mathrm{TIM}]}(\mathbf{x}_{m,k}) \doteq \sum_{\forall \mathbf{x}_{m,k+1}} \alpha_m^{[\mathrm{POS}]}(\mathbf{x}_{m,k+1}) \cdot \beta_m^{[\mathrm{POS}]}(\mathbf{x}_{m,k+1})
$$
$$
\cdot p(\mathbf{x}_{m,k+1}, \hat{\mathbf{x}}_{m,k+1} \mid \mathbf{x}_{m,k}) \cdot \beta_{k+1}^{[\mathrm{TIM}]}(\mathbf{x}_{m,k+1}) \qquad (11.43)
$$

with

$$\alpha_m^{[\text{POS}]}(\mathbf{x}_{m,k}) \doteq \sum_{\forall \mathbf{x}_{m-1,k}} p(\mathbf{x}_{m-1,k}, \hat{\mathbf{x}}_{m-1,k} \mid \mathbf{x}_{m,k}) \cdot \alpha_{m-1}^{[\text{POS}]}(\mathbf{x}_{m-1,k}) \qquad (11.44)$$

$$\beta_m^{[\text{POS}]}(\mathbf{x}_{m,k}) \doteq \sum_{\forall \mathbf{x}_{m+1,k}} p(\mathbf{x}_{m+1,k}, \hat{\mathbf{x}}_{m+1,k} \mid \mathbf{x}_{m,k}) \cdot \beta_{m+1}^{[\text{POS}]}(\mathbf{x}_{m+1,k}) . \qquad (11.45)$$

The upper index in squared brackets indicates whether the main contribution to the reliability $p(\mathbf{x}_{m,k}, \hat{\mathbf{x}}_1^K)$ results from spatial "POS" or temporal "TIM" dependencies. For initialization serve $\alpha_k^{[\text{TIM}]}(\mathbf{x}_{m,k}) = P(\mathbf{x}_{m,k})$, $\beta_k^{[\text{TIM}]}(\mathbf{x}_{m,k}) = 1$, $\alpha_m^{[\text{POS}]}(\mathbf{x}_{m,k}) = 1$, and $\beta_m^{[\text{POS}]}(\mathbf{x}_{m,k}) = 1$.

Finally, with respect to (11.42) to (11.45) near-optimal MMSE parameter estimation (11.17) can be realized with

$$p(\mathbf{x}_m^{(i)}, \hat{\mathbf{x}}_1^K) = \beta_k^{[\text{TIM}]}(\mathbf{x}_m^{(i)}) \cdot \alpha_m^{[\text{POS}]}(\mathbf{x}_m^{(i)}) \cdot \beta_m^{[\text{POS}]}(\mathbf{x}_m^{(i)})$$
$$\cdot \sum_{j=0}^{2^{w_m}-1} p(\mathbf{x}_m^{(i)}, \hat{\mathbf{x}}_{m,k} \mid \mathbf{x}_m^{(j)}) \cdot \alpha_{k-1}^{[\text{TIM}]}(\mathbf{x}_m^{(j)}) . \qquad (11.46)$$

11.6 Illustrative Comparison

A *"jigsaw puzzle"* illustration helps one to understand the key differences between the optimal and near-optimal MMSE parameters estimators. Figure 11.5 shows such a comparison [Adrat et al. 2004].

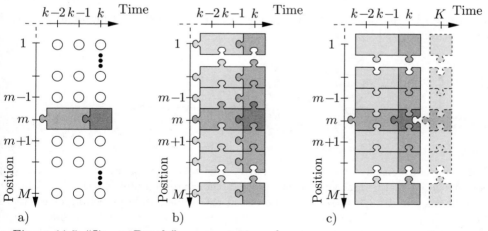

Figure 11.5: "Jigsaw Puzzle" representation of
  a) optimal exploitation of temporal dependencies (SD/AK1)
  b) optimal exploitation of temporal and spatial dependencies
  c) near-optimal exploitation of temporal and spatial dependencies

(SD/AK1) Estimator

The left-most subplot depicts the jigsaw puzzle for the basic extrapolative estimator "(SD/AK1)" introduced in Chap. 10 (see (10.32)). The (SD/AK1) estimator exploits a 1st order Markov property in the time dimension only. Some possibly given spatial dependencies cannot be utilized. The joint PDF $p(\mathbf{x}_{m,k}, \hat{\mathbf{x}}_1^k)$ can be determined from (10.32) as

$$p(\mathbf{x}_{m,k}, \hat{\mathbf{x}}_1^k) = \sum_{\forall \mathbf{x}_{m,k-1}} p(\mathbf{x}_{m,k}, \hat{\mathbf{x}}_{m,k} | \mathbf{x}_{m,k-1}) \cdot \alpha_{k-1}(\mathbf{x}_{m,k-1}). \qquad (11.47)$$

Past received patterns $\hat{\mathbf{x}}_1^{k-1}$ are evaluated in terms of a *forward* recursion $\alpha_{k-1}(\mathbf{x}_{m,k-1})$ (medium gray puzzle piece) while present information is exploited in terms of the innovation $p(\mathbf{x}_{m,k}, \hat{\mathbf{x}}_{m,k} | \mathbf{x}_{m,k-1})$ (dark gray puzzle piece). The interlocking element of both terms is the bit pattern $\mathbf{x}_{m,k-1}$, which is indicated by the specific forms of the two puzzle pieces (see "tabs" and "blanks"). The "tab" of the dark gray puzzle piece (innovation) perfectly fits into the "blank" of the medium gray puzzle piece (*forward* recursion).

The complexity demands of the (SD/AK1) estimator can also roughly be estimated from (11.47). The number of arithmetic operations (MULT, ADD, MAC, etc.) is mainly determined by the sum over all possible V_m realizations of $\mathbf{x}_{m,k-1}$. This summation has to be carried out for each $\mathbf{x}_{m,k} = \mathbf{x}_m^{(i)}$ with $i = 0, \ldots, V_m - 1$. The data memory demands are mainly characterized by the storage demand for the *a priori* parameter statistics $P(\mathbf{x}_{m,k} | \mathbf{x}_{m,k-1})$. In consequence, both terms of complexity (arithmetic operations as well as memory) exhibit demands of the order $O(V_m^2)$.

Optimal MMSE Estimator

The joint PDF $p(\mathbf{x}_{m,k}, \hat{\underline{\mathbf{x}}}_1^k)$ of the optimal MMSE parameter estimator of Sec. 11.4 can be summarized by

$$p(\mathbf{x}_{m,k}, \hat{\underline{\mathbf{x}}}_1^k) = \sum_{\forall \mathbf{x}_{m,k-1}} \Big\{ \beta_m(\mathbf{x}_{m,k}, \mathbf{x}_{m,k-1}) \cdot$$
$$\sum_{\forall \mathbf{x}_{m-1,k}} \sum_{\forall \mathbf{x}_{m-1,k-1}} \Big(p(\mathbf{x}_{m,k}, \hat{\mathbf{x}}_{m,k} | \mathbf{x}_{m,k-1}, \mathbf{x}_{m-1,k}) \cdot$$
$$P(\mathbf{x}_{m,k-1} | \mathbf{x}_{m-1,k-1}, \hat{\underline{\mathbf{x}}}_1^{k-1}) \cdot \alpha_{m-1}(\mathbf{x}_{m-1,k}, \mathbf{x}_{m-1,k-1}) \Big) \Big\}. \qquad (11.48)$$

This equation consists of mainly four parts. The term $p(\mathbf{x}_{m,k}, \hat{\mathbf{x}}_{m,k} | \mathbf{x}_{m,k-1}, \mathbf{x}_{m-1,k})$ is represented by the dark gray puzzle piece in the center subplot of Fig. 11.5. The bit pattern $\mathbf{x}_{m,k}$ under consideration depends on the immediately preceding elements in time $\mathbf{x}_{m,k-1}$ and position $\mathbf{x}_{m-1,k}$. Both dependencies are taken into account by the shape of the puzzle piece (see "tabs" and "blanks"). The medium gray puzzle piece to the left represents $P(\mathbf{x}_{m,k-1} | \mathbf{x}_{m-1,k-1}, \hat{\underline{\mathbf{x}}}_1^{k-1})$. The medium gray and light gray puzzle pieces to the top stand for $\alpha_{m-1}(\mathbf{x}_{m-1,k}, \mathbf{x}_{m-1,k-1})$ and the elements to the bottom for $\beta_m(\mathbf{x}_{m,k}, \mathbf{x}_{m,k-1})$. The thorough interlocking of all puzzle pieces provides a very robust system.

The number of arithmetic operations of optimal MMSE parameter estimation according to (11.48) is mainly determined by the nested summations. For each $\mathbf{x}_{m,k} = \mathbf{x}_m^{(i)}$ with $i = 0, \ldots, V_m - 1$ the nested summation runs over all combinations of $\mathbf{x}_{m,k-1}$ and $\mathbf{x}_{m-1,k}$. Similarly, the *a priori* knowledge $P(\mathbf{x}_{m,k} | \mathbf{x}_{m,k-1}, \mathbf{x}_{m-1,k})$ needs to be stored in a table. Thus, the optimal MMSE parameter estimator exhibits complexity demands of the order $O(V_m^3)$ (assuming that $V_{m-1} \approx V_m$).

Near-Optimal MMSE Estimator

The respective determination rule for the joint PDF of the near-optimal MMSE parameter estimator of Sec. 11.5 reads

$$p(\mathbf{x}_{m,k}, \hat{\underline{\mathbf{x}}}_1^K) = \beta_k^{[\text{TIM}]}(\mathbf{x}_{m,k}) \cdot \alpha_m^{[\text{POS}]}(\mathbf{x}_{m,k}) \cdot \beta_m^{[\text{POS}]}(\mathbf{x}_{m,k})$$

$$\cdot \sum_{\forall \mathbf{x}_{m,k-1}} p(\mathbf{x}_{m,k}, \hat{\mathbf{x}}_{m,k} | \mathbf{x}_{m,k-1}) \cdot \alpha_{k-1}^{[\text{TIM}]}(\mathbf{x}_{m,k-1}). \quad (11.49)$$

This equation consists of mainly five parts that can be interpreted as follows. In the right subplot of Fig. 11.5 the dark gray puzzle piece depicts again the innovation $p(\mathbf{x}_{m,k}, \hat{\mathbf{x}}_{m,k} | \mathbf{x}_{m,k-1})$ for the bit pattern $\mathbf{x}_{m,k}$ under consideration. Notice that $p(\mathbf{x}_{m,k}, \hat{\mathbf{x}}_{m,k} | \mathbf{x}_{m,k-1})$ is only conditioned on $\mathbf{x}_{m,k-1}$ which is precedent in time. Therefore, the puzzle piece has only one "tab". All the other four elements of (11.49) provide extra information for $\mathbf{x}_{m,k}$ from each direction. $\alpha_m^{[\text{POS}]}(\mathbf{x}_{m,k})$ represents the medium gray puzzle pieces to the top and $\beta_m^{[\text{POS}]}(\mathbf{x}_{m,k})$ the respective medium gray puzzle pieces to the bottom. The term $\alpha_{k-1}^{[\text{TIM}]}(\mathbf{x}_{m,k-1})$ stands for all medium gray and light gray puzzle pieces to the left while some possibly given future bit patterns are exploited by $\beta_m^{[\text{POS}]}(\mathbf{x}_{m,k})$ (medium gray and light gray puzzle pieces to the right).

Obviously, if compared with the optimal MMSE parameter estimator, some neighboring puzzle pieces are not connected anymore. This results from the fact that the 1st order Markov properties in time $P(\mathbf{x}_{m,k} | \mathbf{x}_{m,k-1})$ and in position $P(\mathbf{x}_{m,k} | \mathbf{x}_{m-1,k})$ are exploited separately. From this it follows that some robustness of the system might get lost.

Owing to the separate evaluation of the 1st order Markov properties, significant complexity savings become possible. The nested summations of the optimal MMSE parameter estimation process can be resolved, and (11.49) can be considered as a straightforward extension of (11.47). Thus, the complexity demands for both, arithmetic operations and memory, are of the same order $O(V_m^2)$ as for the (SD/AK1) approach. Of course, this complexity needs to be spent for each of the *forward–backward* recursions in time and position.

11.7 Simulation Results

The benefits of the (near-)optimal MMSE estimators over the basic extrapolative estimators introduced in Sec. 10.4.2 for vector sources with spatially and temporally correlated sources will be demonstrated by simulation. For this purpose, the generic source model depicted in Fig. 11.1 is used. The design parameters a and b of this source allow one to adjust auto- and cross-correlation $\rho(1)$ and $\delta(0)$ properties that resemble those of real-world source encoders. Correlation measurements have shown [Heinen 2001b], [Adrat 2003] that the scale factors of audio transform codes as applied in the *Digital Audio Broadcasting* (DAB) system exhibit correlation of $\rho(1) = 0.95$ and $\delta(0) = 0.8$. Differentially encoded *Line Spectral Frequencies* (LSFs) of speech codecs as applied in the *Global System for Mobile communication* (GSM, *Adaptive Multirate Codec* (AMR)) exhibit a residual correlation of $\rho(1) = 0.6$ and $\delta(0) = 0.75$.

For the simulations, the number of parameters $v_{m,k}$ per frame $\underline{v}_k$ is set to $M = 10$. After scalar quantization of each parameter by a $V_m = 16$-level Lloyd–Max quantizer an MMSE optimized index assignment is used [Heinen, Vary 2000], [Heinen 2001b] (see also Chap. 12, Sec. 12.3). A transmission channel serves AWGN with known E_s/N_0.

Figure 11.6 depicts the corresponding simulation results. On the left, the results for the DAB-like settings are shown and, on the right, the results for the GSM-like

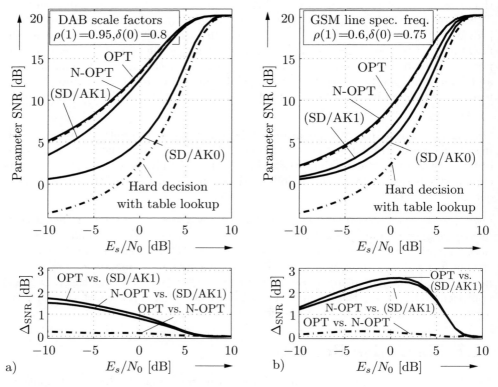

Figure 11.6: Simulation results for
        a) DAB-like settings: auto-corr. $\rho(1) = 0.95$, cross-corr. $\delta(0) = 0.8$
        b) GSM-like settings: auto-corr. $\rho(1) = 0.6$, cross-corr. $\delta(0) = 0.75$

settings. It can be seen that both the optimal (OPT) and the near-optimal (N-OPT) MMSE parameter estimator reveal significant improvements in the *parameter* SNR if compared with the basic estimators *hard decision* by table lookup, *soft decision source decoding* (SDSD) with 0th order *a priori* knowledge (SD/AK0), 1st order *a priori* knowledge (SD/AK1), respectively. In the case of the DAB-like settings the maximum parameter SNR gain of OPT over (SD/AK1) is $\Delta_{SNR} = 1.72$ dB; in the case of the GSM-like settings $\Delta_{SNR} = 2.66$ dB. The maximum loss of N-OPT if compared with OPT is $\Delta_{SNR} = 0.21$ dB (DAB-like), respectively $\Delta_{SNR} = 0.24$ dB (GSM-like), which can be considered as negligibly small.

11.8 Conclusions

In this chapter, we extended the basic concept of *soft decision source decoding* (SDSD). While the basic extrapolative estimators presented in Sec. 10.4, are optimal in the MMSE sense for parameters with temporal correlation, the two new MMSE parameter estimators detailed in this chapter are optimal for vector sources that exhibit temporal and spatial correlation. The respective source model has been introduced and efficient *forward–backward* algorithms for the optimal (OPT) and the near-optimal (N-OPT) MMSE parameter estimation have been derived. The key difference between OPT and N-OPT is the utilization of source redundancy. On the one hand, spatial and temporal correlation are exploited jointly (OPT). On the other, both terms of redundancy are utilized separately (N-OPT). The latter solution offers the potential for significant complexity savings. It has been demonstrated by simulation that both approaches reveal substantial parameter SNR gains over the basic extrapolative estimators. The loss of N-OPT compared with OPT is negligibly small.

In the next chapter, it will be shown that the error correcting capabilities of all SDSD approaches introduced in Chaps. 10 and 11 can be enhanced further if they are used in combination with so called *source optimized channel codes* (SOCC).

Bibliography

Adrat, M. (2003). *Iterative Source-Channel Decoding for Digital Mobile Communications*, PhD thesis. Aachener Beiträge zu digitalen Nachrichtensystemen, vol. 16, P. Vary (ed.), RWTH Aachen University.

Adrat, M.; Picard, J.-M.; Vary, P. (2004). Efficient Near-Optimum Softbit Source Decoding for Sources with Inter- and Intra-Frame Redundancy, *IEEE International Conference on Acoustics, Speech, and Signal Processing (ICASSP)*, Montreal, Canada.

Adrat, M.; Spittka, J.; Heinen, S.; Vary, P. (2000). Error Concealment by Near-Optimum MMSE-Estimation of Source Codec Parameters, *IEEE Workshop on Speech Coding*, Delavan, Wisconsin, USA, pp. 84–86.

Bahl, L. R.; Cocke, J.; Jelinek, F.; Raviv, J. (1974). Optimal Decoding of Linear Codes for Minimizing Symbol Error Rate, *IEEE Transactions on Information Theory*, vol. 20, no. 2, pp. 284–287.

Cover, T. M.; Thomas, J. A. (1991). *Elements of Information Theory*, John Wiley & Sons, Ltd, Chichester.

ETSI, Standard ETS 300 401 (1997). *Radio Broadcasting Systems; Digital Audio Broadcasting (DAB) to Mobile, Portable, and Fixed Receivers.*

ETSI TC-SMG (1999). *Recommendation 06.90: Digital Cellular Telecommunications System (Phase 2+); Adaptive Multi Rate (AMR) Speech Transcoding.*

Fingscheidt, T. (1998). *Softbit-Sprachdecodierung in digitalen Mobilfunksystemen*, PhD thesis. Aachener Beiträge zu digitalen Nachrichtensystemen, vol. 9, P. Vary (ed.), RWTH Aachen University (in German).

Fingscheidt, T.; Hindelang, T.; Cox, R. V.; Seshadri, N. (2002). Joint Source-Channel (De-)Coding for Mobile Communication, *IEEE Transactions on Communications*, vol. 50, no. 2, pp. 200–212.

Fingscheidt, T.; Vary, P. (1997). Robust Speech Decoding: A Universal Approach to Bit Error Concealment, *IEEE International Conference on Acoustics, Speech, and Signal Processing (ICASSP)*, Munich, Germany, vol. 3, pp. 1667–1670.

Fingscheidt, T.; Vary, P. (2001). Softbit Speech Decoding: A New Approach to Error Concealment, *IEEE Transactions on Speech Audio Processing*, vol. 9, no. 3, pp. 240–251.

Heinen, S. (2001a). An Optimal MMSE-Estimator For Source Codec Parameters Using Intra-Frame and Inter-Frame Correlation, *IEEE International Symposium on Information Theory (ISIT)*, Washington, USA.

Heinen, S. (2001b). *Quellenoptimierter Fehlerschutz für digitale Übertragungskanäle*, PhD thesis. Aachener Beiträge zu digitalen Nachrichtensystemen, vol. 14, P. Vary (ed.), RWTH Aachen University (in German).

Heinen, S.; Vary, P. (2000). Source Optimized Channel Codes (SOCCs) for Parameter Protection, *IEEE International Symposium on Information Theory (ISIT)*, Sorrento, Italy.

Hindelang, T. (2001). *Source-Controlled Channel Encoding and Decoding for Mobile Communications*, PhD thesis, VDI-Verlag, Düsseldorf: Fortschritt-Berichte VDI, Reihe 10, Nr. 695, ISBN 3-18-369510-3, Munich University of Technology.

Kliewer, J.; Görtz, N. (2001). Soft-Input Source Decoding for Robust Transmission of Compressed Images Using Two-Dimensional Optimal Estimation, *IEEE International Conference on Acoustics, Speech, and Signal Processing (ICASSP)*, Salt Lake City, Utah, USA.

Lahouti, F.; Khandani, A. K. (2001). Approximating and Exploiting the Residual Redundancies - Applications to Efficient Reconstruction of Speech over Noisy Channels, *IEEE International Conference on Acoustics, Speech, and Signal Processing (ICASSP)*, Salt Lake City, Utah, USA.

Melsa, J. L.; Cohn, D. L. (1978). *Decision and Estimation Theory*, McGray-Hill, New York.

Chapter 12

Source Optimized Channel Codes & Source Controlled Channel Decoding

Stefan Heinen, Thomas Hindelang

12.1 Introduction

Speech, audio, image or video signals are highly correlated and source coding is an essential means to remove redundancy and achieve a high bandwidth efficiency. However, since the source encoder always underlies complexity and delay constraints, redundancy in general can only be imperfectly eliminated. Chapters 10 and 11 coped with the utilization of residual redundancy in the course of *source decoding*. *Channel coding*, which has so far been hidden as a black box in the equivalent channel, is now highlighted. We present two approaches exploiting residual redundancy in channel (de)coding.

First we introduce so-called *Source Optimized Channel Coding* (SOCC). Other than classical channel coding, which primarily aims at a minimum residual bit error rate, the fundamental idea of SOCC is to tailor channel codes entirely with respect to the needs of the source, i.e., to take into account the statistical properties of the source as well as a source-related optimization criterion such as the maximization of the parameter *Signal-to-Noise Ratio* (SNR) after decoding. We will demonstrate in the following that this new design paradigm leads to a new class of powerful non-linear block codes. Related work has been carried out by Zeger and Gersho [Zeger, Gersho 1990] who proposed an algorithm for optimization of non-redundant index assignments and by Skoglund who developed *Channel Constrained Vector Quantization* [Skoglund 1999].

Advances in Digital Speech Transmission
© 2008 John Wiley & Sons, Ltd

Edited by R. Martin, U. Heute and C. Antweiler

Farvardin's *Channel Optimized Vector Quantization* (COVQ) [Farvardin, Vaisham-payan 1987], [Farvardin, Vaishampayan 1991], [Farvardin 1990] will be an issue later in this chapter.

The second approach considered here is *Source Controlled Channel Decoding* (SCCD) published by Hagenauer in 1995 [Hagenauer 1995]. This work describes how con-volutional channel decoding can be improved by exploiting residual redundancy (in terms of *a priori* information) of single parameter bits. Owing to the memory of the channel code not only the parameter itself, and the bits originating from the parameter contribute to the estimation but also the neighboring bits of other param-eters. In contrast to SDSD shown in Chap. 10, here an additional diversity gain can be achieved that allows one to improve the quality of parameters without residual redundancy if their bits are placed close to bits of redundant parameters by a respec-tive interleaver. To improve the quality of signals further, *Unequal Error Protection* (UEP), suitable bit-mapping, and interleaving are applied, which take advantage of the properties of the source or the source code and also of the channel. The start-ing point for investigations in SCCD were channel decoding algorithms, which deliv-ered a reliability of each decoded symbol – so-called Soft-Input/Soft-Output decoders [Bahl et al. 1974], [Battail 1987], [Hagenauer, Höher 1989]. Many publications fo-cused on channel decoding exploiting source statistics to reduce the residual path, symbol, or bit error rate [Boudreau, Dubuc 1998], [Alajaji et al. 1996], [Fazel, Fuja 2000].

In both approaches discussed in this chapter, channel and source decoding are con-sidered under the constraint of typical communication systems where typically delay and complexity constraints limit the length of frames.

This chapter is structured as follows. In Sec. 12.2 we extend the transmission system from Chap. 10 to a system that additionally includes a channel encoder and decoder. Section 12.3 is devoted to the definition, design and analysis of SOCC and a brief comparison with COVQ. In Sec. 12.4 the SCCD approach is presented and extended to operate on parameters (bit groups) rather than single bits. Finally, in Sec. 12.5, the two techniques are compared by applying them to a representative model of a transmission system.

12.2 The Transmission System Used as Reference

Source

Let us assume a speech or an image signal s being source coded on a frame-by-frame basis. The source coder usually generates codec *parameters*, e.g., pitch, spectral co-efficients, etc. In the following, we focus on a parameter vector $\tilde{v} \in \mathbb{R}^M$ generated every frame k as depicted in Fig. 12.1. The M different dimensions of $\tilde{v}$ can basi-cally contain different types of codec parameters (such as, for example, stochastic or adaptive codebook gain factors after source coding), or they can simply describe

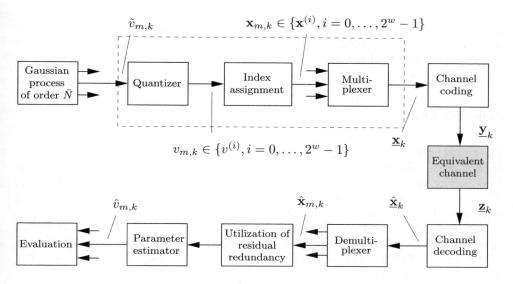

Figure 12.1: A transmission system with a set of Gaussian distributed and temporally correlated parameters used as reference

the same physical signal at different times (e.g., a PCM signal where a definite number M of samples is combined to a frame). If we consider only one arbitrary frame k, we neglect the index k and speak of the parameter vector $\underline{\tilde{v}}$, which contains M parameters $\tilde{v}_m$.

In this reference transmission system we assume all parameters $\tilde{v}_m$ of the parameter vector $\underline{\tilde{v}}$ to be samples of a zero mean and unit variance autoregressive Gaussian process of order $\bar{N} \in \{0, 1\}$, named the AR($\bar{N}$) process of 0th or 1st order as introduced in Sec. 10.4. The parameters are independent of each other ($\tilde{v}_m \neq f(\tilde{v}_i)$ for $i = 1, \ldots, m-1, m+1, \ldots, M$).

Source Coding

Each value continuous parameter $\tilde{v}_m$ is quantized, where we denote the discrete set of values of one quantized parameter with v_m. Each parameter v_m can take V_m levels, which could be represented in binary notation with w_m bits. For the SCCD approach we assume the number of levels V_m being a power of 2 and the same number V_m of levels and thus the same number $w_m = w$ of bits for all parameters. This is not mandatory and especially for the SOCC approach an arbitrary number of levels V_m is applied depending on the respective code design.

Coding and Decoding

The mapping of each quantized parameter v_m onto a bit vector $\mathbf{x}_m$ and also the following multiplexing and coding depends on the chosen approach – SOCC or SCCD – and will be shown in the respective sections.

Also the decoding and the exploitation of *a priori* knowledge differs between the two approaches and will be explained later on. For the SOCC approach even the blocks "Utilization of Residual Redundancy", "Demultiplexer", and "Channel Decoding" can not be seen separately.

Equivalent Channel

The transmission of the channel bits vector $\mathbf{y}$, which comprises a second interleaver, a modulator, a physical channel, a soft demodulation, and a second de-interleaver is described by an equivalent lowpass channel in the same way as in Sec. 10.2.2. However, in between the index assignment and the equivalent channel there is a multiplexing and channel coding unit. For the further decoding approach we assume that the channel delivers – in addition to the received sequence $\underline{\mathbf{z}}_k$ – the reliability information about each received bit either in the form of bit error probabilities as in (10.3) or in the form of log-likelihood ratios [Hagenauer 1995], which allow one to obtain so-called transition probabilities $P(z_k(\kappa)|y_k(\kappa))$.

Parameter Estimation and Evaluation

As long as it is not stated otherwise, the estimated parameter after source decoding is simply the most probable index vector $\mathbf{x}_m$ for each parameter $\hat{v}_m$:

$$\hat{v}_m = \mathcal{M}^{-1}(\arg \max_{\mathbf{x}_m \in \{0,1\}^w} P_d(\mathbf{x}_m)) , \tag{12.1}$$

where P_d denotes the decoding probabilities that will be explained in Sec. 12.4.1 and $\mathcal{M}^{-1}$ is found by a table look up, which was introduced as hard decision decoding in Sec. 10.2.3.

For evaluation, two performance measurements are applied. On the one hand we use the bit error rate (BER) after channel decoding and on the other hand we exploit the parameter SNR according to (10.8).

12.3 Source Optimized Channel Coding (SOCC)

Classical forward error correction coding in general is applied to protect streams of bits and aims at minimizing the residual bit error rate. In contrast, *Source Optimized Channel Coding* (SOCC) deals with *signal* or *parameter* values rather than bits. The key innovation with SOCC is that channel codes are tailored in such a way that the transmission *quality* is maximized. Owing to its practical relevance, but without loss of generality, we use here the parameter SNR as a quality measure. As informal quality assessments show, the SNR of the codec parameters often corresponds well with perceived quality.

12.3.1 Definition

To introduce *Source Optimized Channel Codes* let us briefly recapitulate the considered transmission chain. As a model of the source encoder, we assume a source producing M-dimensional real-valued vectors $\tilde{v}$. The vector elements are either individually or jointly (in terms of a vector quantization) quantized to $\check{V}$ reproduction values $v \in \mathbb{V}$ having distribution $P(v)$. The reproduction values are mapped to binary code vectors $\mathbf{y} = \mathbf{\Gamma}(v)$ and transmitted over a memoryless channel $\mathbf{z} = \mathbf{\Xi}(\mathbf{y})$, which is described by a probabilistic transfer function $p_{\mathbf{z}|\mathbf{y}}(\mathbf{z}|\mathbf{y})$. Finally, the decoder recovers the parameter sets $\hat{v} = \mathbf{f}(\mathbf{z})$ from the disturbed received bit vectors $\mathbf{z}$.

Given this model of a transmission system, we define a *Source Optimized Channel Code* $\mathbb{C}^*$ as the set of codewords

$$\mathbb{C}^* = \{\, \mathbf{y} \,|\, \mathbf{y} = \mathbf{\Gamma}(v),\ v \in \mathbb{V} \,\} = \{\, \mathbf{y}^{(0)}, \ldots, \mathbf{y}^{(\check{V}-1)} \,\}\,, \tag{12.2}$$

which results from solving the optimization problem

$$\min_{\mathbf{\Gamma}} \mathrm{E}\{\, \mathcal{D}[v, \mathbf{f} \circ \mathbf{\Xi} \circ \mathbf{\Gamma}(v)]\,\}\,, \tag{12.3}$$

where the symbol "$\circ$" denotes concatenation of the involved functions and $\mathcal{D}$ some quality measure. In particular, as we aim at maximizing the SNR we get

$$\min_{\mathbf{\Gamma}} \mathrm{E}\{\, \|v - \mathbf{f} \circ \mathbf{\Xi} \circ \mathbf{\Gamma}(v)\|^2 \,\}\,. \tag{12.4}$$

As long as a SOCC is a true subset of the set $\mathbb{Y} = \{0,1\}^B$ of all possible bit combinations of length $B = \sum_{m=0}^{M-1} w_m$, i.e., $\mathbb{C}^* \subset \mathbb{Y}$ the mapping $\mathbf{\Gamma}$ is redundancy increasing. In the case of $\mathbb{C}^* = \mathbb{Y}$ (12.2) degenerates to a non-redundancy increasing index assignment.

We do not impose any constraints concerning the structure of the code, so in general the code will not be linear, i.e., it will not fulfill the formal condition

$$\mathbf{y}^{(i)} \oplus \mathbf{y}^{(j)} \in \mathbb{C}^* \quad \forall\ \mathbf{y}^{(i)}, \mathbf{y}^{(j)} \in \mathbb{C}^*\,, \tag{12.5}$$

where "$\oplus$" denotes bit-wise modulo-2 addition.

In contrast to binary linear codes, whose number of codewords is always a power of two, the code size $\check{V}$ of a SOCC and thus the number of quantizer reproduction values can freely be chosen, e.g., to satisfy a given minimum quality requirement. As will be discussed later in detail, the code size $\check{V}$ is in fact one of the key parameters in SOCC design, which allows us to trade-off quantization accuracy against error-protecting redundancy at a fine level of granularity.

12.3.2 Decoding of Source Optimized Channel Codes

The objective of this section is to derive the optimum decoding algorithm for SOCCs. In the course of the derivation we will notice many similarities with *Soft Decision Source Decoding* (SDSD), which allows us to reuse some of the results from Chap. 10. For this reason we restrict the derivation here to the case of a memoryless parameter source. The extension to the case of correlated sources can straightly be transferred from SDSD to SOCC decoding.

A SOCC decoder is not a channel decoder in the traditional sense. There are no path metrics, error correction or similar. SOCC decoding in fact means estimation of a parameter set $\hat{\underline{v}}$ from the noisy observation $\mathbf{z}$, i.e., $\hat{\underline{v}} = \mathbf{f}(\mathbf{z})$. Since we equated *quality* with a high SNR, the optimal estimator should minimize the mean square error $\mathrm{E}\{\|\tilde{\underline{v}} - \hat{\underline{v}}\|^2\}$. According to estimation theory this estimator is formally given by [Melsa, Cohn 1978]

$$\hat{\underline{v}} = \mathrm{E}\{\tilde{\underline{v}} \,|\, \mathbf{z}\} = \frac{1}{p(\mathbf{z})} \int\limits_{\mathbb{R}^M} \tilde{\underline{v}} \, p(\tilde{\underline{v}}, \mathbf{z}) \, \mathrm{d}\tilde{\underline{v}} \,, \tag{12.6}$$

where $\mathbb{R}^M$ represents the M-dimensional real-valued space. We expand the joint PDF in (12.6) to

$$p(\tilde{\underline{v}}, \mathbf{z}) = \sum_{i=0}^{\check{V}-1} p(\mathbf{z} \,|\, \not{\tilde{\underline{v}}}, \underline{v}^{(i)}) \, P(\underline{v}^{(i)} \,|\, \tilde{\underline{v}}) \, p(\tilde{\underline{v}}) \,, \tag{12.7}$$

where $\tilde{\underline{v}}$ can be omitted in the conditional PDF as knowledge of the quantized parameter set $\underline{v}^{(i)}$ is a sufficient condition. Furthermore, since the channel encoder mapping $\mathbf{y} = \mathbf{\Gamma}(\underline{v})$ is deterministic, the equivalence $p(\mathbf{z} \,|\, \underline{v}^{(i)}) = p(\mathbf{z} \,|\, \mathbf{y}^{(i)})$ holds. Thus, we have expressed the first term of the sum by the known channel statistics.

Let us now consider the remaining two terms. The given quantizer Q establishes the partitioning $\mathbb{R}^M = \mathbb{Q}_1 \cup \mathbb{Q}_2 \cup \cdots \cup \mathbb{Q}_{\check{V}}, \tilde{\underline{v}} \in \mathbb{R}^M$ with the quantization regions $\mathbb{Q}_i$, and without loss of generality we assume that the reproduction values $\underline{v}^{(i)}$ are centroids of their respective quantization cells. Considering that the partitioning implies

$$P(\underline{v}^{(i)} \,|\, \tilde{\underline{v}}) = \begin{cases} 1 & \tilde{\underline{v}} \in \mathbb{Q}_i \\ 0 & \text{else} \end{cases} \tag{12.8}$$

and by substituting (12.7) back into (12.6), we obtain

$$\hat{\underline{v}} = \frac{1}{p(\mathbf{z})} \sum_{i=0}^{\check{V}-1} p(\mathbf{z} \,|\, \mathbf{y}^{(i)}) \int\limits_{\tilde{\underline{v}} \in \mathbb{Q}_i} \tilde{\underline{v}} \, p(\tilde{\underline{v}}) \, \mathrm{d}\tilde{\underline{v}} \,. \tag{12.9}$$

By comparison with the definition of centroids

$$\underline{v}^{(i)} = \mathrm{E}\{\,\tilde{\underline{v}}\,|\,\tilde{\underline{v}} \in \mathbb{Q}_i\,\} = \frac{\int\limits_{\tilde{\underline{v}} \in \mathbb{Q}_i} \tilde{\underline{v}}\,p(\tilde{\underline{v}})\,\mathrm{d}\tilde{\underline{v}}}{P(\underline{v}^{(i)})} \tag{12.10}$$

we can express the remaining integral term by known quantities and finally get the optimum decoder function

$$\hat{\underline{v}} = \mathbf{f}(\mathbf{z}) = \frac{1}{p(\mathbf{z})} \sum_{i=0}^{\check{V}-1} \underline{v}^{(i)}\, p(\mathbf{z}\,|\,\mathbf{y}^{(i)})\, P(\underline{v}^{(i)})\,. \tag{12.11}$$

12.3.3 Design of Source Optimized Channel Codes

We now return to the initial optimization problem (12.3). So far we have considered the estimator function $\mathbf{f}(\mathbf{z})$ and found an optimal solution with respect to the overall optimization criterion. Our actual task, the optimization of the channel encoder $\boldsymbol{\Gamma}$, is still pending. The idea of solving it by an exhaustive search over all possible encoder mapping only works out for codeword lengths up to four bits, since there are quite a lot of possibilities of mapping $\check{V}$ reproduction values to bit vectors of length $B = \sum_{m=0}^{M-1} w_m$ with 2^B combinations. To be precise, in total $2^B!/(2^B - \check{V})!$ mappings exist. Let's give an example: for $\check{V} = 16$ and $B = 5$ we have about $1.26 \cdot 10^{22}$. For comparison, a 32-bit integer can represent numbers up to $4.3 \cdot 10^9$. Even though from the coding point of view some of the mappings are equivalent (e.g., to each mapping there exists a bit-inverted mapping with the same coding properties) the number is still huge.

Therefore, we apply for the design of SOCCs a suboptimal search algorithm, which delivers at least a good local optimum in reasonable time. A powerful means to optimize the encoder mapping is the *Binary Switching Algorithm* (BSA) proposed by Zeger and Gersho [Zeger, Gersho 1990]. This algorithm iteratively approaches a local minimum of (12.4) by repeatedly swapping the code words assigned to a selected pair of reproduction values $\underline{v}^{(\ell)}, \underline{v}^{(j)}$. Let $\boldsymbol{\Gamma}$ and $\boldsymbol{\Gamma}'$ represent the mapping before and after a swap, then one swap iteration is characterized by

$$\boldsymbol{\Gamma}'(\underline{v}^{(i)}) = \begin{cases} \boldsymbol{\Gamma}(\underline{v}^{(i)}) & i \neq \ell,\, i \neq j \\ \boldsymbol{\Gamma}(\underline{v}^{(\ell)}) & i = j \\ \boldsymbol{\Gamma}(\underline{v}^{(j)}) & i = \ell\,. \end{cases} \tag{12.12}$$

Note that swapping a code word $\mathbf{y}^{(k)} \in \mathbb{C}$ with some element from the pool of unused bit combinations $\mathbf{y} \in \mathbb{Y}\backslash\mathbb{C}$ is allowed as well, which enables modifications not only of the mapping $\boldsymbol{\Gamma}$ but also of the code $\mathbb{C}$.

12.3.4 Numerical Aspects of SOCC Design

During the search with the BSA the major computational burden is generated by evaluation of the target function (12.4). Hence, it is crucial to bring it into a form that can be computed with low effort. We substitute the optimum decoder function (12.11) into the expectation (12.4) and obtain after some transformations [Heinen, Vary 2000]

$$\min_{\mathbf{\Gamma}} \mathrm{E}\big\{ \|\underline{v} - \mathbf{f} \circ \mathbf{\Xi} \circ \mathbf{\Gamma}(\underline{v})\|^2 \big\} = \sum_{i=0}^{\check{V}-1} \|\underline{v}^{(i)}\|^2 \, P(\underline{v}^{(i)}) - \int\limits_{\mathbf{z}\in\mathbb{R}^B} \|\mathbf{f}(\mathbf{z})\|^2 \, p(\mathbf{z}) \, \mathrm{d}\mathbf{z} \ . \quad (12.13)$$

As the sum in (12.13) does not depend on $\mathbf{\Gamma}$, a sufficient condition for the optimality of a SOCC is the maximization of the integral term. The integration in (12.13) can be approximated[1] by a sum over a discretized version of $\mathbf{z}$. A closer analysis shows [Heinen 2001] that the error due to this approximation has only a little impact on the performance of the search algorithm. In fact, two discrete levels per vector element of $\mathbf{z}$ are sufficient to achieve maximum performance. This is due to the BSA's property to select a mapping $\mathbf{\Gamma}$ by *relative* comparisons. Although the absolute value of the target function (12.3) is changed by the quantization, the relative relations remain almost the same.

Let $\bar{\mathbf{z}} \in \{\bar{\mathbf{z}}^{(1)}, \ldots, \bar{\mathbf{z}}^{(2^B)}\}$ be the set of discretized reception vectors with elements $\bar{z} \in \{-1, +1\}$. Then, the optimization criterion (12.3) can be reformulated in the numerically evaluable form

$$\max_{\mathbf{\Gamma}} \sum_{j=0}^{2^B-1} \|\mathbf{f}(\bar{\mathbf{z}}^{(j)})\|^2 \, P(\bar{\mathbf{z}}^{(j)}) = \max_{\mathbf{\Gamma}} \sum_{j=0}^{2^B-1} \frac{\big\| \sum_{i=0}^{\check{V}-1} \underline{v}^{(i)} \, P(\bar{\mathbf{z}}^{(j)} \,|\, \underline{\mathbf{y}}^{(i)}) \, P(\underline{v}^{(i)}) \big\|^2}{P(\bar{\mathbf{z}}^{(j)})} \ . \quad (12.14)$$

However, evaluation of (12.14) is still computationally demanding. Since the BSA modifies in each iteration only a small part of the entire mapping $\mathbf{\Gamma}$, reevaluation of (12.14) each time from scratch wastes computing power. Instead, an update-oriented computation can be applied, which leads to the efficient realization of a SOCC design algorithm proposed in [Heinen, Vary 2000].

12.3.5 Bit Allocation between Source and Channel Coding

The optimization (12.4) determines an optimal SOCC $\mathbb{C}^*$ and the corresponding mapping $\mathbf{\Gamma}$ for a given operating point, which is characterized by the number of reproduction values $\check{V}$, the transmission rate $A = B/M$ (number of channel bits in a transmission block / number of parameters per set) and the E_s/N_0 on the channel. In this case $\check{V}$ is a fix constraint for the optimization.

---

[1] Note that this approximation is only necessary for the code search. For the actual estimation at the receiver it is not required as the estimator can cope with continuous $\mathbf{z}$.

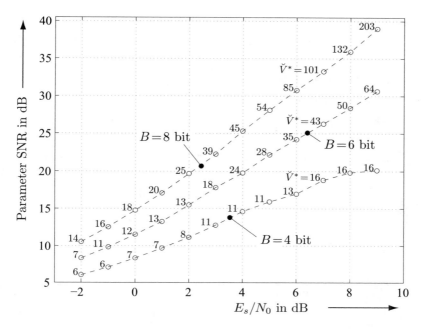

Figure 12.2: SOCC SNR performance and optimal number of reproduction values $\check{V}^*$; individual code design and optimization of $\check{V}$ for each simulation point; white Gaussian parameter source, $\sigma_{\tilde{v}}^2 = 1$; Lloyd–Max quantizer; gross data rate: $A = B/1$ bit per parameter value $\tilde{v}_m$; AWGN channel.
© 2005 IEEE

However, as mentioned earlier, we can also interpret $\check{V}$ as an additional *design* parameter, which is optimized together with $\mathbb{C}^*$ and $\mathbf{\Gamma}$ subject to the criterion (12.4). Then we can write

$$\check{V}^* = \arg\min_{\check{V}} \left[\min_{\mathbf{\Gamma}} \mathrm{E}\{\, \|\underline{\tilde{v}} - \underline{\hat{v}}\| \,\} \right] . \tag{12.15}$$

Figure 12.2 shows the optimization results for the example of a scalar Gaussian distributed parameter $\tilde{v}$ encoded by codewords of lengths $B \in \{4, 6, 8\}$. The figures next to the different considered operating points represent the optimum number of reproduction values / codewords found by (12.15). Without changing the effective code rate, a further SNR improvement can be achieved if two or more parameters are encoded together by one single SOCC. For example $B = 12$ and $M = 2$ would also result in a transmission rate of $A = 6/1$ bits per parameter. But simulations prove that in this case of a joint coding a better SNR can be achieved than with parameter-individual coding as in the middle curve in Fig. 12.2, which results from the improved error protection of the longer channel code. For details refer to [Heinen 2001].

As an interesting side-effect the optimization (12.15) implicitly determines the optimum bit allocation between source and channel coding, since $\log_2 \check{V}^*$ bits are assigned to the source coding while the remaining $B - \log_2 \check{V}^*$ bits contribute to the channel coding. Figure 12.3 displays the resulting coding rates $\log_2 \check{V}^*/B$ as a function of

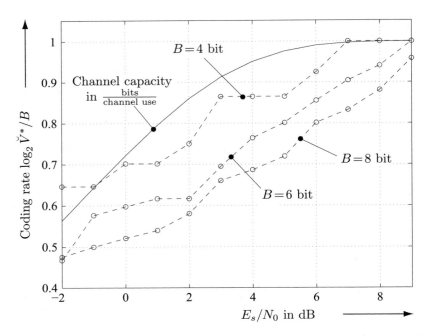

Figure 12.3: SOCC coding rates as function of E_s/N_0; individual code design for
each simulation point; white Gaussian parameter source, $\sigma_{\check{v}}^2 = 1$; AWGN
channel; gross data rate: $A = B/1$ bit per parameter value. © 2005 IEEE

E_s/N_0. For reference, the capacity $\mathcal{C}$ of the AWGN channel is also plotted, which
corresponds to the coding rate of the optimum channel code in the information theoretic sense. It can be observed that with increasing code length the coding rate
shrinks, which means that the more bits per parameter transmission are available the
more of them are "invested" in error protection to achieve the maximum SNR. At
$E_s/N_0 = 3\,\mathrm{dB}$, for example, about 13.5% of the bit rate is used for error protection if
the code length is $B = 4$, while for code length $B = 8$ roughly 34% is consumed.

By protecting several source parameters jointly with one SOCC the coding rate can be
increased compared with a single-parameter SOCC, since channel coding with longer
code words is more effective. According to the coding theorem it must even be possible
to transmit error-free at a coding rate equal to the channel capacity. However, this
would require very long SOCCs.

12.3.6 Relation to Channel Optimized Vector Quantization

A related approach to SOCC is *Channel Optimized Vector Quantization* (COVQ)
as proposed by Farvardin et al. [Farvardin, Vaishampayan 1987], [Farvardin 1990].
COVQ and SOCC are both based on the idea of optimizing source and channel cod-

ing jointly[2]. But in contrast to SOCC, COVQ does no explicit channel coding. Instead, the source encoder is directly optimized with respect to the conditions on the disturbed transmission channel. The error protecting capability of COVQ is a result of shaping the quantizer regions such that an implicit redundancy increase is achieved.

From an engineering point of view the merging of source and channel coding might be seen as a drawback of COVQ. If a source codec based on COVQ has to be adapted to different transmission channels, the quantization scheme has to be changed each time. With SOCC the separation of source and channel codec is preserved, which means that the source codec quantizers can be trained independently from the channel statistics. When adapting to another channel only the SOCC and its encoder mapping $\mathbf{\Gamma}$ has to be redesigned.

Before comparing SOCC and COVQ performance experimentally, we briefly discuss the differences between both approaches in terms of their optimization criteria. The COVQ criterion is given by

$$\min_{Q,\check{V},\mathbf{\Gamma}} \mathrm{E}\big\{ \|\tilde{\underline{v}} - \hat{\underline{v}}\|^2 \big\} \ , \tag{12.16}$$

i.e., it is a true joint optimization of the parameter quantizer Q, the number of reproduction values $\check{V}$ and the mapping $\mathbf{\Gamma}$. Denoting the COVQ centroids by $\underline{v}$ we can decompose (12.16) to

$$\min_{Q,\check{V},\mathbf{\Gamma}} \big[\mathrm{E}\big\{ \|\tilde{\underline{v}} - \underline{v}\|^2 \big\} + \mathrm{E}\big\{ \|\underline{v} - \hat{\underline{v}}\|^2 \big\} \big] \ . \tag{12.17}$$

With SOCC, the total MSE for a given number of reproduction values $\check{V}$ is

$$\min_{Q} \mathrm{E}\big\{ \|\tilde{\underline{v}} - \underline{v}\|^2 \big\} + \min_{\mathbf{\Gamma}} \mathrm{E}\big\{ \|\underline{v} - \hat{\underline{v}}\|^2 \big\} \ . \tag{12.18}$$

If we consider $\check{V}$ as a design parameter the bit allocation between source and channel coding is explicitly determined by a second optimization step

$$\min_{\check{V}} \bigg[\min_{Q} \mathrm{E}\big\{ \|\tilde{\underline{v}} - \underline{v}\|^2 \big\} + \min_{\mathbf{\Gamma}} \mathrm{E}\big\{ \|\underline{v} - \hat{\underline{v}}\|^2 \big\} \bigg] \ . \tag{12.19}$$

Since $\min[\mathrm{E}\{\cdot\} + \mathrm{E}\{\cdot\}] \leq \min \mathrm{E}\{\cdot\} + \min \mathrm{E}\{\cdot\}$ we conclude that only the COVQ criterion ensures a minimization of the total MSE, while the independent design of Q and $\mathbf{\Gamma}$ in the case of SOCC does not necessarily guarantee this.

---

[2]The joint optimization approach at first glance seems to be questionable against the background of Shannon's separation theorem. But in fact there is no contradiction, since the separation theorem is valid for optimum channel and source coding from the information theoretic point of view. In practical transmission systems block lengths and computational complexity of coding algorithms are limited, which leads to suboptimality. In this practical case joint optimization has its justification.

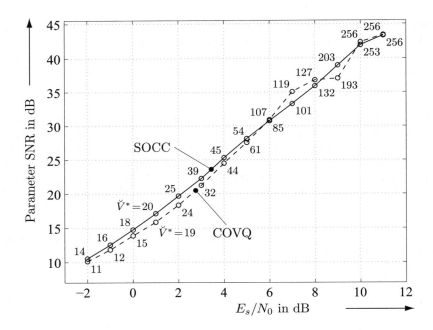

Figure 12.4: Comparison of SOCC and COVQ for gross data rate $A=8$ bit per parameter value; code length $B=8$; individual code design for each simulation point; white Gaussian parameter source, $\sigma_{\tilde{v}}^2=1$; AWGN channel.
© 2005 IEEE

To see how the two approaches compare in practice we trained COVQ codebooks for $A=4$ and $A=8$ and a different E_s/N_0 by applying the original design algorithm proposed by Farvardin [Farvardin 1990]. As our communication model employs a continuous output AWGN channel, the update equations for COVQ training contain integrals over the conditional channel PDF, which cannot be solved analytically. We coped with this by approximating the integrals by sums over a pre-computed set of channel noise realizations. For the highest E_s/N_0 under consideration the COVQ training algorithm was initialized by a source optimized codebook trained with the *split-LBG algorithm* [Linde et al. 1980]. Then, proceeding towards lower E_s/N_0, the COVQ codebook of the previous optimization was applied each time as initialization.

Figure 12.4 shows the comparison of COVQ and SOCC performances. In many cases SOCC slightly outperforms COVQ, which indicates that the applied COVQ training algorithm got stuck in a local optimum. With more sophisticated training methods such as *simulated annealing* it should be possible to achieve at least the SOCC performance.

However, a noteworthy result is that SOCC despite the independent optimization of quantizer and channel encoder provides a performance close to that of COVQ. This means that SOCC allows us to optimize the source encoder independently of the channel without relevant performance loss.

12.4 Source Controlled Channel Decoding (SCCD)

In [Hagenauer 1995] a simple approach was introduced for exploiting *a priori* knowledge on the bit level. Some more details on this can be found in [Hindelang 2001]. Additionally there have been publications where channel decoding with *a priori* knowledge has been applied to audio, image, or video coding. Besides convolutional codes, other codes, such as *Turbo* codes or *Low Density Parity Check* (LDPC) codes, can be used, but they are not the subject of the investigations of this section.

In the following we will exploit *a priori* knowledge on the symbol level in channel decoding. The *a priori* probabilities are obtained from the unequal distribution of parameters and from the temporal correlation. It can be extended to the correlation of parameters within one frame by exploiting the results from Chap. 11.

12.4.1 Channel Coding and Decoding in SCCD

The *Source Controlled Channel Decoding* (SCCD) approach can be applied to any channel decoding approach if a decoder that supports soft-in soft-out decoding is used. In the following we consider linear convolutional codes with rate $r = 1/\Upsilon$.

Figure 12.5 shows the SCCD approach in detail. Each quantized parameter $v_{m,k}$ is mapped to a bit vector $\mathbf{x}_{m,k}$ consisting of the bits $(x_{m,k}(1), x_{m,k}(2), \ldots, x_{m,k}(w))$. Note that, within this section we use the same number w of bits for each bit vector $x_{m,k}$ and we speak of bit mapping instead of index assignment. Once all M parameters belonging to a frame k are quantized and mapped to the respective bit vectors,

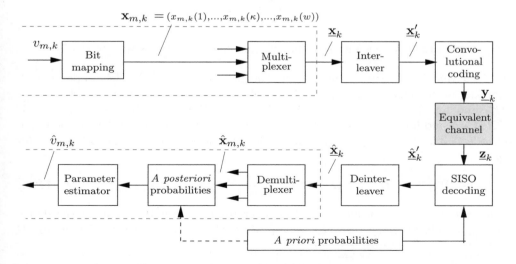

Figure 12.5: Coding and decoding for the SCCD approach

their bits are multiplexed building a frame of $w \cdot M$ source bits denoted as vector $\underline{\mathbf{x}}_k \in \{0,1\}^{w \cdot M}$ with the bits $\underline{x}_k(\kappa)$ and $\kappa = 1, 2, \ldots, w \cdot M$.

The block "Channel Coding" from Fig. 12.1 is split into an interleaver and the convolutional encoder. The interleaving is needed to separate bits belonging to one parameter from each other. This will be explained in detail in Sec. 12.4.5. After interleaving we obtain the bit vector $\underline{\mathbf{x}}'_k$ again with a length of $w \cdot M$ bits.

Convolutional Coding

After interleaving the bit stream $\underline{\mathbf{x}}'$ of each frame k is subject to convolutional encoding with rate r. In this chapter we consider only rate $r = 1/\Upsilon$ codes, i.e., for every bit $x'(\kappa)$ we obtain Υ channel bits $\mathbf{y}(\kappa) = (y((\kappa - 1)\Upsilon + 1), \ldots, y((\kappa - 1)\Upsilon + \Upsilon))$ with $\kappa = 1, \ldots, w \cdot M$. We introduce convolutional coding using an example and follow the notation of [Johannesson, Zigangirov 1999].

The information digits $x'(\kappa)$ from the sequence $\underline{\mathbf{x}}'$ are fed into a shift register with ν stages (in Fig. 12.6, $\nu = 2$). According to [Proakis 2000] the constraint length[3] is then defined as $L = \nu + 1$. The encoder in Fig. 12.6 has $\Upsilon = 2$ linear output sequences $y(1), y(3), \ldots, y((wM - 1)\Upsilon + 1)$ and $y(2), y(4), \ldots, y((wM - 1)\Upsilon + \Upsilon)$. These are serialized to form a single output sequence $\underline{\mathbf{y}} = (\mathbf{y}(1), \mathbf{y}(2), \ldots)$ of length $w \cdot M \cdot \Upsilon$.

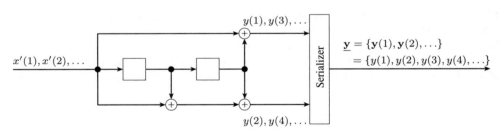

Figure 12.6: An encoder for a binary convolutional code with rate $r = 1/2$

It is convenient to express the coding sequence in terms of the delay operator D. Looking at the example in Fig. 12.6, we obtain for the upper part $G_1(D) = (1 + D^2)$ and for the lower part $G_2(D) = (1 + D + D^2)$ where $G_i, i \in \{1, \ldots, \Upsilon\}$ is called a *generator polynomial* and determines one of the Υ outputs of a convolutional encoder. The individual Υ generator polynomials are merged into one *generator matrix*

$$\mathbf{G}(D) = \left(1 + D^2, 1 + D + D^2\right). \tag{12.20}$$

The output $\mathbf{y}(\kappa)$ is obtained from the generator polynomials by adding the different input bits with the respective delay where the power of the delay operator D

---

[3]In some books, e.g., [Johannesson, Zigangirov 1999], the overall constraint length of a code is defined as the number of stages or shift registers needed to obtain all elements of the output sequence which equals in the case of rate $1/\Upsilon$ codes the memory ν.

denotes the delay. For the used example with the generator matrix from (12.20) we obtain

$$\mathbf{y}(\kappa) = (\, y((\kappa - 1) \cdot \Upsilon + 1)\, ,\ y((\kappa - 1) \cdot \Upsilon + \Upsilon)\,)$$
$$= (\, x'(\kappa) + x'(\kappa - 2)\quad ,\ x'(\kappa) + x'(\kappa - 1) + x'(\kappa - 2)\,)\,, \tag{12.21}$$

where $\Upsilon = 2$ (see Fig. 12.6) and a predefined value, typically zero, is used in the case of an index $i \leq 0$ for the bits $x'(i)$.

The above example is a *non-systematic, non-recursive encoding* matrix. If the input sequence $\underline{x}'$ is mapped onto one, e.g., the upper output sequence, we speak of a *systematic* encoder and if there is a feedback from the shift registers back to the first stage we speak of a *recursive* encoder in analogy to a recursive filter. For simplification the generator matrix is often written in octal notation merging three delays in one octal number starting with the lowest one as the most significant bit, e.g.,

$$\mathbf{G}(D) = (\, 1 + D + D^3\,) = (1 + 1 \cdot D + 0 \cdot D^2 + 1 \cdot D^3) = (15)_8\,. \tag{12.22}$$

For the example from (12.20) we obtain $\mathbf{G} = (5, 7)_8$.

Without going into details, the following properties of convolutional codes will be used in this section.

- Every convolutional generator matrix is equivalent to a recursive systematic encoding matrix. This means that the same output sequences or code words $\mathbf{y}$ are obtained, even if the mapping of $\underline{x}'$ to the possible output sequences may be different. The proof can be found in [Johannesson, Zigangirov 1999].

- Although the distance spectrum of the code and thus the probability of decoding errors is the same, these decoding errors lead to a lower number of bit errors in the case of the equivalent recursive systematic encoding matrix, especially under weak channel conditions.

Considering the generator matrix from (12.20) its equivalent *recursive systematic* generator matrix can be denoted as

$$\mathbf{G}^{[\mathrm{RSC}]}(D) = \left(\, 1, \frac{1 + D + D^2}{1 + D^2}\, \right) \quad \mathbf{G}^{[\mathrm{RSC}]} = (\, 1, \tfrac{7}{5}\,)_8\,. \tag{12.23}$$

Before channel coding tail bits can be added to terminate the convolutional code which increases the overall code rate a little. After channel coding bits can be punctured or repeated to obtain a desired overall code rate or the necessary number of bits per frame.

Decoding

We utilize a *Soft-Input/Soft-Output* (SISO) channel decoder. In the following sections we use the BCJR algorithm [Bahl et al. 1974], which is explained in more detail in Sec. 13.2.2. It delivers the optimal symbol-by-symbol MAP probability as input for the next decoding stage. Additionally, the channel decoder will make use of the available *a priori* knowledge marked by the arrow from the block "*a priori* probabilities" to the SISO channel decoding block in Fig. 12.5.

In contrast to the SDSD approach in Fig. 10.2 where the transition probabilities were introduced as input to the source decoder, in SCCD the transition probabilities are the input to the channel decoder. After channel decoding and de-interleaving, the reliability information $P_d(x_m(\kappa))$ of each bit of a bit vector is obtained where P_d denotes the decoding probability after channel decoding and its content depends on the exploited algorithm. Therefore, we explain P_d in the respective section, in the optimum case it is the MAP probability. As long as every bit is assumed to be statistically independent of the other bits belonging to one parameter, we can derive the probability of each value of the index vector $\mathbf{x}_m$ in one frame k by

$$P_d(\mathbf{x}_{m,k}) = \prod_{\kappa=1}^{w} P_d(x_{m,k}(\kappa)) \,, \qquad (12.24)$$

which serves as input to the source decoder and gives a likelihood for every possibly transmitted bit vector $\mathbf{x}_{m,k}$ similar to the transition probabilities in (10.5). Even though convolutional coding usually yields burst-like errors (path errors), we assume the cascade of channel coder, equivalent channel, and channel decoder to be memoryless. This can be achieved by separating the bits of one bit vector $x_{m,k}(\kappa)$ by at least five times the constraint length of the code. The closer the bits of one bit vector are placed the less exact (12.24) becomes. The interleaver in Fig. 12.5 should therefore be designed such that it complies with this requirement (c.f. [Vary, Martin 2006]).

Example of the Applied Transmission System

To perform simulations for SCCD a concretely defined transmission system on the basis of Figs. 12.1 and 12.5 is used. Each of a set of $M = 20$ Gaussian distributed parameters having AR(1) property with a temporal correlation $\rho = 0.8$ is Lloyd–Max quantized with $V = 8$ levels. The levels are mapped onto a bit vector $\mathbf{x}_{m,k}$ with length $w = 3$ bits. We apply folded binary mapping since it leads to a higher parameter SNR in the case of erroneous transmissions [Hindelang 2001] than the natural binary mapping that was introduced in Sec. 10.4. The quantized values, the assigned bit vectors, and the probabilities for each value are shown in Table 12.1.

After multiplexing we obtain a bit vector $\underline{\mathbf{x}}_k$ of length $M \cdot w = 60$ bits. The interleaver places the *most significant bits* (MSBs) $x_{m,k}(1)$ onto positions 1 to 20, the *second significant bits* (SSBs) $x_{m,k}(2)$ onto positions 21 to 40, and finally the

Table 12.1: Folded binary mapping for $V = 8$ levels

$v_{m,k}$	-2.152	-1.344	-0.756	-0.245	0.245	0.756	1.344	2.152
$\mathbf{x}_{m,k}$	$(0,0,0)$	$(0,0,1)$	$(0,1,0)$	$(0,1,1)$	$(1,1,1)$	$(1,1,0)$	$(1,0,1)$	$(1,0,0)$
$P(\mathbf{x}_{m,k})$	0.0401	0.1071	0.1619	0.1909	0.1909	0.1619	0.1071	0.0401

least significant bits (LSBs) $x_{m,k}(3)$ onto positions 41 to 60 leading to the bit vector $\mathbf{x}'_k$.

Owing to the defined initial state of the convolutional encoder the beginning of each decoded bit stream is less error-prone. The MSBs are therefore located at these better protected positions since they have a larger influence on the quality after source decoding. On the other hand the LSBs are placed at the end of the stream since the error rate of convolutional codes increases to the end of the code due to the truncation of the coding. It has to be mentioned that the effect of truncation can be compensated for by adding so-called tail bits before channel coding. However, due to the different significance of the bits of a quantized symbol the effect of the starting point and the truncation of convolutional codes is used as a scheme for unequal error protection (UEP). To achieve UEP in general, code repetition or puncturing is applied, which will not be considered within this section.

For convolutional coding we use the rate $r = 1/2$, memory $\nu = 6$ recursive systematic convolutional (RSC) encoder with the generator matrix

$$\mathbf{G}(D) = \left(1, \frac{1 + D^2 + D^3 + D^5 + D^6}{1 + D + D^2 + D^3 + D^6} \right) \qquad \mathbf{G} = (1, \tfrac{133}{171})_8$$

leading to $B = w \cdot M \cdot \Upsilon = 120$ channel coded bits.

12.4.2 A Priori Knowledge in Channel Decoding

From Chap. 10 it is known that the redundancy after source coding can be used on the symbol level for source decoding. In the following, the exploitation of *a priori* knowledge on the symbol level for channel decoding is shown. However, as we use binary convolutional codes with rate $1/\Upsilon$ each stage of the convolutional code belongs to one bit and for this reason we have to convert the prediction probabilities to the bit level.

In Fig. 12.7 we show possible realizations of the channel decoder, where AK0 in the upper right box means *a priori* knowledge of 0th order according to a Markov model of 0th order (see Sec. 10.3) and the unequal distribution is exploited. Similarly, AK1 in the lower right box stands for *a priori* knowledge of 1st order and the temporal correlation is exploited. We use the same notation as in Fig. 12.5. The sequence $\mathbf{z}_k$ is the whole received sequence of one frame k from the channel. Furthermore,

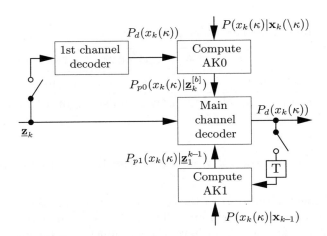

Figure 12.7: Channel decoding using *a priori* knowledge. © 2000 IEEE; © 2002 IEEE

$\underline{\mathbf{z}}_1^k = (\underline{\mathbf{z}}_1, \underline{\mathbf{z}}_2, \ldots, \underline{\mathbf{z}}_k)$ denotes the complete history of frames. For simplification of Fig. 12.7 we do not show the interleaving and multiplexing units that have to be placed before and after the channel decoding modules. They convert the bit vectors $\mathbf{x}_{m,k}$ from the index assignment to the input $\underline{\mathbf{x}}_k'$ of the channel decoder and vice-versa, which is a simple transformation. Therefore, we show the input $\underline{\mathbf{z}}$ from the channel and the different probabilities of the bit vectors $\mathbf{x}_{m,k}$. In the following each parameter $\tilde{v}_{m,k}$ and its bit vector $\mathbf{x}_{m,k}$ is considered separately. Thus, we neglect the index m and we speak of the bit vector $\mathbf{x}_k = (x_k(1), x_k(2), \ldots, x_k(w))$ with its bits $x_k(\kappa)$, which stands for one arbitrary $\mathbf{x}_{m,k}$. Finally, we introduce the residual bit vector

$$\mathbf{x}_k(\backslash \kappa) = (x_k(1), x_k(2), \ldots, x_k(\kappa-1), x_k(\kappa+1), \ldots, x_k(w)) \qquad (12.25)$$

denoting the bit vector $\mathbf{x}_k$ without the κth bit.

In order to get a notation that is independent of the *a priori* knowledge exploited in the channel decoder, we define the decoding probability $P_d(x_k(\kappa))$ which denotes the reliability of a bit $x_k(\kappa)$ at the channel decoder's output and can be used as interface to the source decoder. We will see in the following sections that different types of *a priori* knowledge exploited by the channel decoder lead to different interpretations of $P_d(x_k(\kappa))$.

If both switches in Fig. 12.7 are in "off" position, conventional channel decoding is carried out. If we additionally assume that all source bits are statistically independent and equally distributed, i.e., $P(x_k(\kappa) = 0) = P(x_k(\kappa) = 1) = 0.5$, then no a priori knowledge can be used. We speak of **channel decoding using no *a priori* knowledge** "CD/NAK" and we get the probabilities

$$P_d(x_k(\kappa)) = P(x_k(\kappa)|\underline{\mathbf{z}}_k), \qquad (12.26)$$

i.e., in this case, the decoding probability can be interpreted as *a posteriori* probability to the respective source bits.

12.4.3 Channel Decoding Using Intra-Parameter Correlation

In general, the bits of one bit vector $\mathbf{x}_k$ are statistically dependent if the quantized parameter v_k has a non uniform distribution (see Table 12.1). Therefore, the probability $P(x_k(\kappa))$ denoting the source statistics of a bit being 0 or 1 is dependent on the other bits $\mathbf{x}(\backslash\kappa)$. Since we consider in the following only the unequal distribution of parameters, we call it channel decoding using *a priori* knowledge of 0th order "CD/AK0" following the definition of a Markov model of order $N = 0$.

Without loss of generality we can write

$$P(x_k(\kappa)|\underline{\mathbf{z}}_k) = f(\mathbf{x}_k(\backslash\kappa), \underline{\mathbf{z}}_k), \tag{12.27}$$

indicating that in addition to the dependence on the received channel values there is a dependence on the other bits of a vector belonging to one parameter. Within this section the symbol f states that there is some kind of dependence but it does not mean that these are the only dependencies. Since all probabilities depend on the channel values $\underline{\mathbf{z}}_k$, we speak of *a posteriori* probabilities. For simplification we neglect the index k since we consider only the current frame.

To exploit this dependence we extend the *a posteriori* probability $P(x(\kappa)|\underline{\mathbf{z}})$ by using the symbol *a posteriori* probabilities and sum over all residual bit vectors $\mathbf{x}(\backslash\kappa)$:

$$P(x(\kappa)\,|\,\underline{\mathbf{z}}) = \sum_{i=0}^{2^{w-1}-1} P(\mathbf{x}^{(i)}(\backslash\kappa), x(\kappa)\,|\,\underline{\mathbf{z}})\,. \tag{12.28}$$

For the following derivation we have to separate the received channel sequence into two parts: $\underline{\mathbf{z}} = (\underline{\mathbf{z}}^{[a]}, \underline{\mathbf{z}}^{[b]})$ assuming that $\underline{\mathbf{z}}^{[a]}$ does depend on $x(\kappa)$ but not on $\mathbf{x}(\backslash\kappa)$, and $\underline{\mathbf{z}}^{[b]}$ does depend on $\mathbf{x}(\backslash\kappa)$ but not on $x(\kappa)$. This assumption is fulfilled for nonrecursive codes if we separate the bits of one bit vector within the frame by at least the overall constraint length of the code. For recursive encoders, however, the bits belonging to one bit vector have to be separated by more than five times the memory ν of the convolutional code, since the feedback in the encoder increases the dependence of bits. With a separation of two times ν the degradation due to the violation of the above assumption is negligible. To separate the bits in such a way as to fulfill the above assumptions an interleaver is placed before channel coding.

By applying $\underline{\mathbf{z}} = (\underline{\mathbf{z}}^{[a]}, \underline{\mathbf{z}}^{[b]})$ to (12.28) we obtain

$$P(x(\kappa)|\underline{\mathbf{z}}) = \sum_{i=0}^{2^{w-1}-1} P(\mathbf{x}^{(i)}(\backslash\kappa), x(\kappa)\,|\,\underline{\mathbf{z}}^{[a]}, \underline{\mathbf{z}}^{[b]})\,, \tag{12.29}$$

which can be transformed to

$$P(x(\kappa)|\underline{\mathbf{z}}) = \frac{1}{C} \cdot P(\underline{\mathbf{z}}^{[a]}|x(\kappa)) \cdot \sum_{i=0}^{2^{w-1}-1} \left(P(x(\kappa)|\mathbf{x}^{(i)}(\backslash\kappa)) \cdot P(\mathbf{x}^{(i)}(\backslash\kappa)|\underline{\mathbf{z}}^{[b]}) \right) \tag{12.30}$$

using Bayes' rule. The term $P(\underline{z}^{[a]}|x(\kappa))$ denotes the transition probabilities according to (10.5). Owing to convolutional coding they are obtained in the channel decoding by mapping the channel probabilities $P(z((\kappa-1)\cdot\Upsilon+i)|y((\kappa-1)\cdot\Upsilon+i))$, $i = 1 \ldots \Upsilon$ to the respective transitions in the convolutional decoder, see, e.g., [Bahl et al. 1974]. The constant C is introduced in Sec. 10.4 and applied here in the same way. The detailed derivation can be found in [Hindelang 2001].

Applying (12.30) directly in channel decoding increases the complexity exponentially with each correlated parameter. This is obvious, because the *a posteriori* probabilities $P(x(\kappa)|\underline{z})$ depend directly on $P(\mathbf{x}(\backslash\kappa)|\underline{z}^{[b]})$, which are typically not known at the decoder. This can be solved by parallel paths within the transitions of, e.g., the BCJR algorithm, where the paths denote the different conditional probabilities $P(x(\kappa)|\mathbf{x}(\backslash\kappa))$. However, with every bit of each parameter the number of paths doubles until the last bit of one bit vector is reached within the convolutional code.

To solve now (12.30) in a practicable way, the two-step approach shown in Fig. 12.7 is applied. If the upper branch is switched on and the lower one is switched off, then correlations within the bit vector $\mathbf{x}$ belonging to one parameter are exploited in terms of *a priori* knowledge about one bit $x(\kappa)$ given the $w-1$ other bits $\mathbf{x}(\backslash\kappa)$. Therefore we call it intra-parameter correlation.

The computations are as follows: A first preliminary decoding step is performed as shown in Fig. 12.7, which yields the estimate

$$P(\mathbf{x}(\backslash\kappa)|\underline{z}^{[b]}) = \prod_{i=1,i\neq\kappa}^{w} P(x(i)|\underline{z}^{[b]}) = \prod_{i=1,i\neq\kappa}^{w} P_d(x(i)) \qquad (12.31)$$

with $P(x(i)|\underline{z}^{[b]})$ according to (CD/NAK) (see Sec. 12.4.2). The next step is to calculate the sum from (12.30), the so-called prediction probability

$$P_{p0}(x(\kappa)|\underline{z}^{[b]}) = \sum_{i=0}^{2^{w-1}-1} P(x(\kappa)|\mathbf{x}^{(i)}(\backslash\kappa)) \cdot P(\mathbf{x}^{(i)}(\backslash\kappa)|\underline{z}^{[b]}), \qquad (12.32)$$

where the index $p0$ denotes the fact that *a priori* knowledge of 0th order is exploited. The first part of the sum is given by the source statistics while the second part stems from the first decoding step according to (12.31). In general, a prediction probability denotes some reliability of a symbol or here a bit x based on a part of the received sequence, e.g., $\underline{z}^{[b]}$ (see (12.32)) or $\underline{z}_1^{k-1}$ (see Sec. 10.4.2) which exists only due to some a priori knowledge (e.g., $P(x(\kappa)|\mathbf{x}(\backslash\kappa))$ or $P(\mathbf{x}_k|\mathbf{x}_{k-1})$). The exact derivation depends on the modelling of the source parameter, in (12.32) we assumed a Markov model of order $N = 0$.

The prediction probabilities P_{p0} are now used together with the channel information in the main channel decoder. We obtain decoding probabilities $P_d(x(\kappa)) \approx P(x(\kappa)|\underline{z})$ according to (12.30) depending only on the currently received frame $\underline{z}$. Note that, although these probabilities are formally the same as those delivered by (CD/NAK), they yield in general the better approximation. Now, not only one part $\underline{z}^{[a]}$ of the received sequence from the channel but also a second part $\mathbf{z}^{[b]}$ depending on the other bits of one bit vector is exploited. The approximation is given by the fact that we can not apply (12.30) and thus, we split it into a two-step solution. The *a priori* knowledge given as $P(x(\kappa)|\mathbf{x}(\backslash\kappa))$ can, e.g., be stored in w tables of size 2^{w-1}, which represent a mixed form of *bit* and *bit group* level *a priori* knowledge $P(x(\kappa)|\mathbf{x}(\backslash\kappa))$. These tables can be employed directly in (12.32). Alternatively, $P(x(\kappa)|\mathbf{x}(\backslash\kappa))$ can easily be calculated if the distribution of one parameter v and therefore the distribution of the respective bit vector $\mathbf{x}$ is known. By applying Bayes' rule we obtain

$$P(x(\kappa)|\mathbf{x}(\backslash\kappa)) = \frac{P(\mathbf{x}(\backslash\kappa), x(\kappa))}{P(\mathbf{x}(\backslash\kappa), x(\kappa) = 0) + P(\mathbf{x}(\backslash\kappa), x(\kappa) = 1)} , \qquad (12.33)$$

where each $(\mathbf{x}(\backslash\kappa), x(\kappa))$ leads to one defined bit vector $\mathbf{x}$, and we have to store the distribution $P(\mathbf{x})$ (see, e.g., Table 12.1) with a size of 2^w values and employ (12.33) in (12.32).

12.4.4 Channel Decoding Using Inter-Frame Correlation

In Sec. 12.4.3 the distribution of parameters was considered in the calculation of the prediction probability. In the following, the temporal correlation of symbols is evaluated and converted to the bit level for convolutional decoding. This we call channel decoding using *a priori* knowledge of 1st order "CD/AK1". The principle can be seen in Fig. 12.7 if the lower branch is turned on. Within one frame there may be several parameters $v_{m,k}$ that are correlated to their respective parameters $v_{m,k-1}$ in the previous frame. Again (cf. Sec. 12.4.3) we neglect the index m. The temporal correlation can be expressed in a common way

$$P(\mathbf{x}_k) = f(\mathbf{x}_{k-1}) . \qquad (12.34)$$

In the same manner as in (12.27) we extend (12.34) to *a posteriori* probabilities after channel decoding and convert it to the bit level

$$P(x_k(\kappa)|\underline{\mathbf{z}}_1^k) = f(\mathbf{x}_{k-1}, \underline{\mathbf{z}}_1^{k-1}) , \qquad (12.35)$$

where $P(x_k(\kappa)|\underline{\mathbf{z}}_1^k)$ denotes the dependence of one bit at frame k on the complete history of received frames $0, 1, \ldots, k$.

To exploit (12.35) we split the complete history of received channel values $\underline{\mathbf{z}}_1^k$ into the values of the current frame $\underline{\mathbf{z}}_k$ and all other frames $\underline{\mathbf{z}}_1^{k-1}$. In Sec. 10.4 the symbol *a posteriori* probabilities were derived. For the special case of a 1st order Markov model we modify (10.32), convert it to the bit level, move the prediction probability into a separate part, and obtain

$$P_d(x_k(\kappa)) = P(x_k(\kappa)|\underline{\mathbf{z}}_k, \underline{\mathbf{z}}_1^{k-1}) = \frac{1}{C} \cdot P(\underline{\mathbf{z}}_k|x_k(\kappa)) \cdot P_{p1}(x_k(\kappa)|\underline{\mathbf{z}}_1^{k-1}) . \qquad (12.36)$$

As stated before, we assume a memoryless channel. Again, the bit transition probabilities $P(\underline{\mathbf{z}}_k|x_k(\kappa))$ are given by the equivalent channel. Looking for the prediction probabilities, we obtain

$$P_{p1}(x_k(\kappa)|\underline{\mathbf{z}}_1^{k-1}) = \sum_{i=0}^{2^w-1} P(x_k(\kappa)|\mathbf{x}_{k-1}^{(i)}) \cdot P(\mathbf{x}_{k-1}^{(i)}|\underline{\mathbf{z}}_{k-1}, \underline{\mathbf{z}}_1^{k-2}) \tag{12.37}$$

by considering a Markov model of 1st order. The index $p1$ denotes the prediction probability of 1st order (see Fig. 12.7), which is used together with the current sequence $\underline{\mathbf{z}}_k$ in channel decoding. The *a posteriori* probability $P(\mathbf{x}_{k-1}|\underline{\mathbf{z}}_{k-1}, \underline{\mathbf{z}}_1^{k-2})$ is obtained after the channel decoding of the previous frame on the bit level as $P_d(x_{k-1}(\kappa)) = P(x_{k-1}(\kappa)|\underline{\mathbf{z}}_1^{k-1})$. We assume that these bit level probabilities are statistically independent and apply (12.24). Thus, we obtain the symbol probabilities. Equations (12.36) and (12.37) can be solved in a recursion with one delay element T (see Fig. 12.7). In contrast to (CD/NAK) and (CD/AK) the decoding probability $P_d(x_k(\kappa))$ depends on the history of received frames $\underline{\mathbf{z}}_1^k$.

Since $P_d(\mathbf{x}_{k-1})$ is known from the decoding of the previous frame the prediction probabilities can be calculated before convolutional decoding. The complexity is very low in comparison with that of convolutional decoding and can be almost neglected. For each parameter quantized with w bits, we have to store the probabilities $P(x_k(\kappa)|\mathbf{x}_{k-1})$, which gives a number of $w \cdot 2^w$ elements.

12.4.5 Channel Decoding Using Intra-Parameter and Inter-Frame Correlation

In (SD/AK1) according to Chap. 10 the *a priori* knowledge given by the unequal distribution (SD/AK0) is included. In SCCD, however, the dependence of $x_k(\kappa)$ on the other bits $\mathbf{x}_k(\backslash\kappa)$ is not included in the (CD/AK1) approach due to the conversion to the bit level. Thus, the (CD/AK0) and the (CD/AK1) approaches have to be combined, meaning that both switches in Fig. 12.7 are in the "on" position. We write the prediction probabilities as $P_{p01}(x_k(\kappa)|\underline{\mathbf{z}}_1^k) = P_{p01}(x_k(\kappa)|\underline{\mathbf{z}}_k^{[b]}, \underline{\mathbf{z}}_1^{k-1})$ with the index $p01$ denoting the influence of *a priori* knowledge of 0th and 1st order. If we assume that $\underline{\mathbf{z}}_k^{[b]}$ and $\underline{\mathbf{z}}_1^{k-1}$ are statistically independent the computation is simple because now the (CD/AK1) part according to (12.37) and the (CD/AK0) part according to (12.32) can be calculated separately and multiplied afterwards. We obtain

$$P_{p01}(x_k(\kappa)|\underline{\mathbf{z}}_k^{[b]}, \underline{\mathbf{z}}_1^{k-1}) = \frac{P_{p0}(x_k(\kappa)|\underline{\mathbf{z}}_k^{[b]}) \cdot P_{p1}(x_k(\kappa)|\underline{\mathbf{z}}_1^{k-1})}{P(x_k(\kappa))} . \tag{12.38}$$

The term in the denominator is the unconditioned bit probability given by the unequal distribution of the bit $x_k(\kappa)$ itself. This correction term has to be applied since this probability is considered twice in P_{p0} and in P_{p1}.

Although the channel itself is memoryless, we have a dependence of $\underline{\mathbf{z}}_k$ and $\underline{\mathbf{z}}_{k-1}$ given by the source statistics and thus (12.38) is an approximation. Below we derive the exact solution. Note that for a memoryless channel

$$P(\underline{\mathbf{z}}_k|\mathbf{x}_k, \underline{\mathbf{z}}_{k-1}) = P(\underline{\mathbf{z}}_k|\mathbf{x}_k) \tag{12.39}$$

holds, since the dependence of $\underline{\mathbf{z}}_k$ (or the interesting part $\underline{\mathbf{z}}_k^{[b]}$) on $\underline{\mathbf{z}}_{k-1}$ is completely given by the dependence on the source $\mathbf{x}_k$. This fact was used for the derivation of P_{p1} and we use it in this section likewise. We extend (12.29) to the exploitation of temporal correlation as in (12.37) to obtain the *a posteriori* probabilities for channel decoding using 0th and 1st order *a priori* knowledge "CD/AK0+1"

$$
\begin{aligned}
P_d(x_k(\kappa)) &= P(x_k(\kappa)|\underline{\mathbf{z}}_k^{[a]}, \underline{\mathbf{z}}_k^{[b]}, \underline{\mathbf{z}}_1^{k-1}) \\
&= \sum_{i=0}^{2^w-1} \sum_{j=0}^{2^{w-1}-1} P(x_k(\kappa), \mathbf{x}_k^{(j)}(\backslash\kappa), \mathbf{x}_{k-1}^{(i)}|\underline{\mathbf{z}}_k^{[a]}, \underline{\mathbf{z}}_k^{[b]}, \underline{\mathbf{z}}_1^{k-1}) \\
&= \frac{1}{C} \cdot P(\underline{\mathbf{z}}_k^{[a]}|x_k(\kappa)) \cdot P_{p01}(x_k(\kappa)|\underline{\mathbf{z}}_k^{[b]}, \underline{\mathbf{z}}_1^{k-1}).
\end{aligned}
\tag{12.40}
$$

The exact prediction probability P_{p01} is now a two-dimensional sum over all possible values of $\mathbf{x}_k(\backslash\kappa)$ and $\mathbf{x}_{k-1}$ and is given by

$$P_{p01}(x_k(\kappa)|\underline{\mathbf{z}}_k^{[b]}, \underline{\mathbf{z}}_1^{k-1}) = \tag{12.41}$$

$$\sum_{i=0}^{2^w-1} \sum_{j=0}^{2^{w-1}-1} \left(P(\mathbf{x}_k^{(j)}(\backslash\kappa)|\underline{\mathbf{z}}_k^{[b]}) \cdot P(\mathbf{x}_{k-1}^{(i)}|\underline{\mathbf{z}}_1^{k-1}) \cdot \frac{P(x_k(\kappa), \mathbf{x}_k^{(j)}(\backslash\kappa), \mathbf{x}_{k-1}^{(i)})}{P(\mathbf{x}_k^{(j)}(\backslash\kappa)) \cdot P(\mathbf{x}_{k-1}^{(i)})} \right).$$

The last part of (12.41), which considers only the dependence of the parameters due to the source statistics can be converted to

$$\frac{P(x_k(\kappa), \mathbf{x}_k(\backslash\kappa), \mathbf{x}_{k-1})}{P(\mathbf{x}_k(\backslash\kappa)) \cdot P(\mathbf{x}_{k-1})} = \frac{P(\mathbf{x}_k|\mathbf{x}_{k-1})}{P(\mathbf{x}_k(\backslash\kappa))} \tag{12.42}$$

by merging $x_k(\kappa)$ and $\mathbf{x}_k(\backslash\kappa)$ into $\mathbf{x}_k$ (cf. (12.25)) and applying Bayes' rule. If we compare the exact solution to the approximation in (12.38) it can be shown that they deliver the same result if we assume independence of the two parameters $\mathbf{x}$ at frames k and $k+1$, which is a contradiction to the assumption of temporal correlation. Finally, it has to be mentioned that adapting (10.32) to SCCD and building the sum over the residual bit vector $\mathbf{x}_k(\backslash\kappa)$ leads exactly to the result in (12.41). This summation reflects the conversion to the bit level for one bit $x_k(\kappa)$.

12.4.6 Simulation Results

In Sec. 10.3.2 the residual redundancy was introduced as a measure for the gain in source decoding. These redundancies are converted to the bit level and are split into the redundancy of each bit itself and into the mutual information between each bit and the rest of the bit vector. The first can be derived from (10.17) introducing the index b for the bit level

$$\Delta R_b(\kappa) = H_0 - H(x(\kappa)) = 1 + \sum_{i=0}^{1} P(x(\kappa) = i) \cdot \log_2(P(x(\kappa) = i)) , \qquad (12.43)$$

while the latter can be derived from (10.19) by substituting the history of bit vectors with the dependence to the other bits of the bit vector

$$\Delta R_d(\kappa) = \mathcal{I}(x(\kappa); \mathbf{x}(\backslash\kappa)) \qquad (12.44)$$

$$= \sum_{i=0}^{1} \sum_{j=0}^{2^{w-1}-1} P(x^{(i)}(\kappa), \mathbf{x}^{(j)}(\backslash\kappa)) \cdot \log_2 \frac{P(x^{(i)}(\kappa), \mathbf{x}^{(j)}(\backslash\kappa))}{P(x^{(i)}(\kappa)) \cdot P(\mathbf{x}^{(j)}(\backslash\kappa))} .$$

In the same way the bit level residual redundancy $\Delta R_c(\kappa)$ due to temporal correlation can be derived by converting ΔR_c from (10.19) to the bit level and reducing it to 1st order. In Table 12.2 we show the residual redundancies of the three bits from the exemplary transmission system.

In Fig. 12.8 the bit error rate depending on the bit position is shown for the different approaches of exploiting *a priori* knowledge using the simulation settings of the example in Sec. 12.4.1. Concerning the application of AK0 only, one can see a medium gain for the SSB $x(2)$ interleaved to positions 21 to 40. Comparing the redundancies and the results one can see that the gain of the SSBs $x(2)$ using AK0 can be connected to the redundancies $R_b(\kappa)$ and $R_d(\kappa)$. Note, although there are no redundancies $R_b(\kappa)$ and $R_d(\kappa)$ for the MSB $x(1)$ one can see a gain for them (positions $1 \ldots 20$) due to the mutual influence of bits in convolutional decoding. The reduced bit error rate of the SSBs due to a priori knowledge improves the error rate of the MSBs as well.

Table 12.2: Residual redundancies of the three different bits for a 3 bit Lloyd–Max quantizer of a 1st order Gaussian distributed AR(1) process with temporal correlation $\rho = 0.8$ applying folded binary bit mapping

Bit number	$\Delta R_b(\kappa)$	$\Delta R_d(\kappa)$	$\Delta R_c(\kappa)$
MSB $x(1)$	0.0	0.0	0.373
SSB $x(2)$	0.125	0.022	0.174
LSB $x(3)$	0.027	0.022	0.015

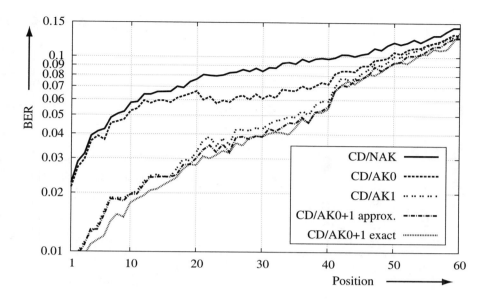

Figure 12.8: BER dependent on the bit position for an AWGN channel at E_s/N_0 of −3 dB exploiting the different approaches for *a priori* knowledge on the symbol level

Looking at AK1 there is a large gain for the MSB $x(1)$ and the SSB $x(2)$ due to the exploitation of $R_b(\kappa)$ and $R_c(\kappa)$ while the extension to AK0+1 gives only a small additional gain due to the relatively small $R_d(\kappa)$.

Owing to the conversion to the bit level the bit mapping plays a central role concerning the residual redundancy and thus different bit mappings lead to different bit error rates. A very interesting fact is that the *a priori* information due to $\Delta R_c(\kappa)$ (given by temporal correlation) and the *a priori* information due to $R_b(\kappa)$ and $R_d(\kappa)$ (both given by unequal distribution) can be converted to some extent to each other. It is possible to find bit mappings that deliver, e.g., a high $R_d(\kappa)$ by reducing the other redundancies $R_b(\kappa)$ and $R_c(\kappa)$. Some more details of the bit mapping together with SCCD can be found in [Hindelang 2001], [Hindelang et al. 2000b].

In a combined source and channel coding system the bit error rate is not that significant. A measure that determines the quality of parameters after source decoding is needed. The parameter SNR as introduced in Sec. 10.1 will be used in the following. Figure 12.9 shows the parameter SNR for the different exploitations of *a priori* knowledge. For parameter estimation a conventional HD source decoding is used as depicted in Fig. 10.1. Summarizing the results of the reference system, there is a large gain by exploiting AK1 since the error rate of the MSBs is mainly reduced by applying the (CD/AK1) approach (compare also to the high $R_c(\kappa)$ in Table 12.2) and errors in the MSBs have the most negative impact on the parameter SNR. A further small gain is obtained with AK0+1, but then two channel decoding steps are necessary. Within this section results have only been shown for the folded binary

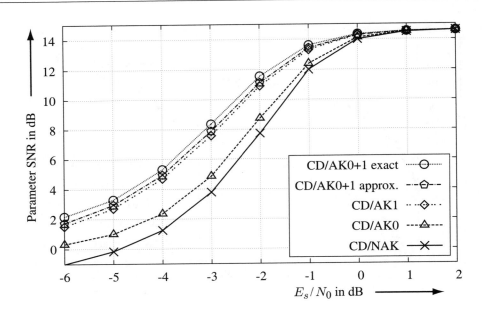

Figure 12.9: Parameter SNR dependent on the E_s/N_0 exploiting the different approaches for *a priori* knowledge on the symbol level

mapping. In general, the results depend strongly on the bit mapping since the parameter SNR may vary for different bit mappings even if the bit error rate is the same.

Thus, in SCCD the bit mapping has influence on both the quality after source decoding (e.g., the parameter SNR) and the channel decoding gain due to residual redundancy (e.g., measurable by the bit error rate) which complicates the search for an optimal bit mapping. The same challenge exists for the iterative source-channel decoding approach as we will see in Chap. 13. In [Hindelang 2001] details on the bit mapping can be found and it turns out that the folded binary mapping is a very good solution for SCCD in the case of Gaussian distributed AR(1) parameters with temporal correlation $\rho \in 0.5 \ldots 0.9$. It is of interest that many parameters in typical speech coding systems have similar properties.

12.4.7 Exploiting A Priori Knowledge in Source and/or Channel Decoding

The previous section has shown the gain of exploiting *a priori* knowledge in channel decoding. In the following, we will show the advantages of exploiting *a priori* knowledge in channel decoding and compare it with SDSD from Chap. 10. Channel decoding suffers from the conversion down to the bit level, but then it gains from

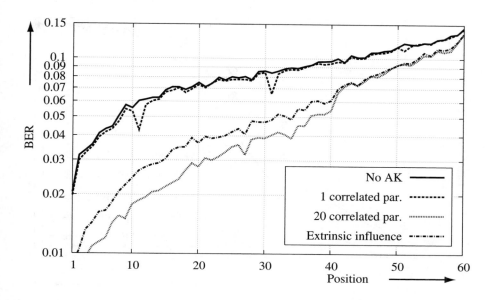

Figure 12.10: BER depending on the bit position for an AWGN channel at an E_s/N_0 of -3 dB for one correlated parameter in comparison to 20 correlated parameters and the gain due to the influence of the other parameters

the properties of the channel codes that lead – by applying respective interleavers – to an interdependence of different parameters (see e.g., the gain of the MSBs with (CD/AK0) in Fig. 12.8).

In Fig. 12.10 the dashed curve shows the gain of applying (CD/AK0+1) to a system where only one parameter stems from a Gauss–Markov source of 1st order. The three bits are placed at positions 11, 31, and 51. All other parameters are equally distributed and uncorrelated. Nevertheless, they gain a little in the bit error rate, especially if they are placed close to the considered parameter, e.g., positions 10 and 12. If we look at the curve with 20 correlated parameters a large gain can be obtained due to the mutual influence of the bits of each parameter, e.g., the bit error rate for the MSB at position 11 is reduced from about 0.042 to less than 0.02. For comparison the gain that stems from the influence of the other 19 parameters to the respective bit of one parameter at each position is shown (dash–dotted line). There, the *a posteriori* probability after channel decoding is divided by the prediction probabilities of each bit. Thus, only the gain due to the a priori knowledge of all other parameters is utilized. We will come back to this so-called gain by *extrinsic* influence later.

Finally, different approaches exploiting *a priori* knowledge either in source or in channel decoding are compared. For a fair comparison we use the MMSE estimator according to (10.7) where simple (SD/NAK) denotes a MMSE estimation without making use of the source statistics of the Gauss–Markov parameter. Looking at Fig. 12.11 the dashed and dash–dotted curves show that due to the conversion to the bit level in

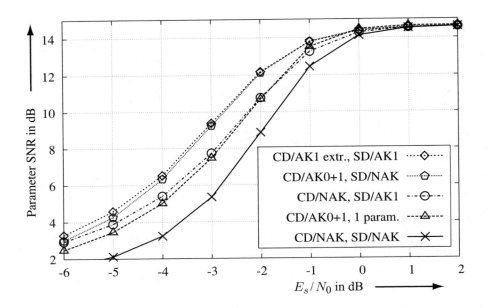

Figure 12.11: Parameter SNR dependent on the E_s/N_0 exploiting the different approaches for *a priori* knowledge on the symbol level

channel decoding the application of *a priori* knowledge in source decoding performs better in most cases. However, if there are 20 correlated parameters the (CD/AK0+1) approach (see dotted line with pentagons) outperforms the (SD/AK1) approach due to the mutual influence of the parameters in convolutional decoding. This mutual influence can not be exploited by the SDSD approach.

Summarizing the results we see that (SD/AK1) gains from the symbol level and the (CD/AK0+1) gains from the mutual influence within convolutional decoding. The question that arises is: can we take advantage of both effects and use *a priori* knowledge in source *and* channel decoding? To answer this question we have to ensure that the same *a priori* knowledge is not used twice. Therefore, we have already shown in Fig. 12.10 the gain by parameters other than those considered and in Sec. 13.2 the so-called *extrinsic* information is introduced in detail. Here, we show only a first result, called (CD/AK1 extr., SD/AK1), where (CD/AK1 extr.) denotes that only the gain due to the extrinsic influence is exploited in (CD/AK1). It delivers a better performance than the (CD/AK0+1, SD/NAK) curve. Note that the complex channel decoding is performed once, if only (CD/AK1) is used. In other simulations it turned out that this low complex approach gains especially if the channel code is weaker, e.g., if a convolutional code with $\nu = 3$ instead of $\nu = 6$ is applied. The application of (CD/AK0+1 extr.) together with (SD/AK1), which makes use of the full *a priori* knowledge both in channel and source decoding, would improve the result further by up to 0.3 dB.

Summarizing the SCCD approach, it has to be mentioned that it can be applied straight forwardly to each standardized system, e.g., video, audio and speech transmission, where residual redundancy is left after source encoding and where a channel coding approach that leads to a dependence between different source bits or symbols is applied, e.g., as in GSM or UMTS. Additionally, the SCCD approach leaves a lot of room for optimization, like quantization of parameters, bit mapping, interleaving, or unequal error protection, to design a system that is well adapted to both the source and the channel still keeping the state-of-the-art source coding (e.g., ACELP) or channel coding (e.g., convolutional codes). The capability and efficiency of SCCD has been shown at the example of the ANSI-136 standard and its modification [Hindelang et al. 2000a].

12.5 Comparison of SOCC versus SCCD

The SOCC and the SCCD approaches were developed between 1998 and 2000 in a scientific competition between the Institute of Communication Systems and Data Processing (IND) at RWTH Aachen University and the Institute for Communications Engineering (LNT) at Munich University of Technology (TUM). The question to be answered was, "Is it better to exploit the source *a priori* knowledge in channel decoding or in source decoding?" The Munich team investigated *source-controlled channel decoding* (SCCD) and the Aachen team investigated *soft decision source decoding* (SDSD) with the extension to *source-optimized channel codes* (SOCC). To allow a comparison a reference system was devised which will be shown in the following. Afterwards, the approaches from Secs. 12.3 and 12.4 will be compared.

System Configuration

The communication model that is mandatory for both was introduced in Sec. 12.2. The source model is designed to approximate the characteristics of the source parameters generated by block based speech coding schemes, such as those used in GSM or UMTS. Therefore, the source produces time discrete vectors $\tilde{v}$ of M elements $\tilde{v}_m \in \mathbb{R}$ where each parameter $\tilde{v}_m$ is modeled by an individual Gauss–Markov process of order one as shown in Sec. 10.2.1. The elements $\tilde{v}_m$ are quantized, mapped and channel coded to $M \cdot A$-dimensional bit vectors $\mathbf{y}$. The decoder receives the channel soft-output vector $\mathbf{z} \in \mathbb{R}^{(M \cdot A)}$ to estimate sample vectors $\hat{v} \in \mathbb{R}^M$, which are delivered to the sink.

To cover different possible system configurations we vary the following parameters of our source model:

- Correlation $\rho \in \{0, 0.75, 0.9\}$

- Maximum allowed bit rate on the channel: $A \in \{4, 6, 8\}$ bits/parameter

- Required source codec quality defined as the parameter $\mathrm{SNR}(\hat{v})$ under noise-free conditions according to Table 12.3.

Table 12.3: SNR requirements

Bit rate A in bits/parameter	4	6	8
Parameter SNR $(\hat{v})$ in dB	9	13	17

The number of bits transmitted per vector $\mathbf{z}$ will be fixed at $B = M \cdot A = 120$, thus the number of dimensions in $\tilde{v}$ takes the values $M \in \{30, 20, 15\}$.

Results

Applying the SCCD approach we employ the transmission system described in Sec. 12.4.1 with the difference that we now quantize each parameter for the three bit rates A (see Table 12.3) with $w = 2, 3, 4$ respectively leading to a block length of $M \cdot w = 60$ for all three cases. The interleaver maps the MSBs of each parameter index to the first part of the input vector $\mathbf{x}'$ for the convolutional encoder, then the SSBs and so on up to the fourth bit. Unequal error protection is achieved simply by not terminating the convolutional code. For short block lengths $(M \cdot w < 100)$ this turned out to be a good scheme for UEP. Owing to the applied code rate $r = 1/2$ in the channel encoder we obtain $B = 120$ transmitted bits. The channel decoder exploits (CD/AK0+1) according to (12.41) and the parameter estimation is made using the MMSE criterion without making use of the source statistics of the Gauss–Markov parameters (SD/NAK).

The SOCC-based transmission model operates at 4 bits per parameter with a two-dimensional ($\tilde{L} = 2$, i.e., pairs of parameters are jointly coded) SOCC with $\check{V} = 36$ levels out of 256 possible bit combinations. Therefore, the code rate is $\log_2(\check{V})/(4 \cdot \tilde{L}) = \log_2(36)/8 \approx 0.65$. With 6 bits per parameter again a two-dimensional SOCC with $\check{V} = 144$ levels is applied. The code rate is $\log_2(\check{V})/(6 \cdot \tilde{L}) = \log_2(144)/12 = 0.60$. Finally, for complexity reasons with 8 bits a one-dimensional ($\tilde{L} = 1$, i.e., all parameters individually coded) SOCC with $\check{V} = 20$ is applied, which leads to a code rate of $\log_2(\check{V})/(8 \cdot \tilde{L}) = \log_2(20)/8 = 0.54$. In all cases the SOCC was optimized for $E_s/N_0 = -1\,\mathrm{dB}$.

Figures 12.12–12.14 depict the simulation results for the given correlation factors ρ and transmission rates. By exploiting *a priori* knowledge either in *source-optimized channel codes* (SOCC) or in *source-controlled channel decoding* (SCCD) a remarkable enhancement of the parameter SNR is achieved for reasonably correlated sources. In Fig. 12.12 the SOCC approach is compared with SCCD for a transmission rate of $A = 4$ bits per parameter. While with $\rho \le 0.75$ there is an SNR range where SOCC performs slightly worse than the SCCD scheme, SOCC outperforms SCCD for all channel conditions if $\rho = 0.9$; a gain in parameter SNR of 1 to 3 dB can be observed. With 6 bits per dimension (Fig. 12.13) the SNR range where SCCD performs better increases. The simulation results for 8 bits are shown in Fig. 12.14. SCCD shows a good performance at a wide range of channel conditions that can be explained by the low SOCC dimension of 1. Nevertheless, SOCCs still achieve gains under very bad

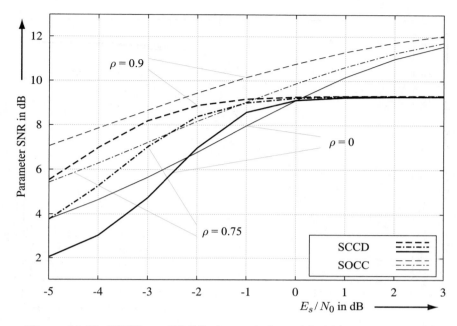

Figure 12.12: SCCD vs. SOCC, transmission with 4 bits per parameter

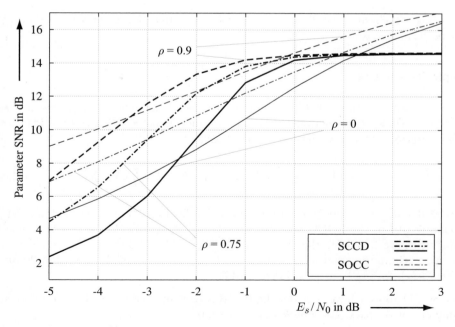

Figure 12.13: SCCD vs. SOCC, transmission with 6 bits per parameter

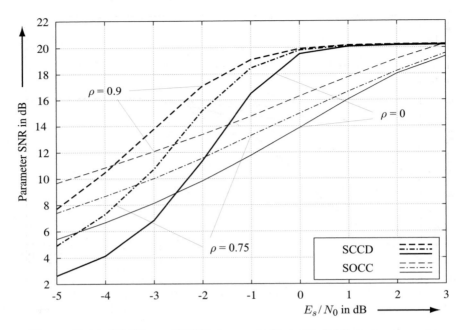

Figure 12.14: SCCD vs. SOCC, transmission with 8 bits per parameter

and very good channel conditions, which raises the question of how to evaluate the total performance of both systems.

Complexity Aspects

The complexity of the SOCC decoder is not shown in detail. Applying a scalar quantizer with $\check{V}$ reproduction levels and $\tilde{L}$-dimensional SOCCs, a straightforward implementation of the decoder has a complexity of

$$CX = \frac{\check{V} \cdot \tilde{L} \cdot M}{\tilde{L}} \left[\left(\frac{M \cdot A \cdot \tilde{L}}{M} + \check{V} \cdot \tilde{L} + 2 \right) \cdot \mathfrak{M} + (\check{V} \cdot \tilde{L} + 1) \cdot \mathfrak{A} \right], \qquad (12.45)$$

where $\mathfrak{A}$ denotes additions and $\mathfrak{M}$ denotes multiplications. For a detailed analysis of the complexity of the SOCC decoder see [Heinen 2001]. In Table 12.4 the decoder complexities are listed for the three coding rates. The MOPS column is calculated under the assumption that a parameter block has to be transmitted within 20 ms as it is in speech frames of, e.g., GSM or UMTS.

In the SCCD approach we consider the BCJR algorithm [Bahl et al. 1974] for decoding of convolutional codes. Its complexity is given by

$$CX_{\mathrm{BCJR}} = (4 \cdot 2^{\nu+1} + 2^{\Upsilon+1} \cdot \Upsilon) \cdot M \cdot w \cdot \mathfrak{M}$$
$$+ \left(2 \cdot 2^{\nu} + (2^{\nu} - 1) \cdot 2\right) \cdot M \cdot w \cdot \mathfrak{A}, \qquad (12.46)$$

Table 12.4: SOCC decoding complexities for one block of 120 transmitted bits

Bits/parameter	Add.	Mult.	Total op's	MOPS
4	19980	24840	44820	2.24
6	208800	227520	436320	21
8	6300	9000	15300	0.765

which can be obtained by summarizing over the calculation of the transition probabilities, the forward and the backward recursion, and the decoding probabilities (for details see [Hindelang 2001]). With the used settings $M \cdot w = 60$ the memory $\nu = 6$, and rate $r = 1/\Upsilon = 1/2$ we obtain $\mathfrak{M} = 31680$ and $\mathfrak{A} = 15240$. Additionally, there is some complexity needed for the calculation of the prediction probabilities P_{p01} and the MMSE source parameter estimation that contributes with less than 10% compared with two times convolutional decoding with the BCJR algorithm. The exact numbers without their derivation are shown in Table 12.5.

Table 12.5 shows that the complexity increases moderately with the number of quantization bits because the main term in the complexity is the one required for decoding the convolutional code. If only AK1 is used, which leads to a very small loss in performance for the folded binary mapping (see Fig. 12.9), the complexity is reduced to less than one half (bottom line in Table 12.5) because convolutional decoding is done only once per block. As a further step to reduce complexity, the BCJR algorithm could be replaced by the simpler soft-output Viterbi algorithm (SOVA). The complexity of SCCD decoding can then be reduced again by one half with only a little performance degradation.

In summary, the complexity of the two approaches is in the same range for the transmission with 4 bits per parameter and a two-dimensional SOCC. With 6 bits the SOCCs become much more complex in the two-dimensional case while they are less complex in the one-dimensional case, which was used for the scheme with 8 bits per parameter.

Table 12.5: The complexity of the SCCD approach using (CD/AK0+1) or (CD/AK1) (bottom line)

Bits/parameter	Add.	Mult.	Total op's	MOPS
4	31440	64320	95760	4.79
6	33360	66240	99600	4.98
8	40080	72960	113040	5.65
(CD/AK1) only, 8	18120	34560	52680	2.63

Summary

Both approaches, SOCC and SCCD, were evaluated in detailed simulations. It turned out that the SCCD approach performs better for a higher number of quantization levels, because of the very complex design of SOCCs for a higher number of bits per parameter. The SCCD approach gains up to 3 dB in the parameter SNR at 8 bits per parameter and $\rho = 0.0$ (Fig. 12.14). The SOCC approach performs better for a higher correlation of parameters since the correlation can be used completely on the symbol level. At 4 bits per parameter and a correlation $\rho = 0.9$ in particular the SOCC approach outperforms the SCCD approach over the whole range of channel SNRs (Fig. 12.12). For all considered transmission rates the SOCC performed better than the SCCD, under very bad and very good channel conditions.

12.6 Conclusions

In this chapter we presented two approaches, *Source Optimized Channel Coding* (SOCC) and *Source Controlled Channel Decoding* (SCCD), which combine the parameter estimation of Chap. 10 with channel coding.

SOCC is based on a new class of non-linear block codes, which are tailored for optimizing the parameter SNR at the receiver. The code optimization takes into account the source quantization as well as source and channel statistics. An interesting aspect of SOCC is that source and channel coding can be optimized separately. Despite this the performance of SOCC is comparable to the "true" joint optimization using COVQ.

SCCD on the other hand takes standard channel codes and makes use of the source statistics in the encoder by applying UEP, bit-mapping, and interleaving and in the decoder by exploiting the residual redundancy in terms of a priori probabilities.

Both approaches perform best if residual redundancy is left, either in terms of unequal distribution, temporal and also spatial correlation as demonstrated in this chapter and Chap. 11. The benchmarking unveiled that both approaches have their strengths and weaknesses depending on the particular operating point and hence the competition of Sec. 12.5 ended with a tie. A very promising option would be the combination of SOCC and SCCD, e.g., in terms of an adaptive multi-rate scheme.

In Sec. 12.4.7 it is shown that SCCD can be combined with SDSD and that *a priori* knowledge can be used twice under some circumstances. This was the starting point for further investigations of source and channel decoding using *a priori* knowledge and will be presented in Chap. 13.

Bibliography

Alajaji, F. I.; Phamdo, N. C.; Fuja, T. E. (1996). Channel Codes that Exploit the Residual Redundancy in CELP-Encoded Speech, *IEEE Transactions on Speech and Audio Processing*, vol. 4, no. 5, pp. 325–336.

Bahl, L. R.; Cocke, J.; Jelinek, F.; Raviv, J. (1974). Optimal Decoding of Linear Codes for Minimizing Symbol Error Rate, *IEEE Transactions on Information Theory*, vol. 20, no. 2, pp. 284–287.

Battail, G. (1987). Pondération des symboles décodés par l'algorithme de Viterbi, *Annales des Télécommunications*, vol. 42, no. 1, pp. 31–38 (in French).

Boudreau, D.; Dubuc, C. (1998). APRI-SOVA-Based Source Controlled Channel Decoding with the ITU-T G.729 Speech Coding Standard, *19th Biennial Symposium on Communications*, Kingston, ON, Canada, pp. 160–163.

Farvardin, N. (1990). A Study of Vector Quantization for Noisy Channels, *IEEE Transactions on Information Theory*, vol. 36, no. 4, pp. 799–809.

Farvardin, N.; Vaishampayan, V. (1987). Optimal Quantizer Design for Noisy Channels: An Approach to Combined Source-Channel Coding, *IEEE Transactions on Information Theory*, vol. 33, no. 6, pp. 827–838.

Farvardin, N.; Vaishampayan, V. (1991). On the Performance and Complexity of Channel Optimized Vector Quantizers, *IEEE Transactions on Information Theory*, vol. 37, no. 1, pp. 155–160.

Fazel, T.; Fuja, T. E. (2000). Joint Source-Channel Decoding of Block-Encoded Compressed Speech, *Conference on Information Sciences and Systems (CISS)*, Princeton, NJ, USA, pp. FA5.1–FA5.6.

Hagenauer, J. (1995). Source-Controlled Channel Decoding, *IEEE Transactions on Communications*, vol. 43, no. 9, pp. 2449–2457.

Hagenauer, J.; Höher, P. (1989). A Viterbi Algorithm with Soft-Decision Outputs and its Applications, *IEEE Global Telecommunications Conference (GLOBECOM)*, Dallas, Texas, pp. 1680–1686.

Heinen, S. (2001). *Quellenoptimierter Fehlerschutz für digitale Übertragungskanäle*, PhD thesis. Aachener Beiträge zu digitalen Nachrichtensystemen, vol. 14, P. Vary (ed.), RWTH Aachen University (in German).

Heinen, S.; Vary, P. (2000). Source Optimized Channel Codes (SOCCS) for Parameter Protection, *IEEE International Symposium on Information Theory (ISIT)*, Sorrento, Italy.

Hindelang, T. (2001). *Source-Controlled Channel Encoding and Decoding for Mobile Communications*, PhD thesis, VDI-Verlag, Düsseldorf: Fortschritt-Berichte VDI, Reihe 10, Nr. 695, ISBN 3-18-369510-3, Munich University of Technology.

Hindelang, T.; Fingscheidt, T.; Seshadri, N.; Cox, R. V. (2000a). A Re-investigation of Scalar Quantization for Mobile Speech Transmission, *IEEE Vehicular Technology Conference (Fall)*, Boston, Massachusetts, vol. 5, pp. 2459–2466.

Hindelang, T.; Heinen, S.; Vary, P.; Hagenauer, J. (2000b). Two Approaches to Combined Source-Channel Coding: A Scientific Competition in Estimating Correlated Parameters, *International Journal of Electronics and Communications (AEÜ)*, vol. 54, no. 6, pp. 364–378.

Johannesson, R.; Zigangirov, K. S. (1999). *Fundamentals of Convolutional Coding*, IEEE Press, Inc., Piscataway, New Jersey.

Linde, Y.; Buzo, A.; Gray, R. M. (1980). An Algorithm for Vector Quantizer Design, *IEEE Transactions on Communications*, vol. 28, no. 1, pp. 84–95.

Melsa, J. L.; Cohn, D. L. (1978). *Decision and Estimation Theory*, McGraw-Hill, New York.

Proakis, J. G. (2000). *Digital Communications*, 4th edn, McGraw-Hill, New York.

Skoglund, M. (1999). On Channel-Constrained Vector Quantization and Index Assignment for Discrete Memoryless Channels, *IEEE Transactions on Information Theory*, vol. 45, no. 7, pp. 2615–2622.

Vary, P.; Martin, R. (2006). *Digital Speech Transmission*, John Wiley & Sons, Ltd, Chichester.

Zeger, K.; Gersho, A. (1990). Pseudo-Gray Coding, *IEEE Transactions on Communications*, vol. 38, December, pp. 2147–2158.

Chapter 13

Iterative Source-Channel Decoding & Turbo DeCodulation

Marc Adrat, Thorsten Clevorn, Laurent Schmalen

13.1 Introduction

In 1993, the Turbo principle was devised by C. Berrou, A. Glavieux, and P. Thiti-majshima for near Shannon limit error correcting decoding [Berrou et al. 1993] with reasonable computational complexity. The key novelty of the Turbo principle was the iterative exchange of so-called *extrinsic* information between two (or more) decoders, which are concatenated by a large bit interleaver [Hagenauer et al. 1996]. Efficient coding close to the Shannon limit also makes the Turbo principle attractive for many other fields of digital signal processing. Two examples are: *Iterative Source-Channel Decoding* (ISCD) and *Turbo DeCodulation* (TDeC). The convergence behavior of Turbo processes can be understood and visualized by applying so-called *EXtrinsic Information Transfer* (EXIT) charts [ten Brink 1999], [ten Brink 2001].

In Sec. 12.4.7 a joint source-channel coding scheme (CD/AK1, SD/AK1, extr.) was outlined that combines the concepts of *Source Controlled Channel Decoding* (SCCD) and of *Soft Decision Source Decoding* (SDSD). A closer look reveals that this combination exhibits some relationships to the Turbo principle. Such a scheme resembles the iterative evaluation of channel code redundancy and natural residual source redundancy. The *main channel decoder* (of Fig. 12.7) exploits some kind of *extrinsic* information, which is gained from the 0th/1st order *a priori* knowledge (AK0/AK1), in order to enhance the soft-outputs of the preliminary *first channel decoder* [Hagenauer 1995], [Hindelang 2001], [Hindelang et al. 2007]. These enhanced soft-outputs

Advances in Digital Speech Transmission   Edited by R. Martin, U. Heute and C. Antweiler
© 2008 John Wiley & Sons, Ltd

are provided to a subsequent *main source decoder*. Thus, the overall scheme is similar to a Turbo process with two iterations.

However, in the combined scheme of SCCD and SDSD some features or components of a Turbo scheme are missing or used in a different sense, e.g., the bit interleaver is primarily designed for unequal error protection and decorrelating errors. In a Turbo scheme the decorrelation of the exchange of *extrinsic* information is the new essential task of the bit interleaver. Moreover, in a (in some suitable sense) properly designed system even more than only two iterations might be profitable. Thus, it is advisable to strictly apply all basic prerequisites of the Turbo principle to the joint source-channel coding problem. This leads to an *Iterative Source-Channel Decoding* (ISCD) system, e.g., [Hindelang et al. 2000], [Görtz 2000], [Adrat et al. 2001] which is the main topic of the first part of the present chapter. The benefits of ISCD have already been demonstrated for the GSM system [Perkert et al. 2001].

Besides a detailed treatment of the relevant terms of *extrinsic* information, the EXIT chart analysis tool is applied to ISCD [Adrat 2003], [Adrat et al. 2003], [Adrat, Vary 2005]. On the one hand, some limiting factors of ISCD schemes can be identified [Adrat et al. 2005a], [Adrat et al. 2006b]. On the other hand, some design guidelines for highly capable ISCD scheme can be derived [Adrat, Vary 2004], [Adrat et al. 2006a], [Adrat et al. 2005b].

Finally, the ISCD scheme will be extended by a third component, namely the demodulator, to incorporate the iterative demodulation scheme *Bit-Interleaved Coded Modulation with Iterative Decoding* (BICM-ID) [Li et al. 2002], [Hanzo et al. 2002]. Such a multiple Turbo process is called *Turbo DeCodulation* (TDeC) [Clevorn et al. 2005b], [Clevorn et al. 2005a], [Clevorn 2006]. TDeC systems and the respective advancements are analyzed in the second part of this chapter.

13.2 The Key of the Turbo Principle: Extrinsic Information

The key element of all decoding processes according to the Turbo principle is the iterative exchange of so-called *extrinsic* information between the constituent decoders [Berrou et al. 1993], [Hagenauer et al. 1996]. *Extrinsic* information can usually be extracted from the *a posteriori* output of a *Soft-Input/Soft-Output* (SISO) decoder. Considering this *extrinsic* information as additional *a priori* input for the other constituent decoder(s) permits stepwise performance improvements. After several iterations the system converges to a steady state. The maximum number of profitable iterations depends on the properties of the constituent decoders as well as the independence of their extractable *extrinsic* information. The latter side constraint can usually be omitted by placing a properly designed interleaver between the constituent decoders. The overall convergence behavior becomes predictable with the *EXtrinsic Information Transfer* (EXIT) chart analysis tool [ten Brink 1999], [ten Brink 2001].

In Sec. 13.2.1 we will briefly review different terms of reliability information. Afterwards, we will discuss in more detail the *extrinsic* terms of information for SISO channel decoding, SDSD, and *Soft Demodulation* (SDM) in Secs. 13.2.2 to 13.2.4. We will introduce the EXIT chart analysis tool in Sec. 13.2.5. All these considerations are of relevance for the ISCD and TDeC schemes being introduced in Secs. 13.3 and 13.4.

13.2.1 Terms of Reliability Information

In Turbo decoding schemes reliability information for single data bits x is processed in several stages of a receiver. Such reliability information can either be expressed in terms of probabilities $P(\cdot)$ or in *log-likelihood* ratios $L(\cdot)$ (or short: L-values). An L-value is the natural logarithm of the probabilities ratio of both alternative realizations of $x \in \{+1, -1\}$. For instance, the *a posteriori* L-value is [Hagenauer et al. 1996]

$$L(x|\mathbf{z}) = \log_e \frac{P(x = +1|\mathbf{z})}{P(x = -1|\mathbf{z})}. \tag{13.1}$$

For the use of x and $\mathbf{z}$ see also Sec. 10.2.2. The sign of an L-value yields the hard decision, $\hat{x} = \text{sign}\{L(\cdot)\}$, and the magnitude $|L(\cdot)|$ represents the reliability of this decision. The L-value in (13.1) is called the *a posteriori* L-value because at the receiver it allows us to decide on the most probably sent bit x given the received sequence $\mathbf{z}$.

Applying Bayes' theorem in *mixed form* and assuming a memoryless transmission channel allows us to separate the *a posteriori* L-value in several additive terms [Hagenauer et al. 1996],

$$
\begin{aligned}
L(x|\mathbf{z}) &= \log_e \frac{p(\mathbf{z}|x = +1) \cdot P(x = +1)}{p(\mathbf{z}|x = -1) \cdot P(x = -1)} \\
&= \log_e \frac{p(z|x = +1) \cdot P(x = +1) \cdot p(\mathbf{z}^{[\text{ext}]}|x = +1)}{p(z|x = -1) \cdot P(x = -1) \cdot p(\mathbf{z}^{[\text{ext}]}|x = -1)} \\
&= \underbrace{L(z|x)}_{\substack{\text{transmission related} \\ \text{information}}} + \underbrace{L(x)}_{\substack{\text{a priori} \\ \text{information}}} + \underbrace{L^{[\text{ext}]}(x) + \dots}_{\substack{\text{different terms of} \\ \textit{extrinsic information}}} .
\end{aligned}
\tag{13.2}
$$

$P(\cdot)$ denotes a discrete probability and $p(\cdot)$ a probability density function. The term $\mathbf{z}^{[\text{ext}]}$ comprises the same elements as the received sequence $\mathbf{z}$, except the particular received value z for the data bit x under consideration. Note that, to emphasize the *extrinsic* nature of $\mathbf{z}^{[\text{ext}]}$, we use the superscript $[\text{ext}]$ instead of the notation introduced in (12.25).

The first term $L(z|x)$ represents transmission related reliability information. It specifies the L-value for receiving a real-valued $z \in \mathbb{R}$ given that $x \in \{+1, -1\}$ has originally been sent. In the case of a transmission channel with Rayleigh fading (fading coefficient a) as well as *Additive White Gaussian Noise* (AWGN) with known E_s/N_0, $L(z|x)$ can be expressed by [Hagenauer et al. 1996]

$$L(z|x) = 4 \cdot a \cdot E_s/N_0 \cdot z \,. \tag{13.3}$$

The second term $L(x)$ in (13.2) denotes bitwise *a priori* information for bit x,

$$L(x) = \log_e \frac{P(x = +1)}{P(x = -1)} \,. \tag{13.4}$$

Both terms of reliability information, $L(z|x)$ and $L(x)$, represent so-called *intrinsic* information about data bit x. Neither of these values is influenced by one of the other elements in the sent sequence $\underline{x}$ or its received counterpart $\underline{z}$.

An additional term can generally be extracted from $L(x|\underline{z})$ for any SISO decoder involved in a Turbo process. This extra term is called *extrinsic* information because it describes the impact of the other elements $\underline{z}^{[\text{ext}]}$ of $\underline{z}$ (i.e., the elements of the received sequence $\underline{z}$ excluding z) on the bit x under consideration,

$$L^{[\text{ext}]}(x) = L(\underline{z}^{[\text{ext}]}|x) \,. \tag{13.5}$$

This impact results either from artificial mutual dependencies that are introduced between the bits x of $\underline{x}$ (e.g., by channel coding) or from natural redundancies between the data bits x of the originally transmitted sequence $\underline{x}$ of bit patterns. The dots in (13.2) indicate that several of such terms of *extrinsic* information exist in a Turbo process where multiple constituent decoders are concatenated.

13.2.2 Extrinsic Information of Channel Decoding

Channel coding introduces artificial dependencies. The inputs and outputs of a SISO channel decoder are depicted in Fig. 13.1. The most popular linear channel coding concept is binary convolutional encoding. A convolutional code can be described by a trellis diagram. In this trellis diagram nodes represent encoder states and branches represent state transitions. If the single data bits $x \in \{+1, -1\}$ of $\underline{x}$ are encoded one after another, the binary input x at time λ causes a state transition from encoder

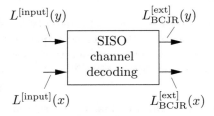

Figure 13.1: Soft-inputs/-outputs of block *SISO channel decoding*

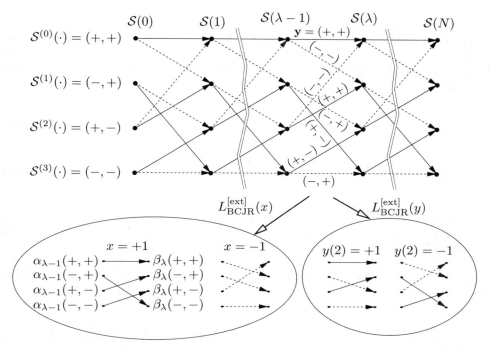

Figure 13.2: Illustration of the *extrinsic* information of a data bit x resulting from channel encoding (solid: $x = +1$, dashed: $x = -1$)

state $\mathcal{S}(\lambda - 1)$ to $\mathcal{S}(\lambda)$. Each state transition is labeled with the specific code word $\mathbf{y}$ being transmitted. The length Υ of $\mathbf{y} = (y(1), \ldots, y(\Upsilon))$ determines the code rate $r = 1/\Upsilon$.

The upper part of Fig. 13.2 depicts an example of a memory $\nu = 2$ (constraint length $L = \nu + 1$, see also Sec. 12.4.1) convolutional code (recursive systematic convolutional code with generator polynomial $\mathbf{G} = \left(1, \frac{5}{7}\right)_8$, compared to (12.23)). If, for instance, the encoder is in state $\mathcal{S}^{(2)}(\lambda - 1) = (+, -)$ then the next transmitted code word can either be $\mathbf{y} = (+, +)$ if $x = +1$ (solid line) or $\mathbf{y} = (-, -)$ if $x = -1$ (dashed line). The code words $(+, -)$ and $(-, +)$ can not follow the state $\mathcal{S}^{(2)}(\lambda - 1)$.

After channel transmission the decoding algorithm at the receiver tries to estimate the transmitted bits x from the noisy received sequence $\underline{\mathbf{z}}$ of $\mathbf{y}$. Owing to the artificial dependencies introduced by channel encoding, the trellis paths to the right and to the left of the bit x under consideration have an influence on x. The reliability gain due to this influence is called *extrinsic* information.

For instance, if the receiver knows the left neighboring state $\mathcal{S}^{(2)}(\lambda - 1) = (+, -)$ and the right neighboring state $\mathcal{S}^{(1)}(\lambda) = (-, +)$, it can conclude from this kind of *extrinsic* information on $\hat{x} = +1$ resp. $\hat{\mathbf{y}} = (+, +)$ (solid line) regardless of any *intrinsic* information given by the received pattern $\underline{\mathbf{z}}$.

The *extrinsic* information resulting from the neighborhood can efficiently be determined using Bahl's *et al.* symbol-by-symbol *Maximum A Posteriori* (MAP) decoder. With deference to the inventors it is also called the BCJR decoder [Bahl et al. 1974]. The BCJR algorithm is based on a *forward–backward* recursive determination rule,

$$\alpha_\lambda(\mathcal{S}^{(j)}(\lambda)) = \sum_{i=0}^{2^\nu-1} \gamma_\lambda(\mathcal{S}^{(i)}(\lambda-1), \mathcal{S}^{(j)}(\lambda)) \cdot \alpha_{\lambda-1}(\mathcal{S}^{(i)}(\lambda-1)) \tag{13.6}$$

$$\beta_\lambda(\mathcal{S}^{(i)}(\lambda)) = \sum_{j=0}^{2^\nu-1} \gamma_{\lambda+1}(\mathcal{S}^{(i)}(\lambda), \mathcal{S}^{(j)}(\lambda+1)) \cdot \beta_{\lambda+1}(\mathcal{S}^{(j)}(\lambda+1)). \tag{13.7}$$

Equations (13.6) and (13.7) specify the reliability of the encoder state $\mathcal{S}(\lambda)$ resulting from preceding resp. succeeding trellis stages. If the convolutional encoder starts in state $\mathcal{S}^{(i)}(0)$, $i = 0, \ldots, 2^\nu - 1$, (13.6) is initialized with $\alpha_0(\mathcal{S}^{(i)}(0)) = 1$ and $\alpha_0(\mathcal{S}^{(j)}(0)) = 0$ for $j = 0, \ldots, 2^\nu - 1$, $j \neq i$. The *backward* recursion (13.7) is initialized similarly. The innovation of each state transition (branch) is introduced by

$$\gamma_\lambda(\mathcal{S}(\lambda-1), \mathcal{S}(\lambda)) = \exp\left(\frac{x}{2} \cdot L^{[\text{input}]}(x)\right) \cdot \exp\left(\sum_{\substack{i=1\\i\neq i_{\text{sys}}}}^{\Upsilon} \frac{y(i)}{2} \cdot L^{[\text{input}]}(y(i))\right). \tag{13.8}$$

At each trellis stage $\lambda = 1, \ldots, N$ the innovation (13.8) needs to be determined for each of the valid $2^\nu \cdot 2$ state transitions from $\mathcal{S}(\lambda - 1)$ to $\mathcal{S}(\lambda)$. If there is no direct connection between $\mathcal{S}(\lambda - 1)$ and $\mathcal{S}(\lambda)$ then $\gamma_\lambda(\mathcal{S}(\lambda - 1), \mathcal{S}(\lambda)) = 0$. The innovation considers the soft-input L-values for both $L^{[\text{input}]}(x)$ for the data bits x as well as $L^{[\text{input}]}(y(i))$ for the parity check bits $y(i)$, $i = 1, \ldots, \Upsilon$ of $\mathbf{y}$. The meaning and usage of these input and output L-values will be explained in detail in Sec. 13.3.1. Note that, if a systematic channel code is used, i.e., if x is explicitly part of $\mathbf{y}$, the addend for $i = i_{\text{sys}}$ is excluded from the summation in (13.8), because we consider it to be already included in $L^{[\text{input}]}(x)$. In general, $L^{[\text{input}]}(x)$ shall contain the systematic information from the channel as well as the *a priori* information on the bit x originating from possible other decoder components.

Finally, using (13.8) to compute the *forward–backward* algorithm (13.6) and (13.7) permits to determine the *extrinsic* L-value of BCJR channel decoding,

$$L_{\text{BCJR}}^{[\text{ext}]}(x) = \tag{13.9}$$

$$\log_e \frac{\displaystyle\sum_{j=0}^{2^\nu-1} \beta_\lambda(\mathcal{S}^{(j)}(\lambda)) \cdot \sum_{i=0}^{2^\nu-1} \gamma_\lambda^{[\text{ext}]}(\mathcal{S}^{(i)}(\lambda-1), \mathcal{S}^{(j)}(\lambda)|x=+1) \cdot \alpha_{\lambda-1}(\mathcal{S}^{(i)}(\lambda-1))}{\displaystyle\sum_{j=0}^{2^\nu-1} \beta_\lambda(\mathcal{S}^{(j)}(\lambda)) \cdot \sum_{i=0}^{2^\nu-1} \gamma_\lambda^{[\text{ext}]}(\mathcal{S}^{(i)}(\lambda-1), \mathcal{S}^{(j)}(\lambda)|x=-1) \cdot \alpha_{\lambda-1}(\mathcal{S}^{(i)}(\lambda-1))}.$$

For the specific trellis stage with the desired bit x under consideration a reduced *extrinsic* innovation

$$\gamma_\lambda^{[\text{ext}]}(\mathcal{S}(\lambda - 1), \mathcal{S}(\lambda)|x) = \exp\left(\sum_{\substack{i=1 \\ i \neq i_{\text{sys}}}}^{\Upsilon} \frac{y(i)}{2} \cdot L^{[\text{input}]}(y(i))\right) \tag{13.10}$$

is taken into account. This reduced *extrinsic* innovation splits the overall set of state transitions of size $2^\nu \cdot 2$ into two subsets of size 2^ν, each for a given $x = \pm 1$ (compare to the lower part on the left of Fig. 13.2). Moreover, it excludes all soft-input L-values $L^{[\text{input}]}(x)$ for the specific bit x under consideration.

In the same way as described by (13.9) and (13.10) *extrinsic* information $L_{\text{BCJR}}^{[\text{ext}]}(y)$ for every code bit y can be determined. For this purpose, in the formulas already mentioned, the term x needs to be replaced by y. The modified separation into the subsets for $y = +1$ (numerator of modified (13.9)) and $y = -1$ (denominator of modified (13.9)) is illustrated for $y(i = 2)$ in the lower part on the right of Fig. 13.2. For the modified reduced *extrinsic* innovation $\gamma_\lambda^{[\text{ext}]}(\mathcal{S}(\lambda - 1), \mathcal{S}(\lambda)|y)$ the addend for the specific $y(i) = y$ under test has to be excluded.

13.2.3 Extrinsic Information of Source Decoding

In the case of source decoding the mutual dependencies between bits x result from the natural residual redundancy of the bit patterns $\mathbf{x}$ representing a codec parameter v. Such natural residual source redundancy typically remains in the bit stream after source encoding due to complexity and delay constraints in the encoding process. The dependencies can be measured in terms of a non-uniform probability distribution $P(\mathbf{x})$ or in terms of $P(\mathbf{x}_k|\mathbf{x}_{k-1})$ if mutual dependencies in time exist. The inputs and output of a SDSD are depicted in Fig. 13.3.

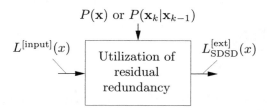

Figure 13.3: Soft-inputs/-outputs of block *utilization of residual redundancy*

Non-Uniform Probability Distribution $P(\mathbf{x})$

Let us illustrate the origin of the *extrinsic* information of SDSD with a comprehensible example. For this purpose, we restrict our first basic considerations to the case where the bit patterns $\mathbf{x}$ exhibit a non-uniform distribution only. In Fig. 13.4-a the probability mass function $P(\mathbf{x})$ of a Gaussian distributed codec parameter v that has been quantized to $2^w = 8$ levels is depicted. Furthermore, we assume that the equally likely bit x under test (left-most bit) is not known, but all the other bits $\mathbf{x}^{[\text{ext}]}$ of $\mathbf{x}$ (two right-most bits) are communicated without any error, i.e., $\mathbf{z}^{[\text{ext}]} = \mathbf{x}^{[\text{ext}]}$. In this case, (13.5) can be approximated by

$$L_{\text{SDSD},P(\mathbf{x}),\text{perf.}}^{[\text{ext}]}(x) = \log_e \frac{P(\mathbf{z}^{[\text{ext}]}|x = +1)}{P(\mathbf{z}^{[\text{ext}]}|x = -1)} = \log_e \frac{P(\mathbf{x}^{[\text{ext}]}, x = +1)}{P(\mathbf{x}^{[\text{ext}]}, x = -1)}. \tag{13.11}$$

If the two right-most bits of $\mathbf{x}$ are supposed to be given as $\mathbf{x}^{[\text{ext}]} = (+, +)$, we can conclude from the probability mass function $P(\mathbf{x})$ that for the left-most bit the realization $x = -1$ is more likely than $x = +1$. As a consequence, the *extrinsic* L-value given by (13.11) will be negative.

Notice, in this example both terms of *intrinsic* information are zero: $L(z|x) = 0$ because bit x is unknown and $L(x) = 0$ because the realizations of the bit x under test are assumed to be equiprobable. Anyhow, a non-zero *extrinsic* resp. *a posteriori* L-value exists due to the mutual dependencies between x and $\mathbf{x}^{[\text{ext}]}$.

In practice, the right-most bits are usually not perfectly known and, therefore, the terms in the numerator and denominator of (13.11) need to be replaced by a weighted

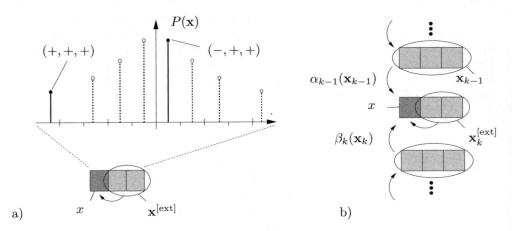

Figure 13.4: Illustrations for the *extrinsic* information of a data bit x
a) Explanation for non-uniform parameter distribution $P(\mathbf{x})$
b) Explanation for redundancies in time $P(\mathbf{x}_k|\mathbf{x}_{k-1})$

sum over all 2^{w-1} possible permutations of $\mathbf{x}^{[\mathrm{ext}]}$,

$$L_{\mathrm{SDSD},P(\mathbf{x})}^{[\mathrm{ext}]}(x) = \log_e \frac{\sum\limits_{i=0}^{2^{w-1}-1} P(\mathbf{x}^{[\mathrm{ext}](i)}, x = +1) \cdot \theta(\mathbf{x}^{[\mathrm{ext}](i)})}{\sum\limits_{i=0}^{2^{w-1}-1} P(\mathbf{x}^{[\mathrm{ext}](i)}, x = -1) \cdot \theta(\mathbf{x}^{[\mathrm{ext}](i)})}. \tag{13.12}$$

In the example, the $2^{w-1} = 4$ permutations for the two right-most bits $x^{[\mathrm{ext}](i)}(\kappa)$, $\kappa = 1,\dots,w-1$ are $\mathbf{x}^{[\mathrm{ext}](i)} \in \{(-,-),(-,+),(+,-),(+,+)\}$. The weights

$$\theta(\mathbf{x}^{[\mathrm{ext}]}) = \exp\left(\sum_{\kappa=1}^{w-1} \frac{x^{[\mathrm{ext}]}(\kappa)}{2} \cdot L^{[\mathrm{input}]}(x^{[\mathrm{ext}]}(\kappa))\right) \tag{13.13}$$

are functions of the soft-input L-values $L^{[\mathrm{input}]}(x^{[\mathrm{ext}]}(\kappa))$.

Mutual Dependencies in Time $P(\mathbf{x}_k|\mathbf{x}_{k-1})$

Besides the non-uniform parameter distribution $P(\mathbf{x})$, source codec parameters determined by real-world source encoders often exhibit mutual dependencies in time k. Such dependencies can be measured in terms of a conditional probability function $P(\mathbf{x}_k|\mathbf{x}_{k-1})$. The generalized determination rule for the *extrinsic* information resulting from $P(\mathbf{x}_k|\mathbf{x}_{k-1})$ reads [Adrat 2003], [Adrat, Vary 2005]

$$L_{\mathrm{SDSD}}^{[\mathrm{ext}]}(x) = \tag{13.14}$$

$$\log_e \frac{\sum\limits_{i=0}^{2^{w-1}-1} \beta_k(\mathbf{x}_k^{[\mathrm{ext}](i)}, x = +1) \cdot \sum\limits_{j=0}^{2^{w-1}} \gamma_k^{[\mathrm{ext}]}(\mathbf{x}_k^{[\mathrm{ext}](i)}, \mathbf{x}_{k-1}^{(j)} | x = +1) \cdot \alpha_{k-1}(\mathbf{x}_{k-1}^{(j)})}{\sum\limits_{i=0}^{2^{w-1}-1} \beta_k(\mathbf{x}_k^{[\mathrm{ext}](i)}, x = -1) \cdot \sum\limits_{j=0}^{2^{w-1}} \gamma_k^{[\mathrm{ext}]}(\mathbf{x}_k^{[\mathrm{ext}](i)}, \mathbf{x}_{k-1}^{(j)} | x = -1) \cdot \alpha_{k-1}(\mathbf{x}_{k-1}^{(j)})}.$$

Equation (13.14) for determining the *extrinsic* L-value of SDSD exhibits many analogies to (13.9) for measuring the $L_{\mathrm{BCJR}}^{[\mathrm{ext}]}(x)$ of channel decoding. The impact of past and some possibly given future bit patterns (see Fig. 13.4-b) can efficiently be considered by a *forward–backward* algorithm,

$$\alpha_k(\mathbf{x}_k^{(j)}) = \sum_{i=0}^{2^w-1} \gamma_k(\mathbf{x}_k^{(j)}, \mathbf{x}_{k-1}^{(i)}) \cdot \alpha_{k-1}(\mathbf{x}_{k-1}^{(i)}) \tag{13.15}$$

$$\beta_k(\mathbf{x}_k^{(i)}) = \sum_{j=0}^{2^w-1} \gamma_{k+1}(\mathbf{x}_{k+1}^{(j)}, \mathbf{x}_k^{(i)}) \cdot \beta_{k+1}(\mathbf{x}_{k+1}^{(j)}) \tag{13.16}$$

with the innovation

$$\gamma_k(\mathbf{x}_k, \mathbf{x}_{k-1}) = P(\mathbf{x}_k|\mathbf{x}_{k-1}) \cdot \exp\left(\sum_{\kappa=1}^{w} \frac{x_k(\kappa)}{2} \cdot L^{[\mathrm{input}]}(x_k(\kappa))\right). \tag{13.17}$$

Table 13.1: Key differences in the determination rules for $L_{\text{BCJR}}^{[\text{ext}]}(x)$ and $L_{\text{SDSD}}^{[\text{ext}]}(x)$

Parameter	BCJR	SDSD
• *Forward–backward* algorithm	(13.6), (13.7)	(13.15), (13.16)
• recursions on ("nodes")	states $\mathcal{S}(\lambda)$	bit patterns $\mathbf{x}_k$
• Innovation	(13.8)	(13.17)
• number of branches ("state transitions")	$2^\nu \cdot 2$ (from every $\mathcal{S}(\lambda)$ via $x = \pm 1$)	$2^w \cdot 2^w$ (from every $\mathbf{x}_{k-1}$ to every $\mathbf{x}_k$)
• branches labeled with	data bit x code word $\mathbf{y}$	parameter v_k bit pattern $\mathbf{x}_k$
• branch specific *a priori* information	$x \cdot L(x)$ (2 values for $x = \pm 1$)	$P(\mathbf{x}_k \vert \mathbf{x}_{k-1})$ ($2^w \cdot 2^w$ values for pairs $\mathbf{x}_k, \mathbf{x}_{k-1}$)

In the *forward–backward* algorithm (13.15), (13.16) the sequence of bit patterns $\mathbf{x}_k$ resembles the sequence of states $\mathcal{S}(\lambda)$ (i.e., the path through the trellis diagram). The key difference is that in the innovation (13.17) "state transitions" from every $\mathbf{x}_{k-1}$ to every $\mathbf{x}_k$ are possible. Each of these $2^w \cdot 2^w$ "state transitions" exhibits a specific probability $P(\mathbf{x}_k \vert \mathbf{x}_{k-1})$.

Similar to (13.10) the particular innovation for the time k under consideration needs to be reduced by all soft-input information for data bit x. For this purpose, $P(\mathbf{x}_k \vert \mathbf{x}_{k-1})$ is divided by the bitwise probability $P(x)$ and the specific addend for $\kappa = \kappa_x$ (i.e., the position of the bit x under consideration in $\mathbf{x}_k$) in the sum of (13.17) is eliminated. The reduced innovation is

$$\gamma_k^{[\text{ext}]}(\mathbf{x}_k^{[\text{ext}]}, \mathbf{x}_{k-1} \vert x) = \frac{P(\mathbf{x}_k \vert \mathbf{x}_{k-1})}{P(x)} \cdot \exp\left(\sum_{\substack{\kappa=1 \\ \kappa \neq \kappa_x}}^{w} \frac{x_k(\kappa)}{2} \cdot L^{[\text{input}]}(x_k(\kappa)) \right) . \quad (13.18)$$

Some of the key differences in the determination rules for $L_{\text{BCJR}}^{[\text{ext}]}(x)$ and $L_{\text{SDSD}}^{[\text{ext}]}(x)$ are summarized in Table 13.1.

13.2.4 Extrinsic Information of Demodulation

In the case of *Soft Demodulation* (SDM) the dependencies that can be used to generate *extrinsic* information are artificially introduced in the modulator by the symbol mapping. The inputs and output of a soft demodulator are depicted in Fig. 13.5.

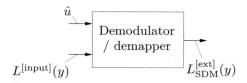

Figure 13.5: Soft-inputs/-outputs of the block *demodulator / demapper*

For this purpose, the channel encoded sequence $\mathbf{y}$ is first partitioned into bit patterns $\mathbf{y} = (y(1), \ldots, y(J))$ of J encoded bits and then mapped to the transmitted modulated symbols u. The *extrinsic* L-value $L_{\text{SDM}}^{[\text{ext}]}(y(j))$ of the demodulator for the encoded bit $y(j)$ at position j of a bit pattern $\mathbf{y}$ can be computed by [Li et al. 2002], [Schreckenbach et al. 2003], [Clevorn 2006]

$$
L_{\text{SDM}}^{[\text{ext}]}(y(j)) = \log_e \frac{\displaystyle\sum_{\check{u} \in \mathbb{U}_{y(j)=+1}} p(\hat{u}|\check{u}) \cdot \exp\left(\sum_{\substack{i=1 \\ i \neq j}}^{J} \frac{\check{y}(i)}{2} \cdot L^{[\text{input}]}(\check{y}(i))\right)}{\displaystyle\sum_{\check{u} \in \mathbb{U}_{y(j)=-1}} p(\hat{u}|\check{u}) \cdot \exp\left(\sum_{\substack{i=1 \\ i \neq j}}^{J} \frac{\check{y}(i)}{2} \cdot L^{[\text{input}]}(\check{y}(i))\right)} , \tag{13.19}
$$

with $\check{u}$ representing the bit pattern $\check{\mathbf{y}}$ under consideration after symbol mapping. Furthermore, the *a priori* L-values $L^{[\text{input}]}(\check{y}(i))$ of the other encoded bits, i.e., $i \neq j$, and the received value $\hat{u}$ from the channel are used, both for the respective bit pattern $\check{\mathbf{y}}$ under consideration. For each j, the set $\mathbb{U}$ of the 2^J possible modulated symbols $\check{u}$ is divided into the two equally sized subsets $\mathbb{U}_{y(j)=+1}$ and $\mathbb{U}_{y(j)=-1}$, which contain the modulated symbols u whose jth bit of the corresponding bit pattern $\check{\mathbf{y}}$ is $\check{y}(j) = +1$ or $\check{y}(j) = -1$, respectively. Thus, the identical exponential term in the nominator and the denominator (which can be considered as *a priori* knowledge or innovation) is weighted by the two channel related conditional probability density functions $p(\hat{u}|\check{u})$ for the two modulated symbols $\check{u} \in \mathbb{U}_{y(j)=+1}$ and $\check{u} \in \mathbb{U}_{y(j)=-1}$.

The benefit of *a priori* information for the *extrinsic* L-value is visualized in Fig. 13.6 using the first bit $\check{y}(1)$ of a Gray symbol mapping for an 8PSK signal constellation set as example. Without *a priori* information, i.e., $L^{[\text{input}]}(\check{y}(i)) = 0$, all distances between $\hat{u}$ and all 2^J possible $\check{u}$ are considered in (13.19) via $p(\hat{u}|\check{u})$. With perfect *a priori* information, i.e., $L^{[\text{input}]}(\check{y}(i)) \in \{\pm\infty\}$, the only remaining distances are the ones to both symbols $\check{u}$, whose bit patterns $\check{\mathbf{y}}$ match the exemplary *a priori* information $L^{[\text{input}]}(\check{y}(2)) \to +\infty$ and $L^{[\text{input}]}(\check{y}(3)) \to -\infty$. This results in an improved *extrinsic* L-value $L_{\text{SDM}}^{[\text{ext}]}(y(j))$. For example, the shortest dashed distance in Fig. 13.6-a does not affect the computation of (13.19) any more in the case of Fig. 13.6-b.

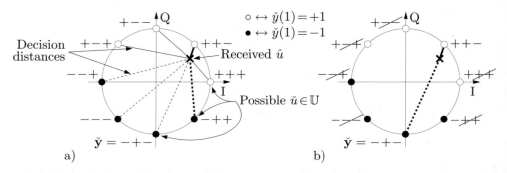

Figure 13.6: Illustrations for the *extrinsic* information of an encoded bit y
    a) Decision distances without *a priori* information, $L^{[\text{input}]}(y(i)) = 0$
    b) Remaining decision distances with perfect *a priori* information,
        $L^{[\text{input}]}(y(2)) \to +\infty$ and $L^{[\text{input}]}(y(3)) \to -\infty$

13.2.5 EXIT Charts

As shown in the previous sections, the determination rules for the *extrinsic* information of BCJR channel decoding $L^{[\text{ext}]}_{\text{BCJR}}(x)$, of SDSD $L^{[\text{ext}]}_{\text{SDSD}}(x)$, and of SDM $L^{[\text{ext}]}_{\text{SDM}}(x)$ are functions of soft-input L-values $L^{[\text{input}]}(x)$. In each decoding approach according to the Turbo principle these soft-input L-values $L^{[\text{input}]}(x)$ contain the *extrinsic* L-value(s) of the other constituent decoder(s) as additive term(s). This is indicated in Figs. 13.8 and 13.13 and will be explained in more detail in Sec. 13.3.1 for *Iterative Source-Channel Decoding* (ISCD) and in Sec. 13.4.1 for *Turbo DeCodulation* (TDeC).

EXIT Characteristics

The transfer of *extrinsic* information from the input $L^{[\text{ext}]}_{\text{In}}(x)$ to the output $L^{[\text{ext}]}_{\text{Out}}(x)$ is specific for each soft-input/soft-output decoding component. The left block in Fig. 13.7-a depicts a generalization of such a block, which represents either BCJR, SDSD, or SDM.

S. ten Brink has shown that this specific transfer can be visualized by an *EXtrinsic Information Transfer* (EXIT) characteristic [ten Brink 1999], [ten Brink 2001]. Such an EXIT characteristic depicts the *mutual information*

$\mathcal{I}^{[\text{ext}]}$ between the originally transmitted bit $x \in \{+1, -1\}$ and the corresponding *extrinsic* L-value $L^{[\text{ext}]}_{\text{Out}}(x)$ at the output as a function of the mutual information

$\mathcal{I}^{[\text{apri}]}$ between x and the *extrinsic* L-value $L^{[\text{ext}]}_{\text{In}}(x)$ at the input.

The *mutual information* measure specifies the amount of information that one random variable contains on average about another random variable [Cover, Thomas 2006].

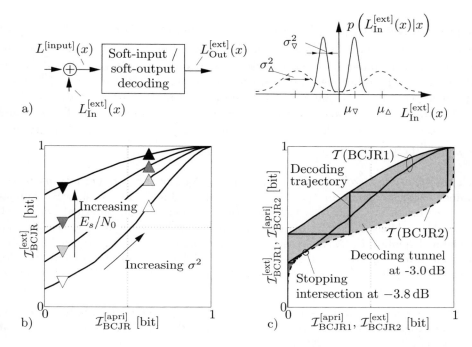

Figure 13.7: Illustrations for the *EXIT charts* of soft-input/soft-output decoders
    a) Modeling of the *extrinsic* input $L_{\mathrm{In}}^{[\mathrm{ext}]}(x)$ by a Gaussian process
    b) EXIT characteristics of a convolutional code for four different E_s/N_0
    c) EXIT chart of two serially concatenated convolutional codes

It can be computed by [ten Brink 2001]

$$\mathcal{I}^{[\cdot]} = \sum_{x=\{+1,-1\}} P(x) \int_{-\infty}^{\infty} p\left(L^{[\mathrm{ext}]}(x)|x\right) \log_2 \frac{p\left(L^{[\mathrm{ext}]}(x)|x\right)}{\sum\limits_{\check{x}=\{+1,-1\}} p\left(L^{[\mathrm{ext}]}(\check{x})|\check{x}\right) P(\check{x})} \, dL^{[\mathrm{ext}]}(x)$$

(13.20)

with $L^{[\mathrm{ext}]}$ being either $L_{\mathrm{In}}^{[\mathrm{ext}]}$ for the calculation of $\mathcal{I}^{[\mathrm{apri}]}$ or $L_{\mathrm{Out}}^{[\mathrm{ext}]}$ for $\mathcal{I}^{[\mathrm{ext}]}$. The bit-wise *a priori* probability $P(x)$ can easily be determined by the marginal distribution of the non-uniform probability mass function $P(\mathbf{x})$. Moreover, ten Brink has observed that for the channel model resulting in (13.3) the conditional probability density $p(L_{\mathrm{In}}^{[\mathrm{ext}]}(x)|x)$ at the input can be approximated by a Gaussian distribution with variance σ^2 and mean $\mu = \sigma^2/2a \cdot x$ [ten Brink 1999], [ten Brink 2001]. Figure 13.7-a illustrates two examples for different values of σ^2 with $\sigma_\nabla^2 < \sigma_\triangle^2$.

With $P(x)$ and the Gaussian approximation of $p\left(L_{\mathrm{In}}^{[\mathrm{ext}]}(x)|x\right)$ the *mutual information* $\mathcal{I}^{[\mathrm{apri}]}$ at the input of the decoder can be determined. If $\sigma^2 = 0$, then $\mathcal{I}^{[\mathrm{apri}]} = 0$ bit. In contrast, if $\sigma^2 \to \infty$, then $\mathcal{I}^{[\mathrm{apri}]}$ approaches the entropy of x.

The probability density $p\left(L_{\text{Out}}^{[\text{ext}]}(x)|x\right)$, which is required to determine $\mathcal{I}^{[\text{ext}]}$, is most conveniently determined by Monte Carlo simulations [ten Brink 2001]. For this purpose, the *extrinsic* input $L_{\text{In}}^{[\text{ext}]}(x)$ is modeled by a Gaussian process with variance σ^2 and mean μ. The other elements contributing to the overall soft-input $L^{[\text{input}]}(x)$ (e.g., some possibly given channel related information $L(z|x)$) are generated similarly as in the respective transmission schemes (see Secs. 13.3.1 and 13.4.1). The L-values $L_{\text{Out}}^{[\text{ext}]}$ at the output are collected in histograms to approximate the probability density $p\left(L_{\text{Out}}^{[\text{ext}]}(x)|x\right)$.

Each combination of σ^2 (for $L_{\text{In}}^{[\text{ext}]}(x)$) and E_s/N_0 (for $L(z|x)$) permits us to determine a specific point in the $\mathcal{I}^{[\text{apri}]}, \mathcal{I}^{[\text{ext}]}$ plot. Figure 13.7-b shows examples for a memory $\nu = 3$, rate $r = 1/2$, recursive, systematic convolutional code with $\mathbf{G} = (1, \frac{13}{15})_8$. The variance σ^2 increases from the left to the right (e.g., $\nabla \rightarrow \triangle$) and the channel quality E_s/N_0 increases from the bottom to the top (with increasing gray scale value: $E_s/N_0 \in \{-5.0, -3.8, -3.0, -2.0\}$ dB). A continuous curve $\mathcal{T}(\cdot)$ for the entire range $0 \le \sigma^2 < \infty$ (i.e., $0 \le \mathcal{I}^{[\text{apri}]} \le 1$) is called the *EXIT characteristic*.

Recently, alternative methods to determine EXIT characteristics analytically, i.e., without histogram measurements, have been proposed [Ashikhmin et al. 2004], [Adrat et al. 2005a], [Kliewer et al. 2006].

EXIT Charts

In a Turbo process the *extrinsic* output $L_{\text{Out}}^{[\text{ext}]}(x)$ of the one decoder serves as additional input $L_{\text{In}}^{[\text{ext}]}(x)$ for the other one and vice versa. From this it follows that the EXIT characteristics of both constituent decoders can be plotted in the same diagram, but with swapped axes. Such a plot is called an *EXIT chart*. It allows us to analyze the convergence behavior. Figure 13.7-c shows an example where two of the above mentioned convolutional codes are serially concatenated. The solid curves $\mathcal{T}(\text{BCJR1})$ represent the first code while the dashed curve $\mathcal{T}(\text{BCJR2})$ describes the second code. Note that only the characteristics of the first code depend on the channel quality E_s/N_0 (for details, see also Sec. 13.3.1 and [ten Brink 1999]).

The area in between both EXIT characteristics $\mathcal{T}(\text{BCJR1})$ and $\mathcal{T}(\text{BCJR2})$ describes the attainable region for the $\mathcal{I}^{[\text{apri}]}, \mathcal{I}^{[\text{ext}]}$ pairs. The step-curves in Fig. 13.7-c are called *decoding trajectories* and they visualize the increase in *mutual information* by the iterations. Each step represents a single iteration. From the example it can be seen that for $E_s/N_0 = -3$ dB the decoding trajectory can reach the upper right corner after $n = 3$ iterations (steps). Thus, perfect reconstruction of the data bits x becomes possible just by knowing the *extrinsic* output $L_{\text{Out}}^{[\text{ext}]}(x)$ resp. input $L_{\text{In}}^{[\text{ext}]}(x)$. However, for $E_s/N_0 = -3.8$ dB there is a *stopping intersection* of the EXIT characteristics. This intersection limits the error correcting capability. More than $n = 2$ iterations cannot provide any further gain in the *mutual information*.

Detailed analyses for the EXIT characteristics and EXIT charts can be found in [ten Brink 2001], [Ashikhmin et al. 2004], [Adrat et al. 2005a], [Kliewer et al. 2006], [Clevorn 2006], [Schmalen et al. 2007].

13.3 Iterative Source-Channel Decoding (ISCD)

Iterative Source-Channel Decoding (ISCD) denotes a Turbo like solution for the joint source-channel decoding problem [Hindelang et al. 2000], [Görtz 2000], [Adrat et al. 2001]. In this solution, the *extrinsic* terms of information resulting from artificial mutual dependencies due to channel coding (Sec. 13.2.2) as well as natural residual redundancies remaining after source encoding (Sec. 13.2.3) are exchanged iteratively according to the Turbo principle.

In Sec. 13.3.1, we will describe the baseband transmission system for ISCD. This description includes details of the Turbo like decoding algorithm. In Sec. 13.3.2, we will present some simulation results as well as the corresponding convergence analysis using EXIT charts. Finally, we will discuss some of the key advancements to ISCD that have recently been proposed in the literature to further improve the error robustness [Adrat, Vary 2005], [Adrat et al. 2005b].

13.3.1 Transmission System and Algorithm

The baseband transmission system for ISCD is shown in Fig. 13.8.

Transmitter

At time k, a parametric source encoder for speech, audio or video signals extracts a set $\tilde{\underline{v}}_k$ of M scalar source codec parameters $\tilde{v}_{k,m} \in \mathbb{R}$, $m = 1, \ldots, M$, from the input signal $\mathbf{s}_k$. For notational convenience we skip the index m in the following. The M codec parameters are individually quantized to $v_k \in \mathbb{V} = \{v^{(i)}, i = 0, \ldots, 2^w - 1\}$. The scalar quantizer codebook contains 2^w reproduction levels $v^{(i)}$. To each realization of v_k a unique bit pattern $\mathbf{x}_k \in \{\mathbf{x}^{(i)}, i = 0, \ldots, 2^w - 1\}$ is assigned. The bit pattern $\mathbf{x}_k$ consists of w data bits $x_k(\kappa) \in \{\pm 1\}$, $\kappa = 1, \ldots, w$. A multiplexer merges the M patterns $\mathbf{x}_k$ representing the input frame $\mathbf{s}_k$ to a sequence of bit patterns $\underline{\mathbf{x}}_k$.

Before channel encoding, the sequence $\underline{\mathbf{x}}_k$ of bit patterns is scrambled by a bit interleaver. Therewith, the single data bits $(x_k(1), \ldots, x_k(w))$ dedicated to a specific bit pattern $\mathbf{x}_k$ are spread far apart from each other over the interleaved sequence of bits. Channel encoding of code rate r expands the sequence $\underline{\mathbf{x}}_k$ of bits x_k by a factor of $1/r$ to a sequence $\underline{\mathbf{y}}_k$ of code bits y_k.

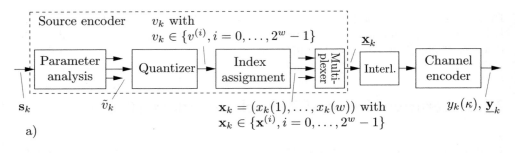

a)

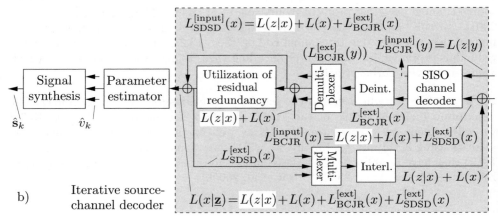

b) Iterative source-channel decoder

Figure 13.8: Block diagram, model and symbol definitions
a) Transmitter: Serially concatenated source and channel encoding
b) Turbo like receiver: Iterative source-channel decoding

Receiver

The aim of the receiver is to minimize the adverse effects of transmission errors on the perceived quality of the reconstructed signal $\hat{s}_k$. To make such an optimization process feasible, the blocks *parameter analysis* and *signal synthesis* are generally excluded from this process. It is most convenient to apply an appropriate quality criterion on the source codec parameters $\tilde{v}_k$. Often, the *parameter Signal-to-Noise Ratio* (SNR) as defined in (10.8) serves as such a quality criterion, i.e.,

$$\text{parameter SNR} = 10 \cdot \log_{10} \frac{\text{E}\{\tilde{v}^2\}}{\text{E}\{(\tilde{v} - \hat{v})^2\}} \quad \text{dB}. \tag{13.21}$$

If the *Minimum Mean Squared Error* (MMSE) is the optimization criterion of the reconstruction process, the individual estimates are determined by [Melsa, Cohn 1978], [Fingscheidt, Vary 2001], [Vary, Martin 2006] (cf. (11.17))

$$\hat{v}_k = \sum_{i=0}^{2^w-1} v^{(i)} \cdot P(\mathbf{x}^{(i)}|\underline{\mathbf{z}}). \tag{13.22}$$

Equation (13.22) is a weighted sum over all 2^w quantizer reproduction levels $v^{(i)}$. The weights are the parameter based *a posteriori* probabilities $P(\mathbf{x}^{(i)}|\mathbf{z})$ for a specific bit pattern $\mathbf{x}^{(i)}$ given the entire received sequence $\mathbf{z}$. The parameter based *a posteriori* probabilities can either be determined by the product of (13.15) and (13.16), or approximated by bit based *a posteriori* L-values $L(x(\kappa)|\mathbf{z})$

$$P(\mathbf{x}^{(i)}|\mathbf{z}) = C_1 \cdot \alpha_k(\mathbf{x}^{(i)}) \cdot \beta_k(\mathbf{x}^{(i)}) \tag{13.23}$$

$$\approx C_2 \cdot \exp\left(\sum_{\kappa=1}^{w} \frac{x_k(\kappa)}{2} \cdot L(x_k(\kappa)|\mathbf{z})\right). \tag{13.24}$$

The constant factors C ensure that the total probability theorem $\sum_{i=0}^{2^w-1} P(\mathbf{x}^{(i)}|\mathbf{z}) = 1$ is fulfilled. Equation (13.24) is an approximation of (13.23) because independence of the bits $x_k(\kappa)$, $\kappa = 1, \ldots, w$ is assumed. According to (13.2), the bit based *a posteriori* L-value can be separated into (up to) four additive terms

$$L(x_k(\kappa)|\mathbf{z}) = L(z|x_k(\kappa)) + L(x_k(\kappa)) + L_{\text{BCJR}}^{[\text{ext}]}(x_k(\kappa)) + L_{\text{SDSD}}^{[\text{ext}]}(x_k(\kappa)). \tag{13.25}$$

The transmission related L-value $L(z|x_k(\kappa))$ can only be separated if a systematic channel code is used. Otherwise, if a non-systematic code is considered, $L(z|x_k(\kappa))=0$ constantly. Both terms of *extrinsic* information are re-calculated iteratively in a Turbo like decoding process.

Turbo like Decoding Algorithm

In the Turbo like decoding process n represents the iteration counter. Before the initial iteration, $n = 1$, all *extrinsic* L-values of SDSD are set to $L_{\text{SDSD}}^{[\text{ext}],n-1=0}(x_k(\kappa)) = 0$ and the transmission related L-values $L(z|y)$ for every sent code bit y resp. $L(z|x_k(\kappa))$ for data bits $x_k(\kappa)$ are determined according to (13.3).

Next, the *inner* decoding step is carried out by computing $L_{\text{BCJR}}^{[\text{ext}],n}(x_k(\kappa))$ according to (13.9) with the soft-inputs (see (13.8), (13.10))

$$L_{\text{BCJR}}^{[\text{input}],n}(y) = L(z|y) \tag{13.26}$$

$$L_{\text{BCJR}}^{[\text{input}],n}(x_k(\kappa)) = L(z|x_k(\kappa)) + L(x_k(\kappa)) + L_{\text{SDSD}}^{[\text{ext}],n-1}(x_k(\kappa)). \tag{13.27}$$

The result for $L_{\text{BCJR}}^{[\text{ext}],n}(x_k(\kappa))$ is considered in the *outer* decoding step as additional *a priori* input information. In this *outer* decoding step $L_{\text{SDSD}}^{[\text{ext}],n}(x_k(\kappa))$ is determined according to (13.14) with the soft-input (see (13.17), (13.18))

$$L_{\text{SDSD}}^{[\text{input}],n}(x_k(\kappa)) = L(z|x_k(\kappa)) + L(x_k(\kappa)) + L_{\text{BCJR}}^{[\text{ext}],n}(x_k(\kappa)). \tag{13.28}$$

Usually, the determined $L_{\text{SDSD}}^{[\text{ext}],n}(x_k(\kappa))$ differ considerably from the initial values $L_{\text{SDSD}}^{[\text{ext}],n-1=0}(x_k(\kappa)) = 0$. Thus, it makes sense to repeat the *inner* decoding step, i.e., to perform iteration $n = n + 1$, where these new *extrinsic* L-values are applied in (13.27). Iteratively executing the *inner* and *outer* decoding steps permits us to increase the reliability of both terms of *extrinsic* information step-by-step.

Reliability improvements are achievable as long as both terms of *extrinsic* information can be considered as being independent of each other. Therefore, a large (de)interleaver is placed between both constituent en-/decoders. Because SDSD gains its *extrinsic* information for bit $x_k(\kappa)$ mainly from the other bits $\mathbf{x}_k^{[\text{ext}]}$ of the same bit pattern $\mathbf{x}_k$ and of some immediately preceding $\mathbf{x}_{k-1}$ resp. succeeding patterns $\mathbf{x}_{k+1}$, those bits need to be spread far apart from each other before BCJR channel decoding. Owing to the iterative interaction, both terms of *extrinsic* information become mutually dependent and thus the iterative process converges to a steady state.

13.3.2 Simulation Examples

For reproducibility matters, instead of using any real-world speech, audio, or video codec with an application specific *parameter analysis* method (see Fig. 13.8) a generic source model for the source codec parameters $\tilde{v}_k$ is used (see Sec. 11.2). For this purpose, each of the M codec parameters $\tilde{v}_k$ is individually modeled by a first order Gauss–Markov process with mean zero and unit variance. The filter coefficient ρ allows us to adjust time dependencies.

In the following, we will present results for two different simulation settings. A comparison of both settings is shown in Table 13.2.

In Configuration A, the codec parameters $\tilde{v}_k$ are at first individually quantized to one out of eight quantizer reproduction levels $v_k \in \mathbb{V} = \{v^{(i)}, i = 0, \ldots, 7\}$ using a scalar *Lloyd–Max Quantizer* (LMQ). Afterwards, each v_k is mapped one-to-one to a unique bit pattern $\mathbf{x}_k$ of length $w_A = 3$ bit/pattern using the *natural binary* index assignment. The sequence $\underline{\mathbf{x}}_k$ of bit patterns is scrambled by a pseudo-random bit interleaver of size $w_A \cdot M$. Finally, the interleaved bit sequence is channel encoded by a memory $\nu = 3$, rate $r_A = 1/2$ *Recursive Systematic Convolutional* (RSC) code with generator polynomial $\mathbf{G}_A = (1, \frac{13}{15})_8$.

Table 13.2: Simulation settings

Configuration	A	B
Quantizer	LMQ	LMQ
Codebook size $|\mathbb{V}|$	8	8
Length w of bit pattern $\mathbf{x}_k$	3	6
Index assignment	Natural binary	EXIT optimized
Channel code	RSC	RNSC
Generator matrix $\mathbf{G}$	$(1, \frac{13}{15})_8$	$(\frac{10}{17})_8$
Code rate r	1/2	1

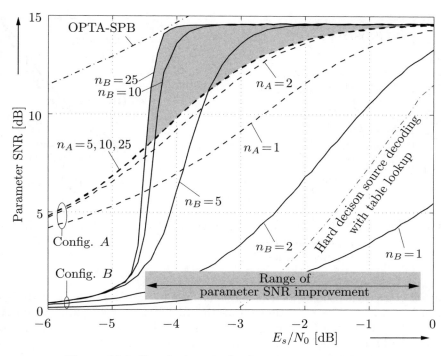

Figure 13.9: Simulation results for $\rho = 0.9$ and $M = 250$

In Configuration B, the same LMQ with codebook size $|\mathbb{V}| = 8$ is used in combination with a redundant *EXIT optimized* index assignment[1]. The bit patterns $\mathbf{x}_k$ are of length $w_B = 6$ bit/pattern. After, pseudo-random bit interleaving of size $w_B \cdot M$ channel encoding is realized by a memory $\nu = 3$, rate $r_B = 1$ *Recursive Non-Systematic Convolutional* (RNSC) code with generator polynomial $\mathbf{G}_B = (\frac{10}{17})_8$.

Notice, the gross bit rate $w \cdot M/r$ on the transmission channel is the same for both settings. Joint source-channel decoding by ISCD is done as described in Sec. 13.3.1. Figure 13.9 depicts the simulation results for the two examples. In addition, the conventional decoding approach (for Configuration A) with BCJR channel decoding and non-iterative source decoding by *hard decision* and *table lookup* is shown as a reference.

The dashed curves show the simulation results with $n_A = 1, 2, 5, 10, 25$ iterations for Configuration A. If compared with the reference, already the first iteration reveals considerable improvements in the end-to-end transmission quality thanks to the extra reliability gain due to SDSD. An additional remarkable parameter SNR gain can be observed for the second iteration. Such a system resembles the SCCD scheme as described in Sec. 12.4. The improvements for higher numbers of iterations are only small.

---

[1]The indices $i = 0, \ldots, 7$ of $v^{(i)}$ are mapped to $\mathbf{x} \in \{52, 10, 17, 04, 77, 27, 01, 40\}_8$, i.e., starting with $i = 0 \mapsto \mathbf{x} = 52_8 = (101010)_2$ up to $i = 7 \mapsto \mathbf{x} = 40_8 = (100000)_2$.

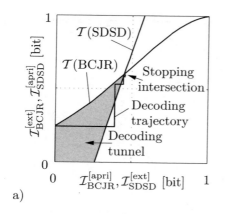
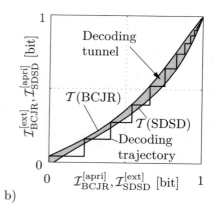

a)　　　　　　　　　　　　　　　　　　b)

Figure 13.10: EXIT charts for $\rho = 0.9$ and $M = 250$ at $E_s/N_0 = -4.0$ dB
　　　　a)　Configuration A
　　　　b)　Configuration B

This quick convergence behavior can be confirmed by an EXIT chart analysis depicted in Fig. 13.10. Figure 13.10-a shows the EXIT chart for Configuration A at $E_s/N_0 = -4.0$ dB. It can be seen that the *decoding trajectory* reaches the intersection of the EXIT characteristics of BCJR channel decoding and SDSD quite closely after $n_A = 3$ iterations. Higher numbers of iterations cannot provide any noteworthy reliability gain.

The solid curves in Fig. 13.9 show the corresponding simulation results for Configuration B. In the initial iterations, substantial quality degradations have to be accepted. But, for numbers of iterations above $n_B = 5$ remarkable gains can be observed in the most interesting range of channel conditions. This range is characterized by the fact that the reconstruction quality starts to drop below its maximum of parameter SNR= 14.4 dB. Up to $n_B = 25$ iterations reveal additional parameter SNR gains. The maximum parameter SNR can (nearly) be guaranteed down to $E_s/N_0 \approx -4.0$ dB. For lower E_s/N_0 values the reconstruction quality decreases rapidly in a waterfall like manner.

The EXIT chart in Fig. 13.10-b confirms the excellent performance of Configuration B. The ISCD approach is not limited by an intersection of the EXIT characteristics and the *decoding trajectory* can pass through a tunnel up to the upper right corner. Therewith, perfect reconstruction becomes possible.

The theoretical limit of an ISCD system can be determined by combining the *channel capacity* and the *rate distortion function* to the so-called *Optimum Performance Theoretically Attainable* (OPTA) limit [Shannon 1959a], [Clevorn et al. 2006b]. When additionally considering the *Sphere Packing Bound* (SPB) [Shannon 1959b] to incorporate the inevitable losses due to a finite block length the OPTA-SPB limit depicted in Fig. 13.9 is obtained. We can observe that we can very closely approach this theoretical bound with Configuration B.

13.3.3 Advancements and Optimizations

The design of first ISCD approaches [Hindelang et al. 2000], [Görtz 2000], [Görtz 2001], [Adrat et al. 2001], [Perkert et al. 2001] resembles those of Configuration A. That means that, e.g., classical index assignments like *natural binary* or *folded binary* [Jayant, Noll 1984] as well as convolutional codes with optimal error correcting capability were used. In recent years, several competing concepts have been investigated in order to improve the error robustness of ISCD systems. A breakthrough was reached when the EXIT chart analysis tool [ten Brink 2001] was applied to the ISCD problem [Adrat et al. 2003], [Adrat 2003], [Adrat, Vary 2005]. The EXIT chart analysis revealed several new design guidelines that ended up in an ISCD system according to Configuration B.

Figure 13.11-a is a copy of the EXIT chart of Fig. 13.10-a. Obviously, the limiting factor of an ISCD system according to Configuration A is the stopping intersection of the EXIT characteristics of BCJR and SDSD at $(0.45, 0.60)$. An ISCD system with improved error robustness can be expected by moving the stopping intersection to the upper right corner of the EXIT chart.

In order to move the stopping intersection most optimizations for the source coding component focus on the *anchor point* of the EXIT characteristic of SDSD at the upper border of the EXIT chart. Such an anchor point is generally given for EXIT characteristics of SDSD when the bit patterns $\mathbf{x}$ being assigned to the full quantizer codebook exhibit a minimal Hamming distance $d_{\mathrm{Ham}} = 1$ [Adrat et al. 2005a]. Figure 13.11-b shows an example where only the index assignment has been modified. The specific index assignment that maximizes the *anchor point* in this case to $(0.80, 1.0)$ is called *EXIT optimized* [Adrat, Vary 2005]. The quantizer reproduction levels can be optimized in a similar way [Adrat et al. 2006a].

The EXIT chart in Fig. 13.11-c shows an example for an optimized channel coding component. It has been demonstrated for instance in [Adrat, Vary 2005] that *Recursive Non-Systematic Convolutional* (RNSC) codes are favorable over *Recursive Systematic Convolutional* (RSC) ones. Notice, changing the channel code from a systematic to non-systematic form also has an impact on the EXIT characteristic of SDSD because the channel related L-value is not given any more (i.e., $L(z|x_k(\kappa)) = 0$ in (13.28)). Anyhow, the stopping intersection can be improved to $(0.75, 0.96)$.

Best performance improvements are achievable if the concepts of *EXIT optimized* index assignment and the RNSC codes are combined with an optimized bit rate allocation between source and channel coding. As first shown by [Ashikhmin et al. 2004] the inner component of a serially concatenated Turbo scheme will be of rate $r = 1$. As a consequence, in an ISCD scheme the full bit budget can be assigned to source coding [Adrat, Vary 2004], [Adrat et al. 2005b]. With this, the anchor point does not have to be a limiting factor for the EXIT characteristic of SDSD any more because bit patterns $\mathbf{x}$ can be assigned to the quantizer codebook such that $d_{\mathrm{Ham}} > 1$. This constraint for the d_{Ham} is in accordance with the findings in [Benedetto et al. 1998]. Figure 13.11-d shows the corresponding EXIT chart.

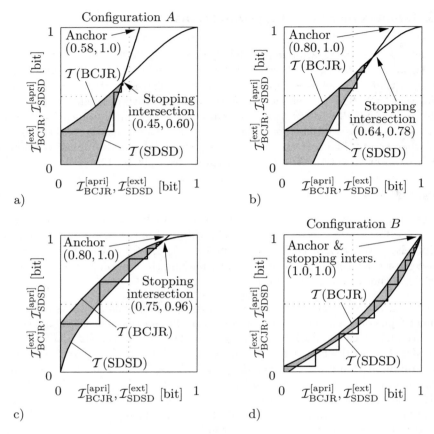

Figure 13.11: EXIT charts at $E_s/N_0 = -4.0\,\mathrm{dB}$
  a) Nat. bin. index assignment and $r_A = 1/2$ RSC code (Config. A)
  b) EXIT opt. index assignment and $r_A = 1/2$ RSC code
  c) EXIT opt. index assignment and $r = 1/2$ RNSC code
  d) EXIT opt. index assignment and $r_B = 1$ RNSC code (Config. B)

Obviously, a stopping intersection can be avoided and error free decoding becomes possible.

An ISCD system design according to Configuration B is the basis for a particularily efficient adaptive multi-mode system [Adrat et al. 2005b]. As mentioned above, when using a channel code of rate $r = 1$, the full bit budget can be assigned to source coding. Here it can be exploited either by the quantizer to reduce the quantization noise or to improve the error robustness by the redundant index assignment (in principle a non-linear block code). Obviously, in order to obtain the highest overall parameter SNR it is beneficial to increase the robustness in bad channel conditions and to reduce the quantization noise in good conditions. The best trade-off can be reached by an adaptive mode switching according to the E_s/N_0 resulting in a *multi-mode envelope*

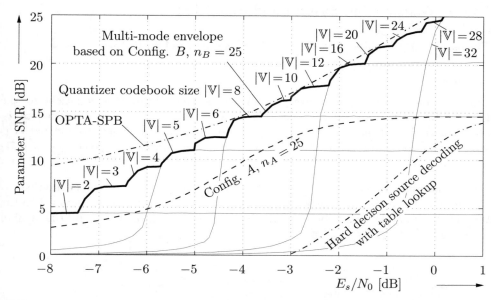

Figure 13.12: Simulation results for multi-mode ISCD with $\rho = 0.9$ and $M = 250$

as shown in Fig. 13.12. This figure depicts a simulation example for a selection of quantizer codebook sizes $|\mathbb{V}|$ ranging from 2 to 32. The size $w_B = 6$ of bit patterns $\mathbf{x}$ after index assignment as well as the channel code with $\mathbf{G}(\frac{10}{17})_8$, $(r = 1)$ are fixed for all cases.

13.4 Turbo DeCodulation (TDeC)

Turbo DeCodulation (TDeC) [Clevorn et al. 2005b], [Clevorn et al. 2005a], [Clevorn 2006] extends ISCD to higher order modulation schemes. TDeC is a multiple Turbo process that comprises the iterative processing of the demodulator, the channel decoder, and the soft decision source decoder. The terminology *DeCodulation* refers to the joint decoding and demodulation [Anderson, Lesh 1981]. Basically, TDeC can be considered as a combination of ISCD introduced in Sec. 13.3 and the iterative demodulation scheme BICM-ID [Li et al. 2002], [Hanzo et al. 2002]. Thus, the iterative refinement of the *extrinsic* information available by artificial mutual dependencies due to the symbol mapping of the modulation (Sec. 13.2.4) and channel coding (Sec. 13.2.2) as well as natural residual redundancies remaining after source encoding (Sec. 13.2.3) is accomplished by TDeC.

In the following, the extension of the ISCD baseband transmission system in Sec. 13.3.1 to a TDeC system will be described in Sec. 13.4.1. In Sec. 13.4.2, we will present some simulation results including the EXIT chart based convergence analysis. Finally, in Sec. 13.4.3, we present recent advancements proposed in the literature.

13.4.1 Transmission System and Algorithm

The baseband transmission system for TDeC is shown in Fig. 13.13.

Transmitter

The first part of the transmitter up to the channel decoder is identical to the ISCD transmitter detailed in Sec. 13.3.1. M scalar source codec parameters $\tilde{v}_k$ are extracted at time k from the input signal $\mathbf{s}_k$ and assigned to bit patterns $\mathbf{x}_k$. After outer interleaving the channel encoder generates a sequence $\underline{\mathbf{y}}_k$ of code bits $y_k(\kappa)$.

Before modulation, the code bits $y_k(\kappa)$ are scrambled by a different inner interleaver. In the modulator bit patterns $\mathbf{y} = (y(1), \dots, y(J))$ of J code bits are mapped to the transmitted modulated symbol u in the complex signal space, $u \in \mathbb{U}$, by the symbol mapping.

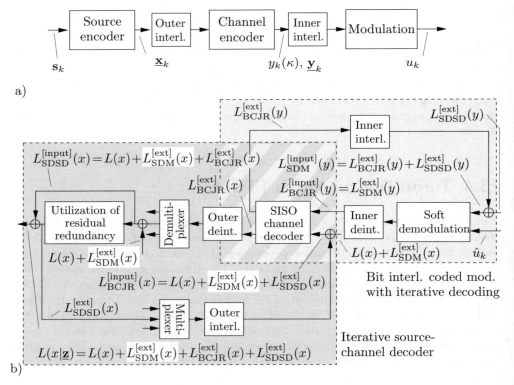

Figure 13.13: Block diagram, model and symbol definitions
  a) Transmitter: Serially concatenated source and channel encoding and modulation
  b) Turbo like receiver: Interaction of ISCD and BICM-ID

Receiver

The receiver of TDeC resembles a multiple Turbo process that evaluates three types of *extrinsic* information, originating from the three components. The basic structure can be considered as a serial concatenation of an ISCD scheme and a BICM-ID scheme via a common channel decoder. In the Turbo process of a BICM-ID receiver the *extrinsic* information $L_{\mathrm{SDM}}^{[\mathrm{ext}]}(y_k(j))$ (see Sec. 13.2.4) and $L_{\mathrm{BCJR}}^{[\mathrm{ext}]}(y_k(j))$ (see Sec. 13.2.2) on the code bits $y_k(j)$ provided by demodulator and channel decoder is refined in an iterative manner. More details on the ISCD receiver can be found in Sec. 13.3.1. Note, the ISCD Turbo process works on the data bits x while the BICM-ID Turbo process works on the code bits y.

One advantage of TDeC is that in the case of a systematic channel code *extrinsic* information on the systematic bits can be exchanged between the demodulator and the SDSD [Clevorn 2006], [Clevorn et al. 2005a]. Owing to the demodulation this potentially available additional *extrinsic* information $L_{\mathrm{SDM}}^{[\mathrm{ext}],n}(x_k(\kappa))$ replaces the transmission related L-value $L(z|x_k(\kappa))$ in the ISCD Turbo loop, transforming (13.27) and (13.28) to

$$L_{\mathrm{BCJR}}^{[\mathrm{input}],n}(x_k(\kappa)) = L(x_k(\kappa)) + L_{\mathrm{SDM}}^{[\mathrm{ext}],n}(x_k(\kappa)) + L_{\mathrm{SDSD}}^{[\mathrm{ext}],n-1}(x_k(\kappa)) \qquad (13.29)$$

$$L_{\mathrm{SDSD}}^{[\mathrm{input}],n}(x_k(\kappa)) = L(x_k(\kappa)) + L_{\mathrm{SDM}}^{[\mathrm{ext}],n}(x_k(\kappa)) + L_{\mathrm{BCJR}}^{[\mathrm{ext}],n}(x_k(\kappa)) . \qquad (13.30)$$

For the BICM-ID Turbo loop we obtain, respectively,

$$L_{\mathrm{BCJR}}^{[\mathrm{input}],n}(y_k(j)) = L_{\mathrm{SDM}}^{[\mathrm{ext}],n}(y_k(j)) \qquad (13.31)$$

$$L_{\mathrm{SDM}}^{[\mathrm{input}],n}(y_k(j)) = L_{\mathrm{BCJR}}^{[\mathrm{ext}],n-1}(y_k(j)) + L_{\mathrm{SDSD}}^{[\mathrm{ext}],n-1}(y_k(j)) \qquad (13.32)$$

and the *a posteriori* L-value in the parameter estimation (see (13.25)) becomes

$$L(x_k(\kappa)|\underline{\mathbf{z}}) = L(x_k(\kappa)) + L_{\mathrm{SDM}}^{[\mathrm{ext}]}(x_k(\kappa)) + L_{\mathrm{BCJR}}^{[\mathrm{ext}]}(x_k(\kappa)) + L_{\mathrm{SDSD}}^{[\mathrm{ext}]}(x_k(\kappa)) . \qquad (13.33)$$

When a non-systematic channel code is used, no *extrinsic* information can be exchanged between the demodulator and the SDSD, i.e., $L_{\mathrm{SDM}}^{[\mathrm{ext}],n}(x_k(\kappa)) = 0$ and $L_{\mathrm{SDSD}}^{[\mathrm{ext}],n-1}(y_k(j)) = 0$ in (13.29)–(13.33) and in Fig. 13.13.

With three components taking part in the Turbo process the order of their execution is of high importance. For example, doing all iterations of the BICM-ID loop first and then executing the iterations of the ISCD loop would not fully exploit the potential of the TDeC scheme. The demodulator cannot profit from the *extrinsic* information refinement of the SDSD since it is not activated any more when the ISCD processing started. The sequential processing of demodulator, channel decoder and SDSD with wrap around has been proven to be quite beneficial [Clevorn 2006]. This corresponds to an alternating processing of the two Turbo loops with a joint channel decoder call. The iteration indices in (13.29)–(13.32) reflect this execution order and it will be used in the following simulation example.

13.4.2 Simulation Examples

For the TDeC simulation example we use the same generic source and source encoding as for ISCD in Sec. 13.3.2. Depending on the configuration, either the *natural binary* or the *EXIT optimized* index assignment is applied. A *Recursive Systematic Convolutional* (RSC) code with rate $r = 1/2$, memory $\nu = 3$, and generator polynomial $\mathbf{G} = (1, \frac{13}{15})_8$ serves as channel coding scheme for all configurations, again taken from Sec. 13.3.2. In the modulator an 8PSK signal constellation set as depicted in Fig. 13.6 is used, i.e., a unique bit pattern $\mathbf{y}$ consisting of $J = 3$ code bits is mapped to each different modulated symbol u. The options for the symbol mapping are the *Gray* (see Fig. 13.6) and the *set-partitioning* symbol mapping [Caire et al. 1998], [Li et al. 2002].

Table 13.3 gives an overview of the different settings described in the following.

Configuration A' represents a setup for a classic non-iterative system. As seen in Sec. 13.3.2 the *natural binary* index assignment yields good results in the first iteration, i.e., in a non-iterative system, but it can provide only a limited improvement when Turbo processing with several iterations is applied. The same holds for the *Gray* symbol mapping, which is the optimum symbol mapping in the non-iterative case [Caire et al. 1998]. However, *Gray* symbol mapping is not suited for iterative demodulation, because the possible gain in *extrinsic* information due to increased decision distances (see Sec. 13.2.4) is rather small [Li et al. 2002], [Clevorn 2006].

In Configuration B' the index assignment and the symbol mapping are adapted to the Turbo processing. The *EXIT optimized* index assignment, which has shown a good performance with ISCD (see 13.3.2), is applied and *set-partitioning* serves as symbol mapping. *Set-partitioning* symbol mapping can provide a significant gain in schemes with iterative demodulation. Note, *set-partitioning* may not be the optimum symbol mapping in a pure BICM-ID system that uses the bit error rate as a quality measure. Here, symbol mappings with an even better asymptotic performance can be designed [Li et al. 2002], [Schreckenbach et al. 2003], [Clevorn et al. 2004], [Clevorn

Table 13.3: Simulation settings

Configuration	A'	B'	BICM-ID	ISCD
Quantizer	LMQ	LMQ	LMQ	LMQ
Codebook size $|\mathbb{V}|$	8	8	8	8
Length w of bit pattern $\mathbf{x}_k$	3	3	3	3
Index assignment	Nat. bin.	EXIT opt.	Nat. bin.	EXIT opt.
Channel code	RSC	RSC	RSC	RSC
Generator matrix $\mathbf{G}$	$(1, \frac{13}{15})_8$	$(1, \frac{13}{15})_8$	$(1, \frac{13}{15})_8$	$(1, \frac{13}{15})_8$
Code rate r	1/2	1/2	1/2	1/2
Signal constellation set $\mathbb{U}$	8PSK	8PSK	8PSK	8PSK
Symbol mapping	Gray	Set-part.	Set-part.	Gray

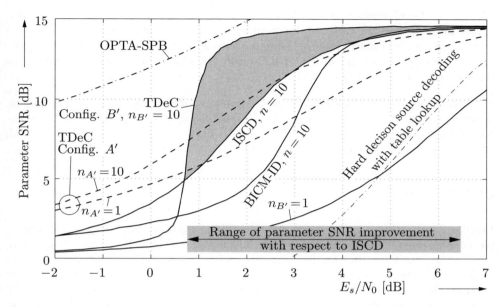

Figure 13.14: Simulation results for $\rho = 0.9$, $M = 250$, and 8PSK modulation

2006]. Nevertheless, *set-partitioning* symbol mapping yields the best parameter SNR results in the investigated TDeC system. To enhance the comparability and emphasize the effects of using two Turbo loops the channel code is not changed. Of course, as shown in Sec. 13.4.3 many more possible optimizations exist, including all the optimizations for ISCD described in Sec. 13.3.3.

For comparison we additionally define a BICM-ID and an ISCD configuration. These configurations apply Turbo processing only to the respective iterative loop. The index assignments and the symbol mappings are chosen such that they match Configuration A' or B', depending on whether the respective component takes part in the Turbo processing or not.

Figure 13.14 depicts the simulation results for the different configurations and the conventional approach with soft demodulation, BCJR channel decoding and non-iterative source decoding by *hard decision* and *table lookup*. Compared with this reference, all investigated configurations show a significant improvement.

The dashed curves show the simulation results for the TDeC Configuration A' with $n_{A'} = 1$ and $n_{A'} = 10$ iterations. A noticeable gain can be observed. However, at high parameter SNRs the BICM-ID and the ISCD system outperform Configuration A', despite performing Turbo processing only in a single loop. BICM-ID and ISCD use an optimized component in their respective iterative loop, yielding their superior performance.

When examining the simulation results for Configuration B' we observe a very poor performance in the first iteration, $n_{B'} = 1$. Nevertheless, with $n_{B'} = 10$ iterations,

TDeC with Configuration B' exhibits a steep slope and shows an impressive performance in the interesting range of the parameter SNR. The gain in E_s/N_0 with respect to the best non-iterative system (Configuration A' with $n_{A'} = 1$) often exceeds the sum of the respective gains of BICM-ID and ISCD, whose serial concatenation forms the structural basis of TDeC.

With its two Turbo loops allowing the exploitation of redundancy in source coding, channel coding, as well as in modulation, TDeC can operate quite close to the theoretical bound for this scenario, the OPTA-SPB limit [Clevorn 2006], [Clevorn et al. 2006b] (see also Sec. 13.3.2).

EXIT Chart Analysis

Despite having three Turbo components a TDeC system obeys the Turbo principle of exchanging and refining *extrinsic* information. Thus, EXIT charts can still serve as an excellent tool for the analysis and optimization. However, with three Turbo components taking part in the Turbo process, the EXIT charts have to be three-dimensional [Clevorn 2006], [Clevorn et al. 2005a]. In Fig. 13.15, EXIT charts for Configurations A' and B' of the simulation results in Fig. 13.14 are depicted.

The areas in Fig. 13.15 are the multi-dimensional EXIT characteristics $\mathcal{T}$ of the Turbo components. Note that the EXIT characteristic of the SDSD is only outlined but not drawn between the two other EXIT characteristics to allow the view on the decoding trajectory. With the utilized systematic channel code, all Turbo

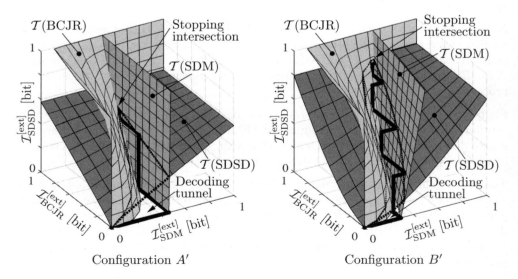

Figure 13.15: EXIT charts of Fig. 13.14 at $E_s/N_0 = 2.8\,\text{dB}$

components have two types of *a priori* information available. Thus, each EXIT characteristic depends on the *extrinsic* information provided by both other components.

The EXIT characteristics can be determined again by the histogram method. But, when the two types of *a priori* information refer to the same input port of the Turbo component under consideration, which is the case for the SDSD and the SDM, the EXIT characteristic can also be interpolated with good accuracy. Combined *a priori* information can be computed using *mutual information* combining [Land et al. 2005], [Clevorn et al. 2005a], [Clevorn 2006].

When comparing Configurations A' and B' in Fig. 13.15 it can be observed that with Configuration B' the decoding tunnel under $\mathcal{T}(\text{SDSD})$ and between $\mathcal{T}(\text{BCJR})$ and $\mathcal{T}(\text{SDM})$ is narrower than for Configuration A'. But, the final stopping intersection occurs at significantly higher values of $\mathcal{I}_{\text{SDM}}^{[\text{ext}]}$, $\mathcal{I}_{\text{BCJR}}^{[\text{ext}]}$, and $\mathcal{I}_{\text{SDSD}}^{[\text{ext}]}$. This is similar to the comparison of the classic Configuration A and the optimized Configuration B in the ISCD example in Sec. 13.3.2. The decoding trajectories in Fig. 13.15, which now of course are also three-dimensional, demonstrate that the TDeC systems fully exploit the decoding tunnel. They first advance along the $\mathcal{I}_{\text{SDM}}^{[\text{ext}]}$ axis, followed by the $\mathcal{I}_{\text{BCJR}}^{[\text{ext}]}$ axis and then the $\mathcal{I}_{\text{SDSD}}^{[\text{ext}]}$ axis, which complies with the iterative processing order outlined in Sec. 13.4.1.

13.4.3 Advancements and Optimizations

There are a manifold of possibilities for the optimization of TDeC. Since TDeC can be considered to be a combination of BICM-ID and ISCD, obviously most of the advancements for these subsystems can be directly applied to TDeC. Amongst others, these comprise the optimizations for ISCD presented in Sec. 13.3.3 as well as the various optimized symbol mappings for BICM-ID proposed, for example, in [Li et al. 2002], [Schreckenbach et al. 2003], [Clevorn et al. 2004]. However, as mentioned before, the effect of a modification on the joint TDeC system has to be studied carefully to avoid an overall degradation by a resulting impairment on the remaining parts.

A particularly interesting interpretation of the TDeC paradigm, iterative joint source-channel decoding and demodulation, is the so-called *block coded* TDeC proposed in [Clevorn et al. 2006a], [Clevorn 2006]. Figure 13.16 compares the transmitter structures of the so far discussed convolutional coded TDeC and this block coded TDeC. The transmitter of convolutional coded TDeC follows the typical design of today's communications systems.

In contrast, block coded TDeC consequently implements the ideas of some recent optimizations of the ISCD and BICM-ID subsystems. This results in the removal of a dedicated channel code. Instead, the channel coding task of adding artificial redundancy is split and integrated into the source coding and the modulation.

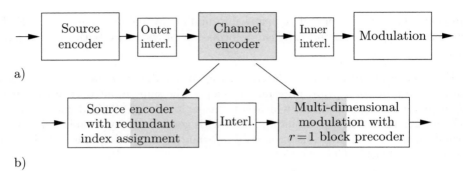

a)

b)

Figure 13.16: Comparison of TDeC transmitter structures
a) Convolutional coded TDeC
b) Block coded TDeC

For the index assignment in the source coding block the optimized bit rate allocation scheme presented in Sec. 13.3.3 (Configuration B' of the ISCD example) is applied. Instead of the EXIT optimized redundant index assignments of Sec. 13.3.3 ($d_{\mathrm{Ham}} > 1$), the block coded index assignments proposed in [Clevorn et al. 2006a], [Clevorn 2006] can be used. Then, the index assignment is based on a simple and short block code.

The modulation is extended to a multi-dimensional modulation [Wei 1987], [Simoens et al. 2004]. A joint symbol mapping to several modulated symbols is applied based on a short $r = 1$ block precoder. This increases the signal space diversity to more than the two dimensions of a single modulated symbol, which improves the iterative demodulation.

Simulation experiments have shown a comparable performance of the two variants of TDeC [Clevorn et al. 2006a], [Clevorn 2006] despite block coded TDeC possessing only a single Turbo loop. The parallel usage of the small block codes allows a high degree of parallelization, which might be advantageous for implementation. Furthermore, the adaptive multi-mode extension suggested for ISCD [Adrat et al. 2005b] can easily be applied to block coded TDeC [Clevorn et al. 2006a].

13.5 Conclusions

In this chapter, the concepts of *Source Controlled Channel Decoding* (SCCD) presented in Sec. 12.4 and the Turbo principle have been combined to *Iterative Source-Channel Decoding* (ISCD). ISCD describes the iterative evaluation of natural residual source redundancy and artificial channel coding redundancy. It turns out that ISCD outperforms conventional non-iterative transmission schemes with respect to the signal-to-noise ratio between the original and the reconstructed signal.

On the one hand, the EXIT chart analysis allows us to analyze the convergence behavior of ISCD and, on the other, it reveals some design guidelines to improve the overall system performance. It has been demonstrated that the inner channel coding component should be recursive, non-systematic, and of rate $r = 1$. In addition, by a proper adjustment of the SDSD EXIT characteristic, the quantization and the (redundant) index assignment of the source encoder can be optimized. The combination of both results in a powerful and highly flexible multi-mode ISCD scheme.

Turbo DeCodulation (TDeC) combines the ideas of ISCD and the iterative demodulation scheme *Bit Interleaved Coded Modulation with Iterative Decoding* (BICM-ID) to a single system, where all three receiver components, i.e., soft demodulator, BCJR channel decoder, and SDSD, act jointly. It has been shown that the performance gain of TDeC can be higher than the sum of the gains of the sub-systems BICM-ID and ISCD. EXIT charts, which are extended to three dimensions in this case, can again serve for analysis and optimization.

Bibliography

Adrat, M. (2003). *Iterative Source-Channel Decoding for Digital Mobile Communications*, PhD thesis. Aachener Beiträge zu digitalen Nachrichtensystemen, vol. 16, P. Vary (ed.), RWTH Aachen University.

Adrat, M.; Antweiler, M.; Clevorn, T.; Korall, B.; Vary, P. (2006a). EXIT-Optimized Quantizer Design for Iterative Source-Channel Decoding, *4th International Symposium on Turbo Codes and Related Topics*, Munich, Germany.

Adrat, M.; Brauers, J.; Clevorn, T.; Vary, P. (2005a). The EXIT-Characteristic of Softbit-Source Decoders, *IEEE Communications Letters*, vol. 9, no. 6, pp. 540–542.

Adrat, M.; Clevorn, T.; Brauers, J.; Vary, P. (2006b). Minimum Terms of Residual Redundancy for Successful Iterative Source-Channel Decoding, *IEEE Communications Letters*, vol. 10, no. 11, pp. 778–780.

Adrat, M.; v. Agris, U.; Vary, P. (2003). Convergence Behavior of Iterative Source-Channel Decoding, *IEEE International Conference on Acoustics, Speech, and Signal Processing (ICASSP)*, Hongkong, China.

Adrat, M.; Vary, P. (2004). Iterative Source-Channel Decoding with Code Rates near $r = 1$, *IEEE International Conference on Communications (ICC)*, Paris, France.

Adrat, M.; Vary, P. (2005). Iterative Source-Channel Decoding: Improved System Design Using EXIT Charts, *EURASIP Journal on Applied Signal Processing (Special Issue: Turbo Processing)*, vol. 2005, no. 6, pp. 928–941.

Adrat, M.; Vary, P.; Clevorn, T. (2005b). Optimized Bit Rate Allocation for Iterative Source-Channel Decoding and its Extension towards Multi-Mode Transmission, *14th IST Mobile and Wireless Communications Summit*, Dresden, Germany.

Adrat, M.; Vary, P.; Spittka, J. (2001). Iterative Source-Channel Decoder Using Extrinsic Information from Softbit-Source Decoding, *IEEE International Conference on Acoustics, Speech, and Signal Processing (ICASSP)*, Salt Lake City, UT, USA.

Anderson, J. B.; Lesh, J. R. (1981). Guest Editors' Prologue, *IEEE Transactions on Communications*, vol. 29, no. 3, pp. 185–186.

Ashikhmin, A.; Kramer, G.; ten Brink, S. (2004). Extrinsic Information Transfer Functions: Model and Erasure Channel Properties, *IEEE Transactions on Information Theory*, vol. 50, no. 11, pp. 2657–2673.

Bahl, L. R.; Cocke, J.; Jelinek, F.; Raviv, J. (1974). Optimal Decoding of Linear Codes for Minimizing Symbol Error Rate, *IEEE Transactions on Information Theory*, vol. 20, no. 2, pp. 284–287.

Benedetto, S.; Divsalar, D.; Montorsi, G.; Pollara, F. (1998). Serial Concatention of Interleaved Codes: Performance Analysis, Design, and Iterative Decoding, *IEEE Transactions on Information Theory*, vol. 44, no. 3, pp. 909–921.

Berrou, C.; Glavieux, A.; Thitimajshima, P. (1993). Near Shannon Limit Error-Correcting Coding and Decoding, *IEEE International Conference on Communications (ICC)*, Geneva, Switzerland.

Caire, G.; Taricco, G.; Biglieri, E. (1998). Bit-Interleaved Coded Modulation, *IEEE Transactions on Information Theory*, vol. 44, no. 3, pp. 927–946.

Clevorn, T. (2006). *Turbo DeCodulation: Iterative Joint Source-Channel Decoding and Demodulation*, PhD thesis. Aachener Beiträge zu digitalen Nachrichtensystemen, vol. 24, P. Vary (ed.), RWTH Aachen University.

Clevorn, T.; Adrat, M.; Vary, P. (2006a). Turbo DeCodulation using Highly Redundant Index Assignments and Multi-Dimensional Mappings, *4th International Symposium on Turbo Codes and Related Topics*, Munich, Germany.

Clevorn, T.; Brauers, J.; Adrat, M.; Vary, P. (2005a). EXIT Chart Analysis of Turbo DeCodulation, *IEEE International Symposium on Personal Indoor and Mobile Radio Communications (PIMRC)*, Berlin, Germany.

Clevorn, T.; Brauers, J.; Adrat, M.; Vary, P. (2005b). Turbo DeCodulation: Iterative Combined Demodulation and Source-Channel Decoding, *IEEE Communications Letters*, vol. 9, no. 9, pp. 820–822.

Clevorn, T.; Godtmann, S.; Vary, P. (2004). PSK versus QAM for Iterative Decoding of Bit-Interleaved Coded Modulation, *IEEE Global Telecommunications Conference (GLOBECOM)*, Dallas, TX, USA.

Clevorn, T.; Schmalen, L.; Vary, P.; Adrat, M. (2006b). On the Optimum Performance Theoretically Attainable for Scalarly Quantized Correlated Sources, *International Symposium on Information Theory and its Applications (ISITA)*, Seoul, Korea.

Cover, T. M.; Thomas, J. A. (2006). *Elements of Information Theory*, 2nd edn, John Wiley & Sons, Ltd, Chichester.

Fingscheidt, T.; Vary, P. (2001). Softbit Speech Decoding: A New Approach to Error Concealment, *IEEE Transactions on Speech and Audio Processing*, vol. 9, no. 3, pp. 240–251.

Görtz, N. (2000). Iterative Source-Channel Decoding using Soft-In/Soft-Out Decoders, *IEEE International Symposium on Information Theory (ISIT)*, Sorrento, Italy.

Görtz, N. (2001). On the Iterative Approximation of Optimal Joint Source-Channel Decoding, *IEEE Journal on Selected Areas in Communications*, vol. 19, no. 9, pp. 1662–1670.

Hagenauer, J. (1995). Source-Controlled Channel Decoding, *IEEE Transactions on Communications*, vol. 43, no. 9, pp. 2449–2457.

Hagenauer, J.; Offer, E.; Papke, L. (1996). Iterative Decoding of Binary Block and Convolutional Codes, *IEEE Transactions on Information Theory*, vol. 42, no. 2, pp. 429–445.

Hanzo, L.; Liew, T. H.; Yeap, B. L. (2002). *Turbo Coding, Turbo Equalisation and Space-Time Coding for Transmission over Fading Channels*, John Wiley & Sons, Ltd, Chichester.

Hindelang, T. (2001). *Source-Controlled Channel Encoding and Decoding for Mobile Communications*, PhD thesis, VDI-Verlag, Düsseldorf: Fortschritt-Berichte VDI, Reihe 10, Nr. 695, ISBN 3-18-369510-3, Munich University of Technology.

Hindelang, T.; Adrat, M.; Fingscheidt, T.; Heinen, S. (2007). Joint Source and Channel Coding: From the Beginning until the 'Exit', *European Transactions on Telecommunications (ETT)*, vol. 18, no. 8.

Hindelang, T.; Fingscheidt, T.; Seshadri, N.; Cox, R. V. (2000). Combined Source/Channel (De-)Coding: Can A Priori Information Be Used Twice?, *IEEE International Symposium on Information Theory (ISIT)*, Sorrento, Italy.

Jayant, N. S.; Noll, P. (1984). *Digital Coding of Waveforms: Principles and Applications to Speech and Audio*, Prentice-Hall, Englewood Cliffs, New Jersey.

Kliewer, J.; Ng, S. X.; Hanzo, L. (2006). On the Computation of EXIT Characteristics for Symbol-Based Iterative Decoding, *4th International Symposium on Turbo Codes and Related Topics*, Munich, Germany.

Land, I.; Huettinger, S.; Hoeher, P. A.; Huber, J. B. (2005). Bounds on Information Combining, *IEEE Transactions on Information Theory*, vol. 51, no. 2, pp. 612–619.

Li, X.; Chindapol, A.; Ritcey, J. A. (2002). Bit-Interleaved Coded Modulation with Iterative Decoding and 8PSK Signaling, *IEEE Transactions on Communications*, vol. 50, no. 8, pp. 1250–1257.

Melsa, J. L.; Cohn, D. L. (1978). *Decision and Estimation Theory*, McGraw-Hill, New York.

Perkert, R.; Kaindl, M.; Hindelang, T. (2001). Iterative Source and Channel Decoding for GSM, *IEEE International Conference on Acoustics, Speech, and Signal Processing (ICASSP)*, Salt Lake City, UT, USA.

Schmalen, L.; Vary, P.; Adrat, M.; Clevorn, T. (2007). On the EXIT Characteristics of Feed Forward Convolutional Codes, *IEEE International Symposium on Information Theory (ISIT)*, Nice, France.

Schreckenbach, F.; Görtz, N.; Hagenauer, J.; Bauch, G. (2003). Optimized Symbol Mappings for Bit-Interleaved Coded Modulation with Iterative Decoding, *IEEE Global Telecommunications Conference (GLOBECOM)*, San Francisco, CA, USA.

Shannon, C. E. (1959a). Coding Theorems for a Discrete Source with a Fidelity Criterion, *IRE National Convention Records*, vol. 4, pp. 142–163.

Shannon, C. E. (1959b). Probability of Error for Optimal Codes in a Gaussian Channel, *The Bell Systems Technical Journal*, vol. 38, pp. 611–656.

Simoens, F.; Wymeersch, H.; Moeneclaey, M. (2004). Spatial Mapping for MIMO Systems, *IEEE Information Theory Workshop (ITW)*, San Antonio, TX, USA.

ten Brink, S. (1999). Convergence of Iterative Decoding, *IEE Electronics Letters*, vol. 35, no. 13, pp. 806–808.

ten Brink, S. (2001). Convergence Behavior of Iteratively Decoded Parallel Concatenated Codes, *IEEE Transactions on Communications*, vol. 49, no. 10, pp. 1727–1737.

Vary, P.; Martin, R. (2006). *Digital Speech Transmission - Enhancement, Coding and Error Concealment*, John Wiley & Sons, Ltd, Chichester.

Wei, L.-F. (1987). Trellis Coded Modulation with Multidimensional Constellations, *IEEE Transactions on Information Theory*, vol. 33, no. 4, pp. 483–501.

V

Speech Processing in Hearing Instruments

Chapter 14

Binaural Signal Processing in Hearing Aids: Technologies and Algorithms

Volkmar Hamacher, Ulrich Kornagel, Thomas Lotter, Henning Puder

14.1 Introduction

The history of electronic hearing aids begins with analog systems, which mainly provided a rudimentary concept of hearing loss compensation. These systems only offered a limited degree of freedom to account for the various types of individual hearing loss. A critical breakthrough was the transition from analog to digital systems. The primary advantage of digital systems is their high flexibility. With some limitations, they can be programmed like a computer and thus digital signal processing methods can be integrated. This is the prerequisite for many algorithms that are tailored to the requirements of hearing aid wearers.

Usually, hearing aids are worn bilaterally, i.e., one hearing aid on each ear. One further breakthrough was the introduction of a wireless link which exchanges data between both hearing aids. It offers an additional degree of freedom to develop more powerful algorithms since it is possible to make use of additional spatial information and synchronize state information. Thus, the development of binaural algorithms became possible and a pair of hearing aids that are connected by the binaural link can be seen as one binaural system. The technology of the binaural link and the algorithms utilizing it are the subject of this chapter.

To give an impression of how modern monaural hearing aids (i.e., hearing aids without a binaural link) work, Sec. 14.1.1 introduces the algorithms that are usually applied. Section 14.1.2 then shows how several algorithms can profit from a binaural link. Finally, Sec. 14.1.3 describes how the remainder of this chapter is organized.

Advances in Digital Speech Transmission   Edited by R. Martin, U. Heute and C. Antweiler
© 2008 John Wiley & Sons, Ltd

14.1.1 Monaural Hearing Aids - State of the Art

The signal processing components of a modern digital hearing aid system can be divided into two principal categories:

- algorithms that fulfill audiological purposes, and

- algorithms and modules with technical motivation.

Figure 14.1 shows the signal processing modules and the binaural extension that will be explained in Sec. 14.1.2.

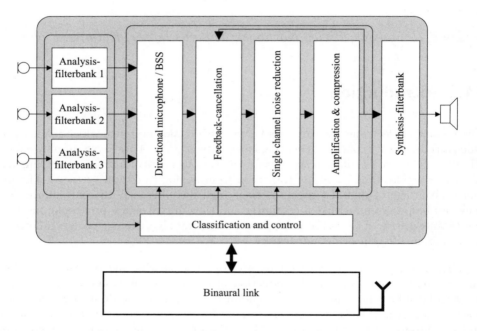

Figure 14.1: Processing stages of a high-end hearing aid. The three input signals are analysed by three separate analysis-filterbanks with usually eight to twenty frequency bands each. The subsequent module can be a classic directional microphone, a binaural beamformer or a blind source separation algorithm. All the following modules, the feedback-cancellation module, the single channel noise reduction module and the amplification and compression module, work in frequency bands, as well. As the last step in the signal processing chain, the resulting band signals are summed up by means of the synthesis-filterbank. The classification system as a superordinated system controls all signal processing modules. The binaural link provides the wireless connection to the other hearing aid and the remote control

Audiological Algorithms

The audiological algorithms have the goal of compensating for the hearing loss with the main focus on speech intelligibility. The most prominent examples are the frequency dependent amplification concept and the compression algorithm, which compensate for the increased hearing threshold. Further examples are single channel noise reduction algorithms and directional microphone algorithms.

The compression algorithm is needed to compensate for the recruitment phenomenon which is typically observed as a by-product of sensorineural hearing loss [Dillon 2001]. This phenomenon can roughly be described as follows. Low level sounds are inaudible because of the increased hearing threshold, while louder sounds are nearly normally perceived. That is, the application of linear gain is not completely effective. Instead, a level dependent gain has to be applied, which compresses the dynamic range of the input signal. At low signal levels, more gain is necessary than at higher signal levels. Since the recruitment phenomenon is frequency dependent, it is advantageous to apply a compression system within a subband concept.

These compression systems are called automatic gain control (AGC). The major component of AGC is a level meter with different release and attack time constants. AGC with time constants of the order of seconds are often referred to as automatic volume control (AVC). The resulting gain is then able to adapt to varying listening environments. AGC with time constants of the order of milliseconds are called "syllabic compression" as they are able to follow the temporal level changes of vowels and consonants within a syllable. Combinations of both types of AGC are known as "dual compression" AGC.

Single channel noise reduction algorithms use the signal of only one input source, e.g., one microphone. The separation of the desired signal from the noisy mixture is done based on *a priori* knowledge of both the desired signal and the noise signal. One example of *a priori* knowledge is the stationarity behavior of speech (desired signal) and noise: It is assumed that noise is more stationary than speech. Examples of single channel noise reduction algorithms are the long-term smoothed modulation frequency-based noise reduction, Wiener-filter-based short-term smoothed noise reduction methods and Ephraim–Malah-based short-term smoothed noise reduction methods [Hamacher et al. 2005].

Another noise reduction approach based on more than one microphone uses the spatial distribution of the desired signal and the noise signals. Often, one can assume that the desired signal comes from the front direction, and any signal that does not fall in some front-angle range is defined as noise. This spatial separation can be achieved via directional microphone algorithms. Modern hearing aids are designed with up to three microphones per device to capture the desired front signal.

Technical Algorithms

One very prominent example that falls into this category is the feedback cancellation system. The underlying problem is caused by the acoustic feedback path from the hearing aid receiver back to the microphone (or microphones, if a directional microphone exits). In combination with the forward amplification path of the hearing aid, the feedback path completes a closed loop. If the amplification of the loop exceeds a certain value, the system becomes unstable and feedback (whistling) occurs. One possible solution to this problem is the reduction of the loop gain exactly at the frequency of the feedback. A more powerful approach is related to the echo compensation approach in hands-free equipment. The acoustic feedback path is estimated and inversely mimicked within the digital hearing aid to compensate for the acoustic feedback. [Hamacher et al. 2006] explains this algorithm in more detail.

Another example of technically motivated algorithms is the classification system. It is desirable since most algorithms have specific optimized parameter settings for different listening situations in everyday life. The system analyses the acoustic environment and switches to the most appropriate setting.

14.1.2 Binaural Hearing Aids

The binaural link has two impacts on hearing aid algorithms and concepts. On the one hand, it allows the extension and improvement of existing monaural algorithms, as summarized in Sec. 14.1.1. On the other hand, it allows the realization of completely new algorithms and concepts. Many of these must wait for future implementation because they cannot be realized due to the high demand of computational power and data rate of the binaural link. But some of them are already implemented in modern binaural high-end hearing aids.

The technical constraints of a binaural link within hearing aids are very challenging. The analog components have to be as small as possible to accommodate space requirements in hearing aids. Additionally, the power consumption has to be very low. Depending on the application, the binaural link may need to be active at all times. The only power source available is a small battery within the hearing aid, which needs to power the binaural link as well as all other components for at least a day.

One application of a low data rate binaural link addresses the ease of use of bilaterally worn hearing aids. Examples are the volume control and the program switch, which can be manipulated at only one hearing aid. The synchronization of the other hearing aid is done via the binaural link, which transmits the control information. Another currently existing application is the binaural classification system. Since modern hearing aids adapt to the current hearing situation, it is important that both hearing aids maintain uniform classification on both sides. Again, this demands a binaural link, which exchanges the respective information.

More demanding for the data rate of the binaural link is the binaural beamformer. We see the beginning approach in the monaural use at the directional microphone. Additionally, the assemblage of microphones of both hearing aids must be viewed as a large microphone array. Since there is only one algorithm computing all microphone signals, the binaural link has to transmit the microphone inputs at full sampling frequency.

A further expansion and enhancement of the beamformer concept is the blind source separation (BSS) approach. This offers a more flexible and intelligent choice of desired source. With this approach all sound sources are initially captured, regardless of being desired sources or noise sources. The choice of the desired source can be determined in a separate step by means of an intelligent algorithm or a hearing aid wearer command. The selected source has to be presented binaurally, for the relevant binaural cues for localization to be reconstructed. Since the BSS algorithm processes the microphone signals of both hearing aids and produces a stereo signal, again, a binaural link with high data rate becomes a mandatory component.

14.1.3 Organization of this Chapter

In Sec. 14.2 we introduce the physical and technical aspects of the binaural link. As outlined, it is the prerequisite for all binaural applications. The concepts that increase the ease of use are described at the end of Sec. 14.2, and the binaural classification system is explained in Sec. 14.3. The two applications that need a high data rate for binaural link are the binaural beamformer, as addressed in Sec. 14.4, and the blind source separation as explained in Sec. 14.5. Finally, conclusions are given in Sec. 14.6.

14.2 Wireless System for Hearing Aids

People with hearing loss can benefit from the introduction of a wireless system into their hearing aids in various ways. The usability of the hearing aid can be improved by synchronizing the hearing aids and introducing a wireless remote control. Additionally, speech intelligibility can be raised in difficult acoustic situations by applying a wireless interconnection between the binaural hearings aids. However, the unique requirements of the hearing aid world call for a wireless connection system specific for hearing instruments. Siemens introduced a wireless system for hearing aids in 2004. In the following, background about this system is given.

14.2.1 Comparison of Wireless Systems

Wireless systems like GSM, UMTS and Personal Area Network (PAN) systems like Bluetooth or WLAN are to date ubiquitous. However, they are not suitable for integration into a hearing aid. The power consumption requirements of the hearing aid are significantly more strict than those of a mobile communication device, especially due to the limited space for the battery. This impedes the use of high transmission frequencies in the area of 1 GHz as used for many mobile communication systems, as the corresponding analog and digital parts of the transceivers consume far too much power. Also, for hearing instrument applications, a transmission range covering the distance between the ears is sufficient.

Table 14.1: Comparison of properties of different wireless systems

	GSM, UMTS	PANs	RFID (HF)	Hearing aid
Frequency	900/1800/1900 MHz	2.4 GHz	13.56 Mhz	120 kHz
Near/Far field limit	2–5 cm	2 cm	3.5 m	400 m
Range	up to 35 km	10–100 m	up to 150 cm	25 cm

The comparably high range of mobile communication systems or PANs is reached by electromagnetic transmission with the receiver being positioned in the far field of the transmitter. Assuming free space propagation, the field strengths, e.g., the magnetic field strength H_{EM}, can be considered as decaying in proportion to the distance d between receiver and transmitter:

$$H_{\mathrm{EM}} \sim d^{-1}. \tag{14.1}$$

The digital wireless system realized in the Siemens hearing aids is realized at a much lower transmission frequency and lower transmission range. This is achieved by an inductive transmission system, in which the receiver is positioned in the near field of the transmitter. The term near field means that the receiver is placed at a distance from the transmitter that is small compared with the transmitted wavelength λ, i.e.,

$$d << \frac{\lambda}{2\pi} = \frac{c}{2\pi f_{\mathrm{trans}}}. \tag{14.2}$$

In the near field, the magnetic field strength at the receiver position decays by

$$H_{\mathrm{MAG}} \sim d^{-3}. \tag{14.3}$$

Table 14.1 gives an overview of transmission ranges and frequencies of the wireless inductive system for the Siemens hearing aids compared with other well known wireless systems.

Given the lower transmission frequency of the digital hearing aids, the digital parts of the transceivers can be realized with significantly less power consumption compared with other wireless systems due to the much lower system clock.

14.2.2 Functional Description of the Wireless System for Hearing Aids

Each of the binaural hearing aids may operate as both transmitter and receiver. In addition to the hearing aids, a remote control may act as wireless transceiver. Figure 14.2 shows a block diagram of the wireless system transmitter and receiver.

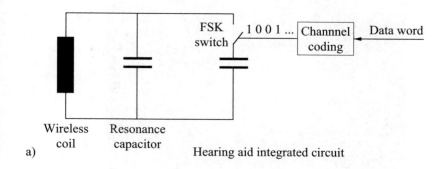

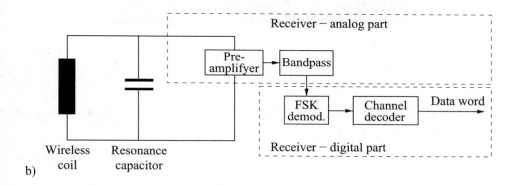

Figure 14.2: Block diagram of the hearing aid wireless
a) transmitter
b) receiver

On the transmitter side, the digital data symbols are protected by channel coding and are split into bits, which are transmitted via Frequency Shift Keying (FSK) modulation. An external resonance circuit of resonance frequency $f = 120\,\text{kHz}$ is applied, which consists of a wireless coil (inductivity) and a capacitor. The FSK modulation is then realized by changing the resonance frequency with the respective data bit by switching an integrated capacitor on or off. As only control sequences are exchanged between the hearing aids, the overall data rate of the currently realized system is quite low, i.e., $200\,\text{bit/s}$, which allows the system to operate with very little power.

On the receiver side, the signal is captured using the same resonance circuit and is processed in the analog world via pre-amplification and bandpass filtering. The bandpass filtering improves the SNR by attenuating out-of-band noises. In-band noises that potentially disturb the hearing aid receiver are primarily emissions from digital circuits in close proximity to the hearing aid. After bandpass filtering, the

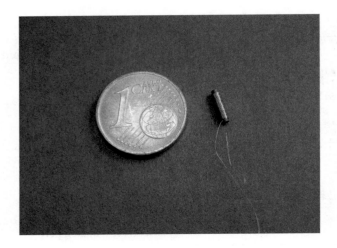

Figure 14.3: Wireless receiver coil inside hearing aid

received signals are digitized in the FSK demodulator and channel decoded to extract the transmitted data symbol.

To reach the transmission range of the binaural hearing aids, i.e., 25 cm, the design of wireless transceiver coil is crucial to the performance of the system.

Figure 14.3 shows the wireless coil used in the Siemens hearing CIC (completely-in-the-canal) hearing aids compared with a 1 cent coin. The wireless coil is composed of several hundred windings at a very small diameter. The coil inductivity needs to be very accurate to exactly match the resonance frequency of the circuit.

Table 14.2 provides an overview of the properties of the wireless system for Siemens hearing aids.

Table 14.2: Properties of the Siemens wireless system for hearing aids

Transmission frequency	120 kHz
Modulation mechanism	FSK
Data rate	200 bit/s
Power consumption	90 μW
Transmission range from hearing aid	25 cm
Transmission range from remote control	100 cm

The power consumption is below 10 % of the typical power consumption of the whole hearing aid. Compared with other wireless systems like, e.g., Bluetooth, the hearing aids' wireless system consumes less than 1 % of the power. The overall data rate is limited to 200 bit/s, which allows the applications described in the following sections.

14.2.3 Applications of the Wireless System for Hearing Aids

Figure 14.4 shows an overview of the applications provided by the wireless system. The two hearing aids interchange low rate data between the ears, while the external remote control can be applied to control functional states of the hearing aids or to read out and display functional states to the end-user in an integrated display. The range of the interconnection between the hearing aids is limited to the distance between the two ears. The transmitting range of the remote control must be 100 cm to accommodate arm length and this is possible due to the higher battery capacity in the remote control. For applications, in which the remote control acts as receiver, the application is limited to the 25 cm transmission range.

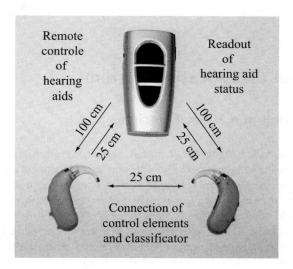

Figure 14.4: Applications of the wireless system for hearing aids

Binaural Interaction

The applications of the binaurally interacting hearing aids can be classified in technically motivated functions and usability motivated functions. The technical functions aim at optimizing the speech intelligibility in noisy environments. For that purpose, the classification units of both hearing aids are synchronized via the wireless data exchange to avoid consequences of detection of different hearing situations as described later in detail. Usability comfort concepts comprise the synchronization of the hearing programs at both ears and the volume controls of the two hearing aids. It is sufficient to change the program or volume via only one hearing aid as the other device will immediately follow this program or volume switch. The synchronization also enables the building of smaller binaural hearing aids, as control elements can be omitted from the second device or distributed between the two hearing aids. The left hearing aid can, e.g., include a volume control element while the right hearing aid includes a program change switch.

Remote Control

The remote control can operate all the wireless functionalities of the hearing aid, such as switching hearing programs or volume controls and may also reset the state of the hearing aids. Moreover, it enables one to read out the functional states of the hearing aids and displays these to the customer: besides hearing programs or volume setting, this includes the battery charge status as indication when the primary hearing aid cell must be exchanged or the secondary cell must be recharged again.

14.3 Binaural Classification Systems

14.3.1 Motivation and Basic Principle

Modern hearing aids offer a variety of signal processing algorithms and specialized parameter configurations to cope with the different listening situations in everyday life. Most algorithms and their parameter configurations were designed to address specific acoustic situations, in which they are beneficial for the hearing aid users. However, these settings would have no or even negative effects in the other conditions. For example, the single channel noise reduction algorithms described in Sec. 14.1.1, which suppress stationary background noise efficiently would worsen the sound quality of music and, therefore, should be disabled here. Even if the optimal signal processing algorithms for any relevant situation were available, the problem would remain of activating them specifically as well as reliably. However, many hearing aid users are not willing or even able to monitor the acoustic environment continuously, recognize the specific acoustic scenes and activate the related set of algorithms and parameter configurations ("hearing aid programs"). Moreover, elderly people often do not have the manual dexterity to easily use the miniature push buttons of the hearing aids to switch between the different programs. As shown in the results of a previous field study [Cord et al. 2002], the manual activation of a simple directional microphone is already a severe problem for many hearing aid users. Consequently, the proper activation of the different algorithms is a fundamental requirement needed to change the technical capabilities into the substantial hearing improvement that is expected by the hearing aid users in their everyday lives.

A solution to this problem is a superordinated classification-based control systems as depicted in Fig. 14.1. These systems continuously analyze and classify the acoustic listening environment and activate the appropriate hearing aid processing automatically [Powers, Hamacher 2002], [Buechler et al. 2000], [Kates 1995], [Ostendorf et al. 1998]. Such classification-based control systems are already key elements in today's hearing aids and they will become even more important in the future, since they facilitate the beneficial application of the growing portfolio of highly specialized algorithms without overextending the dexterity and mental capabilities of the hearing aid users.

The classification-based control systems of today's hearing aids are able to detect classes of situation such as "speech in quiet", "speech in noise", "noise" and "music" [Powers, Hamacher 2002], [Buechler et al. 2000]. As depicted in Fig. 14.1, they mainly control all major blocks of the hearing aid processing, such as the amplification and dynamic compression, the noise-reduction and directional microphone algorithms, as well as the feedback reduction. Today's systems comprise different functional stages. First, "features" are extracted from the microphone signal. A feature is a certain property of the signal that is as different as possible for the different situation classes. In the literature, several spectral and temporal features have been proposed, e.g., the profile and temporal changes of the frequency spectrum [Buechler et al. 2000], [Kates 1995], [Feldbusch 1998], the statistical distribution of signal amplitudes, and the modulation frequencies of the signal envelope [Ostendorf et al. 1998].

The adaption of the hearing aid signal processing to the detected situation is divided into two steps. The stage "selection of algorithm and parameters" contains an "action matrix", which defines the appropriate algorithms and parameter configurations for each situation class. The action matrix is derived empirically from perceptual studies with hearing impaired subjects in which the effect of each particular algorithm and configuration in the different situations was systematically investigated. The following stage generates the "on/off"-control signals. Since sudden switches of the signal processing components between "off" and "on" can be irritating, appropriate fading mechanisms are applied, for example, by simple low-pass filtering of the control signals. In this way, a smooth transition between successive operation states of the hearing aid processing is ensured.

As described in [Hamacher et al. 2006], detection rates in the range of 75–90 % for the situation classes given above can be achieved. These rates turn out to be sufficient for the reliable and beneficial control of the various hearing aid signal processing algorithms. The impact of the misdetections are reduced to a negligible level by nonlinear temporal averaging of the classification results. In addition, the smooth transition described above conceals misdetections, as long as their duration is shorter than the time constant of the transition process.

14.3.2 Binaural Classification

Since a hearing loss normally affects both ears, hearing aids are fitted bilaterally in most cases. If hearing aids with classification-based control are used, the problem must be considered that different situation classes can be detected in the left and right hearing aid resulting in significantly different processing schemes. Such differences in classification results are mainly caused by head shading effects in asymmetrical hearing situations. For example, a music source on one side of the head of the hearing aid user and a speech source on the other side could lead to local classification decisions dominated by the ipsilateral source. The contralateral source is shaded, i.e., attenuated, by the head. Figure 14.5 shows the result of real-life evaluations with a KEMAR and bilaterally fitted Siemens Acuris BTE instruments. In 43 different hearing situations of everyday-life, the classification results of both hearing aids were recorded and the temporal percentage of asymmetrical decisions was calculated. Obviously, in nearly 50 % of the encountered hearing situations, the two classification systems came to different results more than 15 % of the time.

Processing differences between both hearing aids, e.g., if the directional microphone is activated only on one side, can temporarily reduce the sound quality as well as the speech intelligibility. An example for the effect on speech understanding is the study of Hornsby and Ricketts [Hornsby, Ricketts 2007]. They tested speech understanding of 16 hearing impaired subjects in a complex condition with speech arriving from the front and interfered by babble noise provided by five uncorrelated sources, spatially arranged as illustrated in Fig. 14.6-a. The subjects were bilaterally fitted with advanced hearing aids offering omnidirectional and directional microphone modes. The speech understanding was measured with the directional microphone activated in both instruments (D/D) and additionally in two mixed conditions, in which only one instrument was in directional mode and the other in omnidirectional mode (D/O, O/D). For each of the three conditions, the directional benefit

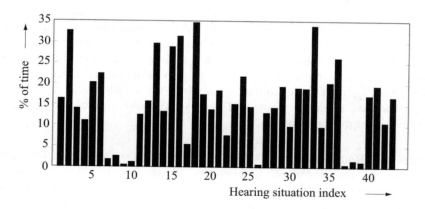

Figure 14.5: Temporal percentage of asymmetrical classification results in 43 different hearing situations

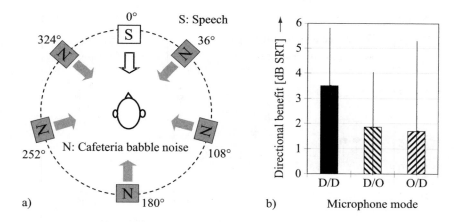

a)

b)

Figure 14.6: Study of Hornsby and Ricketts [Hornsby, Ricketts 2007]
    a) Test condition
    b) Average directional benefit (as defined in [Hornsby, Ricketts 2007])
       and standard deviation for three different microphone modes

was averaged across the subjects (Fig. 14.6-b). It can be seen that both asymmetrical fittings (D/O, O/D) cause a significant degradation in speech understanding performance compared with the symmetrical condition with both instruments in directional mode (D/D).

In addition to that, different signal processing schemes in both hearing aids can introduce artificial interaural time and level differences, reducing the localization ability of the hearing impaired listener, which is mainly based on analyzing these signal cues. An example of this effect is described by Keidser et al. [Keidser et al. 2006], who investigated the impact of compression, noise reduction, and the directional microphone on the localization performance of 24 bilaterally fitted hearing aid wearers. The tests regarding the impact of the microphone mode were performed in two symmetrical and two asymmetrical conditions: 1) symmetrical with both hearing aids in omnidirectional mode (O/O), 2) symmetrical with both hearing aids in cardioid-directional mode (DC/DC), 3) asymmetrical with one hearing aid in figure-8 directional mode and the other in cardioid directional mode (D8/DC), 4) asymmetrical with one hearing aid cardioid-directional and the other omni-directional (DC/O). Details on the impact of each microphone setting on the interaural time- and level-difference cues can be found in [Keidser et al. 2006]. The results of the localization tests performed after two weeks wearing time and two months wearing time are given in Fig. 14.7. It can be seen that the average localization error is significantly larger in both asymmetrical conditions than in the symmetrical conditions. Therefore, asymmetrical microphone settings should be avoided.

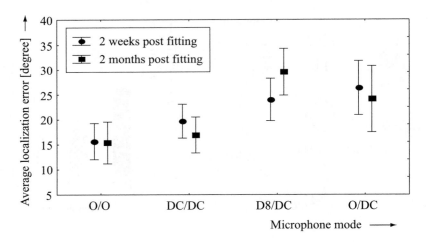

Figure 14.7: The effect of the directional microphone on horizontal localization performance in hearing aid users [Keidser et al. 2006]

As an efficient solution for the described problems, a binaural synchronization of the classification systems based on the bidirectional wireless link described in the previous chapter has recently been introduced. In this realization, both hearing aids first analyze the sound field independently and then exchange information about the local classification results (Fig. 14.8). With this information they then follow exactly the same rules in parallel to determine the global "binaural" class. These rules are defined by the "binaural decision matrix". An example of this matrix is shown in Fig. 14.8-b. The decision matrix is defined empirically based on extensive analysis of the classification system and the processing settings preferred by hearing aid wearers in numerous real-life situations. Finally, both hearing aids are adapted synchronously to the signal processing and parameter settings prescribed for the common class. Doing so, the above mentioned disadvantages in asymmetrical hearing situations are avoided.

The current classification systems rely only on statistical information derived from one microphone signal. However, spatial information has not been incorporated into the classification process up to now. Since spatial attributes would provide a more accurate description of the "acoustic scene" they will be a major key to providing larger and more specific sets of classes in future hearing aid generations. Additionally, any knowledge about the spatial distribution of the sound sources will increase the consistency between the decisions of the classification system and the hearing aid user regarding the desired source. For example, in ambiguous situations, e.g., conversation in a music cafe, the classification system can decide in favor of the source to which the user is turning his or her head, which is probably the desired source. Consequently, the further-development of the classification systems for hearing aids will be strongly connected to the availability and performance of localization algorithms.

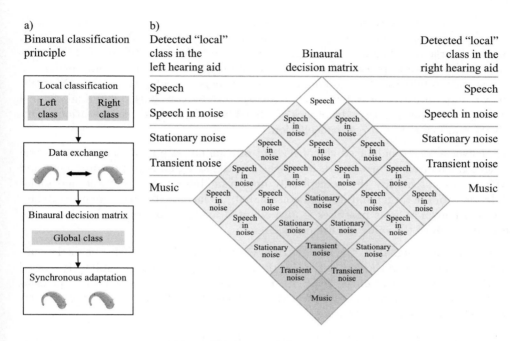

Figure 14.8: Binaural classification system
a) Basic principle
b) Example of a binaural decision matrix

14.4 Binaural Beamformer

By transmitting the actual audio signals between the hearing aids instead of low bit-rate data, binaural algorithms can be applied to improve the speech intelligibility of the hearing aid wearer. Binaural beamformers are candidate algorithms for that purpose.

Speech enhancement by beamforming uses spatial diversity of desired speech and interfering speech or noise sources by combining multiple noisy input signals. Beamformer realizations can be classified into fixed and adaptive. A fixed beamformer combines the noisy signals of multiple microphones by a time-invariant filter-and-sum operation. The combining filters can be designed to achieve constructive superposition towards a desired direction (delay-and-sum beamformer) or to maximize the SNR improvement (superdirective beamformer), e.g., [Bitzer, Simmer 2001]. Adaptive beamformers commonly consist of a fixed beamformer steered towards a desired direction and a time varying branch that adaptively steers spatial nulls towards interfering sources. Among various adaptive beamformers, the Griffiths–Jim beamformer [Griffiths, Jim 1982], or extensions of it, e.g., [Hoshuyama, Sugiyama 2001] are more widely known.

While classic beamformer applications output one enhanced signal given multiple noisy observations, a binaural speech enhancement system, e.g., [Wittkop et al. 1997], must deliver a dual channel output signal, preferably without modification of the inter-aural amplitude and phase differences to maintain the original spatial impression. In this section, a fullband, binaural input–output array is presented that applies the well-known superdirective beamformer as core structure [Lotter 2004]. The dual-channel system thus comprises the advantages of a fixed beamformer, i.e., computational sim-plicity and low risk of target cancellation.

To deliver a stereo enhanced signal instead of a mono output, an adaptive spectral weight calculation is introduced, in which the desired signal is passed unfiltered and which does not modify the perceptually important interaural amplitude and phase differences of the target and residual noise signal. To increase the performance further, the well-known Wiener postfilter is also adapted for the binaural application under consideration of the same requirements.

The microphone signals at the left and right ear not only differ in their phases depend-ing on the position of the source relative to the head, but also in their intensity caused by the shadowing effect of the head. At discrete Fourier transform (DFT) frequencies ω_μ with frequency bin index μ, given the DFT spectrum $S(\omega_\mu)$ of a source, the left and right ear signal spectra are given by

$$Y_l(\omega_\mu) = D_l(\omega_\mu)S(\omega_\mu) \quad \text{and} \quad Y_r(\omega_\mu) = D_r(\omega_\mu)S(\omega_\mu). \tag{14.4}$$

The shadowing effect of the head might be described for a source impinging from an angle θ_S in the horizontal plane by angle and frequency-dependent amplitude factors $\alpha_l(\mu, \theta_S), \alpha_r(\mu, \theta_S)$ for the left and right ear sides, respectively time delays $\tau_l(\theta_S), \tau_r(\theta_S)$ characterize the propagation time from the source to the left and right ear:

$$\boldsymbol{D}(\mu, \theta_S) = [\alpha_l(\mu, \theta_S)e^{-j\omega_\mu \tau_l(\theta_S)}, \; \alpha_r(\mu, \theta_S)e^{-j\omega_\mu \tau_r(\theta_S)}]^T. \tag{14.5}$$

14.4.1 Dual Channel Input–Output Beamformer Design

As a special case of the general superdirective beamformer, the binaural beamformer can be realized by summing the input DFT coefficients after complex multiplication with superdirective coefficients,

$$Z(\mu) = W_l^*(\mu)Y_l(\mu) + W_r^*(\mu)Y_r(\mu). \tag{14.6}$$

The objective of the superdirective design of the weight vector $\boldsymbol{W}$ is to maximize the output SNR. This can be achieved by minimizing the output energy with the constraint of an unfiltered signal from the desired direction leading to the design rule

$$\mathbf{W}(\mu, \theta_S) = \frac{\boldsymbol{\Phi}_{22}^{-1}(\mu)\boldsymbol{D}(\mu, \theta_S)}{\boldsymbol{D}^H(\mu, \theta_S)\boldsymbol{\Phi}_{22}^{-1}(\mu)\boldsymbol{D}(\mu, \theta_S)}. \tag{14.7}$$

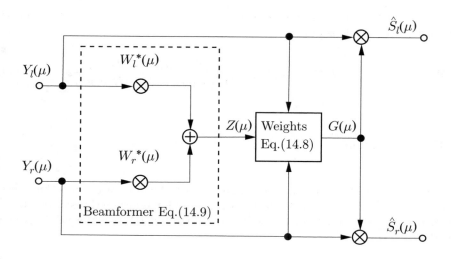

Figure 14.9: Superdirective binaural input–output beamformer

The 2×2 cross power spectral density matrix $\mathbf{\Phi}_{22}(\mu)$ can be calculated using the head related coherence function obtained from measurements from a binaural signal model.

A beamformer that outputs a monaural signal would, however, be unacceptable, because the benefit in terms of noise reduction is consumed by the loss of spatial hearing. Therefore, the beamformer output is used for the calculation of spectral weights. Figure 14.9 shows a block diagram of the proposed superdirective binaural input–output beamformer in the frequency domain.

The enhanced Fourier coefficients $Z(\mu)$ can serve as reference for the calculation of weight factors $G(\mu)$, which output binaural enhanced spectra $\hat{S}_1(\mu)$, $\hat{S}_r(\mu)$ via multiplication with the input spectra $Y_1(\mu)$, $Y_r(\mu)$. Regarding the weight calculation method, it is advantageous to determine a single real-valued gain for both left and right ear spectral coefficients. By doing so, the interaural time and amplitude differences will be preserved in the enhanced signal. Consequently, distortions of the spatial impression will be minimized in the output signal. Real-valued weight factors $G_{\mathrm{super}}(\mu)$ are desirable in order to minimize distortions from the frequency domain filter. In addition, a distortionless response for the desired direction should be guaranteed, i.e., $G_{\mathrm{super}}(\mu, \theta_S) \overset{!}{=} 1$.

To fulfill the demand of just one weight for both the left and right ear sides, the weights are calculated by comparing the spectral amplitudes of the beamformer output with the sum of both input spectral amplitudes,

$$G_{\mathrm{super}}(\mu) = \frac{|Z(\mu)|}{|Y_1(\mu)| + |Y_r(\mu)|}. \qquad (14.8)$$

To fulfil the distortionless response of the desired signal with (14.8) the following beamformer design rule is applied:

$$\mathbf{W}(\mu,\theta_S) = (\alpha_l(\mu,\theta_S) + \alpha_r(\mu,\theta_S)) \cdot \frac{(\mathbf{\Phi}_{22}^{-1}(\mu))\mathbf{D}(\mu,\theta_S)}{\mathbf{D}^H(\mu,\theta_S)(\mathbf{\Phi}_{22}^{-1}(\mu))\mathbf{D}(\mu,\theta_S)}. \tag{14.9}$$

Directivity Evaluation

The performance of the dual-channel beamformer can be illustrated with the directivity pattern $\Psi(\mu,\theta_s,\theta)$, which is defined as the squared magnitude transfer function for a signal that arrives from a certain spatial direction θ if the beamformer is designed for angle θ_s.

Figure 14.10 shows the directivity pattern for the desired direction $\theta_s = 0°$, where $\theta_S = 0°$ corresponds to a broadside look direction. The achieved directivity is comparably low at low frequencies. At higher frequencies, the phase difference generated by a lateral source becomes significant and leads to a higher degree of directivity. Figure 14.11 shows the directivity for $\theta_S = -60°$. The directivity increases especially for low frequencies and the main lobe becomes more narrow as the amplitude differences become more important.

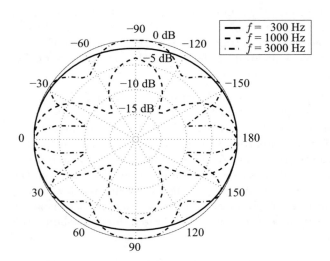

Figure 14.10: Beam pattern $\Psi(\mu,\theta_s = 0°,\theta)$ of superdirective binaural input–output beamformer for DFT bins μ corresponding to $f = 300\,\text{Hz}$, $f = 1000\,\text{Hz}$ and $f = 3000\,\text{Hz}$

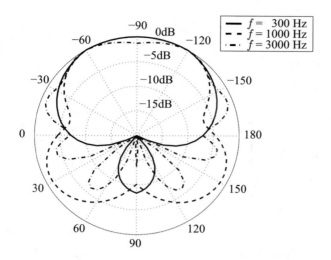

Figure 14.11: Beam pattern $\Psi(\mu, \theta_s = -60°, \theta)$ of superdirective binaural input–output beamformer for DFT bins μ corresponding to $f = 300\,\text{Hz}$, $f = 1000\,\text{Hz}$ and $f = 3000\,\text{Hz}$

14.4.2 Multichannel Postfilter

The superdirective beamformer produces the best possible signal-to-noise ratio for a narrowband input by minimizing the noise power subject to the constraint of a distortionless response for a desired direction [Monzingo, Miller 1980]. It can be shown [Simmer et al. 2001] that the best possible estimate in the minimum mean square error (MMSE) sense is the multi-channel Wiener filter, which can be factorized into the superdirective beamformer followed by a single-channel Wiener postfilter.

The dual-channel input–output beamformer can be extended by also adapting the formulation of the multichannel Wiener postfilter according to [Simmer et al. 2001] into the spectral weighting framework. Again, only one postfilter weight for both left and right ear spectral coefficients is applied to maintain the original spatial impression, i.e., the interaural amplitude and phase differences. Secondly, a source from a desired direction θ_S should pass unfiltered, i.e., the spectral postfilter weight for a signal from that direction should be one.

In analogy to the optimal MMSE estimate, postfilter weights $G_{\text{post}}(\mu)$ are multiplicatively combined with the beamformer weights $G_{\text{super}}(\mu)$ according to (14.8) to the resulting weights $G(\mu)$

$$G(\mu) = G_{\text{super}}(\mu) \cdot G_{\text{post}}(\mu).$$
(14.10)

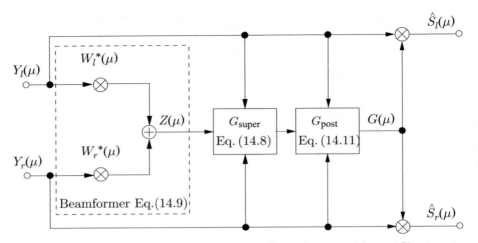

Figure 14.12: Superdirective input–output beamformer with postfiltering

The condition $G_{\text{post}}(\mu) = 1$ from a source impinging from θ_S gives

$$G_{\text{post}}(\mu) = \frac{|Z(\mu)|^2}{|Y_l(\mu)|^2 + |Y_r(\mu)|^2} \cdot \frac{(\alpha_l(\mu, \theta_S))^2 + (\alpha_r(\mu, \theta_S))^2}{(\alpha_l(\mu, \theta_S) + \alpha_r(\mu, \theta_S))^2} . \tag{14.11}$$

Figure 14.12 shows a block diagram of the resulting system with the stereo input–output beamformer plus Wiener postfilter in the DFT domain. After the dual channel beamformer processing, the postfilter weights are calculated according to (14.11) and are multiplicatively combined with the beamformer gains according to (14.10). The dual-channel output spectral coefficients $\hat{S}_l(\mu)$, $\hat{S}_r(\mu)$ are generated by a multiplication of left and right side input coefficients $Y_l(\mu)$, $Y_r(\mu)$ with the same final weight $G(\mu)$.

14.4.3 Performance Evaluation

The performance of the dual channel input–output beamformer with postfilter is evaluated in a multi-talker situation in a conference room (reverberation time $T_{60} \approx 800\,\text{ms}$) with a dummy head wearing binaural hearing aids. In the experiments, a desired speech source s_1 arrives from angle θ_{S_1} towards which the beamformer is steered and an interfering speech signal s_2 arrives from angle θ_{S_2}. To judge the benefit of the frequency dependent noise reduction, a speech intelligibility weighted noise reduction gain is applied, which measures the attenuation of the unwanted source relative to the desired source frequency weighted with the intelligibility contribution of the respective frequency band as given in [ANSI-S3.5 1997].

Figure 14.13 plots the performance of the superdirective binaural input–output beamformer in terms of the speech intelligibility weighted gain for a desired speech source from $0°$ and speech interferers from variable directions.

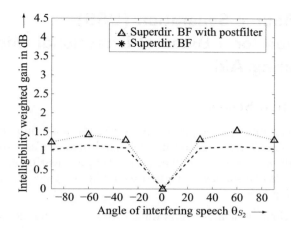

Figure 14.13: Intelligibility weighted gain of superdirective binaural input–output beamformer with and without postfilter for speech from $\theta_{S_1} = 0°$ and speech interferer from other directions, θ_{S_2}

For $\theta_{S_1} = 0$, the binaural input–output superdirective beamformer delivers about 1 dB intelligibility weighted improvement, which is further improved by 0.5 dB with the postfilter.

Higher directivity gains can be achieved for lateral directions. Figure 14.14 plots the performance of the superdirective binaural input–output beamformer when the desired signal arrives from $\theta_{S_1} = -60°$. Here, the algorithm delivers up to 4.5 dB gain, depending on the position of the interfering source.

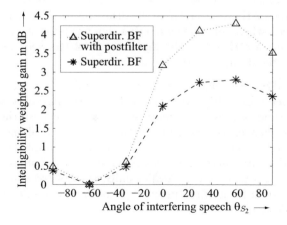

Figure 14.14: Intelligibility weighted gain of superdirective binaural input–output beamformer with and without postfilter for speech from $\theta_{S_1} = -60°$ and speech interferer from other directions, θ_{S_2}

14.5 Blind Source Separation (BSS): An Application for a Binaural Directional Microphone Array in Hearing Aids

14.5.1 Application Scenario

Currently, directional microphone processing is still the predominant approach to increase the signal-to-noise ratio (SNR) in noisy environments and to enhance speech intelligibility for normal hearing as well as for people with hearing impairment. Therefore, research effort is continuously invested to further enhance the current methods, especially for hearing aids.

These current solutions typically exhibit differential processing methods of monaural, small aperture arrays with two or three microphones. Differential processing methods are appropriate since spatial aliasing is not relevant for such small microphone distance settings. However, these approaches suffer from a strong microphone noise gain that increases with decreasing microphone distance and with an increasing number of array microphones.

The physical limitations of such approaches are the motivation to develop alternatives, especially binaural approaches. The targets of these approaches are to outperform current monaural solutions and to offer possibilities for completely new applications such as directional microphone processing for small In-The-Ear (ITE) custom hearing aid devices. Here, due to space limitations in these small custom devices, only one microphone can be integrated. Currently, directional processing has yet to be integrated in such binaural systems. One solution of a binaural beamformer has been described in Sec. 14.4. This beamformer has to be designed for a certain direction of the desired source, which has to be assumed. However, as described there, especially for the desired sources arriving from the front direction (0° azimuth) the gain in SNR is rather limited, even when only one competing signal is present. The limits of classic binaural beamformers are therefore the limited gain of the SNR and the required *a priori* knowledge of the desired source.

Approaches for Blind Source Separation (BSS) as described in publications of Buchner and Makino [Buchner et al. 2005], [Makino 2003] have the potential to overcome these limits. These blind approaches are especially attractive because they do not require *a priori* knowledge such as the position of the array microphones or the location of sources. This is advantageous for binaural microphone applications since, due to different head widths, the microphone positions and their distance from the input, respectively, are not *a priori* known. Additionally, the desired signal sources are not required to be directly in front of the listener. This offers more degrees of freedom for the beam pattern design.

Theoretical aspects of BSS with the classic application scenario and free field microphones have been described in detail in [Buchner et al. 2005], [Makino 2003]. Here, we will focus on hearing aid applications for BSS and the specific problems and solutions required.

14.5.2 Specific Hearing Aid Challenges and Solutions

Specific problems arise for the classic signal separation task mainly due to the specific hearing aid setup, i.e., head shading and head movements. Head shading occurs due to the microphone positioning next to the head. Head movements require fast adaptation procedures to cope with the tracking of moving sources.

Additionally, solutions are required to select the desired output since blind separation methods provide all separated signals. People with BSS integrated in their hearing aids should be relieved from the task of selecting their desired source. An intelligent system has to cope with this task.

Also, a solution has to be found for providing the hearing aid user with a binaural output. One approach has been presented in [Takatami et al. 2005]. BSS methods provide only monaural output signals for each separated source signal. The desired monaural source signal has to be further processed for generating the desired binaural output, which allows the hearing aid user to localize the source position correctly within the acoustic environment.

These three tasks are illustrated in Fig. 14.15 and have already been discussed [Puder 2005]. It is shown that for hearing aid applications, the signal separation stage has to be followed by the source selection algorithm and a procedure for generating binaural output signals.

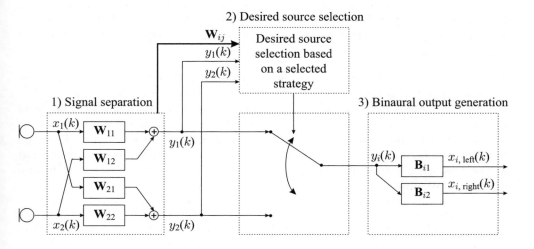

Figure 14.15: Three stages are necessary for an integration of BSS in hearing aids. Here, $\mathbf{W}_{ij}$ and $\mathbf{B}_{ij}$ with $i, j \in \{1, 2\}$ denote the frequency vector of the filters, i.e., $\mathbf{W}_{ij} = [W_{ij}(0), W_{ij}(1), \ldots, W_{ij}(\mu), \ldots, W_{ij}(M-1),]^T$ where M is the number of frequency bins

14.5.3 Signal Separation with Hearing Aid Constraints

Whatever the different approaches for solving the BSS problem, all of them are theoretically able to separate a number of sound sources in the acoustic environment equivalent to the number of microphones used.

For the above described binaural hearing aid setup, using one microphone on each side of the head, two sources can be separated. Consequently, the classic BSS question arises for the hearing aid applications. How robust is the system in the case of the presence of more environmental sources or even in the case of diffuse background noise? Here, the BSS systems for hearing aids have to provide answers just like other BSS applications.

Therefore, we would like to focus on specific problems occurring for the hearing aid application:

- head shading,

- head movements, and

- own voice presence.

Head shading describes the effects when microphones are positioned next to the head and not in free-field. For sources arriving from the opposite side from the head, the shadowed microphone signal is more attenuated compared with the microphone turned towards the signal source. The BSS system has to compensate for this effect. Experiments in real applications showed that BSS approaches, already proven in free-field, can cope with this problem and separate the signals comparably in the presence of head shading.

Head movements challenge the tracking performance of BSS algorithms since the entire array is completely repositioned, not only the position of single sources. The result of vast experiments is that BSS is generally able to continuously track head movements and suppress non-desired sources. However, in a specific setup, tracking may fail. Where there are two sources in a certain angle from the listener who wears the BSS system, and who alternates focus between these sources. After several head movements, the system is no longer able to separate the sources. The system freezes in a local optimum.

In one approach to overcoming this problem successful results were obtained by implementing a "shadow" system in parallel with the normal system with shorter filters to reduce the computational effort [Wehr et al. 2007]. Its output signals are only calculated for comparing the signal separation performance of the shadow filter system with the actual system. In order to avoid freezing of the shadow system, its filter coefficients are periodically reset. Comparing the signal separation performance of both BSS systems continuously, a freezing of the actual system can be detected. In this case, the filter coefficients of the shadow filter system are copied to the current system and the freezing is released.

The *own voice problem* describes the setup that in addition to the desired and non-desired external sources the lister's own voice is present as a third source which can disturb the adaptation of the system. Since one's voice is rather symmetric for both hearing aids, its signal properties are comparable to external sources located in the front direction. Thus, the real *three source problem* only occurs in the case when none of the external sources are located in the front direction.

14.5.4 Output Signal Selection

Referring to Fig. 14.15, the second stage for the BSS application in hearing aids is the stage for the desired source selection. So that the listener does not need to select the desired source manually, an automatic system should perform this task according to a certain strategy. Here, several different approaches are envisaged.

The most obvious one is to select the signal arriving from the front direction as desired or the signal closest to the direction of sight. In this case BSS, would work comparably to classic directional microphone systems.

However, this approach would not completely utilize the potential offered by BSS, especially to track sources during head movements.

The following approaches are useful:

- selecting only speech signals and excluding other signals from the selection, independent of their location;

- prioritizing signals of known speakers;

- prioritizing signals with the closest distance to the listener; and

- allowing the listener to track sources selected with a hold button, independent of their location.

The first two options consider the signal content. First, only those signals from speech are considered, and secondly, only preferred speakers are selected. Here, of course, combinations with speech classification systems or speaker recognition systems, respectively, are required. In the second case, this includes training to specific characteristics of the selected speakers.

The third approach also considers the distance to the sources. Here, the idea is to estimate the distance between the listener and the separated sources by estimating the amount of reverberation of these signals, which is related to the distance: The greater the distance, the smaller is the relation of the direct signal path to the reverberation components of the signal. The aim is to select the signal with the shortest distance.

The fourth method takes user interaction into account. Once the system has selected a specific signal, then the user can decide to track this source independently of the future location by pressing a "hold" button on the hearing instrument or the remote control. The source will then be tracked until a "release" button is pressed, a certain time passed, or the system loses the tracking ability, e.g., because the source has completely disappeared.

14.5.5 Binaural Output Generation

Finally, the third stage required for BSS in hearing aids is a unit for the generation of a binaural signal in order to allow the hearing impaired to localize the source signal positions correctly within the acoustic environment.

Here, from among different possibilities, one based on the filters which were identified within the BSS system, $\mathbf{W}_{ij}$, is described. The BSS method determines these filters $\mathbf{W}_{ij}$ such that the matrix equation (14.12) is fulfilled in each frequency bin μ, i.e., the cross-correlation at the output is minimized and the signals are separated.

Figure 14.16 shows a more general setup compared with Fig. 14.15. Here the signal propagation model from the sources to the microphones is included. However, no signal selection stage is shown. Instead, to be more general, a binaural output is generated for both of the separated sources

$$
\begin{bmatrix} Y_1(\mu) \\ Y_2(\mu) \end{bmatrix} = \begin{bmatrix} W_{11}(\mu) & W_{12}(\mu) \\ W_{21}(\mu) & W_{22}(\mu) \end{bmatrix} \cdot \begin{bmatrix} H_{11}(\mu) & H_{12}(\mu) \\ H_{21}(\mu) & H_{22}(\mu) \end{bmatrix} \cdot \begin{bmatrix} S_1(\mu) \\ S_2(\mu) \end{bmatrix}
$$
$$
\overset{!}{=} \begin{bmatrix} c_1(\mu) & 0 \\ 0 & c_2(\mu) \end{bmatrix} \cdot \begin{bmatrix} S_1(\mu) \\ S_2(\mu) \end{bmatrix}. \tag{14.12}
$$

The target for determining the filters $B_{ij}(\mu)$ is that for each source two output signals $Y_l(\mu)B_{ij}(\mu)$ are generated that correspond to the components of the respective signals

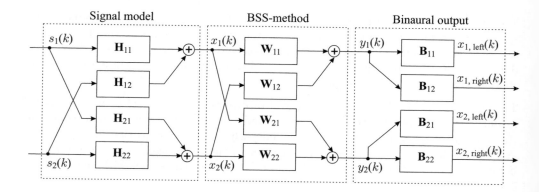

Figure 14.16: BSS system including the signal propagation model

at each microphone $S_l(\mu) H_{ij}(\mu)$, i.e., include the propagation path of the acoustic environment $H_{ij}(\mu)$. Based on (14.12) these filters can be determined as given in (14.13)–(14.16). The beneficial property of this approach is that the unknown transfer functions $c_i(\Omega)$, which are degrees of freedom for the BSS system, are eliminated in this setup. Therefore, no undesired signal distortion provoked by these unknown transfer functions should be present at the binaural output

$$S_1(\mu) H_{11}(\mu) \overset{!}{=} Y_1(\mu) B_{11}(\mu) = \frac{Y_1(\mu) W_{22}(\mu)}{W_{11}(\mu) W_{22}(\mu) - W_{21}(\mu) W_{12}(\mu)} \tag{14.13}$$

$$S_1(\mu) H_{21}(\mu) \overset{!}{=} Y_1(\mu) B_{12}(\mu) = \frac{Y_1(\mu) W_{21}(\mu)}{W_{21}(\mu) W_{12}(\mu) - W_{11}(\mu) W_{22}(\mu)} \tag{14.14}$$

$$S_2(\mu) H_{12}(\mu) \overset{!}{=} Y_2(\mu) B_{21}(\mu) = \frac{Y_2(\mu) W_{12}(\mu)}{W_{21}(\mu) W_{12}(\mu) - W_{11}(\mu) W_{22}(\mu)} \tag{14.15}$$

$$S_2(\mu) H_{22}(\mu) \overset{!}{=} Y_2(\mu) B_{22}(\mu) = \frac{Y_2(\mu) W_{11}(\mu)}{W_{11}(\mu) W_{22}(\mu) - W_{21}(\mu) W_{12}(\mu)} \,. \tag{14.16}$$

14.5.6 Concluding Remarks

In this section, a solution for a binaural directional microphone for hearing aids has been described based on Blind Source Separation. It has been shown that for a complete solution applicable for hearing aids, besides the classic and often described signal separation stage, additional stages for signal selection and for the generation of a binaural output are necessary.

For each of the three stages, the requirements were described, showing specific properties and constellations of hearing aids and different solution approaches were presented.

Surely, all of the described methods represent first steps for an application in hearing aids. To obtain solutions that perform robust and reliably in daily life environments, considerably work remains to be done.

14.6 Conclusions

This chapter dealt with the design of the binaural link technology for hearing aids and its impact on hearing aid algorithms and concepts. It was shown that the binaural link allows the extension and improvement of current monaural algorithms. In this context, the binaural classification system was explained as a low data rate application that already exists in modern hearing aids. It guarantees that both hearing aids work synchronously and adapt to the current hearing situation.

Another low data rate feature discussed addresses the ease of use of bilaterally worn hearing aids. With these, the binaural link allows adjustment of the volume control and the program switch by only one hearing aid.

More demanding for the data rate of the binaural link is the binaural beamformer that is applied to an assemblage of microphones for both hearing aids. As outlined, this approach uses spatial diversity of the input signals to capture the desired signal. The extension of this idea leads to the blind source separation approach, which again requires a high data rate link. As explained, it can be seen as a binaural adaptive beamformer which is flexible in choosing a desired source.

Since both the power of hearing aid processors and the efficiency of the binaural link will increase in the future, more powerful signal processing features for digital hearing aids can be expected over time.

Bibliography

ANSI-S3.5 (1997). Methods for Calculation of the Speech Intelligibility Index, *ANSI S3.5-1997*.

Bitzer, J.; Simmer, K. U. (2001). Superdirective Microphone Arrays, *in* M. Brandstein; D. Ward (eds.), *Microphone Arrays*, Springer Verlag, pp. 19–38.

Buchner, H.; Aichner, R.; Kellermann, W. (2005). A Generalization of Blind Source Separation Algorithm for Convolutive Mixtures Based on Second-Order Statistics, *IEEE Transactions on Speech and Audio Processing*, vol. 13, no. 1, pp. 120–134.

Buechler, M.; Dillier, N.; Allegro, S.; Launer, S. (2000). Klassifizierung der akustischen Umgebung fuer Hoergeraete-Anwendungen, *Proceedings of German Annual Conference on Acoustics (DAGA)*, pp. 282–283.

Cord, M. T.; Surr, R. K.; Walden, B. E.; Olson, L. (2002). Performance of Directional Microphone Hearing Aids in Everyday Life, *Journal of the American Academy of Audiology*, vol. 13, no. 6, pp. 295–307.

Dillon, H. (2001). *Hearing Aids*, Thieme, New York, Stuttgart, Boomerang Press.

Feldbusch, F. (1998). Geraeuscherkennung mittels Neuronaler Netze, *Zeitschrift fuer Audiologie*, vol. 1, pp. 30–36.

Griffiths, L. J.; Jim, C. W. (1982). An Alternative Approach to Linearly Constrained Adaptive Beamforming, *IEEE Transactions on Antennas and Propagation*, vol. AP-30, pp. 27–34.

Hamacher, V.; Chalupper, J.; Eggers, J.; Fischer, E.; Kornagel, U.; Puder, H.; Rass, U. (2005). Signal Processing in High-End Hearing Aids: State of the Art, Challenges, and Future Trends, *Signal Processing in High-End Hearing Aids: State of the Art, Challenges, and Future Trends*, vol. 18, pp. 2915–2929.

Hamacher, V.; Fischer, E.; Kornagel, U.; Puder, H. (2006). *Applications of Adaptive Signal Processing Methods in High-End Hearing Aids, Topics in Acoustic Echo and Noise Control*, Springer, Berlin, Heidelberg, New York.

Hornsby, B.; Ricketts, T. A. (2007). Effects of Noise Source Configuration on Directional Benefit Using Symmetric and Asymmetric Directional Hearing Aid Fittings, *Ear and Hearing*, vol. 28, no. 2, pp. 177–186.

Hoshuyama, O.; Sugiyama, A. (2001). Robust adaptive beamforming, *in* M. Brandstein; D. Ward (eds.), *Microphone Arrays*, Springer Verlag, pp. 87–109.

Kates, J. M. (1995). Classification of Background Noises for Hearing-Aid Applications, *Journal of the Acoustical Society of America*, vol. 97, no. 1, pp. 461–70.

Keidser, G.; Rohrseitz, K.; Dillon, H.; V. Hamacher, L. C.; Rass, U.; Convery, E. (2006). The Effect of Multi-Channel Wide Dynamic Range Compression, Noise Reduction, and the Directional Microphone on Horizontal Localization Performance in Hearing Aid Wearers, *International Journal of Audiology*, vol. 45, no. 10, pp. 563–79.

Lotter, T. (2004). *Single and Multichannel Speech Enhancement for Hearing Aids*, PhD thesis. Aachener Beiträge zu digitalen Nachrichtensystemen, vol. 18, P. Vary (ed.), RWTH Aachen University.

Makino, S. (2003). *Blind Source Separation of Convolutive Mixtures of Speech*, Benesty, J., Huang, Y. (eds.), Adaptive Signal Processing: Application to Real-World Problems, Springer, pp. 195-226.

Monzingo, R.; Miller, T. (1980). *Introduction to Adaptive Arrays*, John Wiley & Sons, Ltd.

Ostendorf, M.; Hohmann, V.; Kollmeier, B. (1998). Klassifikation von akustischen Signalen basierend auf der Analyse von Modulationsspektren zur Anwendung in digitalen Hoergeraeten, *Proceedings of German Annual Conference on Acoustics (DAGA)*, pp. 402–3.

Powers, T. A.; Hamacher, V. (2002). Three Microphone Instrument is Designed to Extend Benefits of Directionality, *The Hearing Journal*, vol. 55, no. 10, pp. 38–45.

Puder, H. (2005). Challenges when Integrating BSS Methods in Hearing Aids, *International Forum for Hearing Instrument Developers*, Oldenburg.

Simmer, K. U.; Bitzer, J.; Marro, C. (2001). Post-Filtering Techniques, *in* M. Brandstein; D. Ward (eds.), *Microphone Arrays*, Springer Verlag, pp. 39–60.

Takatami, T.; Ukai, S.; Nishikawa, T.; Saruwatari, H.; Shikano, K. (2005). Evaluation of SIMO Separation Methods for Blind Decomposition of Binaural Mixed Signals, *International Workshop on Acoustic Echo and Noise Control (IWAENC)*, Eindhoven, Netherlands.

Wehr, S.; Lombard, A.; Buchner, H.; Kellermann, W. (2007). Shadow BSS for Blind Source Separation in Rapidly Time-Varying Acoustic Scenes, *7th International Converence on Independent Component Analysis and Signal Separation*, London.

Wittkop, T.; Albani, S.; Hohmann, V.; Peissig, J.; Woods, J.; Kollmeier, B. (1997). Speech Processing for Hearing Aids: Noise Reduction Motivated by Models of Binaural Interaction, *Acustica united with Acta Acustica*, vol. 84, no. 4, pp. 684–699.

Chapter 15

Auditory-profile-based Physical Evaluation of Multi-microphone Noise Reduction Techniques in Hearing Instruments

Koen Eneman, Arne Leijon, Simon Doclo, Ann Spriet, Marc Moonen, Jan Wouters

15.1 Introduction

During recent years significant progress has been made in the design of hearing aid and cochlear implant devices thanks to the incorporation of digital signal processing techniques. Development has among other things been concentrated on the design of multi-microphone solutions with advanced signal enhancement capabilities, such as noise reduction and feedback cancellation. Thanks to these novel features, hearing impaired people show improved abilities to function and to interact in formerly adverse listening conditions such as conversations on a street corner, in a restaurant, or during a cocktail party.

Of the many digital signal enhancement techniques that have been proposed during the past decades, only a limited number have been effectively implemented and integrated in commercial hearing instruments. In fact, the customization of a signal processing scheme towards the implementation in a hearing aid or cochlear implant device makes strong demands in terms of computational complexity and processing delay, and requires a profound performance assessment through physical and perceptual validation tests. For the perceptual evaluation, typically, a large number of time-consuming listening tests with hearing-impaired subjects are required.

Advances in Digital Speech Transmission   Edited by R. Martin, U. Heute and C. Antweiler
© 2008 John Wiley & Sons, Ltd

In the frame of the European HearCom[1] project a number of representative signal enhancement schemes are evaluated for future usage in hearing instrument devices. Through advanced physical evaluation based on speech-intelligibility-weighted performance measures incorporating the auditory profile of the hearing aid user, a better performance assessment of the algorithms can be made solely based on simulation experiments. In this way the number of subjective listening tests can be restricted. In this chapter the proposed physical performance measures are presented and, by way of illustration, they are applied to a state-of-the-art as well as to a more recently developed multi-microphone noise cancellation approach.

Subjective Performance Assessment

As part of the HearCom project several advanced state-of-the-art and novel signal enhancement solutions are studied and implemented on a common real-time hardware platform. With a view to the integration of these techniques in future hearing aid devices a profound evaluation of the proposed schemes is required under various realistic test conditions. To limit the number of listening tests with normal hearing and hearing impaired subjects a set of physical performance measures has been defined to quantify the expected speech distortion and speech intelligibility improvement offered by the different algorithmic approaches. These measures build upon a functional auditory model that incorporates several aspects of normal and hearing impaired listening that are included in the so-called auditory profile of the hearing aid user. The auditory profile has been defined within the HearCom project to be able to characterize the auditory impairment profile of an individual in a comparable way across Europe. The auditory profile includes results on a number of diagnostic tests assessing audibility, loudness perception, frequency resolution, and temporal acuity, speech perception in noise, spatial listening, subjective judgments and communication, listening effort, and cognitive abilities.

Performance assessment of the envisaged signal enhancement approaches using the proposed evaluation measures allows reliable benchmarking between different algorithms and makes it possible to perform initial parameter tuning entirely through simulation. In this way the number of time-consuming subjective listening tests can be limited.

Computational Complexity

Apart from a profound performance assessment focusing on speech intelligibility and signal distortion, the designer also needs to keep an eye on other, more implementation related parameters such as the computational complexity of the algorithm.

---
[1] The work presented in this chapter has been supported by grants from the European Union FP6 Project 004171 HearCom. The information in this document is provided as is and no guarantee or warranty is given that the information is fit for any particular purpose. The user thereof uses the information at its sole risk and liability.

Hearing aid instruments are battery-powered devices mostly relying on 1.4-V zinc–air batteries. To ensure sufficiently long battery autonomy hearing aid devices operate at a low voltage (around 1 V) and run at low clock frequencies. Hence, if the integration in a commercial hearing aid device is aimed at, the algorithm complexity figures need to be carefully monitored and the required number of operations per second should not exceed the computational capabilities of typical current or near-future hearing aid processor technology.

Finally, one should realize that the signal enhancement module taking care, e.g., of the noise suppression is just one of the many functional blocks in the signal processing chain, and can therefore claim only a (small) part of the available execution time and power consumption.

Memory Requirements

To operate properly, signal enhancement algorithms require a certain amount of memory, which is needed to store both the intermediate algorithm results and the algorithm program code. As memory banks occupy relatively large silicon areas and consume a non-negligible part of the scarcely available power, dedicated hardware devices such as hearing aids typically dispose of a limited amount of memory. As a consequence, the amount of memory that is directly available to the algorithm is often considerably restricted. The algorithm might therefore need to undergo structural changes, which, as a side effect, could increase the overall computational complexity of the approach.

Signal Delay

Finally, also the total signal delay that is introduced by the hearing aid processing needs to be carefully monitored. The total signal delay typically is a combination of an interface delay (due to A/D and D/A conversion), a delay caused by the block processing (a technique used by many signal processing algorithms) and a group delay introduced on purpose inside the signal enhancement approach itself.

Most hearing-instrument users receive processed sound together with unprocessed sound leaking directly into the ear canal. At low frequencies these components may have similar amplitudes. Interference then may cause noticeable effects if the processed signal is delayed more than about 5–10 ms with respect to the unprocessed sound [Dillon 2001]. Listeners with a severe hearing loss hardly perceive the unprocessed sound leaking directly into the ear canal. However, they might suffer from the asynchrony between the perceived speech sounds and visual information such as lip movements. In that case, delays of up to a few tens of milliseconds are acceptable. Similar results were obtained through subjective disturbance assessment tests in [Stone, Moore 2005].

Often there is a trade-off between delay and computational complexity. For example, several algorithm approaches can be equivalently implemented in the frequency domain, leading to solutions that are computationally more efficient than the corresponding time-domain realizations, however typically at the cost of a longer signal delay [Shynk 1992].

Organization of the Chapter

The chapter is organized as follows. In Sec. 15.2 an overview is given of selected multi-microphone noise reduction techniques that are suited for hearing instrument applications. Classical state-of-the-art as well as novel schemes are reviewed. Section 15.3 defines a number of perceptually weighted performance measures that incorporate aspects of normal and impaired human hearing. These measures are intended to simulate perceptual evaluation with different types of signal enhancement algorithms under realistic test conditions. By way of illustration two multi-microphone noise reduction algorithms are compared based on the proposed performance measures. The selected test conditions are described in Sec. 15.4. Sections 15.5 and 15.6 present and discuss the simulation results. Finally, some conclusions are formulated in Sec. 15.7.

15.2 Multi-microphone Noise Reduction in Hearing Instruments

15.2.1 Classical Solutions

Classical, analog single-microphone devices have dominated the hearing aid market for several decades, solely providing basic functionalities such as frequency-dependent amplification and compression according to the auditory profile of the hearing aid user. The first fully digital hearing aids were introduced in the mid-1990s. Following the exponential growth in silicon and microprocessor technology digital hearing aid solutions now systematically replace classical analog devices. Moreover, through the use of digital signal processing techniques an increased number of functionalities can be offered to the end user for a given amount of power consumption. Additionally, more advanced functions can be provided that are much harder to realize with classical analog technology.

Hardware Directional Microphone

In the early 1970s hardware directional microphones were brought to market. Thanks to their angle-dependent sensitivity, hardware directional microphones can reduce unwanted background noise, resulting in an improvement in signal-to-noise ratio (SNR), up to several dBs in the case where the jammer sounds are spatially separated from the frontal look direction. This typically leads to a clearly noticeable increase in speech intelligibility. In that respect, it was found that a rise of 1 dB in signal-to-noise ratio roughly offers an increase of about 10% in speech understanding [Dillon 2001].

Disadvantages of hardware directional (pressure gradient) microphone technology are the inherent highpass response [Eargle 2001] and the larger component size compared to omnidirectional solutions. The highpass characteristic can however be compensated for by using analog lowpass filtering techniques, typically at the expense of extra noise insertion in the low and mid-frequency range and additional hardware requirements for the lowpass filter implementation.

Software Directional Microphone

Through an electrical combination of outputs from several (omnidirectional) microphones mounted in the same device a software controllable angular response can be obtained, leading to the software directional microphone. Software directional microphones offer more flexibility than their hardware directional counterparts as they can selectively be activated depending on the listening situation, providing noise suppression in noisy environments and offering an omnidirectional response under less challenging noise conditions. Unfortunately, software directional microphones are sensitive to changes in microphone characteristics and microphone placement, and are therefore typically less robust than mechanical hardware directional microphones.

Acoustic Beamforming

The idea of combining signals coming from several microphones is a well-known approach commonly referred to as acoustic beamforming. By using several microphones the sound field around the hearing aid can in fact be sampled in the spatial domain, making it possible to process signals not only in the spectral domain, but also in a direction or position dependent way. This principle forms the basis for multi-microphone signal enhancement techniques such as noise suppression, dereverberation, and source separation.

As far as acoustic beamforming is concerned, different kinds of realization can be distinguished. In its most simple form microphone signals are delayed and summed together, leading to the delay-and-sum beamformer structure. If the delays are appropriately set, noise can be reduced as the desired speech components in the microphone signals will be added in phase, whereas noise contributions fail to be added in phase, and are hence attenuated. More advanced topologies first perform a filtering operation on each of the microphone channels and then sum all channels together, realizing a so-called filter-and-sum beamformer. Owing to the small size of the hearing aid device beamforming techniques appear to be most effective in the higher range of the audio spectrum. For a more detailed discussion on beamforming read [Van Veen, Buckley 1988] or Chap. 12 in [Vary, Martin 2006].

Standard beamforming solutions use fixed filter or delay settings and are therefore data independent. As a consequence, they offer limited performance and flexibility. Nevertheless, often significant improvements in speech understanding can be achieved. This was shown by [Luts et al. 2004] and [van der Beek et al. 2007], who evaluated a commercial hearing instrument device that is based on beamforming technology.

Digital hearing aid technology not only paved the way for the integration of powerful multi-microphone signal enhancement techniques into commercial hearing aid devices, but also offered a platform to realize single-microphone noise reduction schemes such as the Wiener filter or spectral subtraction based methods that are described in [Vary, Martin 2006], Chap. 11. Whereas single-microphone noise suppression techniques are ideally suited to combat diffuse noise, multi-microphone techniques utilize the spatial diversity of the setup and are therefore good candidates to suppress point-like noise sources. As they use angular-dependent discrimination techniques to "zoom in" on the desired source the expected noise suppression will decrease if the reverberation time of the recording room increases. Indeed, signals that come from a desired or a competing source reach the hearing aid basically from all directions. Hence, discrimination based on the direction of arrival is no longer reliable. Nevertheless, multi-microphone techniques are also capable of reducing diffuse noise, albeit to a lesser extent. To further reduce the residual noise in diffuse and highly reverberating environments single-microphone noise suppression techniques are often employed as a postprocessing stage acting on the output of a multi-microphone algorithm [Simmer et al. 2001].

15.2.2 Generalized Sidelobe Canceler

Through the incorporation of adaptivity the performance of classical fixed beamforming based solutions can be improved. In this way, additional noise suppression can be obtained and the algorithm is given the ability to adapt its settings to a specific environmental scenario. The first solution of this type has been presented by [Frost 1972]. Later on, Griffiths and Jim proposed an improved scheme that is known nowadays as the Generalized Sidelobe Canceler (GSC) [Griffiths, Jim 1982]. In fact, many state-of-the-art multi-microphone noise suppression techniques that are used in hearing instruments nowadays are based on this principle.

The Generalized Sidelobe Canceler (GSC) consists of a fixed spatial preprocessor, i.e., a fixed beamformer with blocking matrix, and an adaptive stage, as shown in Fig. 15.1. The fixed beamformer, which in the most general case acts as a multichannel filter $\mathbf{A}(k)$, creates a so-called speech reference $y_0(k) = x_0(k) + v_0(k)$, where $x_0(k)$ and $v_0(k)$ are the speech and noise components, in $y_0(k)$, respectively. Similarly, the blocking matrix $\mathbf{B}(k)$ creates $M - 1$ noise references $y_m(k) = x_m(k) + v_m(k)$, $m = 1 : M - 1$, where M is the number of microphones. In general, it holds that with $M - 1$ noise

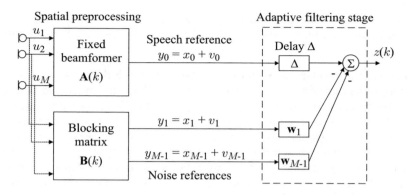

Figure 15.1: Structure of the Generalized Sidelobe Canceler (GSC)

references $M - 1$ noise sources can be removed. The noise references are created by spatially suppressing sounds arriving from the direction from where the desired speaker sound is assumed to be coming. One thereby assumes that the speaker is in front of the microphone array (broadside direction) or in the direction along the axis of the array (endfire direction). In many practical realizations both $\mathbf{A}(k)$ and $\mathbf{B}(k)$ are (low-order) multichannel FIR filters.

The goal of the adaptive stage is to make an estimate of the noise component $v_0(k)$ in the speech reference $y_0(k)$ and to subtract this noise from $y_0(k)$ to create an enhanced output signal $z(k)$. To estimate the noise component optimal filter weights $\mathbf{w}_1, \ldots, \mathbf{w}_{M-1}$ are computed that minimize the cost function

$$J_{\mathrm{GSC}}(\mathbf{w}(k)) = \mathrm{E}\left\{|v_0(k - \Delta) - \mathbf{w}^T(k)\,\mathbf{v}(k)|^2\right\}, \tag{15.1}$$

with

$$\mathbf{w}(k) = \left(\mathbf{w}_1^T(k), \ldots, \mathbf{w}_{M-1}^T(k)\right)^T,$$
$$\mathbf{v}(k) = \left(\mathbf{v}_1^T(k), \ldots, \mathbf{v}_{M-1}^T(k)\right)^T,$$
$$\mathbf{v}_m(k) = \left(v_m(k), \ldots, v_m(k - L + 1]\right))^T.$$

Parameter L is the length of the adaptive filters $\mathbf{w}_m(k)$ and $\mathrm{E}\{\cdot\}$ symbolizes the expectation operator. Observe that the cost function minimizes the residual noise energy, so that the Generalized Sidelobe Canceler is primarily focused on noise suppression only. Typically, the minimum of the cost function is computed online using adaptive filtering techniques [Widrow, Stearns 1985]. For proper convergence the filters are adapted during speech pauses only. During speech periods the filters are kept constant, assuming that the environmental conditions do not significantly change meanwhile. Speech pauses can be identified using a voice activity detector (VAD).

Thanks to the adaptivity of the approach the filter coefficients are data dependent, giving the Generalized Sidelobe Canceler an environmentally specific spatial sensitivity, and self-learning and noise tracking capabilities. In this way, better noise suppression and more flexibility can be obtained than with classical, fixed beamforming solutions.

Unfortunately, the proper operation of the Generalized Sidelobe Canceler relies on some assumptions that are often violated in practice. As indicated above, it is assumed that the desired speaker sound arrives from a specific direction with respect to the array. In this way, noise references can be created that solely contain contributions of the undesired background noise. In practice however, the position of the speaker varies slightly and the characteristics of the microphones tend to deviate from their ideal values. On top of that, in a highly reverberating environment signals arrive at the array basically from any direction. As a consequence, the noise references do contain contributions of the desired speech, which eventually distorts the desired speech signal. This phenomenon is called speech leakage. Hence, whereas in general high noise suppression figures can be obtained with the Generalized Sidelobe Canceler, the quality of the desired signal can be severely compromised. Performance is further reduced if the voice activity detection fails, as then the adaptive filters fail to converge to the desired solution. More information about the Generalized Sidelobe Canceler can be found in [Vary, Martin 2006], Sec. 12.8.

Many extensions to the Generalized Sidelobe Canceler have been proposed, such as [Nordebo et al. 1994], [Herbordt, Kellermann 2003] and [Gannot et al. 2001]. Also in the HearCom project two variants on the Generalized Sidelobe Canceler have been considered: the adaptive two-stage beamforming (A2B) approach and the spatially preprocessed speech-distortion-weighted multichannel Wiener filter (SDW-MWF). Both solutions will be briefly discussed in Sec. 15.2.3 and 15.2.4, respectively. More detailed information on the algorithms can be found in the references that are given in these sections.

15.2.3 Adaptive Two-stage Beamforming Approach

A first multi-microphone noise cancellation algorithm that has been evaluated in the frame of the HearCom project is the Adaptive Two-stage Beamforming (A2B) approach [Wouters, Vanden Berghe 2001] [Wouters et al. 2002] [Maj et al. 2004] [Maj 2004]. This two-microphone adaptive noise cancellation algorithm can be considered as a special case of the Generalized Sidelobe Canceler (GSC). Algorithmic variants of this approach have recently been integrated in commercial hearing instruments [Spriet et al. 2007].

It is observed that the A2B adaptive noise canceler of Fig. 15.2 consists of three stages. First, the outputs of two omnidirectional microphones are combined to create a software directional microphone. Then, the second microphone output has to be appropriately delayed with respect to the first microphone signal. Next comes a fixed first beamforming stage, which is implemented using a 10-taps FIR filter. The first stage creates a speech and a noise reference that are input into an adaptive second stage. The 30-taps adaptive filter makes use of normalized least-means squares (NLMS)

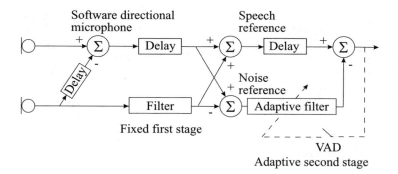

Figure 15.2: Structure of the Adaptive Two-stage Beamforming (A2B) approach

optimization [Widrow, Stearns 1985] [Vary, Martin 2006], and is controlled by an energy-based voice activity detector (VAD) [Maj 2004].

The A2B algorithm uses rather small filter lengths to be able to fit in standard hearing aid processors. The A2B scheme has been included in our evaluation to act as a reference for state-of-the-art multi-microphone noise suppression technology that is currently used in commercial hearing instruments.

15.2.4 Spatially Preprocessed Speech-distortion-weighted Multichannel Wiener Filtering

A more recently derived variant of the Generalized Sidelobe Canceler is the spatially preprocessed speech-distortion-weighted multichannel Wiener filtering algorithm (SDWMWF), which is based on work described in [Doclo et al. 2004] [Doclo et al. 2005] [Spriet et al. 2005] [Spriet et al. 2004] [Spriet 2004]. The structure of the SDWMWF approach is shown in Fig. 15.3. Observe that the algorithm perfectly copies the structure of the Generalized Sidelobe Canceler, the main difference being in the adaptation of the second-stage filters $\mathbf{w}_1, \ldots, \mathbf{w}_{M-1}$. In the SDWMWF approach the optimal filter weights $\mathbf{w}_1, \ldots, \mathbf{w}_{M-1}$ are computed by minimizing

$$J_{\mathrm{SDWMWF}}(\mathbf{w}(k)) = \mathrm{E}\left\{|v_0(k-\Delta) - \mathbf{w}^T(k)\,\mathbf{v}(k)|^2\right\} + \frac{1}{\mu}\mathrm{E}\left\{|\mathbf{w}^T(k)\,\mathbf{x}(k)|^2\right\}, \quad (15.2)$$

where

$$\mathbf{x}(k) = \left(\mathbf{x}_1^T(k), \ldots, \mathbf{x}_{M-1}^T(k)\right)^T,$$
$$\mathbf{x}_m(k) = \left(x_m(k), \ldots, x_m(k-L+1)\right)^T.$$

Observe that the cost function minimizes the weighted sum of the residual noise energy and the speech distortion energy. Parameter $\mu \in [0, \infty[$ provides a trade-off between noise reduction and speech distortion. If $\mu \to \infty$ speech distortion is

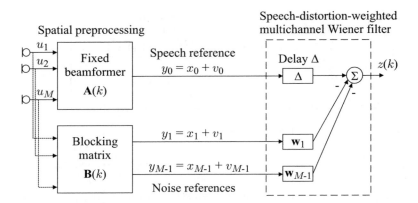

Figure 15.3: Structure of the Spatially Preprocessed Speech-Distortion-Weighted Multichannel Wiener Filter (SDWMWF)

completely ignored and the algorithm approaches the Generalized Sidelobe Canceler (GSC) (compare (15.2) with (15.1)). The SDWMWF algorithm can therefore be considered as an extension of the GSC. Thanks to the extra term in the cost function the SDWMWF makes a trade-off between noise suppression and speech distortion, making the algorithm more robust against speech leakage than the standard Generalized Sidelobe Canceler [Spriet et al. 2004].

Assuming that the speech and noise components are independent, the solution to the cost function (15.2) can be expressed as

$$\mathbf{w}(k) = \mathbf{R}(k)^{-1} \cdot \mathrm{E}\left\{\mathbf{v}(k)\, v_0(k - \Delta)\right\}, \tag{15.3}$$

with

$$\mathbf{R}(k) = \mathrm{E}\left\{\mathbf{v}(k)\, \mathbf{v}^T(k)\right\} + \frac{1}{\mu}\left(\mathrm{E}\left\{\mathbf{y}(k)\, \mathbf{y}^T(k)\right\} - \mathrm{E}\left\{\mathbf{v}(k)\, \mathbf{v}^T(k)\right\}\right), \tag{15.4}$$

and $\mathbf{y}(k)$ defined similarly to $\mathbf{x}(k)$. Hence, $\mathbf{w}(k)$ can be computed based on the noise correlation matrix $\mathrm{E}\left\{\mathbf{v}(k)\mathbf{v}^T(k)\right\}$, which is updated during noise-only-periods, and the speech correlation matrix $\mathrm{E}\left\{\mathbf{y}(k)\mathbf{y}^T(k)\right\}$, which is adapted during speech periods. To make a distinction between speech and noise periods the SDWMWF algorithm is complemented with a log-energy-based voice activity detector [Van Gerven, Xie 1997]. Efficient stochastic-gradient algorithms have been derived to update the filter weights (15.3) in an efficient way [Spriet et al. 2005] [Doclo et al. 2005].

In the evaluation (Sec. 15.5) a three-microphone version of the algorithm is considered that relies on a frequency-domain variant of cost function (15.2) and uses efficient correlation matrix updating. The first, non-adaptive stage consists of 48-taps FIR filters. Block length L in the Wiener filtering stage is set to 32 and trade-off parameter $1/\mu$ is set to 0.5. In this way, compared with the standard Generalized Sidelobe Canceler, more emphasis is put on signal distortion at the expense of less noise suppression.

15.3 Auditory-profile-based Physical Evaluation

For a proper validation of signal enhancement algorithms in a hearing aid context dedicated evaluation measures are required that can accurately predict algorithm performance for a number of representative hearing loss profiles. However, a reliable performance assessment would require intensive speech intelligibility testing, listening effort assessment and quality scoring with a large number of test subjects under several realistic environmental conditions. This type of procedure is very time consuming, needs a dedicated and well equipped test site and requires access to a large number of hearing-impaired subjects with different hearing loss profiles.

Owing to the lack of familiarity of many algorithm designers with the testing of hearing-impaired listeners, and the huge amount of time that is required to set up and perform such tests, many novel signal enhancement approaches are solely evaluated under rather academic conditions. In that respect, often, merely a number of signal theoretic performance measures are applied, which typically fail to incorporate important aspects of normal and impaired human hearing. As a consequence, many promising signal enhancement algorithms eventually fail to be competitive with existing solutions once they are evaluated in a hearing aid context under real-life conditions. Furthermore, they often exceed the computational and memory capabilities of the hardware or cannot meet the delay constraints imposed by the application.

Taking this into account, there is clearly a need for advanced physical evaluation measures that incorporate aspects of human hearing and that can reliably predict algorithm performance through simulation experiments only. In a first phase of our research we relied on a number of physical evaluation measures that incorporate basic aspects of normal human hearing, such as the intelligibility-weighted signal-to-noise ratio (SNR), the segmental intelligibility-weighted SNR, the segmental SNR and a frequency-weighted log-spectral signal distortion measure. However, given the hearing aid application we have in mind, more advanced evaluation measures need to be derived that take into account aspects of impaired hearing as well. With this aim, a number of physical performance measures have been proposed that assess various aspects of user-perceived signal quality, such as speech intelligibility, signal distortion, and relative loudness of desired and undesired signal components. In order to evaluate algorithm performance across different auditory profiles the proposed measures make use of an auditory functional model that takes into account aspects of normal as well as impaired hearing. The auditory functional model and the physical evaluation measures will be presented in Sec. 15.3.1 and 15.3.2, respectively.

Of course, no physical performance measure can perfectly predict performance in real life, in particular with hearing-impaired users. Although the established measures that will be presented in Sec. 15.3.2 relate to some important aspects of user-perceived signal quality they have to be used with this restriction in mind.

15.3.1 Simulation of Hearing-impaired Perception

In this section an auditory functional model is proposed that accounts for some of the most fundamental aspects of normal and impaired hearing. The auditory model is used for the performance measures that will be presented in Sec. 15.3.2.

Functional Auditory Model

The functional auditory model on which the objective performance measures are based accounts for normal auditory functions and for some fundamental effects of hearing impairment, as there are loss of audibility at low input levels, loudness recruitment (i.e., reduced dynamic range caused by loss of normal non-linear, compressive outer-hair-cell amplification at low input levels), reduced frequency resolution (also caused by dysfunction of outer hair cells) and reduced ability to extract supra-threshold speech cues for speech recognition.

The simulation includes the following main steps. First, the (unprocessed or algorithm-processed) input signal is segmented and transformed into a sequence of short-time power spectra, where the time resolution can vary between evaluation measures. Each short-time spectrum is then modified by the head-related transfer function for frontal incidence, by the simulated hearing-aid insertion-gain frequency response, and by the middle-ear transmission gain (which was assumed to be normal, because all the simulated auditory profiles include only sensorineural loss). The resulting input spectra are then transformed by filtering and compression to simulate effects of inner-ear processing, as described below, for each auditory profile. All inner-ear processing is simulated in the frequency domain.

Our functional auditory model, which is explained in more detail in [Leijon 2007], closely corresponds to that of [Moore et al. 1997] and [Moore, Glasberg 2004], who used much earlier established principles from [Fletcher, Munson 1937] and [Zwicker 1958]. The core function of the model is to transform the sound pressure at the input of the ear into a two-dimensional auditory excitation pattern $E(z, t)$ as a function of time t and of place (position) z along the basilar membrane in the cochlea (inner ear). The excitation E is a power-like quantity representing the output from the non-linear auditory filtering process. The excitation at each position z is assumed to cause afferent neural activity in auditory neurons tonotopically connected at that position along the array of inner hair cells on the basilar membrane. Furthermore, the excitation pattern $E(z, t)$ always includes a minimal spontaneous excitation representing internal physiological noise.

Neurons are most sensitive to input frequencies near the center frequency of an auditory bandpass filter that is associated with the position of the neurons along the basilar membrane. The auditory filtering is the result of interaction between at least two physiological mechanisms: 1) passive mechanical filtering of the traveling wave along the inner-ear basilar membrane, and 2) active mechanical processes involving outer hair cells. The passive filtering is approximately linear, i.e., independent of input signal, whereas the outer hair cell contribution is known to be non-linear and compressive. Outer hair cells provide high gain at low levels and no gain at high levels, and this amplification is effective only for input frequencies in a narrow frequency range near the center frequency of the auditory bandpass filter that is associated with the position of the outer hair cells along the basilar membrane.

The auditory filter bandwidth is related to the traditional concept of auditory Critical Bands (CB). However, extensive masking experiments have revealed that the Equivalent Rectangular Bandwidths (ERB) of auditory filters [Moore, Glasberg 2004] are slightly smaller than CB bandwidths [Zwicker, Terhardt 1980]. Therefore, we used the ERB estimates in our model. In this way, the auditory place scale z of the model [Moore, Glasberg 2004], is very similar, but not identical to the traditional Bark scale [Zwicker, Terhardt 1980].

The obtained excitation pattern can be used to predict the loudness of a sound in a quiet background and the partial loudness of a sound that is partially masked by another simultaneous sound. The difference between logarithms of excitation patterns for two sounds is used to predict the listener's ability to discriminate between the sounds. Similarly, the difference between the log-excitation pattern caused by an external sound and the spontaneous log-excitation pattern determines the detectability of a sound in quiet.

The discrimination ability is limited by the inherent random variability in the neural data reaching the brain and in the brain's decision processes. Signal-detection theory defines a discrimination index d' indicating the effective perceptual "distance" between two sounds with stationary excitation patterns $E_1(z)$ and $E_2(z)$, as

$$d' = \sqrt{\int_0^\infty \frac{(\log E_1(z) - \log E_2(z))^2}{\sigma_L^2(\Delta)} \, dz} . \tag{15.5}$$

Here, Δ is the duration of the sound, and $\sigma_L^2(\Delta)$ represents the total underlying variance in the auditory process of observing log-excitation patterns and using them for detection or discrimination. The log-excitation domain is used here, because then the variance can be approximated simply as a level-independent constant

$$\sigma_L^2(\Delta) = c/\min(\Delta, \Delta_{\max}) , \tag{15.6}$$

where constant c is chosen to reproduce empirical results on intensity discrimination for broadband noise. Discrimination improves with stimulus durations up to an approximate maximal duration $\Delta_{\max}=0.2$ s.

The implemented model includes the following transformation steps, just like [Moore et al. 1997] and [Moore, Glasberg 2004]: 1) a fixed filter for the transfer from a specified sound field to the eardrum, 2) a fixed filter representing middle-ear transmission, and 3) linear and non-linear filtering at the outer hair cells level to mimic auditory frequency resolution. The present implementation deviates slightly from that of [Moore et al. 1997], but was verified to be at least as much in agreement with empirical loudness-balance data as the version validated by [Moore et al. 1997].

It should be noted however that the model attempts to simulate only the most fundamental and reasonably well-known effects of auditory processing. In particular, temporal masking effects and specific deficits in the binaural integration are not included in the model.

Model-simulated Auditory Profiles

For the evaluation of the signal enhancement algorithms presented in Sec. 15.5 typical auditory profiles have been selected based on a broad study of audiometric data from a large number of hearing impaired listeners including individual pure-tone air-conduction thresholds, speech recognition, and results of categorical loudness judgments. In this way six common categories ranging from mild over moderate to severe hearing loss have been considered. The audiograms of the selected auditory profiles are shown in Fig. 15.4. Normal hearing was also added as a seventh reference profile.

Hearing-aid Amplification and Spectral Shaping

In addition to the special signal enhancement methods that are to be evaluated, any hearing aid also presents all sounds with amplification and spectral shaping, individually adjusted for each user. In practice, the hearing aid settings are fine-tuned according to individual listener preferences. In this work, to allow full automatic evaluation we relied on the NAL-RP prescription rule [Byrne, Dillon 1986] [Byrne et al. 1990] [Byrne et al. 1991] to set the amplification and spectral shaping for each profile. The NAL-RP prescription rule defines a single non-adaptive frequency response, which simulates a hearing aid with slow automatic gain control (AGC) that adjusts the frequency response to the acoustic sound environment, but does not adapt rapidly to every short segment in the input signal.

15.3.2 Physical Evaluation Measures

In this section four physical evaluation measures are introduced that take into account aspects of normal and impaired hearing. They are intended to be used to assess the performance of signal enhancement algorithms for hearing aid applications.

The performance measures require separate estimates of the desired signal (speech) component and of the competing signal (noise) component at the input and the output of the algorithm. The separation of the single-channel output signal into a speech

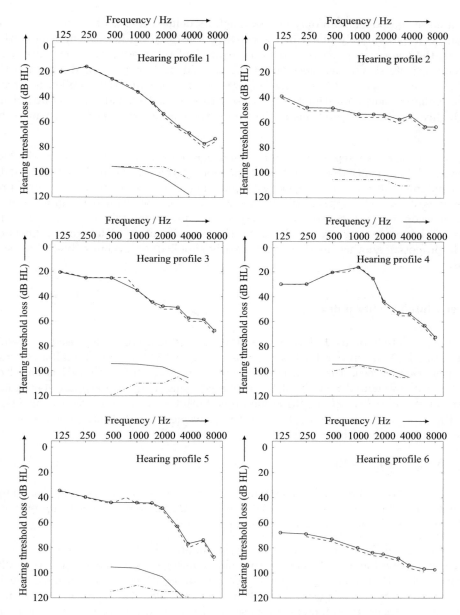

Figure 15.4: Audiograms for the selected auditory profiles, showing audiometric pure-tone air-conduction hearing threshold loss (unmarked dashed curves) and corresponding detection thresholds (at $d' = 1$) for the auditory model (marked with circles). Audiometric levels of discomfort (dash–dot curves) may be compared with model pure-tone levels yielding a calculated loudness of 64 sone (unmarked solid curves). 0 dB HL on the vertical scales represents the normal hearing threshold

and a noise component can only be meaningfully defined if the algorithm signal processing is approximately linear in a short-time sense. Regardless of implementation details, the algorithm processing must be equivalent to an adaptive linear filter that varies relatively slowly, i.e., does not change considerably within the duration of the impulse response of the filter. This requirement is approximately fulfilled for many signal enhancement algorithms. Of course this does not exclude the algorithm using highly non-linear operations to determine the characteristics of the adaptive filter.

In practice, the two separate output signal components are obtained by shadow-filtering. This means that first the adaptation of the algorithm is determined by the combined (desired + competing) input signal, and that then the same processing is separately applied to the desired and to the competing input signal. This method is most convenient in algorithm simulations where the internal details of the processing are available to the experimenter. [Hagerman, Olofsson 2004] proposed another method that can be applied to evaluate black-box systems.

Speech Intelligibility Index

The Speech Intelligibility Index (SII) is standardized and commonly used to predict speech intelligibility in non-fluctuating noise [ANSI-S3.5 1997]. For broadband external noise that exceeds the hearing threshold at all frequencies, the SII is based on the frequency-weighted SNR (in dB), calculated using long-term average speech and noise power spectra. Once the SII is computed both for the unprocessed signal (SII_{in}) and the processed signal (SII_{out}), the SII improvement can be determined, which is achieved by the algorithm:

$$\Delta \text{SII} = \text{SII}_{out} - \text{SII}_{in} . \tag{15.7}$$

The SII is always a number between 0 and 1. For normal-hearing listeners, this range corresponds to SNR values from $-15\,\text{dB}$ to $+15\,\text{dB}$. Therefore, an SII improvement of 0.1 corresponds to a real SNR improvement of $3\,\text{dB}$.

It is well known that the standard SII overestimates the speech-recognition performance of hearing-impaired listeners, especially in noisy environments. Various modifications of the SII have been proposed to account for additional suprathreshold deficits in impaired ears [Pavlovic et al. 1986] [Ching et al. 1998]. Our present implementation uses "desensitization factors" proposed and validated for noisy environments by [Pavlovic et al. 1986] and [Magnusson 1996].

Segmental Speech Intelligibility Index

The SII standard does not claim to account for the effects of fluctuating noise. Additionally, the frequency-weighted long-term SNR may obscure some segmental effects introduced by the noise reduction algorithms. Therefore, a slightly modified procedure is used to derive a segmental SII measure (segSII). The SII is first calculated for each short-time segment of 50 ms, and is then averaged over the full duration of the test signal. The segSII measure has not been empirically validated and is used here tentatively as a complement to the SII measure. If an algorithm improves only the SII, but not the segSII, it is questionable whether the algorithm will effectively improve real speech recognition.

Signal-to-noise Loudness Level Difference

Using the auditory excitation model for each simulated hearing loss, the partial loudness of both speech and competing signals can be calculated, including the masking effect of noise on the speech loudness, and vice versa. A large number of different loudness estimation procedures were evaluated by [Skovenborg, Nielsen 2004], but these procedures did not include the effects of impaired hearing. Therefore, the partial loudness values for the desired (speech) and the competing (noise) signal have been computed in a similar way as in [Moore et al. 1997] and [Moore, Glasberg 2004].

Preliminary instantaneous partial loudness density patterns $N'_{ps}(z, t)$ for the desired signal, and $N'_{pn}(z, t)$ for the competing signal, are calculated as

$$N'_{ps}(z, t) = N_0\big((E_s(z, t) + E_n(z, t) + E_a(z))^\alpha - (E_n(z, t) + E_a(z))^\alpha\big) \qquad (15.8)$$

$$N'_{pn}(z, t) = N_0\big((E_n(z, t) + E_s(z, t) + E_a(z))^\alpha - (E_s(z, t) + E_a(z))^\alpha\big), \qquad (15.9)$$

where z is the auditory place scale along the basilar membrane and t denotes the block index of the analyzed signal segment. $E_s(z, t)$ and $E_n(z, t)$ are the stimulus-related excitation components for the desired and the competing signal, respectively, $E_a(z)$ is the fixed internal spontaneous excitation, and α is the loudness-growth exponent, adapted to provide a 16-fold increase of loudness for a 1000 Hz tone when the presentation level is increased from 40 to 80 dB SPL. N_0 is a scale factor set to give a final loudness value of 1 sone for a 1000 Hz tone at 40 dB SPL for normal-hearing listeners. To represent the reduced partial loudness of speech in the presence of noise, and vice versa, the loudness density is further reduced smoothly towards zero, as

$$p_s(z, t) = \frac{N'_{ps}(z, t)}{N'_{ps}(z, t) + N'_{pn}(z, t)} \qquad (15.10)$$

$$N'_s(z, t) = N'_{ps}(z, t)\big(p_s(z, t)(2 - p_s(z, t))\big)^\beta, \qquad (15.11)$$

with symmetric expressions for $N'_n(z, t)$. This operation makes a considerable difference only when one of the signal components is much weaker than the other. This method is slightly different from the procedure suggested by [Moore et al. 1997]. The exponent was set to $\beta = 2$ to achieve agreement with empirical data. The instantaneous loudness function is then calculated by numerical integration over the auditory place scale z along the basilar membrane:

$$N_s(t) = \int_0^{z_{\max}} N_s'(z,t) \, dz \, . \tag{15.12}$$

Finally, the partial-loudness estimates N_s and N_n are calculated by three steps of non-linear smoothing of the instantaneous loudness functions $N_s(t)$ and $N_n(t)$ [Glasberg, Moore 2002].

Loudness levels (in phon) of the desired and competing signal are calculated from the loudness in sone simply as

$$LL_s = \max(0, 40 + 10 \log_2 N_s) \tag{15.13}$$
$$LL_n = \max(0, 40 + 10 \log_2 N_n) \, . \tag{15.14}$$

The loudness-level difference between desired and competing signal is obtained as

$$SNLL = LL_s - LL_n \, . \tag{15.15}$$

Thus, if the loudness N_n of the competing signal is very small, the SNLL can never exceed the loudness level of the desired signal alone.

Signal Excitation-level Distortion (SED)

The Signal Excitation-level Distortion (SED) is a measure of the spectral deviation between the unprocessed and processed desired signal. This measure is calculated as a root-mean-square average of excitation-level differences between the desired signal component in the unprocessed sound $(E_{s,in}(z,t))$ and the desired signal component in the processed sound $(E_{s,out}(z,t))$:

$$SED = \sqrt{\frac{T_m}{\Delta} \frac{1}{n(T_{SN})} \sum_{t \in T_{SN}} \int_0^{z_{\max}} w(z,t) \frac{(\log E_{s,in}(z,t) - \log E_{s,out}(z,t))^2}{\sigma_L^2(\Delta)} \, dz} \, , \tag{15.16}$$

where

$$w(z,t) = \begin{cases} 1 & \text{if } \frac{E_{s,in}(z,t)}{E_{n,in}(z,t)} \geq \delta_E \ \vee \ \frac{E_{s,out}(z,t)}{E_{n,out}(z,t)} \geq \delta_E \\ 0 & \text{otherwise} \, . \end{cases} \tag{15.17}$$

$E_{n,in}(z,t)$ and $E_{n,out}(z,t)$ represent the signal excitation patterns corresponding to the competing signal component in the unprocessed and the processed sound, respectively. To avoid the influence of speech pauses in the test material, signal segments are included in the calculation only if the segmental power signal-to-noise ratio is larger than -15 dB for either the unprocessed or the processed signal. Indices of these signal segments are included in the set T_{SN}, containing $n(T_{SN})$ index elements. Furthermore, to avoid including distortion elements that are completely masked by noise, the binary-valued function $w(z,t)$ allows non-zero contributions only at those auditory places z and time segment indices t where the signal-to-noise excitation ratios exceed δ_E. This parameter is set to correspond to an SNR of $-10\,\text{dB}$ in normal hearing. At lower signal-to-noise excitation-level differences, any spectral deviations in the desired signal are assumed to be masked by the noise.

The SED is closely related to the discrimination index d' for auditory spectral and intensity discrimination, defined in (15.5). The index value is calculated using the internal perceptual variance $\sigma_L^2(\Delta)$ of the log-excitation, estimated to predict empirical intensity discrimination, as shown in Sec. 15.3.1. This variance represents the auditory discrimination limit for each short sound segment. The segment duration was $\Delta = 0.02$ s in these calculations. However, the listener can remember signal features and improve discrimination by accumulating evidence by "multiple looks" over many segments. Therefore, the final result is scaled to give the average discrimination index for an integration time constant T_m, regardless of the actual duration of the test sound or the duration of signal segments. This time constant represents the memory span for which the listener can effectively accumulate perceptual information. The value has been set to $T_m = 1$ s, somewhat arbitrarily. The exact value is not critical, as this constant is merely a scale factor that does not change the qualitative comparison across different auditory profiles.

15.4 Test Conditions

Apart from the perceptually and intelligibility-weighted performance measures that were presented in Sec. 15.3 realistic acoustic test conditions need to be defined for a reliable performance assessment.

Recording Database

To simulate realistic acoustic test conditions we relied on a database with real-life audio recordings provided by Siemens Audiologische Technik, Erlangen, Germany for use in the HearCom project. The database contains recordings with different kinds of audio signals in a number of representative recording rooms. All test material was recorded by small microphones mounted in a behind-the-ear hearing aid case that was placed on an artificial head-and-torso manikin. The distance between the three microphones of the hearing aid device was about 10 mm. All signals were simultaneously recorded at a sampling rate of 16 kHz.

Environmental Conditions

Two representative recording rooms have been selected as follows.

Living room. The reverberation time T_{60} of this room ranges from about 0.3 s to 0.4 s. A music (classical piano concerto) point source was presented with azimuth angle 60 degrees at 150 cm distance from the recording manikin, where azimuth angle

+90 degrees refers to a signal source directly pointing to the right ear of the recording manikin.

Cafeteria. This large cafeteria has a reverberation time T_{60} of about 2 s between 250 and 2000 Hz. The competing source in this scenario was the natural diffuse noise that was recorded at lunch time, consisting of fairly stable babble sounds and other background noise in the cafeteria.

We remark that the selected subset of test material includes both point-source material and diffuse-like jammer sounds. In all cases the desired speech signal was coming from the frontal direction, corresponding to 0 degrees.

Several other environmental conditions were considered during a more profound analysis, such as a low-reverberant room, a car cabin, and a street corner.

Test Signals

The actual test signals are generated by additive mixing of the desired (speech) signal with the competing signals at specified signal-to-noise ratios (SNR). The desired signal and the competing signal are thus always separately available. In all test conditions the overall long-term sound pressure level of the desired (speech) signal is fixed at 70 dB SPL, and the level of the competing signal is varied to achieve the desired SNR value. Nominal presentation levels and SNRs are defined by long-term equivalent levels of electrical signals recorded by the most frontal microphone of the hearing aid. This frontal microphone signal is used during the computation of the evaluation measures as the reference representing unprocessed sound.

15.5 Simulation Results

In this section simulation results are presented that have been obtained with the multi-microphone noise suppression algorithms of Secs. 15.2.3 and 15.2.4 under the test conditions specified in Sec. 15.4. All test conditions have been evaluated for six SNR ratios: $-5, 0, 5, 10, 15$, and 20 dB. Furthermore, for every test condition four simulations were performed, each time with a different speaker (two male German, a female German and a female American English speaker). Afterwards, the results were averaged over these four test runs.

For the simulations with the SDWMWF algorithm of Sec. 15.2.4 all three microphone channels were input to the algorithm. The test results obtained with the A2B approach of Sec. 15.2.3 were obtained based on the front and the rear microphone signal of the three-microphone hearing aid device, as this combination was expected to deliver highest performance.

Both algorithms were evaluated with the functional-auditory-model-based performance measures of Sec. 15.3.2, for the six different auditory profiles shown in Fig. 15.4 and for the normal-hearing profile. For each simulation the algorithm started from

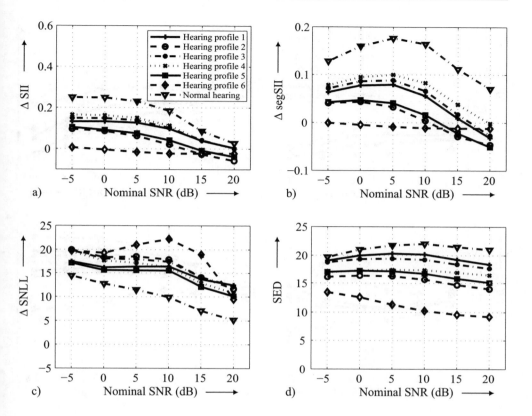

Figure 15.5: Living room with music source at 60°: A2B performance for a number
of representative hearing loss profiles
a) Speech Intelligibility Index improvement (Δ SII)
b) Segmental Speech Intelligibility Index improvement (Δ segSII)
c) Signal-to-Noise Loudness Level difference improvement (Δ SNLL)
d) Signal Excitation-level Distortion (SED)

its default state and was then allowed to adapt and converge for 29 seconds. The
performance measures were computed based on the output over the last 20 seconds.

The simulation results obtained with the A2B algorithm in the living room and in
the cafeteria are presented in Fig. 15.5 and Fig. 15.6, respectively. The corresponding
figures for the SDWMWF algorithm are Fig. 15.7 and Fig. 15.8. Each figure consists
of four subplots, which show the speech intelligibility index improvement (Δ SII),
the segmental speech intelligibility index improvement (Δ segSII), the improvement
in signal-to-noise loudness level difference (Δ SNLL) and the signal excitation-level
distortion (SED), respectively. Note that improvements are shown with respect to
the reference microphone (front hearing aid microphone), rather than the absolute
(seg)SII or SNLL values.

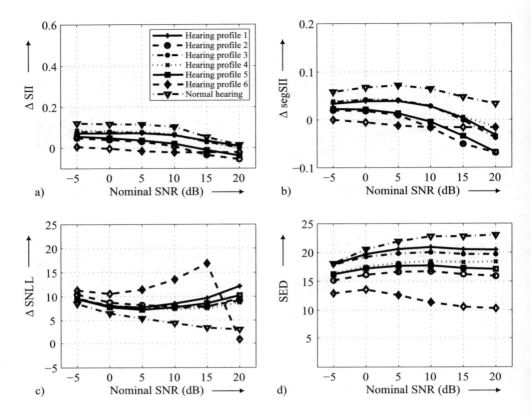

Figure 15.6: Cafeteria with diffuse babble noise: A2B performance for a number of
representative hearing loss profiles
a) Speech Intelligibility Index improvement (Δ SII)
b) Segmental Speech Intelligibility Index improvement (Δ segSII)
c) Signal-to-Noise Loudness Level difference improvement (Δ SNLL)
d) Signal Excitation-level Distortion (SED)

15.6 Discussion

Based on Figs. 15.5–15.8 a number of conclusions can be drawn with respect to the expected improvement in intelligibility and speech distortion for each algorithm. In this
way, the different algorithm variants can easily be compared.

A first observation that follows from an inspection of Figs. 15.5–15.8 is that for a
given hearing profile in general a clear trade-off occurs between noise suppression
and speech distortion: higher noise suppression implies higher speech distortion. For
the SDWMWF algorithm this trade-off is even one of the key design parameters, as
explained in Sec. 15.2.4: More emphasis can be put on either speech distortion or
noise suppression by means of trade-off parameter $1/\mu$.

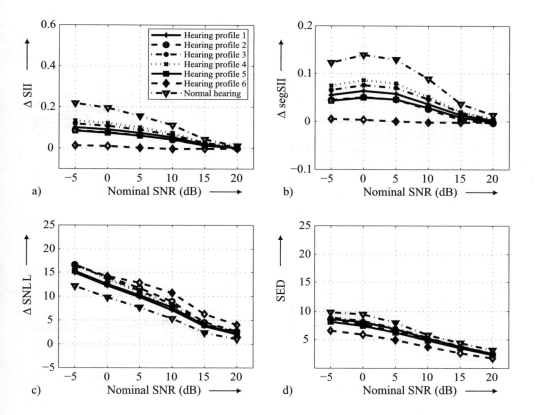

Figure 15.7: Living room with music source at 60°: SDWMWF performance for a
number of representative hearing loss profiles
  a) Speech Intelligibility Index improvement (Δ SII)
  b) Segmental Speech Intelligibility Index improvement (Δ segSII)
  c) Signal-to-Noise Loudness Level difference improvement (Δ SNLL)
  d) Signal Excitation-level Distortion (SED)

The A2B approach, unlike the SDWMWF algorithm, does not explicitly take speech
distortion into account, and leads to (slightly) higher (seg)SII and SNLL improvements, but more speech distortion than the SDWMWF algorithm. The high distortion figures that are observed for the A2B approach are partly due to the presence of
a software directional microphone in the signal flow graph (see Fig. 15.2), which acts
as a highpass filter, and considerably distorts the speech. It can hence be concluded
that the SDWMWF algorithm provides comparable SNR improvements as the A2B
approach, but a significantly lower distortion.

Also note that for some hearing profiles at high SNR ratios, the A2B algorithm shows
negative SII and segmental SII improvements, a phenomenon which is much less
prominent in the case of the SDWMWF algorithm.

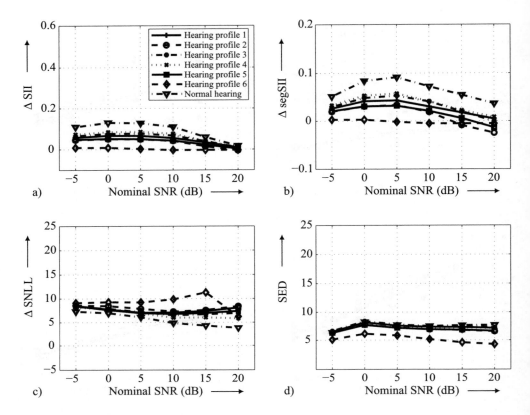

Figure 15.8: Cafeteria room with diffuse babble noise: SDWMWF performance for a
number of representative hearing loss profiles
   a) Speech Intelligibility Index improvement (Δ SII)
   b) Segmental Speech Intelligibility Index improvement (Δ segSII)
   c) Signal-to-Noise Loudness Level difference improvement (Δ SNLL)
   d) Signal Excitation-level Distortion (SED)

The effect of the room is also clearly noticeable. Generalized Sidelobe Canceler based
solutions are expected to show high noise suppression in low-reverberant rooms due to
the directionality of the sources and the limited amount of reverberation. However, in
more highly reverberating rooms with a point source disturbance, such as the living
room environment (see Figs. 15.5 and 15.7), SII scores are clearly reduced. The cafeteria scenario is most difficult as the algorithms have to cope with diffuse jammer sources
like babble noise, and experience even more reverberation. In this more challenging environment the SDWMWF algorithm offers slightly better SII and segmentally-
weighted SII figures than the A2B approach thanks to the longer filters that are used
and the additional microphone that is available.

As far as the influence of the input SNR is concerned, it is observed that the intelligibility improvement (Δ SII and Δ segSII) generally decreases with increasing

signal-to-noise ratio, especially in the higher SNR range. This can be understood as follows: The higher the input SNR, the more difficult it is to obtain additional enhancement, leading to moderate or low SNR improvements and hence reduced Δ SII and Δ segSII values at high input SNRs.

As far as the different auditory profiles are concerned, the normal hearing profile gives the highest SII and segmental SII improvements, followed by profiles 4, 3, 1, 5 and 2. Profile 6 (severe loss) is clearly worst. As a conclusion, it can be stated that the algorithms are expected to improve speech intelligibility for all hearing profiles, except for severe hearing loss, as long as the input SNR is lower than about 15 dB.

Both algorithms have also been evaluated based on another widely used hearing aid prescription, the ConstSL rule. The same tendencies and dependencies were observed as with the NAL-RP rule. Nevertheless, corresponding performance numbers sometimes differed significantly. It is therefore worthwhile putting sufficient effort into choosing the right hearing aid and defining appropriate device settings.

Furthermore, more profound comparison tests have been performed in a car cabin and street environment and in a low-reverberant room. The results obtained under these acoustic conditions are not included in this chapter. It was observed that the SDWMWF algorithm in general shows better performance than the A2B approach, especially in more realistic, complex scenarios, introducing less speech distortion, and typically offering similar or even higher segmental intelligibility improvements. The A2B algorithm however uses only two microphones and shorter filters, and therefore has a smaller computational complexity and introduces less delay than the SDWMWF algorithm.

15.7 Conclusions

In an attempt to meet the increasing demand for improved listening comfort in adverse listening conditions such as speech understanding amidst disturbances and environmental noise (cocktail-party effect), modern digital hearing aid and cochlear implant devices now standardly make use of advanced signal processing schemes such as multimicrophone noise suppression techniques.

Given the increased complexity of the proposed algorithmic solutions profound *a priori* testing is required before an integration onto a real-time hearing aid hardware platform can be taken into consideration. Owing to the inherent time-consuming character of subjective listening tests there is a great demand for automated quantitative test procedures to assess algorithm performance. Through the incorporation of fundamental aspects of normal and impaired human hearing, a number of perceptually meaningful evaluation measures have been obtained to quantify speech intelligibility improvement and speech distortion introduced by signal enhancement algorithms, and this for a number of representative hearing profile groups. The effectiveness of the proposed measures has been illustrated through the evaluation of a representative state-of-the-art and a recently developed multi-microphone noise reduction algorithm. By following this procedure major trends and dependencies can be highlighted from simulation results only, as shown by Figs. 15.5–15.8. In this way, promising algorithms variants can be ranked and selected and rough parameter tuning can be performed through simulation, which reduces the number of time-consuming perceptual tests with hearing-impaired subjects. Some of these results are now being validated with psychophysical speech reception tests.

Bibliography

ANSI-S3.5 (1997). *American National Standard Methods for the Calculation of the Speech Intelligibility Index*, American National Standards Institute, New York.

Byrne, D.; Dillon, H. (1986). The National Acoustic Laboratories New Procedure for Selecting the Gain and Frequency Response of a Hearing Aid, *Ear and Hearing*, vol. 7, pp. 257–265.

Byrne, D.; Parkinson, A.; Newall, P. (1990). Hearing Aid Gain and Frequency Response Requirements for the Severely/profoundly Hearing Impaired, *Ear and Hearing*, vol. 11, pp. 40–49.

Byrne, D.; Parkinson, A.; Newall, P. (1991). Modified Hearing Aid Selection Procedures for Severe/profound Hearing Losses, *in* G. Studebaker; F. Bess; L. Beck (eds.), *The Vanderbilt Hearing Aid Report II*, York Press, Parkton, Maryland, pp. 295–300.

Ching, T. Y. C.; Dillon, H.; Byrne, D. (1998). Speech Recognition of Hearing-impaired Listeners: Predictions from Audibility and the Limited Role of High-frequency Amplification, *Journal of the Acoustical Society of America*, vol. 103, no. 2, pp. 1128–1140.

Dillon, H. (2001). *Hearing Aids*, Boomerang Press, Sydney.

Doclo, S.; Spriet, A.; Moonen, M. (2004). Efficient Frequency-domain Implementation of Speech Distortion Weighted Multi-channel Wiener Filtering for Noise Reduction, *Proc. of the European Signal Processing Conference (EUSIPCO)*, Vienna, Austria, pp. 2007–2010.

Doclo, S.; Spriet, A.; Wouters, J.; Moonen, M. (2005). Speech Distortion Weighted Multi-channel Wiener Filtering Techniques for Noise Reduction (Chapter 9), *in* J. Benesty; S. Makino; J. Chen (eds.), *Speech Enhancement*, Springer, pp. 199–228.

Eargle, J. (2001). *The Microphone Book*, Butterworth-Heinemann, Woburn MA.

Fletcher, H.; Munson, W. (1937). Relations between Loudness and Masking, *J. Acoust. Soc. Amer.*, vol. 9, pp. 1–10.

Frost, O. (1972). An Algorithm for Linearly Constrained Adaptive Array Processing, *Proceedings of the IEEE*, vol. 60, no. 8, pp. 926–935.

Gannot, S.; Burshtein, D.; Weinstein, E. (2001). Signal Enhancement Using Beamforming and Non-Stationarity with Applications to Speech, *IEEE Trans. Signal Processing*, vol. 49, no. 8, pp. 1614–1626.

Glasberg, B. R.; Moore, B. C. J. (2002). A Model of Loudness Applicable to Time-varying Sounds, *J. Aud. Eng. Soc.*, vol. 50, no. 5, pp. 331–342.

Griffiths, L.; Jim, C. (1982). An Alternative Approach to Lineraly Constrained Adaptive Beamforming, *IEEE Transactions on Antennas Propagation*, vol. 30, January, pp. 27–34.

Hagerman, B.; Olofsson, A. (2004). A method to Measure the Effect of Noise Reduction Algorithms using Simultaneous Speech and Noise, *Acustica – Acta Acustica*, vol. 90, no. 2, pp. 356–361.

Herbordt, W.; Kellermann, W. (2003). Adaptive Beamforming for Audio Signal Acquisition (Chapter 6), *in* J. Benesty; Y. Huang (eds.), *Adaptive Signal Processing: Applications to Real-World Problems*, Springer, pp. 155–194.

Leijon, A. (2007). Predicting the Benefit of Noise Suppression in Hearing Aids. In preparation.

Luts, H.; Maj, J.; Soede, W.; Wouters, J. (2004). Better Speech Perception in Noise with an Assistive Multimicrophone Array for Hearing Aids, *Ear and Hearing*, vol. 25, no. 5, pp. 411–420.

Magnusson, L. (1996). Predicting the Speech Recognition Performance of Elderly Individuals with Sensorineural Hearing Impairment, *Scandinavian Audiology*, vol. 25, pp. 215–222.

Maj, J. (2004). *Adaptive Noise Reduction Algorithms for Speech Intelligibility Improvement in Dual Microphone Hearing Aids*, PhD thesis, Leuven, Belgium, Katholieke Universiteit Leuven.

Maj, J.; Wouters, J.; Moonen, M. (2004). Noise Reduction Results of an Adaptive Filtering Technique for Dual-Microphone Behind-the-Ear Hearing Aids, *Ear and Hearing*, vol. 25, no. 3, pp. 215–229.

Moore, B. C.; Glasberg, B. R. (2004). A Revised Model of Loudness Perception Applied to Cochlear Hearing Loss, *Hearing Research*, vol. 188, pp. 70–88.

Moore, B. C. J.; Glasberg, B. R.; Baer, T. (1997). A Model for the Prediction of Thresholds, Loudness, and Partial Loudness, *J. Aud. Eng. Soc.*, vol. 45, no. 4, pp. 224–240.

Nordebo, S.; Claesson, I.; Nordholm, S. (1994). Adaptive Beamforming: Spatial Filter Designed Blocking Matrix, *IEEE Journal of Oceanic Engineering*, vol. 19, no. 4, pp. 583–590.

Pavlovic, C.; Studebaker, G.; Sherbecoe, R. (1986). An Articulation Index Based Procedure for Predicting the Speech Recognition Performance of Hearing-impaired Individuals, *J. Acoust. Soc. Amer.*, vol. 80, pp. 50–57.

Shynk, J. (1992). Frequency-Domain and Multirate Adaptive Filtering, *IEEE Signal Processing Magazine*, vol. 9, no. 1, pp. 15–37.

Simmer, K. U.; Bitzer, J.; Marro, C. (2001). Post-Filtering Techniques (chapter 3), *in* M. S. Brandstein; D. B. Ward (eds.), *Microphone Arrays: Signal Processing Techniques and Applications*, Springer, pp. 39–60.

Skovenborg, E.; Nielsen, S. H. (2004). Evaluation of Different Loudness Models with Music and Speech Material, *Audio Engineering Society 117th Convention*, San Fransisco, CA, pp. 1–34.

Spriet, A. (2004). *Adaptive Filtering Techniques for Noise Reduction and Acoustic Feedback Cancellation in Hearing Aids*, PhD thesis, Leuven, Belgium, Katholieke Universiteit Leuven.

Spriet, A.; Moonen, M.; Wouters, J. (2004). Spatially Pre-processed Speech Distortion Weighted Multi-channel Wiener Filtering for Noise Reduction, *Signal Processing*, vol. 84, no. 12, pp. 2367–2387.

Spriet, A.; Moonen, M.; Wouters, J. (2005). Stochastic Gradient Based Implementation of Spatially Pre-processed Speech Distortion Weighted Multi-channel Wiener Filtering for Noise Reduction in Hearing Aids, *IEEE Transactions on Signal Processing*, vol. 53, no. 3, pp. 911–925.

Spriet, A.; Van Deun, L.; Eftaxiadis, K.; Laneau, J.; Moonen, M.; van Dijk, B.; van Wieringen, A.; Wouters, J. (2007). Speech Understanding in Background Noise with the Two-microphone Adaptive Beamformer BEAM in the Nucleus Freedom Cochlear Implant System, *Ear and Hearing*, vol. 28, no. 1, pp. 62–72.

Stone, M. A.; Moore, B. C. J. (2005). Tolerable Hearing-Aid Delays. IV. Effects on Subjective Disturbance During Speech Production by Hearing-Impaired Subjects, *Ear and Hearing*, vol. 26, no. 2, pp. 225–235.

van der Beek, F.; Soede, W.; Frijns, J. (2007). Evaluation of the Benefit for Cochlear Implantees of Two Assistive Birectional Microphone Systems in an Artificial Diffuse Noise Situation, *Ear and Hearing*, vol. 28, no. 1, pp. 99–110.

Van Gerven, S.; Xie, F. (1997). A Comparative Study of Speech Detection Methods, *Proc. European Conference on Speech Communication and Technology*, Rhodos, Greece, vol. 3, pp. 1095–1098.

Van Veen, B.; Buckley, K. (1988). Beamforming : A Versatile Approach to Spatial Filtering, *IEEE Magazine on Acoustics, Speech and Signal Processing*, vol. 36, no. 7, pp. 953–964.

Vary, P.; Martin, R. (2006). *Digital Speech Transmission – Enhancement, Coding and Error Concealment*, John Wiley & Sons, Ltd.

Widrow, B.; Stearns, S. (1985). *Adaptive Signal Processing*, Prentice Hall, Englewood Cliffs, New Jersey.

Wouters, J.; Vanden Berghe, J. (2001). Speech Recognition in Noise for Cochlear Implantees with a Two-microphone Monaural Adaptive Noise Reduction System, *Ear and Hearing*, vol. 22, no. 5, pp. 420–430.

Wouters, J.; Vanden Berghe, J.; Maj, J. (2002). Adaptive Noise Suppression for a Dual-microphone Hearing Aid, *International Journal of Audiology*, vol. 41, no. 7, pp. 401–407.

Zwicker, E. (1958). Über Psychologische und Methodische Grundlagen der Lautheit, *Acustica*, vol. 8, pp. 237–258.

Zwicker, E.; Terhardt, E. (1980). Analytical Expressions for Critical-band Rate and Critical Bandwidth as a Function of Frequency, *J. Acoust. Soc. Amer.*, vol. 68, no. 5, pp. 1523–1525.

VI

Speech Processing for Human–Machine Interfaces

Chapter 16

Automatic Speech Recognition in Adverse Acoustic Conditions

Hans-Günter Hirsch

16.1 Introduction

Automatic recognition of speech can be applied in a lot of practical application scenarios. However, the recognition performance of most recognition systems deteriorates considerably in the presence of adverse acoustic conditions such as noise and reverberation. The improvement of recognition rates in the case of adverse acoustic conditions is still one of the major research topics in this area [Junqua 2000], [Peinado, Segura 2006].

To approach this subject, a short overview of the signal processing and pattern matching techniques in the field of speech recognition is given. Most of today's speech recognition systems are based on a cepstral analysis scheme for representing short-term spectral speech features. Furthermore, the statistical approach of modeling speech units by Hidden Markov Models (HMMs) and the application of the Viterbi algorithm are state-of-the-art for pattern recognition in this field [Rabiner, Juang 1993], [Jelinek 1998].

The acoustic conditions will be analyzed as they occur during the speech input in practical applications. The major effects that influence the characteristics of speech will be discussed.

To investigate the influence of different distortion effects, noisy speech data are needed that have been either recorded in noisy situations or have been artificially created. We will present a signal processing tool that allows the simulation of various acoustic

Advances in Digital Speech Transmission   Edited by R. Martin, U. Heute and C. Antweiler
© 2008 John Wiley & Sons, Ltd

conditions. A new database has been created with this simulation tool. Furthermore a set of recognition experiments has been defined for the new data. Both are publicly available [Hirsch, Pearce 2006]. This database allows investigations on the same task so that the achieved recognition performance can be compared with other approaches.

The quantitative deterioration in the presence of different distortion effects is shown in Sec. 16.3.2. This allows a rating of different distortion effects with respect to the degree of performance degradation.

The existing approaches for improving the recognition performance can be roughly separated into techniques that determine robust acoustic features as part of the front-end processing and techniques that modify the pattern recognition process, e.g., by adapting the parameters of the reference patterns. As representative of the front-end processing approach we briefly describe a few details of a robust feature extraction scheme as it has been standardized by ETSI (European Telecommunication Standards Institute) [ETSI 2003a]. We present a new method for adapting certain HMM parameters in adverse conditions. This new approach can be especially used to adapt to the condition of a hands-free speech input in a room. Furthermore, it is shown that this technique can be combined with an adaptation to stationary background noise and unknown frequency characteristics. We demonstrate the efficiency of the new approaches by evaluating the results for recognizing the distorted data in the newly created database.

16.2 Structure of Speech Recognition Systems

The principal structure of a speech recognition system is shown in Fig. 16.1. The speech signal is analyzed by a short-term spectral analysis method, which is usually called feature extraction or front-end processing. The output is a set of acoustic features. The set of features is taken to build the components of a vector. Usually, about 100 feature vectors per second are created as the output of the analysis stage.

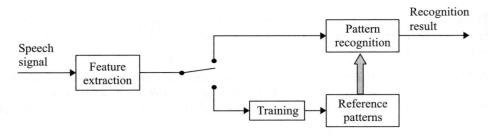

Figure 16.1: Structure of speech recognition system

During a training phase the stream of feature vectors is used to estimate the parameters of the reference patterns. A reference pattern can model a whole word, a single phoneme or some other speech unit. Depending on the recognition task the training consists of a more or less complex processing. For speaker dependent recognition there might be only a single or a few utterances for each word to be trained. But in the case of speaker independent recognition there are usually thousands of utterances that contain the signal of the desired reference pattern. These utterances are usually part of a speech database that has been collected in advance. Extracting the representative features and building a reference model can be a time consuming and complex task.

During recognition, the sequence of feature vectors is compared with the reference patterns. The likelihood is calculated that the vector sequence can be represented by a reference pattern or a sequence of reference patterns. Most often some type of Viterbi algorithm is applied for an efficient calculation of the likelihoods [Jelinek 1998]. The pattern or the sequence of patterns with the highest likelihood are presented as recognition result.

Nowadays, most often some type of Mel filterbank in combination with a transformation of the Mel spectrum to the cepstral domain is used as the feature extraction scheme. This type of processing will be explained in the next section. A processing scheme will be described that has been standardized by ETSI as a front-end technique [ETSI 2003b]. Hidden Markov Models (HMMs) are used most often as reference patterns. The characteristics are presented in Sec. 16.2.2. Furthermore a method is shown to visualize the acoustic features that define an HMM. The visualization is performed in the spectral domain.

16.2.1 Mel Frequency Cepstral Analysis

The block diagram of the feature extraction scheme is shown in Fig. 16.2 as standardized by ETSI [ETSI 2003b]. This technique can be seen as a typical representative for an extraction of the so called Mel frequency cepstral coefficients (MFCCs) [Rabiner, Juang 1993]. The speech signal is filtered with a high-pass filter at a very low cut-off frequency to remove any DC offset that might have been introduced by an Analog-to-Digital converter. This filtering is especially needed to realize a correct framewise calculation of the short-term energy. Then, the signal is segmented into frames of 25 ms duration. The energy as well as the cepstral parameters are calculated every $d_{\text{shift}} = 10\,\text{ms}$ by shifting the frame window. Since humans perceive loudness on a non-linear scale, the logarithm of the energy is calculated. The logarithmic frame energy is taken as one component of the feature vector.

A further high-pass filtering is applied as a so called preemphasis to enhance the components in the higher frequency region where speech has less energy. The short-term spectrum is calculated by weighting the samples of each frame with a Hamming window and by applying a Fast Fourier Transformation (FFT). The magnitude spectrum is taken because the short-term phase does not contain useful information about the

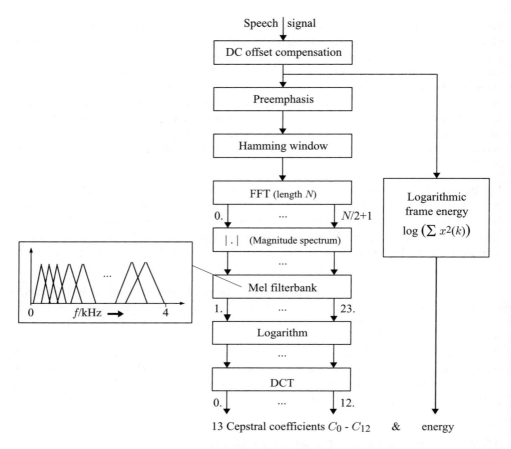

Figure 16.2: Block diagram of the cepstral analysis scheme as standardized by ETSI

contents of speech. Derived from the knowledge that the human auditory system accumulates the energies in critical bands, a so called Mel-scale filterbank is applied [Zwicker, Fastl 1999]. The filterbank consists of 23 subbands. The spectral FFT components are weighted with a triangular function and accumulated in the desired frequency region of a specific subband to form a Mel spectral component. The width of the frequency regions increases for increasing frequency according to the relation of linear and Mel frequency. As for the energy the logarithm of the Mel spectrum is calculated. Neighboring Mel frequency components are fairly correlated. To get feature components that are statistically more independent of each other, a Discrete Cosine Transformation (DCT) is applied to the logarithmic Mel spectrum. This statistical independence is of advantage for modeling the speech characteristics in the reference models and calculating the likelihoods in the pattern matching process.

In the case of the front-end as standardized by ETSI, 13 cepstral coefficients are calculated including the zeroth cepstral coefficient. The zeroth cepstral coefficient represents the mean of the logarithmic Mel spectrum. Thus, this value is closely related to the frame energy. Usually either the logarithmic frame energy as calculated from the time signal or the zeroth cepstral coefficient is used as a parameter in the recognition process. The feature vectors for the recognition often contain the logarithmic frame energy and the 12 cepstral coefficients C_1 to C_{12}. To apply the adaptation techniques that are presented in Sec. 16.4, we also need the zeroth cepstral coefficient C_0. Therefore C_0 is especially extracted for the training data so that C_0 becomes an HMM parameter. Thus, the set of cepstral coefficients in the reference patterns can be transformed back to the Mel spectrum. But C_0 is not used for the pattern recognition.

The acoustic parameters mentioned so far are called static parameters because they are calculated only from the speech signal of a short 25 ms frame. It turned out that the recognition performance can be increased by adding further dynamic features. This is often realized by looking at the contour of each static parameter over time and calculating the derivative of this contour. The parameters calculated this way are called delta coefficients. The first derivative $\Delta C_i(k)$ of the cepstral coefficient C_i is estimated according to

$$\Delta C_i(k) = \frac{\sum_{j=1}^{N_\Delta} j \cdot [C_i(k+j) - C_i(k-j)]}{2 \cdot \sum_{j=1}^{N_\Delta} j^2} . \tag{16.1}$$

This approach is used in many speech recognition systems, e.g., in the software package HTK [Young, et. al. 2005]. A value of 3 is a typical choice for N_Δ. In this case the delta coefficients are calculated from seven frames. Thus, they contain information about the dynamic behavior in a segment of about 85 ms. In the same way the second derivative can be estimated by applying (16.1) to the contour of the first derivative. These parameters are called delta–delta coefficients. Usually the time span for calculating the second derivative is less than the one for estimating the first derivative. In total the delta–delta coefficients are calculated from a segment of about 150 ms. The delta and delta–delta coefficients are added to the static parameters to form the final feature vector. Thus, the typical feature vector will consist of 39 components including the logarithmic frame energy and the 12 cepstral coefficients C_1 to C_{12} as static parameters.

Another feature extraction scheme has been standardized by ETSI [ETSI 2003a], [Machoa et al. 2002], [Peinado, Segura 2006]. This is an extension of the front-end shown in Fig. 16.2. Two further processing blocks are added as visualized in Fig. 16.3. This scheme aims at an extraction of acoustic features that are robust in the presence of background noise and unknown transfer function. The noise reduction consists of a two-stage Wiener filter that is applied to the noisy speech signal. The characteristics of the Wiener filter is estimated in the frequency domain. The filtering itself is done in the time domain after transforming back the smoothed estimated filter characteristic to the time domain. A further SNR dependent waveform processing is applied to the filtered signal. The noise reduced signal is taken as the input signal to a cepstral analysis scheme as described before. The output of the cepstral analysis stage are again

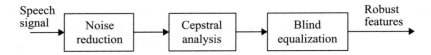

Figure 16.3: Block diagram of the robust ETSI feature extraction scheme

13 cepstral coefficients including the zeroth coefficient and one logarithmic energy co-efficient per frame. To compensate for the influence of an unknown transfer function a further processing block has been introduced. This contains a blind equalization scheme that is based on a comparison of the speech spectra with a "flat" spectrum and applying the LMS (least mean square) algorithm for adapting an equalization filter [Mauuary 1998].

16.2.2 Modeling Speech Units as HMMs

Most speech recognition systems are based on modeling speech segments as Hidden Markov Models (HMMs). A segment can be, e.g., a whole word or a phoneme. An HMM consists of a chain of states as shown in Fig. 16.4 for the word "six". The transitions between the states are defined in a simple left-to-right topology derived from the natural behavior of uttering the speech segments one after each other along time. The transition to a succeeding state or the remaining in a certain state are described by transition probabilities. The probability to remain in a state contains the information about the average duration of a speech segment.

In the simple example shown in Fig. 16.4 each state is related to a phoneme whereas in practical implementations each phoneme is usually modeled by a sequence of at least three states. Each state contains a set of parameters that mainly describe the spectral and energy characteristics of the corresponding segment. In the simple case of a speaker dependent model the set of parameters can correspond to the vector of features. The components of this vector can be estimated, e.g., as the average of all feature vectors that are mapped on this HMM state during the training procedure. In the more complex case of a speaker independent recognition each state contains information about the statistical distribution of each acoustic parameter. Modeling the spectral and energy characteristics by means of a distribution function accounts for the effect that the pronunciation of a speech segment varies for different speakers.

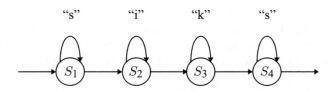

Figure 16.4: Structure of an HMM in speech recognition

Based on the assumption of statistically independent acoustic features each parameter can be described by an individual one-dimensional distribution function. Usually a Gaussian distribution or a weighted mixture of Gaussian distributions is taken. The characteristics of these distribution functions are estimated in the training phase by a large number of utterances from different speakers.

For research purposes it is of interest to visualize the spectral characteristics that are contained in an HMM. Furthermore, we are interested in visualizing the spectral modifications that are introduced by the adaptation of HMMs. Therefore, for each HMM state S_n the vector $\bar{\mathbf{C}}(S_n)$ of static MFCCs is transformed back to the linear Mel spectral domain according to (16.2). This is based on the availability of the zeroth cepstral coefficient as described in the previous section. The output is the vector $|\bar{\mathbf{X}}(S_n)|$ containing the magnitude Mel spectrum in K_{mel} bands. A typical value for K_{mel} is 24. The Mel spectrum $|\bar{\mathbf{X}}(S_n)|$ represents the spectral information of the speech segment as modeled by HMM state S_n

$$\bar{\mathbf{C}}(S_n) \quad \underset{\text{IDCT}}{\Longrightarrow} \quad \log_e(|\bar{\mathbf{X}}(S_n)|) \quad \underset{\exp}{\Longrightarrow} \quad |\bar{\mathbf{X}}(S_n)| \qquad \text{for all states } S_n . \tag{16.2}$$

The MFCCs of vector $\bar{\mathbf{C}}(\mathbf{S_n})$ are the means of individual Gaussian distribution functions in the case of modeling with a single distribution only. When modeling with a mixture of distributions, vector $\bar{\mathbf{C}}(\mathbf{S_n})$ represents the weighted average over the means of all N_{mix} distribution functions. The weighted average $\bar{C}_i(S_n)$ of a single cepstral coefficient is calculated according to

$$\bar{C}_i(S_n) = \sum_{j=1}^{N_{\text{mix}}} w_j \cdot C_i(S_n, j), \qquad \sum_{j=1}^{N_{\text{mix}}} w_j = 1 \tag{16.3}$$

applying the mixture weight w_j to the cepstral coefficient $C_i(S_n, j)$ of the Gaussian distribution with index j. Besides the spectral characteristics of the speech segment we need the temporal information at which point in time it occurs and how long the segment is. The average duration $d(S_n)$ of a speech segment can be estimated from the conditional probability $p(S_n|S_n)$ of remaining in the corresponding HMM state S_n according to (16.4). d_{shift} describes the time shift of the analysis window as applied during feature extraction

$$d(S_n) = \frac{1}{1 - p(S_n|S_n)} \cdot d_{\text{shift}} . \tag{16.4}$$

Then, the Mel spectrum $|\mathbf{X}(S_n)|$ of HMM state S_n can be positioned with respect to its point in time $t(S_n)$ at the middle of each segment

$$t(S_n) = \sum_{j=1}^{n-1} d(S_j) + \frac{d(S_n)}{2} . \tag{16.5}$$

Furthermore, a spline interpolation is individually applied to the contour of the magnitude spectral values in each Mel band. Thus, the magnitude Mel spectrum is recreated at a frame rate as it is defined by the window shift in the feature extraction

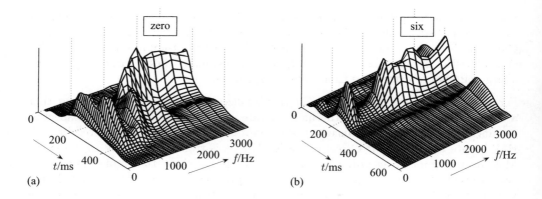

Figure 16.5: 3-D visualization of the spectrograms derived from two HMMs

process. Two spectrograms that have been derived as described before are visualized in a three-dimensional representation in Fig. 16.5. They represent the characteristics of the HMMs for the words "zero" and "six" where the HMMs are the result of training with the male training subset of the TIDigits data [Leonard 1984].

The spectrogram for the word "zero" in Fig. 16.5-a contains components at higher frequencies that exist at the beginning of the word due to the fricative. Furthermore, the characteristics of the vowel "e" become visible with a second formant at a frequency of about 2 kHz. The formants of the vowel "o" at lower frequencies are visible towards the end of the word. Figure 16.5-b contains the formant spectrum of the vowel in the middle for the word "six" as well as the spectral components at higher frequencies due to the fricatives at the beginning and at the end. The valley between the vowel and the fricative at the end is due to the plosive "k" that manifests as a short pause in the spectrum.

16.3 Acoustic Scenarios during Speech Input

An overview of the distortion effects as they occur in practical applications of speech recognition systems is given in Fig. 16.6.

The application of automatic speech recognition makes especially sense in situations where the user does not have his or her hands available for controlling a device by keyboard input or mouse movements. In many cases it would be desirable to allow speech input to a recognition system without the need to wear a close-talking microphone. But this hands-free speech input leads to a modification of the speech signal due to multiple reflections of the sound inside a room, i.e., reverberation.

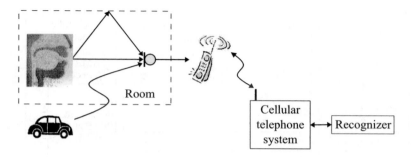

Figure 16.6: Acoustic scenarios at the speech input to a recognition system

Besides reverberation, background noise is present in almost all applications. The presence of noise leads to a superposition of speech and noise signals. Noise can be classified into two categories when considering its influence on speech recognition. It can be a stationary signal such as the noise of an engine or a fan or it can be non-stationary such as a knocking at a door. In most applications both types are present. The worst case is the presence of speech or music in the background because it leads to the task of separating a desired speech signal from a competing speech signal.

Furthermore, in a lot of applications the recognizer is not at the same location as the microphone. Most information retrieval systems are located at a remote position somewhere in a telephone or data network. In this case the speech signal has to be transmitted over an analog or digital telephone line or a digital data line. The transmission can take place over a fixed or a mobile network. The speech signal is encoded before the transmission and decoded later on. This causes a minor degradation of the speech signal dependent on the type of speech coding [Hirsch 2002]. Additional distortion is introduced in mobile networks when the speech is transmitted over a noisy cellular channel.

16.3.1 Simulation of the Acoustic Environment

The focus of this work is on situations as they are of interest for practical applications of speech recognition systems. Three situations are considered as being of great interest:

1. Hands-free speech input while driving a car;

2. Hands-free speech input at a desk in an office with the intention of controlling the phone itself or using it for information retrieval from a remote system; and

3. Hands-free speech input in a living room with the intention of controlling, e.g., audio or video devices.

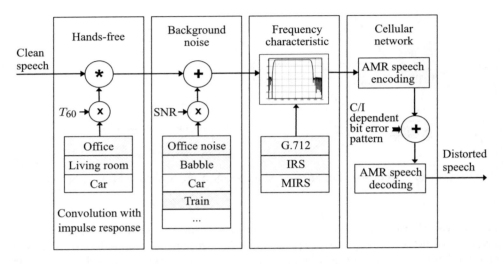

Figure 16.7: Processing scheme for the simulation of acoustic environments

In this section we describe a processing scheme to simulate the different acoustic scenarios as they might occur during speech input to a recognition system. An overview about the processing options is given in Fig. 16.7.

We model the hands-free speech input by convolving the speech signal with an impulse response reflecting the transmission between the speaker's mouth and the microphone in the desired acoustic environment. In practice the impulse response is usually a time variant function when the speaker moves in the room or the room configuration is modified by, e.g., opening or closing doors or windows. A time-invariant response is used assuming stationary conditions during an individual utterance as speech input to the recognition system. However, the given impulse responses can be modified dependent on the desired reverberation time T_{60}. Thus, we can model slight changes of the acoustic environment. We artificially created impulse responses for the two room conditions while taking a measured impulse response for modeling hands-free communication in a car.

The presence of noise in the background can be simulated by adding a recorded noise signal to the speech signal at a desired SNR. The estimation of speech and noise levels is done according to ITU recommendation P.56 [ITU 1993]. The same approach was applied in earlier investigations where noisy speech data were generated for the evaluations inside the ETSI working group AURORA [Hirsch, Pearce 2000]. The speech and noise signals are filtered with the G.712 telephone frequency characteristic [Campos-Neto 1999] to estimate their levels. Thus, the SNR is related to the energy of the signals in the perceptually important frequency range from about 300 to 3400 Hz.

Certain frequency characteristics can be applied to simulate the recording with a telephone device and the transmission over telephone networks. The applied frequency

responses have been defined by ITU [Campos-Neto 1999] and are known by their abbreviations G.712, IRS and MIRS. In all cases frequency components below 300 Hz
and above 3400 Hz are considerably attenuated. The G.712 filtering has a flat characteristic in the range between 300 and 3400 Hz, whereas the frequency responses of IRS
and MIRS show an increasing trend in this range with a slightly higher attenuation
at low frequencies.

The usage of mobile phones is a typical scenario while accessing a speech dialog system for information retrieval. The encoding and decoding of the speech as well as
the transmission over the noisy cellular channel degrade the speech signal [Vary, Martin 2006]. The influence of these distortions could be avoided by the approach of
a distributed speech recognition system where the acoustic features are extracted in
the terminal and transmitted as digital data with high error protection. However,
so far this approach is not seen in practical applications. Thus, it seems to be useful to simulate the transmission over voice channels in cellular networks. Here, the
AMR (adaptive multi rate) coding schemes are applied for considering the influence of
speech encoding and decoding [ETSI 2000]. These schemes are mainly used for speech
transmission in GSM and UMTS networks. There are two sets of coding schemes,
one for encoding speech in the narrow-band frequency range up to about 3.4 kHz and
one in the wide-band range up to about 7 kHz. The AMR-NB (narrow-band) codec
includes eight coding modes with data rates between 4.75 and 12.2 kBit/s. The AMR-
WB (wide-band) coding scheme includes nine coding modes with data rates between
6.6 and 23.85 kBit/s.

The influence of the transmission over GSM and UMTS channels is simulated by applying bit error patterns to the data stream between speech encoding and decoding.
These error patterns have been derived by simulating channel encoding and decoding
together with the typical error patterns that are applied between channel encoding
and decoding. Error patterns exist for different transmission scenarios as, e.g., driving in a car. An advantage of creating versions of the typical bit error patterns that
can be directly applied to the data stream after speech encoding, is that no channel
encoding and decoding blocks are needed in the simulation tool. In the case of GSM
transmission, error patterns exist for all AMR coding modes that are designated for
their usage in GSM networks. For each speech coding mode there are patterns for
different C/I (carrier-to-interference) ratios. The value of the C/I in dB describes the
quality of the cellular channel. For example, a value of 4 dB describes the communication at the border of a radio cell whereas a value of 16 dB describes the situation in the
center of the radio cell. In the case of CDMA based transmission in UMTS networks
the quality of the cellular channel is defined by the frame error rate that describes the
percentage of erroneous frames. Error patterns are available for all AMR coding
modes and frame error rates of 0.5 %, 1 %, and 3 %.

A graphical user interface is available in the World Wide Web as access to this tool.
It allows the definition of a desired speech input scenario and can be used to upload
and process speech files [Finster 2005].

A noisy speech database has been created by means of the simulation tool for simulating the influence of different distortion effects. This database has been made publicly

available by distributing it via the European Languages Resource Association ELRA under the name "Aurora-5". Thus, it can be used by researchers and developers to determine the recognition performance of their own algorithms and compare their results with others. The principal setup is similar to the one of the "Aurora-2" database that was created some time ago as a first approach for comparative investigations in the field of robust recognition.

The well known TIDigits speech database [Leonard 1984] is the basis for the creation of noisy data where the "clean" TIDigits data have been downsampled to a sampling frequency of 8 kHz. Only the recordings of the adult American speakers are used. These contain sequences of English digits with a maximum of 7 digits per utterance. The data are separated into two sets. One has been designated for training a recognition system and the other one can be used for testing.

All available 8700 test utterances with a total of about 8700 digits are used to create test sets for different acoustic conditions. We focus on the two scenarios of applying a speech recognition system:

1. Inside a car to control devices in the car or to retrieve information from a remote speech server somewhere in a telephone network;

2. Inside a room to enable the hands-free control of telephone, audio or video equipment.

For the car environment we created test sets at SNRs of 0 to 15 dB by:

- adding car noise only,

- simulating a hands-free speech input and adding car noise,

- simulating a hands-free speech input, adding car noise and simulating a transmission over the GSM network.

This set-up allows the comparison of different speech input and transmission conditions.

To investigate the influence of a speech input in a noisy room environment we created test sets at SNRs of 0 to 15 dB by:

- adding interior noise only,

- simulating the hands-free speech input in an office room and adding interior noise,

- simulating the hands-free speech input in a living room and adding interior noise.

This enables the comparison of a close-talk and a hands-free input in noisy conditions.

We used a set of different noise recordings for the creation of a single test set. These recordings reflect the desired condition of the test set. For the Aurora-2 experiment

only a single recording was taken. Thus, we increase the variance of noise characteristics within each test set in comparison with Aurora-2. As for Aurora-2, we defined
experiments for training a recognition system and investigating the speaker independent recognition of connected digits with the freely available software package HTK
[Young, et. al. 2005].

Recognition results on this new database will be presented in Sec. 16.4.5 after introducing different techniques for improving the robustness of recognition systems.

16.3.2 Recognition Results for Different Distortion Effects

With the availability of the simulation tool we study the influence of different distortion effects on the performance of a recognition system. The goal is a comparative
rating of the different effects with respect to the deterioration of the recognition performance.

As for the creation of the new database, the speaker independent recognition of connected English digits is considered by creating modified versions of the TIDigits. The
robust analysis scheme as described in Sec. 16.2.1 is taken for the extraction of acoustic features [ETSI 2003a]. We use this front-end here because it can be considered
as representing a state-of-the-art feature extraction as applied in most recognition
systems today. It includes a certain robustness against noise. The processing scheme
provides twelve cepstral coefficients and one energy coefficient as acoustic parameters
for describing a short speech segment of 25 ms duration. A feature vectors contains 39
components including the corresponding delta and delta–delta coefficients. A feature
vector is determined every 10 ms.

The training of the recognition system is done with clean data only. The corresponding
tools of HTK are applied for training and recognition [Young, et. al. 2005]. The
training set of the TIDigits is taken for the training of whole word HMMs. Gender
dependent models are determined for the 11 digits including the two versions "zero"
and "oh" for the digit "0". Each digit HMM consists of 16 states where each acoustic
parameter is modeled with a mixture of two Gaussian distributions. A further one
state model with a mixture of eight Gaussian distributions is trained as representation
of the speech pauses. We achieve a word error rate of 0.69 % on the recognition of the
clean TIDigits test data including deletion and insertion errors. This value represents
the baseline performance for this recognition task.

The word error rates are shown in Fig. 16.8-a for recognizing the TIDigits after applying a simulation where the speech data are recorded in hands-free mode.

Word error rates are presented for the two rooms as they were defined in the simulation tool. We refer to them as "office" and "living" room. Furthermore we vary the
reverberation time in a certain range. Error rates are shown for a variation of the
reverberation time of between 0.2 and 0.9 seconds. Error rates considerably increase
for increasing reverberation time. Another result is plotted as a star in Fig. 16.8-a.
It was achieved by applying the measured impulse response inside a real room used

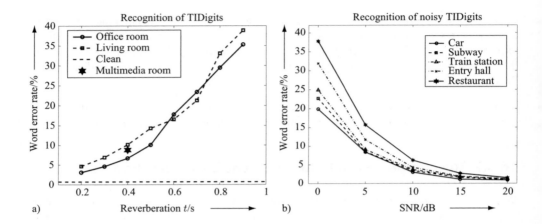

a)   b)

Figure 16.8: Word error rates applying the robust ETSI front-end for speech input in hands-free mode and in the presence of background noise

for multi-media presentations. This room has a reverberation time of about 0.4 s. The result is very close to the results achieved with the simulated rooms, providing evidence that the room simulation is reasonably realistic.

In an additional experiment we investigated the hands-free speech input in a car. Compared with the clean and anechoic speech we observed an increase in the error rate by about a factor of five from 0.69 % to 3.16 %. As expected, the deterioration is much less than in the rooms.

We study the presence of background noise as another distortion effect. Word error rates are shown in Fig. 16.8-b for the recording in five different noise scenarios. Noise signals have been added at SNRs in the range 0 to 20 dB. The results for these five noises represent a typical range of word error rates. The general characteristic of increasing error rates for lower SNRs is well known from many investigations. The difference between the noise signals is due to the characteristics of the individual noise scenarios. It depends on the presence and the length of non-stationary segments, where the noise level and the spectrum of the noise signals changes. The curve with the lowest error rates is achieved for the recording inside a car, where the noise is fairly stationary. Results are worst for the recording in a restaurant with music and people chatting in the background. One has to keep in mind that a robust feature extraction scheme is applied. This scheme contains a processing block for the reduction of stationary noise. Error rates would be much higher in the case of applying, e.g., a standard cepstral analysis scheme as described in Sec. 16.2.1.

Word error rates are listed in Table 16.1 showing the performance in the presence of a fixed spectral weighting as occurs in telephony. The digit recordings are filtered with typical telephone frequency characteristics as they have been defined by ITU and as they are available in the simulation tool.

Table 16.1:   Word error rates of TIDigits data for different filter characteristics

Filter characteristic	No filter	G.712	MIRS	IRS
Word error rate/%	0.69	0.72	0.79	1.07

Only a small deterioration of the recognition performance can be seen when comparing this type of distortion with the effect of reverberation or background noise. This is due to the processing block of the feature extraction scheme that performs a blind estimation and equalization of unknown frequency characteristics. Without such a compensation scheme the influence of an unknown frequency characteristic is much higher. This was a major problem in older recognition systems.

The use of a mobile phone for accessing a recognition system is studied as the last distortion effect. The word error rates are listed in Table 16.2 for the eight coding modes of the AMR scheme [ETSI 2000].

Table 16.2:   Word error rates of TIDigits data for AMR speech en(de)coding

AMR-mode (kBit/s)	No coding	12.2	10.2	7.95	7.4	6.7	5.9	5.15	4.75
Word error rate/%	0.69		0.98	1.01	1.23	1.32	1.38	1.34	1.84	1.82

We can see that the error rate increases for coding modes with a lower data rate. But the deterioration is relatively small in comparison with the other distortion effects. The word error rates shown in Fig. 16.9 also take into account the transmission over the noisy cellular channel of a GSM network.

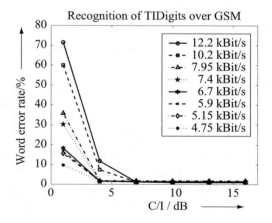

Figure 16.9: Word error rates of TIDigits data after transmission over GSM networks

The curves show a higher deterioration for coding modes with a higher date rate at low values of C/I because the higher data rate for the source coding comes along with a lower data rate for the channel coding. The lower data rate for the channel coding causes a lower percentage of detectable and correctable errors, which results in a lower quality of speech. The AMR coding schemes have been introduced to improve the speech quality in mobile communication by switching to a mode with a lower data rate in case of a bad cellular channel. Keeping this in mind, the influence of using a mobile phone as speech input to a recognition system is also relatively small in comparison with the influences of reverberation or background noise.

As a conclusion of these investigations it turns out that reverberation and background noise are the effects that cause major deterioration of recognition performance. Techniques should be introduced to compensate for the influence of background noise and reverberation.

16.4 Improving the Recognition Performance in Adverse Conditions

The existing approaches for improving the recognition performance in adverse conditions can be roughly separated into two classes as visualized in Fig. 16.10.

One class contains the techniques to extract robust features from the speech signal that should be independent of the acoustic input conditions. We presented an example of a robust front-end in Sec. 16.2.1.

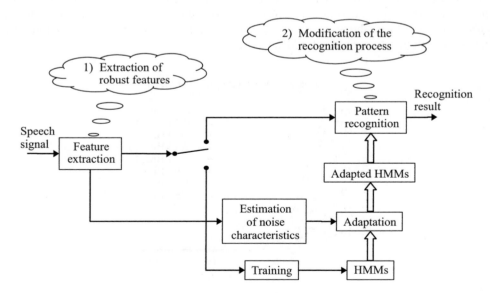

Figure 16.10: Approaches for improving the recognition performance in adverse conditions

The second class consists of methods to modify the pattern recognition process dependent on the acoustic conditions. This can be done in different ways. As a first approach the recognition system can be trained on data that have been recorded in adverse conditions. In most cases the highest performance can be achieved when applying reference patterns that have been trained on data recorded in exactly the same conditions as in the recognition phase. But the conditions have to be known in advance. Therefore, the performance deteriorates as soon as the conditions during testing become different from the ones used for training. As a compromise, the reference patterns can be trained on data from all the expected acoustic input conditions. The term "multi-condition" is used for this type of training. However, again the conditions have to be known in advance, and it needs a great deal of effort in recording or artificially creating data for all conditions.

Besides the training of reference patterns on adverse conditions the recognition process can be modified by omitting speech segments or, more precisely, certain spectral regions of these segments that are marked as unreliable. This approach is known as the "missing feature" technique [Cooke et al. 2001]. The assessment of speech segments or spectral regions as unreliable can be, for example, based on the estimation of the local SNR in the corresponding spectral regions. In the case of estimating a low SNR, these spectral regions are not used at all or get a lower weight for the calculation of the probabilities in the pattern recognition process. The probabilities describe the likelihood that the sequence of feature vectors can be modelled by an HMM or a sequence of HMMs. A disadvantage of the missing feature technique so far is its restriction to the use of spectral features so that it can not easily be applied to recognition systems that are based on cepstral features.

A third possibility of modifying the pattern recognition is an adaptation of the references based on an estimation of the acoustic conditions. This is indicated in Fig. 16.10. There are different approaches to realizing the adaptation, e.g., [Leggeter, Woodland 1995], [Woodland 2001], [Gauvain, Lee 1994], [Sankar, Lee 1996], [Gales, Young 1996] and [Minami, Furui 1996]. We will present our approach to adapting the energy and spectral parameters as they are contained in HMMs to a reverberant environment. Furthermore, we introduce a new approach for adapting the delta and delta–delta coefficients. Recognition results are presented for the recognition of reverberant signals. Then, we show how this adaptation technique can be combined with an adaptation to background noise and unknown frequency characteristics. The efficiency of the adaptation scheme is demonstrated by presenting results on the recognition of the distorted data from the new database "Aurora-5", which was described in Sec. 16.3.1.

16.4.1 Adapting HMMs to Reverberation

To derive the approach for adapting the HMMs, we first analyze the modifications that are introduced by reverberation with respect to the usage of a cepstral analysis scheme and the modelling with HMMs as applied in the field of speech recognition. Based on the results of this analysis we present the new method with its mathematical details for adapting the static parameters of HMMs to the reverberant signals of a hands-free speech input. The only parameter that is needed for performing the adaptation is an estimate of the reverberation time T_{60}.

Modeling the Influence of a Hands-free Speech Input

Ideally, the multiple reflections of sound in a room can be described by an exponential decay of the acoustic energy, which has been the result of early investigations in room acoustics [Kuttruff 2000]. This leads to a room impulse response (RIR) $h(t)$ with an exponentially decaying envelope according to

$$h^2(t) \sim e^{-\frac{6 \cdot ln(10)}{T_{60}} \cdot t} . \tag{16.6}$$

The only parameter for defining the exponential decay is the reverberation time T_{60}, which takes values in the range of about 0.2 to 0.4 seconds for smaller rooms and of about 0.4 to 0.8 seconds for larger rooms. It can take values above 1 second for very large rooms such as naves. The reverberation time depends on the interior equipment in the room and the individual absorption characteristics of the walls.

The RIR can be transformed to the room transfer function by means of a Fourier transform. The room transfer function has a contour that changes very fast with frequency. Usually only the envelope of the room transfer function is of interest when looking at the filterbank approaches that are applied for extracting acoustic features in speech recognition. This effect can be covered by an adaptation to an unknown frequency characteristic. More important for the frame-based analysis in speech recognition is the influence on the contour of the short-term energy over time. The energy contours of a speech signal are shown in Fig. 16.11-a before and after the transmission in a

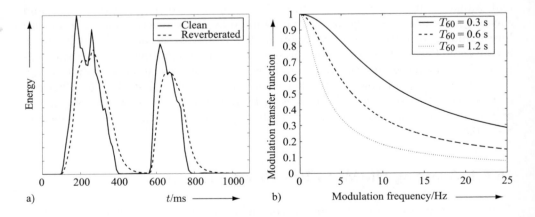

Figure 16.11: Short-term energy contours of a speech signal and modulation transfer functions

room. The energy is usually estimated as short-term energy in frames of about 25 ms duration. It can be seen that the reverberation leads to an extension of each sound contribution. This extension occurs as the so called reverberation tail with the exponentially decaying envelope of the RIR. The same effect can also be seen when looking at the energy contours in single subbands of a Mel-scale filterbank, which is usually applied in the front-end of a speech recognition system.

Transforming such energy contours to the so called modulation spectrum by means of a Fourier transform leads to the estimation of the modulation transfer function $m(F)$ [Houtgast et al. 1980], which can be mathematically described by

$$m(F) = \frac{1}{\sqrt{1 + (2 \cdot \pi \cdot F \cdot \frac{T_{60}}{6 \cdot ln(10)})^2}}. \tag{16.7}$$

Fig. 16.11-b shows the low pass characteristic of the modulation transfer function for different values of T_{60}. The cut-off frequency of the low pass characteristic shifts to lower values of the modulation frequency for increasing values of T_{60}. This corresponds to longer reverberation tails for higher values of T_{60}. The extension of sound contributions can lead to masking in the acoustic parameters of low energy sounds by the parameters of a preceding sound with higher energy.

Adaptation of Static Parameters

In reverberant conditions it can be expected that the acoustic excitation described by the parameters of a single state will also occur in succeeding states with some attenuation when clean speech HMMs are used. Figure 16.12 visualizes this effect. Each state S_n of an HMM describes a speech segment with a certain average duration. This duration $d(S_n)$ can be derived according to (16.4) from the probability of remaining in this state. The speech segment contains a certain acoustic excitation with a defined energy. Owing to the reverberation with its exponentially decaying RIR, this energy will be spread over time. Looking at the visualization in Fig. 16.12 an energy contribution due to the excitation in state S_1 will also appear in state S_3. The factor $\alpha_{n,1}$ describing the energy contribution in a succeeding state S_n due to the excitation in state S_1 can be calculated according to

$$\alpha_{n,1} = \frac{\int\limits_{t_s(S_n)}^{t_e(S_n)} h^2(t)\, dt}{\int\limits_{0}^{\infty} h^2(t)\, dt} \quad \text{with} \quad t_s(S_n) = \sum_{j=1}^{n-1} d(S_j) \tag{16.8}$$

$$t_e(S_n) = t_s(S_n) + d(S_n).$$

Given an estimate of the reverberation time T_{60}, the contribution factors can be individually calculated for all states of all HMMs.

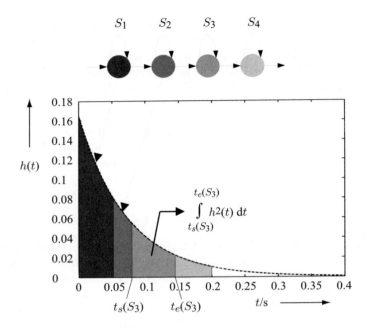

Figure 16.12: The distribution of energy at state S_1 due to reverberation

The mean of the energy parameter at an individual state S_n can be adapted by adding the energy contributions of the state itself and the preceding states according to

$$\tilde{E}(S_n, j) = \alpha_{n,n} \cdot E(S_n, j) + \alpha_{n,n-1} \cdot \bar{E}(S_{n-1}) + \alpha_{n,n-2} \cdot \bar{E}(S_{n-2}) + \ldots \quad (16.9)$$

$$= \alpha_{n,n} \cdot E(S_n, j) + \sum_{l=1}^{n-1} \alpha_{n,l} \cdot \bar{E}(S_l), \quad 1 \le j \le N_{\text{mix}}, \quad (16.10)$$

where N_{mix} denotes the number of mixture components. The adaptation is individually applied to each mixture component with index j. For the preceding states $\bar{E}$ denotes the weighted average that is calculated over the mixture of Gaussians in the same way as defined for the cepstral coefficients in (16.3).

In the same way the means of the power spectral density can be adapted. When using MFCCs, the cepstral coefficients have to be transformed back to the spectral domain according to (16.2). The power spectral density value at state S_n in the Mel bin with index μ can be adapted by adding the contributions of the preceding states according to

$$|\tilde{X}_\mu(S_n, j)|^2 = \alpha_{n,n} \cdot |X_\mu(S_n, j)|^2 + \sum_{l=1}^{n-1} \alpha_{n,l} \cdot |\bar{X}_\mu(S_l)|^2, \quad 1 \le \mu \le K_{\text{mel}}. \quad (16.11)$$

The adapted spectra have to be transformed to the cepstral domain according to

$$|\tilde{\mathbf{X}}(S_n, j)|^2 \underset{\text{sqrt\&log}}{\Longrightarrow} \log_e(|\tilde{\mathbf{X}}(S_n, j)|) \cdot \underset{\text{DCT}}{\Longrightarrow} \tilde{\mathbf{C}}(S_n, j). \tag{16.12}$$

In practice, mainly two to three preceding HMM states have an influence on the current state. This depends on the reverberation time and on the average durations of the HMM states.

The variances of the HMM are not adapted. It turned out in earlier investigations that the modification of the variances has only a minor influence on the improvement of the recognition performance [Gales 1995], [Hirsch 2001].

The effects of this adaptation approach are visualized in Fig. 16.13 in the spectral domain by comparing the spectral characteristics as they can be derived for three different HMMs, which was described in Sec. 16.2.2. The three plots show different HMM versions of the word "six". The spectrogram of the clean HMM is shown in Fig. 16.13-

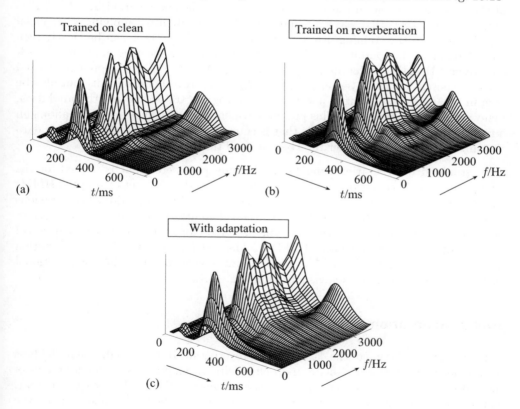

Figure 16.13: The spectral characteristics of different HMMs for the word "six"

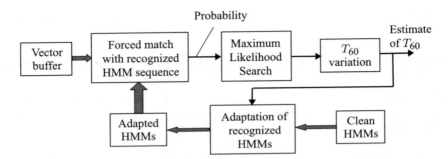

Figure 16.14: Estimation of T_{60} by an iterative search of the maximum likelihood

a as it was trained with the utterances of the TIDigits database. Only the utterances containing a single digit were taken for the training. The contributions of the fricatives at the end can be clearly seen in the high frequency region while the formants of the vowel are visible in the middle of the word. The spectrogram in Fig. 16.13-b represents the HMM that was trained on the TIDigits data after applying an artificial reverberation to the training utterances. The reverberation tails can be clearly seen when looking at the contours of individual Mel bands. The spectrogram in Fig. 16.13-c represents the HMM after adapting the clean HMM with the new approach. A fixed value is chosen for the reverberation time T_{60}. The reverberation tails can also be seen in this figure. Comparing it with the spectrogram trained on reverberated data, many similarities are visible. This provides evidence that the new approach allows an adaptation of the static parameters that is comparable to training the HMM on data that have been recorded under reverberant conditions.

The estimated reverberation time T_{60} is the only parameter that is needed for the adaptation. The recognition of an utterance is achieved with a set of adapted HMMs where the applied value of T_{60} has been estimated from the recognition of the previous utterance. T_{60} is estimated after the recognition of an utterance by a search for this set of adapted HMMs that leads to a maximum likelihood for another forced recognition of the already recognized sequence of HMMs. The restriction to the forced recognition of the already recognized HMM sequence is introduced to limit the computational costs. This iterative process is visualized in Fig. 16.14.

16.4.2 Adaptation of Delta Parameters

Comparing the contours of the clean and the reverberant HMM at individual Mel bins it becomes obvious that the delta and delta–delta parameters as time derivatives of the static parameters are also modified by the influence of the hands-free speech input. This can be seen, for example, in Fig. 16.13-a where a "valley" is visible between the vowel and the succeeding phoneme for the clean HMM. This "valley" is filled by the reverberation tails for the reverberant HMM versions. This indicates that the time derivatives will also be different in this region. The calculation of the delta parameters in the feature extraction is described in Sec. 16.2.1. We estimate the delta parameters of the reverberant speech by looking at the contours of the adapted static parameters from all HMM states. The average logarithmic frame energies of all states are considered. A spline interpolation according to (16.13) is applied to recreate the average energy contour at the frame rate of the feature extraction

$$\{\log_e \bar{E}_{ada}(S_1), \ldots, \log_e \bar{E}_{ada}(S_{n-1}), \log_e \bar{E}_{ada}(S_n), \log_e \bar{E}_{ada}(S_{n+1}), \ldots\} \underset{\text{Spline}}{\Longrightarrow}$$

$$\{\log_e \bar{E}_{ada}(0), \ldots, \log_e \bar{E}_{ada}[(k-1) \cdot d_{shift}], \log_e \bar{E}_{ada}[k \cdot d_{shift}],$$
$$\log_e \bar{E}_{ada}[(k+1) \cdot d_{shift}], \ldots\}. \quad (16.13)$$

In the same way an interpolated version of the average energy contour can be calculated for the average frame energies of the clean speech HMM. Applying (16.1) we get a set of average logarithmic delta energies $\Delta \log_e \bar{E}_{clean}[k \cdot d_{shift}]$ for the clean HMM as well as the set $\Delta \log_e \bar{E}_{ada}[k \cdot d_{shift}]$ for the adapted HMM. The number of energy values is dependent on the total length as modeled by the individual HMM. But the number is equal for the clean and the adapted HMM. Thus the differences between the clean and the adapted delta energies can be calculated with

$$\Delta \log_e \bar{E}_{diff}[k \cdot d_{shift}] = \Delta \log_e \bar{E}_{ada}[k \cdot d_{shift}] - \Delta \log_e \bar{E}_{clean}[k \cdot d_{shift}] \quad (16.14)$$

for all k and each frame at time $k \cdot d_{shift}$. These values describe the average differences between the delta logarithmic energies of the adapted and the clean HMM at each frame. By means of a spline interpolation the average differences are calculated for all HMM states according to

$$\{\Delta \log_e \bar{E}_{diff}(0), \ldots, \Delta \log_e \bar{E}_{diff}[(k-1) \cdot d_{shift}], \Delta \log_e \bar{E}_{diff}[k \cdot d_{shift}],$$
$$\Delta \log_e \bar{E}_{diff}[(k+1) \cdot d_{shift}], \ldots\} \underset{\text{Spline}}{\Longrightarrow}$$

$$\{\Delta \log_e \bar{E}_{diff}(S_1), \ldots, \Delta \log_e \bar{E}_{diff}(S_{n-1}), \Delta \log_e \bar{E}_{diff}(S_n),$$
$$\Delta \log_e \bar{E}_{diff}(S_{n+1}), \ldots\}. \quad (16.15)$$

A weighted version of these average differences is added to the corresponding delta parameters as contained in the clean HMM to create a set of adapted delta parameters according to

$$\Delta \log_e \tilde{E}[S_n, j] = \Delta \log_e E_{clean}[S_n, j] + \beta \cdot \Delta \log_e \bar{E}_{diff}[S_n], \quad 1 \leq j \leq N_{mix}. \quad (16.16)$$

This is done individually for each state S_n and for each mixture component with index j. A factor β is introduced for the weighted summation of the differences. During recognition experiments we found a value of 0.7 for β to achieve the highest performance.

The delta cepstral parameters can be adapted in the same way. The average logarithmic Mel spectral values are taken as the basis as they can be calculated by (16.2) from the average cepstral coefficients for each HMM state. The spline interpolation

$$\{\log_e |\bar{X}_\mu(S_1)|, \ldots, \log_e |\bar{X}_\mu(S_{n-1})|, \log_e |\bar{X}_\mu(S_n)|, \log_e |\bar{X}_\mu(S_{n+1})|, \ldots\} \underset{\text{Spline}}{\Longrightarrow}$$

$$\{\log_e |\bar{X}_\mu(0)|, \ldots, \log_e |\bar{X}_\mu[(k-1) \cdot d_{\text{shift}}]|, \log_e |\bar{X}_\mu[k \cdot d_{\text{shift}}]|,$$
$$\log_e |\bar{X}_\mu[(k+1) \cdot d_{\text{shift}}]|, \ldots\}, \quad 1 \leq \mu \leq K_{\text{mel}} \quad (16.17)$$

is applied to recreate the contour of the logarithmic Mel magnitude spectral components $\log_e |\bar{X}_\mu[k \cdot d_{\text{shift}}]|$ in each Mel bin μ and for each frame with index k at the frame rate of the feature extraction. The logarithmic spectral domain seems to be the right domain for applying the spline interpolation even though the interpolation could also be immediately applied to the average cepstral parameters. The interpolated average logarithmic spectrum $\log_e(|\bar{\mathbf{X}}[k \cdot d_{\text{shift}}])$ is transformed to the cepstral domain for each frame with index k

$$\log_e(|\bar{\mathbf{X}}[k \cdot d_{\text{shift}}]|) \underset{\text{DCT}}{\Longrightarrow} \bar{\mathbf{C}}[k \cdot d_{\text{shift}}]. \quad (16.18)$$

The delta coefficients $\Delta\bar{\mathbf{C}}[k \cdot d_{\text{shift}}]$ are calculated for the contour of each individual average cepstral coefficient $\bar{C}_i$ and all frames. This is repeated for the clean as well as for the adapted HMM so that the difference between these two versions can be estimated

$$\Delta\bar{\mathbf{C}}_{\text{diff}}[k \cdot d_{\text{shift}}] = \Delta\bar{\mathbf{C}}_{\text{ada}}[k \cdot d_{\text{shift}}] - \Delta\bar{\mathbf{C}}_{\text{clean}}[k \cdot d_{\text{shift}}] \quad \text{for all } k. \quad (16.19)$$

These values describe the average differences between the delta cepstral coefficients of the adapted and the clean HMM at each frame. By means of a spline interpolation the average differences are calculated individually for each cepstral coefficient $\bar{C}_{\text{diff}_i}(S_n)$ for all HMM states according to

$$\{\Delta\bar{C}_{\text{diff}_i}(0), \ldots, \Delta\bar{C}_{\text{diff}_i}[(k-1) \cdot d_{\text{shift}}], \Delta\bar{C}_{\text{diff}_i}[k \cdot d_{\text{shift}}],$$
$$\Delta\bar{C}_{\text{diff}_i}[(k+1) \cdot d_{\text{shift}}], \ldots\} \underset{\text{Spline}}{\Longrightarrow}$$

$$\{\Delta\bar{C}_{\text{diff}_i}(S_1), \ldots, \Delta\bar{C}_{\text{diff}_i}(S_{n-1}), \Delta\bar{C}_{\text{diff}_i}(S_n), \Delta\bar{C}_{\text{diff}_i}(S_{n+1}), \ldots\}. \quad (16.20)$$

The adapted cepstral coefficients can be estimated by adding the differences as contained in vector $\Delta\bar{\mathbf{C}}_{\text{diff}}[S_n, j]$ to the delta coefficients of the clean HMM according to

$$\Delta\tilde{\mathbf{C}}[S_n, j] = \Delta\mathbf{C}_{\text{clean}}[S_n, j] + \beta \cdot \Delta\bar{\mathbf{C}}_{\text{diff}}[S_n], \quad 1 \leq j \leq N_{\text{mix}}. \quad (16.21)$$

This can be done individually for each HMM state S_n and each mixture component with index j. The value of β is the same as in (16.16) for the adaptation of the energy coefficients.

The delta–delta parameters can be adapted in the same way as the delta parameters. As described in Sec. 16.2.1 the delta–delta parameters are calculated from the delta parameters in the same way the delta parameters are determined from the static parameters. Thus the average delta–delta parameters can be calculated from the average delta energy and the average cepstral coefficients as they are determined in (16.13) and (16.18). Otherwise the adaptation of the delta–delta parameters is achieved as described by (16.14)–(16.16) and (16.19)–(16.21) just by substituting delta by delta–delta.

16.4.3 Recognition Experiments on Hands-Free Speech Input

Recognition experiments have been performed to validate the applicability of the new adaptation approaches and to quantify the improvements that can be achieved. A cepstral analysis scheme is applied that is similar to the one described in Sec. 16.2.1. Each feature vector contains the 13 cepstral coefficients C_0 to C_{12} and the logarithm of the energy parameter. C_0 is only needed to transform back the cepstral coefficients to the spectral domain as part of the adaptation process. C_0 is not used for the recognition. Furthermore, the delta and delta–delta parameters of C_1 to C_{12} and the energy are part of each feature vector. The TIDigits database is considered again. Gender dependent HMMs are trained for each word as described in Sec. 16.3.2. The adaptation is individually applied to each speech utterance when detecting the beginning of speech. The applied VAD (voice activity detector) is based on a detection of changes in the Mel magnitude spectrum [Hirsch, Ehrlicher 1995].

The word error rates are shown in Fig. 16.15 for the recognition of the clean TIDigits as well as two further versions where the recording in two rooms has been simulated. The results are presented for four conditions that differ in the type of feature extraction or the adaptation mode. The first condition is based on the application of the robust ETSI front-end [ETSI 2003a]. The cepstral analysis scheme as described before is applied for the three other conditions in combination with and without adaptation. Word error rates are presented for the three cases where the recognition is done:

- without any adaptation, or

- with adaptation of the static parameters only, or

- with adaptation of the static and the delta and delta–delta parameters.

The error rates for the robust ETSI front-end are higher in comparison to applying the cepstral analysis scheme without any adaptation. For the condition of a hands-free speech input in reverberant environments it looks as if the ETSI front-end does not work as efficiently as it does in the presence of background noise.

The application of the proposed adaptation methods leads to a considerable reduction of the error rates compared with the ETSI front-end as well as compared with the cepstral analysis scheme. As mentioned before, the adaptation is individually applied to each utterance when detecting the beginning of speech in this utterance. We observe that the additional adaptation of the delta parameters is very efficient and results in a further gain in recognition performance.

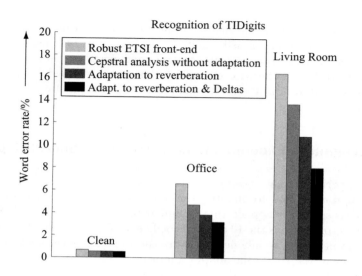

Figure 16.15: Word error rates for the recognition of the TIDigits in hands-free mode

The experiments include the recognition of utterances containing single digits and utterances containing sequences of digits. For the case of recognizing isolated words the adaptation approach as described in the previous section is correct and applicable without restrictions. But for the recognition of connected words the adaptation will not be perfect when the words are uttered without pauses between words. In this case the acoustic information at the beginning of a word is modified by the acoustic information at the end of the preceding word due to the reverberation. These "inter-word" modifications occur especially when sequences of words are spoken fluently with coarticulation effects. Looking at the sequences of the TIDigits the speaking rate varies considerably between speakers. Analyzing the recognition errors in more detail, it turns out that about half of the errors are due to deletions when recognizing the living room data with adaptation. In this case the adaptation of the first states of an HMM would need a knowledge of the acoustic information contained in the final states of the preceding word. But knowledge about the preceding word is not available in advance when recognizing sequences of digits. Because of this our approach as described in the previous section does not cover these "inter-word" interference.

Thinking about a phoneme based recognition using triphone models, some knowledge about the preceding phoneme model is available due to the property of triphones to model a phoneme dependent on the preceding and the succeeding phoneme. In this case the adaptation of a phoneme model can also include acoustic knowledge about the preceding and the succeeding phoneme. We could show by some further experiments that the modified adaptation can be successfully applied to the phoneme based recognition with triphone models.

16.4.4 Combined Adaptation to All Distortion Effects

The hands-free speech input in a room comes along with the recording of background noise as it is present in almost all applications of speech recognition systems. Furthermore, the spectrum of the speech is modified by the frequency characteristics of the microphone and of an additional transmission channel, e.g., when transmitting the speech via telephone to a remote recognition system. This creates the need to compensate also for these distortion effects. We developed an adaptation scheme [Hirsch 2001] that is based on the well known PMC (parallel model combination) approach [Gales, Young 1996]. This scheme consists of an adaptation of the static Mel frequency cepstral coefficients. The cepstral coefficients are transformed back to the Mel spectral domain where the adaptation can be realized by a multiplication with a frequency weighting function as an estimate for the frequency characteristics and by adding the estimated noise spectrum. The cepstral coefficients of all HMMs are individually adapted for each speech utterance when the beginning of speech is detected. Furthermore, the energy parameter can be adapted with an estimate of the noise energy. We present a short overview of the techniques for estimating the spectrum of the background noise and the frequency weighting function in the next section. Having obtained these estimates as well as an estimation of T_{60}, it will be shown that the earlier adaptation approach can be combined with the new method of adapting the spectra to a hands-free speech input.

Estimation of Distortion Parameters

The Mel spectrum of the background noise is estimated by looking at a smoothed version of the Mel magnitude spectrum $\mathbf{X}(k \cdot d_{\text{shift}})$ as calculated for the frame with index k in the feature extraction. The contour of the spectral magnitude values is smoothed in each Mel subband by applying a first order recursive filtering according to

$$|\mathbf{X}_{\text{smooth}}(k \cdot d_{\text{shift}})| = (1 - \alpha) \cdot |\mathbf{X}(k \cdot d_{\text{shift}})| + \alpha \cdot |\mathbf{X}_{\text{smooth}}((k-1) \cdot d_{\text{shift}})|. \quad (16.22)$$

α takes a value of 0.7 in our realization. A VAD (voice activity detector) is applied that takes the Mel spectra $\mathbf{X}(k \cdot d_{\text{shift}})$ as input. The onset of speech is detected when the estimated signal-to-noise ratios exceed an adaptive threshold in several subbands for a certain number of frames [Hirsch, Ehrlicher 1995], [Hirsch 2001]. When the start of speech is detected the noise spectrum is estimated as the smoothed spectrum of the last analysis frame with index k_{last} that is marked as a pause frame

$$|\hat{\mathbf{N}}| = |\mathbf{X}_{\text{smooth}}(k_{\text{last}} \cdot d_{\text{shift}})|. \quad (16.23)$$

Furthermore, the energy of the noise is estimated as the energy of the last pause frame

$$\hat{E}_{\text{noise}} = E(k_{\text{last}} \cdot d_{\text{shift}}). \quad (16.24)$$

The detection of speech onsets also triggers the adaptation of all HMMs. For the simulation experiments we take the acoustic parameters of all frames from a recorded utterance as input for the Viterbi recognition. For the real-time version of the recognizer as it is applied in a speech dialog system, we start the recognition process five frames before the first frame marked as speech. Thus, the Viterbi calculation can run almost in parallel with the feature extraction.

The frequency weighting function is estimated after the recognition of an utterance. It is applied for the recognition of the next utterance. This is based on the assumption that the frequency characteristics of the transmission from the speaker's mouth to the input of the recognizer will not change rapidly. Usually the microphone and the other transmission conditions do not change during a recognition session. The weighting function is estimated by comparing the long-term spectra of the noisy input speech with that of the clean speech. The sequence of HMM states is considered as it is available after the Viterbi match by backtracking the path with the highest likelihood. The long-term spectrum $\mathbf{X}_{\text{long}}$ of the noisy input speech is calculated for all K_{speech} analysis frames that are mapped on speech HMMs excluding the frames that are mapped on the pause model

$$|X_{\text{long}}| = \frac{1}{K_{\text{speech}}} \cdot \sum_{\text{speech frames}} |\mathbf{X}(k \cdot d_{\text{shift}})| \,. \tag{16.25}$$

In a similar way the long-term spectrum of the clean speech is estimated by looking at the spectral information contained in the HMM states on the path with highest likelihood. A set of adapted HMMs is used for the recognition. But for the estimation of the clean spectrum the spectral information is extracted from the corresponding clean HMMs. The cepstral coefficients of the corresponding clean HMM states are transformed back to the Mel spectral domain according to (16.2). In the case of HMMs with multiple mixture components, the spectrum of this mixture component with the smallest spectral distance to the corresponding spectrum of the input signal is taken. The long-term spectrum of the clean speech is estimated as the sum of all K_{speech} clean Mel spectra $|\hat{\mathbf{S}}(j)|$ from the corresponding HMM states on the best path

$$|\hat{\mathbf{S}}_{\text{long}}| = \left[\frac{1}{K_{\text{speech}}} \cdot \sum_{\text{best path}} |\hat{\mathbf{S}}(j)| \right] - |\mathbf{N}_{\text{sil}}| \,. \tag{16.26}$$

$|\mathbf{N}_{\text{sil}}|$ is the Mel spectrum that can be derived from the single state pause model. It contains the spectral information of the background noise that was present during the recording of the training data. In the case of clean training data, the spectrum $|\mathbf{N}_{\text{sil}}|$ takes only small values. It is subtracted here to compensate for its presence in the spectral parameters of all HMMs. In the rare case of getting a negative value after the subtraction the result is set to a fixed small positive value.

Subtracting the estimated noise spectrum $|\hat{\mathbf{N}}|$ as determined in (16.23) from the long-term spectrum of the noisy input speech, the frequency weighting function can be estimated by comparing the noise reduced input spectrum with the estimated clean spectrum as defined by

$$\mathbf{W} = \frac{|\mathbf{X}_{\text{long}}| - |\hat{\mathbf{N}}|}{|\hat{\mathbf{S}}_{\text{long}}|}.$$
(16.27)

It turned out in earlier investigations that this way of estimating the spectral difference between the input signal and the clean HMMs works well [Hirsch 2001]. By comparing the spectral information from the input signal and the clean HMMs, the weighting function not only contains the spectral characteristics of the recording equipment and the transmission line but also the frequency characteristics of the individual speaker to some extent.

In the same way the difference between the energy contours of the input speech and the best HMM sequence can be calculated. A weighting factor can be calculated that describes the average energy difference between the input signal and the energies contained in the sequence of HMM states on the best path. This factor contains information about the loudness of the individual speaker in comparison with the average energy contained in the HMMs.

Combined Adaptation Scheme

Having estimates for the noise spectrum, the frequency weighting function and the reverberation time, the Mel spectra of the clean HMMs are adapted as shown in Fig. 16.16.

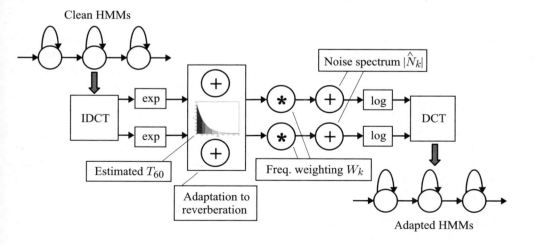

Figure 16.16: Scheme for adapting HMMs to all distortion effects

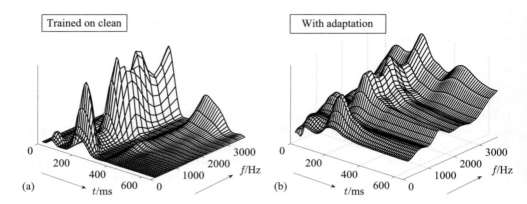

Figure 16.17: Spectral characteristics of the clean and the adapted HMMs for the word "six"

The cepstral coefficients of each state and mixture component are transformed back to the linear Mel spectrum for all clean HMMs. The Mel spectra are adapted to the estimated reverberation condition as described in Sec. 16.4.1. The estimated weighting function and the estimated noise spectrum are applied for further adaptation. This is done for each state S_n and each mixture component

$$|\hat{\mathbf{X}}(S_n, j)| = \mathbf{W} \cdot |\tilde{\mathbf{X}}(S_n, j)| + |\hat{\mathbf{N}}|. \tag{16.28}$$

The adapted Mel spectra $\hat{\mathbf{X}}(S_n, j)$ are transformed to the cepstral domain again. In the same way the energy parameter is adapted to reverberation first. Then it is adapted to the loudness of the individual speaker and the noise energy of the acoustic environment.

The adaptation to reverberation and noise is visualized by the three-dimensional spectral plots in Fig. 16.17 representing the spectral characteristics as they can be derived from two HMMs. The spectra shown in Fig. 16.17-a are calculated from the HMM of the word "six" trained on clean data. In Fig. 16.17-b the adapted version of this HMM is visualized. The adapted HMM was extracted during the recognition of artificially distorted TIDigits data. These data have been created from a simulation of the hands-free recording in a noisy living room environment. The noise spectrum as it is estimated for the individual input utterance becomes visible as a shift of the complete spectrogram. The reverberation tails can also be seen when looking at the contours along time in individual subbands.

16.4.5 Recognition Experiments on Hands-Free Speech Input in Noisy Environments

The new database "Aurora-5" as described in Sec. 16.3.1 is used for the evaluation of the combined adaptation approach. This database has been designed to contain the combinations of distortion effects as they might occur in practical applications. Thus it is well suited for investigations on the efficiency of the combined adaptation to several distortion effects. The results with and without the application of the adaptation are compared against the results when applying the robust ETSI front-end instead or when applying the well known MLLR (maximum likelihood linear regression) approach as alternative adaptation technique [Leggeter, Woodland 1995]. MLLR estimates a set of transformations for the mean and variance parameters of HMMs based on the availability of adaptation data. The goal is to obtain a reduction of the mismatch between an initial HMM set and the adaptation data. We apply the MLLR technique here in a similar way to that in our own adaptation method. This means that only the preceding utterance or several preceding utterances are applied for an unsupervised adaptation where it is not known what has been spoken.

The cepstral analysis scheme and gender dependent HMMs are applied as described in Sec. 16.4.3. The word error rates are presented in Fig. 16.18 for the three different versions containing car noise.

Looking at the condition with additive noise only, shown in Fig. 16.18-a, the expected improvement can be seen when comparing the results for the robust ETSI front-end against the results for a conventional cepstral analysis. Further small improvements are achieved when adapting the HMMs to all distortion effects. Furthermore, the error rates are shown for the unsupervised HMM adaptation with MLLR as it is available as part of the HTK Viterbi recognizer. An incremental MLLR is performed after each utterance. We observed a worse recognition performance when applying MLLR every two or more utterances. The adaptation is performed on the HMMs containing the features of the cepstral analysis so that the results can be immediately compared with the new adaptation approach. The error rates for MLLR are only a little bit worse when looking at the condition of additive noise only.

The improvement, comparing the new adaptation approach against the ETSI front-end, is greater when looking at the condition of a hands-free speech input in the noisy car environment. This is shown in Fig. 16.18-b. The reverberation time is fairly small in a car in comparison with rooms. The major impact of the hands-free recording inside a car is a modification of the frequency characteristics. The MLLR adaptation seems to compensate for these effects to a higher extent than the robust feature extraction except for the low SNR of 0 dB. The error rates for MLLR are again slightly worse.

The adaptation scheme shows its usability also for the case of an additional transmission over the GSM cellular network as shown by the results in Fig. 16.18-c. In this case the speech is further modified by the encoding and decoding and the transmission errors on the cellular channel. The adaptation technique seems to cover this type of distortion considerably better than the robust ETSI front-end. The performance of MLLR is extremely low for the SNR of 0 dB. This has been observed in several experiments where the performance without adaptation was already quite low. MLLR seems to be unable to find the right feature mapping in such cases and it seems to adapt the features in the wrong direction.

The cases with car noise do not include the major effects of a hands-free speech input in a reverberant room environment. The word error rates presented in Fig. 16.19 do include such effects. These experiments investigate recordings of speech inside a noisy room environment.

In the case of additive noise only, shown in Fig. 16.19-a, the new adaptation scheme leads to similar error rates to those for the robust front-end. In general the recognition performance is lower in comparison with the case with car noise because the interior noise signals contain more non-stationary segments. A considerable improvement is observed when comparing the new adaptation technique against the robust front-end for the cases of a hands-free speech input in an office or a living room as shown in Fig. 16.19-b and Fig. 16.19-c. The additional adaptation to reverberation causes this improvement.

MLLR adaptation leads to worse results, especially for SNRs below 15 dB. It looks as if the mapping on the basis of a linear regression is not able to completely compensate

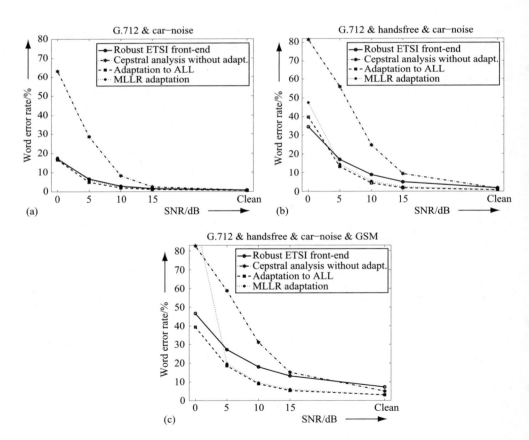

Figure 16.18: Word error rates for different recording conditions inside a car

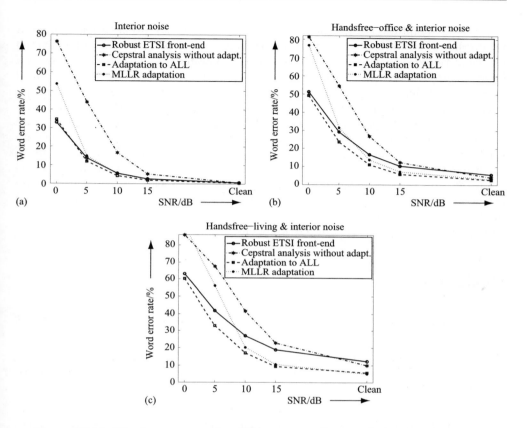

Figure 16.19: Word error rates for different recording conditions inside rooms

for the sum of spectral modifications caused by background noise and reverberation. While additive noise and a frequency weighting can be modeled as a stationary modification of each frame, reverberation includes modifications along the time axis. Both effects can not be completely compensated for with a linear mapping. As already observed for the car noise conditions, MLLR seems to adapt into the wrong direction in the case of a low performance without adaptation.

16.5 Conclusions

The problem of achieving a high recognition performance in adverse conditions has been addressed in this chapter. The speech input scenarios have been analyzed where speech recognition systems are applied. A simulation tool has been developed to simulate the different acoustic input conditions. The tool was used to create a new database for performing comparative recognition experiments on data that contain the distortion effects of realistic input scenarios. After a short overview about the existing approaches for improving the robustness a new method has been introduced that is based on an adaptation of HMMs. Especially new is the adaptation to a speech input in reverberant conditions. The static spectral and energy parameters can be adapted as well as the corresponding delta and delta–delta parameters. The new technique can be applied to the case of a hands-free speech input in rooms as well as to a combined occurrence of different distortion effects. An estimate of the stationary background noise, of the unknown frequency characteristic as well as of the reverberation time are needed as adaptation parameters. The background noise is estimated from the pause segment before the speech onset whereas some type of blind estimation techniques are used for the two other parameters. We could demonstrate with several recognition experiments on data from the newly created database that the recognition can be improved with this technique in comparison to applying a widely used robust feature extraction scheme. We could verify our results on artificially distorted data by running further experiments on data that have been recorded under noisy conditions, e.g., in hands-free mode in a reverberant environment.

Bibliography

Campos-Neto, S. F. (1999). The ITU-T Software Library, *International Journal of Speech Technology*, pp. 259–272.

Cooke, M.; Green, P.; Josifowski, L.; Vizinho, A. (2001). Robust Automatic Speech Recognition with Missing and Unreliable Acoustic Data, *Speech Communication*, vol. 34, pp. 267–285.

ETSI (2000). GSM 06.90: Digital Cellular Telecommunications Systems (Phase2+); Adaptive Multi Rate (AMR) Speech Transcoding, ETSI standard document EN 301 712 v7.2.1.

ETSI (2003a). Speech Processing, Transmission and Quality Aspects (STQ); Distributed Speech Recognition; Advanced Front-End Feature Extraction Algorithm; Compression Algorithm, ETSI standard document ES 202 050 v1.1.3.

ETSI (2003b). Speech Processing, Transmission and Quality Aspects (STQ); Distributed Speech Recognition; Front-End Feature Extraction Algorithm; Compression Algorithm, ETSI standard document ES 201 108 v1.1.3.

Finster, H. (2005). *Web Interface to Experience the Simulation of Acoustic Scenarios*, http://dnt.kr.hsnr.de/sireac.html.

Gales, M. J. F. (1995). *Model Based Techniques for Noise Robust Speech Recognition*, PhD thesis, University of Cambridge, Great Britain.

Gales, M. J. F.; Young, S. J. (1996). Robust Continuous Speech Recognition Using Parallel Model Combination, *IEEE Transactions on Speech and Audio Processing*, vol. 4, pp. 352–359.

Gauvain, J. L.; Lee, C. H. (1994). Maximum a Posteriori Estimation for Multivariate Gaussian Mixture Observations of Markov Chains, *IEEE Transactions on Speech and Audio Processing*, vol. 2, pp. 291–298.

Hirsch, H. G. (2001). HMM Adaptation for Applications in Telecommunication, *Speech Communication*, vol. 34, pp. 127–139.

Hirsch, H. G. (2002). The Influence of Speech Coding on Recognition Performance in Telecommunication Networks, *Proceedings of the International Conference on Spoken Language Processing (ICSLP)*, pp. 1877–1880.

Hirsch, H. G.; Ehrlicher, C. (1995). Noise Estimation Techniques for Robust Speech Recognition, *Proceedings of the IEEE International Conference on Acoustics, Speech, and Signal Processing (ICASSP)*, pp. 153–156.

Hirsch, H. G.; Pearce, D. (2000). The Aurora Experimental Framework for the Performance Evaluation of Speech Recognition Systems under Noisy Conditions, *Proceedings of the ISCA workshop ASR2000*, Paris, France.

Hirsch, H. G.; Pearce, D. (2006). *The Aurora Project*, http://aurora.hsnr.de.

Houtgast, T.; Steeneken, H. J. M.; Plomp, R. (1980). Predicting Speech Intelligibility in Rooms from the Modulation Transfer Function, I. General Room Acoustics, *Acustica*, vol. 46, pp. 60–72.

ITU (1993). Telephone Transmission Quality Objective Measuring Apparatus: Objective Measurement of Active Speech Level, ITU-T recommendation P.56.

Jelinek, F. (1998). *Statistical Methods for Speech Recognition*, MIT Press.

Junqua, J. (2000). *Robust Speech Recognition in Embedded Systems and PC Applications*, Kluwer Academic Publisher.

Kuttruff, H. (2000). *Room Acoustics*, 4th edn, Spon Press.

Leggeter, C.; Woodland, P. (1995). Maximum Likelihood Linear Regression for Speaker Adaptation of Continuous Density Hidden Markov Models, *Computer Speech and Language*, vol. 9, pp. 171–185.

Leonard, R. (1984). A Database for Speaker-Independent Digit Recognition, *Proceedings of the IEEE International Conference on Acoustics, Speech, and Signal Processing (ICASSP)*, vol. 3.

Machoa, D.; Mauuary, L.; Pearce, D.; et. al. (2002). Evaluation of a Noise Robust DSR Front-end on Aurora Databases, *Proceedings of the International Conference on Spoken Language Processing (ICSLP)*, pp. 17–20.

Mauuary, L. (1998). Blind Equalization in the Cepstral Domain for Robust Telephone Based Speech Recognition, *Proceedings of the European Signal Processing Conference (EUSIPCO)*, pp. 359–362.

Minami, Y.; Furui, S. (1996). Adaptation Method Based on HMM Composition and EM Algorithm, *Proceedings of the IEEE International Conference on Acoustics, Speech, and Signal Processing (ICASSP)*, pp. 327–330.

Peinado, A.; Segura, J. (2006). *Speech Recognition over Digital Channels; Robustness and Standards*, John Wiley & Sons, Ltd.

Rabiner, L.; Juang, B. H. (1993). *Fundamentals of Speech Recognition*, Prentice Hall.

Sankar, A. J.; Lee, C. H. (1996). A Maximum Likelihood Approach to Stochastic Matching for Robust Speech Recognition, *IEEE Transactions on Speech and Audio Processing*, pp. 190–201.

Vary, P.; Martin, R. (2006). *Digital Speech Transmission. Enhancement, Coding and Error Concealment*, John Wiley & Sons, Ltd.

Woodland, P. C. (2001). Speaker Adaptation for Continuous Density HMMs: A Review, *Proceedings of the Int. Workshop on Adaptation Methods for Speech Recognition*, Sophia Antipolis, France.

Young, S.; et. al. (2005). *The HTK Book (Version 3.3)*, Cambridge University Engineering Department, http://htk.eng.cam.ac.uk.

Zwicker, E.; Fastl, H. (1999). *Psychoacoustics*, 2nd edn, Springer Verlag, Berlin.

Chapter 17

Speaker Classification for Next-Generation Voice-Dialog Systems

Felix Burkhardt, Florian Metze, Joachim Stegmann

17.1 Introduction

Customer Relationship Management (CRM) is a growing business factor for medium and large enterprises. For cost reduction, the automation of business processes in call centers based on Interactive-Voice-Response (IVR) systems has been introduced in many companies. In state-of-the-art IVR systems automation based on automatic speech recognition (ASR) is mainly used for pre-qualifying of customers' requests with subsequent skill-based routing to a human agent or complete automation of simple business processes such as checking of account balances or tariff changes.

These automated voice-dialog systems are currently not adapted to the preferences or needs of specific user groups. For market success of new voice-controlled value-added services as well as for increased usability and efficiency in process automation, it is important to divide customers into specific target groups with a tailored adaptation of the respective voice dialogs. However, personalization of voice dialogs in state-of-the-art IVR systems can be performed only if the caller is known and has been authenticated by the system. In many applications and services, the information about the identity of the caller is not available. Additionally, time-variant features such as the emotional state of the caller are not known and can not be utilized by the system.

Advances in Digital Speech Transmission   Edited by R. Martin, U. Heute and C. Antweiler

One solution is to introduce automatic speaker classification into voice-dialog systems. In a first step, time-invariant or slowly time-varying features such as gender and age can be detected from the first utterances of the caller in a voice dialog. This information can then be used to assign the caller to a specific target group and switch the dialog parameters accordingly. Additionally, time-varying features of the caller such as anger can be monitored during the entire dialog. In the case of any problems within the dialog, the system can help the customer by either offering the assistance of human operators or trying to react with appropriate de-escalating dialog strategies.

This chapter is organized as follows: Section 17.2 explains the basics of speaker classification, gives an overview of the relevant classification algorithms, and describes methods for their evaluation. Then, Sec. 17.3 discusses the classification of age and gender in detail, while Sec. 17.4 focuses on anger detection. In Sec. 17.5 examples for applications in the area of telecommunications are described and results of corresponding usability evaluations are discussed. Finally, Sec. 17.6 gives a conclusion and shows possible directions for future work.

17.2 Speaker Classification

17.2.1 Overview

Speaker *classification* is concerned with assigning every test speaker to a given class or group of speakers. In our case, the decision is taken on the basis of sample acoustic data [Müller, Schötz 2007]. To support the decision, knowledge about the characteristics of different classes needs to be learned from training data and compiled into rules or models.

Similar problems are given by the task of speaker *identification*, which tries to identify an individual single speaker (and not only assign him or her to a group of speakers), while speaker *verification* tries to ascertain a given individual's identity [Reynolds 2002]. These techniques, although important in practice, will not be discussed here.

As the aim of this chapter is to demonstrate the most important principles of speaker classification, we content ourselves with providing pointers to other related work such as the automatic detection of stress and fraud [Board for Professional and Occupational Regulation 2003], intoxicated speech [Tanner, Tanner 2004], or multi-lingual speech technologies [Schultz, Kirchhoff 2006].

Figure 17.1 shows example criteria employed for speaker classification and attempts to structure them according to the degree of variability over time: the result of a classification can be time-invariant, constant within a dialog, or variable within a dialog.

Speaker classification as the task of assigning measurements of speaker characteristics to a class can therefore be seen as a pattern recognition problem: following [Duda et al.

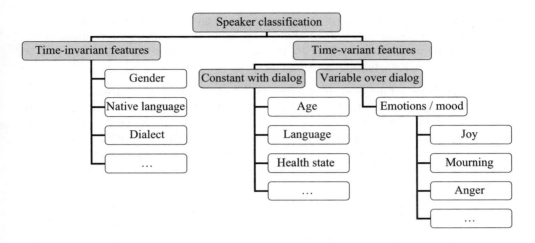

Figure 17.1: Taxonomy of classification criteria for speaker classification: speaker properties can be immutable (e.g., *gender*), constant within an interaction (e.g., *age*), or change rapidly (e.g., *emotion*)

2000], this necessitates the steps of 1) data collection, 2) feature selection, 3) model selection, 4) training, and 5) evaluation. Section 17.2.2 will therefore discuss data collection and particularly feature selection, Sec. 17.2.3 will present the most relevant model selection and training algorithms, and Sec. 17.2.4 will give an overview on techniques allowing us to compare or evaluate systems.

17.2.2 Feature Extraction

Algorithms are usually based on the transformation of the digital input signal into a parametric representation that is better suited for classic machine learning techniques than a digitized wave-form.

Following [Reynolds et al. 2003], the features used to describe speech signals can be looked at from different levels of abstraction if one follows a development of longer temporal segments and higher linguistic abstraction as in the following list: acoustic – prosodic – phonotactic – idiolectal – dialogic – semantic.

Of course, some categories are better suited for certain classifications than others; for example, [Shafran et al. 2003] report that, in contrast to emotion detection, gender, age and dialect detection was not enhanced by regarding word-based features, while acoustic- and pitch-based features were useful.

Generally spoken, these features can be used in two ways: On the one hand in comparison with a "world model", i.e., a classifier trained with samples from the desired groups, on the other hand as a deviation from an assumed "neutral" state determined from the user beforehand. The latter possibility, of course, only makes sense for time-variant criteria such as, for example, emotional state, and assumes that the speaker is known beforehand. For unknown speakers comparison with a dialog situation where the user is supposed to be neutral, e.g., the beginning of a dialog, can be used.

Acoustic features: The basis of computing acoustic features is usually the computation of the Mel frequency cepstral coefficients (MFCCs), which are primarily known as the basis for automatic speech recognition. A MFCC vector encodes the spectral properties, computed by a fast Fourier transformation (FFT), of a small frame of speech (about 10–20 ms) and translated to the Mel frequency scale in order to account for the near logarithmic resolution of human hearing [Rabiner 1978].

The acoustic features can be measured on a sub-phonemic level and measure different features such as, for example, jitter (the micro variation of fundamental frequency), and shimmer (the micro variation of amplitude or the Harmonics-to-Noise ratio (HNR), which gives the proportion between harmonic and random signal parts [Müller 2005]). Also measures connected with the position of formants fall into that category.

Prosodic features: On a higher level, pitch, duration, and intensity features can be subsumed as prosodic because they are related to the rhythmic and melodic structure of the speech. Just like the acoustic features, they can be the basis for global values such as the mean, maximum, minimum, or standard deviation. Also values that describe their contour, such as regression coefficients or the position of a maximum or a minimum on the time axis, are used. If no phonetic analysis is done, duration measures are often computed for successive voiced respectively, unvoiced parts of speech. One example is the work described in [Burkhardt et al. 2005a] where the anger detector is based primarily on prosodic features.

Phonotactic features: A phonotactic analysis presupposes a phonetic classification. Based on that the probability of the speech sample belonging to the phonetic set as well as pronunciation variants can be computed. These set of features are primarily known to be used with language or dialect detection for obvious reasons but might also be useful for other criteria like, e.g., emotion detection where a change in articulation is a known effect [Kienast et al. 1999].

Idiolectal features: These features are based on word recognition and can make predictions on the use of certain words by speaker groups. Although they are obviously used by humans in order to assign a speaker to a certain age or social group, automatic detection is difficult because a very-large-vocabulary speaker-independent speech recognizer is not yet realistic.

An example for the analysis of idiolectal features given a limited vocabulary is described by [Lee, Narayanan 2005]. The reported computation of the "emotional salience" of each recognized word helped emotion detection in that case.

Dialogic features: On the dialogic level, features regarding the interaction between human and machine can be measured, e.g., the average length of turn, the frequency of interruption, or even dialog specific issues such as the use of help mechanisms, etc. Such features are described extensively in [Walker et al. 2002].

Semantic features: Semantic models, finally, are used to detect what the speaker's intention is. The features of this layer may also be used to support speaker classification. In the context of an automated voice portal, a scenario for the use of semantic features would be, for example, to take an interest in a certain product as input for the classifier. Another example is described in [Ang et al. 2002]. In this study, one of the features under investigation consisted of the number of requests for repetition during a dialog.

17.2.3 Classification Algorithms

A classifier can formally be described as a set of *discriminant functions* $g_i(x)$, $i = 1, \ldots, c$, with c being the number of classes [Duda et al. 2000]. A classifier $x \mapsto c(x)$ can then be described as a network in which c discriminant functions are being evaluated and the input vector x is being assigned to the class i resulting in the highest discriminant function:

$$x \mapsto \operatorname*{argmax}_{i} g_i(x) \,. \tag{17.1}$$

The "Bayes error" is the lowest possible probability of a misclassification and is reached by an optimal classifier. Virtually all current approaches use data-driven approaches, in which training data is used to first build a statistical model for the domain (training phase). The model is then used to classify unknown data into one of the target classes (test or evaluation phase). In many cases, a successful approach will rely on a combination of individual classifiers on different features. To classify men and women, one would collect labeled speech data from both sexes, train independent classifiers on several features that can be extracted from the speech signal, for example F_0 (fundamental frequency) and the harmonics-to-noise ratio, and then train a "meta-classifier" to combine the decision of the individual classifiers into a final decision on a held-out data set. For most applications, this is easier and more flexible than training a single multi-dimensional classifier.

Classifiers can be defined for items of *fixed* length (e.g., the fundamental frequency value for a given point in time) or of *variable* length (e.g., a whole utterance). In the latter case, the overall decision can be reached either by computing some statistical derivative of the observation (e.g., the mean) over the duration of the utterance, or by using more elaborate classifiers, which take into account interdependencies of individual measurements (classifications), for example, temporal or sequential structures. Only if observations are known to be independent, can they be combined using a naive Bayes classifier [Domingos, Pazzani 1997]. In this chapter, we shall restrict ourselves to presenting algorithms using the following types of classifiers.

Principal Component Analysis (PCA) and *Linear Discriminant Analysis* [Fukunaga 1990] are frequently used for simple problems, or as a pre-processing step for more elaborate classifiers, as they improve the separation of classes without significant extra cost during recognition, for example, by rotating the input vectors so that their components become linearly independent. This improves the "separability" for other classification algorithms, by aligning the class separation or "decision boundary" with the model's assumptions.

k-Nearest Neighbor (kNN) classifiers [Fukunaga 1990]: An input vector x is assigned to class c if c is the most frequent class label among the k training samples nearest to the incoming data sample x. While simple to realize, this algorithm is often impractical, because it requires the storage of a large number of individual training samples, instead of compacting them into some kind of statistical model.

Gaussian Mixture Model (GMMs) [Redner, Walker 1984] classifiers build a statistical model for the data using mixtures of Gaussian probability density functions (PDFs). The advantage is that one does not need to store a large selection of data points as with kNN, but only a statistical model of the data. This usually generalizes better than kNN but, of course, the model has to be learned during the training phase. A decision is taken by computing the class-specific PDFs on a test vector and assigning the sample vector to the class with the highest probability, possibly also taking into account a prior distribution over the classes.

Artificial Neural Networks (ANNs) [Duda et al. 2000], for example, Multi-Layer Perceptrons (MLPs), can be used for classification just as for GMMs, but they offer a variety of training methods inspired by nature, which can incorporate discriminative information very efficiently. As some of the training procedures can learn and enhance discriminative features, ANNs are a very powerful tool. Neural networks are often used to combine individual, simpler classifiers and the resulting structure is sometimes referred to as a "hierarchy of experts".

Support Vector Machines (SVMs) [Schölkopf, Smola 2002] have recently gained much interest as a powerful type of classifiers using "kernel" functions to allow one to use virtually unconstrained decision boundaries for classification. Kernel functions provide for a non-linear mapping of the training data into a higher-dimensional feature space, in which it is possible to separate the data using simple classifiers, e.g., a simple linear hyperplane.

Classification and Regression Trees (CART) can be applied to individual features or for combining classifiers [Breiman et al. 1984]. Basically, classification trees define a tree structure in which questions about properties are asked and the answers are used to put the input data into bins. This frequently also allows for visual analysis of dependencies between features and diagnosis of problems.

(Continuous Density) Hidden Markov Models ((CD-)HMMs) [Rabiner 1989] provide for an appropriate stochastic model, if the individual measurements form a time series with well-defined transition probabilities between individual states of the sequence. They offer a principled approach to model dependencies between specific states. Derivatives thereof, namely phoneme-recognition-based classifiers, e.g., Parallel Phone Recognizers (PPRs), are usually used in combination with Viterbi decoding [Jelinek 1998]. These allow one to model individual data points (for example using GMMs as

"continuous densities") together with their temporal or physical structure as a series of stochastic events with defined transition probabilities.

Dynamic Bayesian Networks (DBNs) are a standard approach to model dependencies between certain observations in the decision process in a principled way and have been used successfully in many speech-related applications [Zweig, Russell 1998]. The approach is to explicitly model known or learned assumptions about feature dependencies instead of assuming features to be independent.

The best choice of classifier(s) for a given problem depends on the maximum amount of training data available, the time and memory available during evaluation of data sets, and the experience of the experimenter.

17.2.4 Evaluation of Classifiers

For evaluation purposes, independent test data from the same domain, i.e., additional data that was collected under identical conditions as the training data and processed identically, needs to be available. In some cases, particularly for classifier combination approaches, if test data were used to optimize certain parameters of the classification, the true performance of the system can only be determined on a cross-validation set, which is entirely independent of all the other sets.

Experimental results can be tabulated in a confusion matrix. The confusion matrix is a full table of all possible outcomes of an experiment. It contains all information on how many items of each input category were assigned to what output category. As an example, the diagonal elements in Table 17.1 contain the "good" results (hits), while the off-diagonal elements contain confusions. Good classifiers have low counts in the off-diagonal elements.

In the literature, classification performance is often given as *recall* and *precision* values. The recall Rec_A of a class A means the ratio of correctly identified occurrences C_A to the number of instances T_A in the reference, while the precision is given as the ratio of the number of correctly predicted cases C_A to the total number P_A of occurrences predicted. If the experiment is set so that $\Sigma_i T_i = \Sigma_i P_i := N$ and $\Sigma_i C_i := C$,

Table 17.1: An example confusion matrix: all four lemons are correctly identified, while one pear and three apples are mis-classified as lemons

Reference	Hypothesis			Sum
	Apple	Pear	Lemon	
Apple	10	5	3	18
Pear	2	8	1	11
Lemon	0	0	4	4
Sum	12	13	8	

the overall ratio of correctly identified cases to the number of experiments is usually referred to as the *accuracy* Acc:

$$\text{Rec}_A = \frac{C_A}{T_A} \tag{17.2}$$

$$\text{Prec}_A = \frac{C_A}{P_A} \tag{17.3}$$

$$\text{Acc} = \frac{C}{N}. \tag{17.4}$$

In the above example,

$$\text{Prec}_{\text{Apple}} = 10/(10 + 5 + 3) = 0.56, \tag{17.5}$$

while

$$\text{Rec}_{\text{Apple}} = \frac{10}{10 + 2 + 0} = 0.83. \tag{17.6}$$

The overall *accuracy* of the classifier is given by

$$\text{Acc} = \frac{10 + 8 + 4}{12 + 13 + 8} = \frac{10 + 8 + 4}{18 + 11 + 4} = 0.67 \tag{17.7}$$

for $i \in \{\text{Apple}, \text{Pear}, \text{Lemon}\}$. This number, however, may not be very meaningful if rare classes need to be detected reliably as well.

To express precision and recall as a single number or *figure of merit*, the so-called *F-measure* [van Rijsbergen 1979] can be employed. It corresponds to the weighted harmonic mean of precision and recall, and is defined as

$$F_\alpha = (1 + \alpha)\frac{\text{Prec}_A \cdot \text{Rec}_A}{\alpha \cdot \text{Prec}_A + \text{Rec}_a} \tag{17.8}$$

for $\alpha > 0$. We then have $F_\alpha \in [0, 1]$. If $\alpha = 1$, this number is known as the traditional *F*-measure or balanced *F*-score. This is also known as the F_1 measure, because recall and precision are evenly weighted. Two other commonly used *F* measures are the F_2 measure, which weights recall twice as much as precision, and the $F_{0.5}$ measure, which weights precision twice as much as recall. $\alpha \neq 1$ will be used, if different types of error are differently severe or "expensive". For example, it may be more acceptable to classify an adult as a child and connect him or her to an operator, than to admit a child to adult-only services.

Depending on the amount of context (visual information, semantic information, etc.) available, age and gender classification based on acoustic information alone can be very challenging for people, too. For the work presented in this chapter, the annotation of data can therefore be a difficult problem on its own: inter-labeler agreement (i.e., the degree to which the annotation achieved by one individual will be reproduced by another annotator) can be low and the Human baseline for a task can have a significant error rate, too. Section 17.3 provides a comparison of Human performance with automatic performance on the age/ gender task. In this case, ground truth was established by asking the speakers during data collection for their age and gender.

17.3 Detection of Age and Gender

17.3.1 Background

This section will give an overview about age and gender classification algorithms that can be used in telephony-based applications, present two different approaches to classification of age and gender using Human speech over telephone channels, evaluate them, and compare their performance with a Human baseline.

While research on the general influence of speaker age on voice characteristics has been carried out since the late 1950s [Mysak 1959], [Linville 2001], the first systems that could automatically estimate the age and gender of a speaker have been developed only recently [Müller et al. 2003], [Minematsu et al. 2002], [Shafran et al. 2003], [Schötz 2004]. The quality of these systems is difficult to compare, however, as the type of speech material used as well as the age classes considered are not consistent in the literature.

As studies show that the dialog strategies employed in IVR systems can be adapted to age and gender of the caller [Hempel 2006], there is currently increasing interest in these algorithms in order to improve overall service quality.

The results presented in this section were achieved in an evaluation experiment that used controlled procedures in order to compare several approaches to age and gender recognition under fair conditions [Metze et al. 2007]. The numbers reported were achieved after the systems were optimized on common training and development test data sets in a one month time window.

The systems use the following seven age and gender groups and labels:

- children: ≤ 13 years (C),

- young people: 14–19 years, male (YM) and female (YF),

- adults: 20–64 years, male (AM) and female (AF),

- seniors: ≥ 65 years, male (SM) and female (SF).

While somewhat arbitrary, these classes stem from an IVR application currently under development. Data were taken from the German SpeechDat II corpus [Höge et al. 1999], which is annotated with age and gender labels as given by callers at the time of recording. This database consists of 4000 native German speakers, who called a recording system over the telephone and read a set of numbers, words and sentences. For each age/gender group, the data of about 80 individuals were selected for training and about 20 for testing purpose. Children and senior groups where slightly under-represented but all in all a weighted age and gender structure was achieved. Training data consisted of the whole utterance set of each person, up to 44 utterances.

For further analysis, the test set was partitioned into a sub-set of short utterances "SpeechDat_short" (SpeechDat II corpus identifiers "a" and "o": command words and city names) and another set of longer sentences "SpeechDat_long" (identifier "s").

17.3.2 Algorithms

While many different types of algorithms have been tried on the task of detection of age and gender, we choose to present two exemplary algorithms because they performed best in the above-mentioned evaluation and also because they demonstrate how similar results can be achieved using very different approaches.

Segment-based approach

This system was derived from an existing approach to age and gender classification with an overall development effort of approximately three years. The system architecture is shown in Fig. 17.2.

For classification of audio data, a 17-dimensional feature vector is computed on whole segments of speech. The parameters included in this vector are 1) **jitter** (micro-variations of the F0 frequency); 2) **shimmer** (micro-variations of the F0 amplitude), for each of which multiple algorithms were used including the Relative Average Perturbation (RAP) and the Period Perturbation Quotient (PPQ) for jitter as well as the three-, five- and eleven-point Amplitude Perturbation Quotient (APQ) for shimmer [Baken, Orlikoff 2000]; 3) the mean and the standard deviation of the

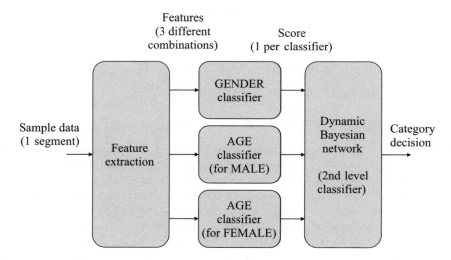

Figure 17.2: Example age/gender detection system based on segmental features and a Dynamic Bayesian Network to combine individual decisions

harmonics-to-noise-ratio [Ferrand 2002]; 4) several statistical derivatives of the **fundamental frequency** (F_0) including mean, standard deviation, and mean average slope (MAS).

A 17-dimensional Gaussian model was trained for each individual class and the resulting distributions were analyzed manually to rate their respective discriminative power. On the basis of this analysis, three classifiers were constructed that could determine

- the gender (male or female) of the speaker,
- the age class under the assumption that the speaker is male, and
- the age class under the assumption that the speaker is female.

These classifiers C_1, C_2, C_3 constitute the *first layer* of a two-level structure and were each trained on a different combination of initial features. Multi-layer Perceptron Networks (MLPs) with one hidden layer and sigmoid activation functions were used. The number of hidden units N corresponded to $N = (f + c + 1)/2$, where f is the number of input units (features) and c is the number of output units (classes) [Müller 2005].

The *second layer* performs post processing on the initial classification results using Dynamic Bayesian Networks (DBNs).

The DBN is primarily used to model the classification-inherent uncertainty by introducing: 1) three observable nodes O_1, O_2, O_3, representing the result of one classifier C_i with states corresponding to the classes (e.g., MALE and FEMALE); 2) the nodes AGE and GENDER representing the actual speaker class; 3) the links between O_1, O_2, O_3 and AGE/ GENDER representing a causal relationship. In this structure, the uncertainty of that relationship is modeled in terms of the conditional probability table (CPT) attached to the nodes O_i. The CPT-values were optimized on a cross-validation set of the respective classifier C_i.

At the same time, the DBN fuses the results of the classifiers by letting the nodes AGE and GENDER both be parents of each O_i. Appropriate CPTs then provide the precedence of one age classifier over the other depending on the result of the gender classifier.

In real-life applications, the third function of the second-layer DBN is to successively improve the model when sequentially processing multiple utterances of the same speaker. This is provided by $1 : 1$ transitions from AGE_{n-1} and GENDER_{n-1} to AGE_n and GENDER_n nodes. This feature, however, could not be employed in the experiment presented here, as no speaker IDs were available to the system.

Frame-based approach

The underlying system was originally developed for Automatic Speech Recognition (ASR) and automatic acoustic Language Identification (LID). It is based on

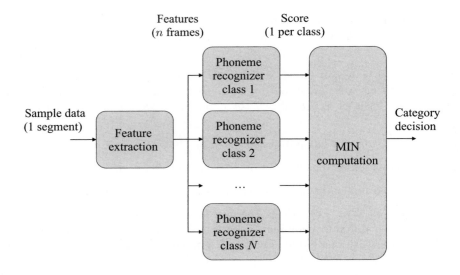

Figure 17.3: Example age/gender system based on Parallel Phoneme Recognizers (PPR) for three categories, e.g., the "frame-based" approach. The "MIN" decision selects the category pertaining to the phoneme recognizer yielding the best (lowest) score

Parallel Phoneme Recognizers (PPR) using Continuous Density Hidden Markov Models (CD-HMMs) and phoneme bi-grams to model the transition probabilities between individual states of the HMM. Feature extraction consists of the computation of Mel frequency cepstral coefficients (MFCCs) and a linear transformation based on Linear Discriminant Analysis (LDA), retaining 24 components for the final feature vectors. Figure 17.3 explains the system architecture.

During recognizer development, a specific phoneme recognizer for each of the seven age/gender categories with class-specific HMM and phoneme bi-gram was trained on the respective sub-set of the training data. To build the PPR system, we first created category-specific mono-phone HMMs using maximum likelihood estimation as used for standard ASR system generation. For the following LDA, the category specific mono-phone HMM states served as the LDA classes. Based on the retained LDA matrix optimized for age/ gender classification, the final category specific mono-phone HMMs were built. In a final step the phoneme recognizers were applied to the training material to estimate category-specific phoneme bi-grams also based on the maximum likelihood criterion.

During recognition, negative log-likelihood scores for each category are computed using a Viterbi decoder. In a final step, the classified category is determined by choosing the category with the minimum score once all the frames have been evaluated. This structure is shown in Fig. 17.3.

17.3.3 Results

The results reached with these approaches in our evaluation are shown in Table 17.2. Accuracy on SpeechDat II ranges between 40 % and 54 % when distinguishing all seven classes, while recall is between 52 % and 55 %.

The "frame-based" recognizer using class-specific phone recognizers reaches the best performance and also shows the most balanced confusion matrix. Performance, however, drops for the "short" utterances, presumably due to the temporal structure realized in the phone bi-grams. The approach based on a combination of features, on the other hand, is based on multiple prosodic features computed on the entire segment, and its accuracy shows very little dependence on the length of the utterance.

The results of a baseline experiment involving Human listeners labeling the data using the same classes are shown in Table 17.3. For this experiment, over 54 members of five different speech research laboratories listened to 100 randomly chosen audio files each over headphones and annotated them, covering about 85 % of the evaluation corpus.

The overall classification accuracy on the (near complete) SpeechDat II evaluation set is 55 %, with a precision of 69 % (see Table 17.2). Comparing automatic and Human results, the performance of the best automatic system is not too far behind the Human performance, although the recall is significantly lower. The difference between long and short sentences also exists for Human labelers, although Human labelers do not perform that much worse on short sentences. The F-measure for the Human baseline experiment is $F_1 = 0.61$, while the automatic approaches reach $F_{1,Segment} = 0.45$ and $F_{1,Frame} = 0.54$.

Assuming that age estimation should be robust across languages [Braun, Cerrato 1999], these results can be compared with other results on telephony speech [Cerrato et al. 2000], and the same "centralization" trend for the perceived age and a similar performance of our Human labelers on longer utterances can be found, even though the average sentence length of SpeechDat_long utterances is below the 40 s measured in [Cerrato et al. 2000].

Table 17.2: Precision (left) and recall (right) on the different data sets for the individual systems. The "frame-based" system is based on CD-HMMs and Parallel Phone Recognizers, while the "segment-based" system uses a Neural Network to combine several segment-level features

Approach	SpeechDat II		SD_short		SD_long	
Frame-based	54 %	55 %	45 %	46 %	61 %	61 %
Segment-based	40 %	52 %	38 %	51 %	42 %	62 %
Human baseline	55 %	69 %	51 %	67 %	60 %	73 %

Table 17.3: Confusion matrix of Human comparison experiment on SpeechDat II. Class symbols are defined in Sec. 17.3.1; columns contain hypothesized (estimated) classes, rows contain reference (true) classes

Reference	Hypothesis							Sum
	C	YM	YF	AM	AF	SM	SF	
C	164	69	139	33	44	2	1	452
YM	5	295	2	411	12	4	3	732
YF	38	34	413	4	238	0	5	732
AM	0	31	0	631	1	59	0	722
AF	5	2	33	4	658	0	54	756
SM	0	10	0	342	4	268	5	629
SF	2	2	13	10	538	5	287	857
Sum	214	443	600	1435	1495	338	355	

A "majority voting" combination approach did not improve the performance on SpeechDat II data. This may be an indication that, despite the different approaches implemented, the systems' errors are highly correlated, resulting in similar confusion matrices. However, the segment-based approach can be improved by adding more individual features, and initial experiments in that direction show promising results.

17.4 Detection of Anger

17.4.1 Background

Emotion-aware voice portals are one of the most prominent application ideas for the monitoring of emotional speech. Voice portals could use detection of negative feelings such as anger to appease the users by mirroring their expressions or to collect statistical data for quality measurement [Burkhardt et al. 2005a], [Yacoub et al. 2003], [Shafran et al. 2003]. In the context of customer care voice portals, it can be helpful to detect potential problems that arise from an unsatisfactory course of interaction in order to help the customer by either offering the assistance of a human operator or trying to react with appropriate dialog strategies. Some of these strategies are described in [Burkhardt et al. 2005a].

The number of studies that deal with emotional speech detection has increased significantly in the last few years. Despite the general progress in human–machine dialog systems, studies that deal with real-life telephone data, which is the typical outcome of automated human–machine voice portal dialogs, are still rare.

[Petrushin 1999] investigated 56 voice messages containing acted emotional expression spoken by 18 people. With an Artificial Neural Network (ANN) trained on acoustic features such as pitch, formants, energy and duration, he achieved about 23 % error rate for binary classification (agitated vs. calm).

[Devillers et al. 2002] reported on explorations based on about 5000 turns from customer-agent dialogs from a stock-service voice portal, i.e., they didn't use human–computer dialogs but human–human interaction. They investigated the separation of the utterances into four emotion related states solely based on words with a uni-gram topic tracker classification algorithm, which had originally been developed to see whether a document concerned a specified topic. They achieved a, comparably with the literature, low error rate of 32 % for the four target emotions and the neutral state, which can be partially explained by the fact that the labeling as well as the classification was based solely on spoken words, i.e.: the use of key words triggered the emotion detection.

[Ang et al. 2002] reported an investigation based on data coming from a faked air-travel arrangement system. The data consisted of 830 dialogs with more than 20000 turns, 75 % of which was used for training and 25 % for validation. They investigated three classes of features: acoustic-features (duration based on phonemes, spectral tilt, F0, energy), words based on automatic speech recognition (ASR), and a manually labeled "speaking-style" distinguishing between "hyper-articulating", "pausing" and "raised voice". It is not very surprising that classification of the material based on speaking style resulted in a low error rate. The use of a Classification And Regression Tree (CART) approach resulted in about 15 % error rate for a binary decision (nega-tive/else), and about 30 % error rate for ternary (annoyed, frustrated, else) decisions based solely on the acoustic features. The classification based on words resulted in about a 25 % error rate.

[Walker et al. 2002] used, like many other investigations, a subset of AT&T's How-May-I-help-You (HMIHY) database. The noteworthiness of this study is that they are not interested in the classification of a single utterance but in the detection of whole dialogs. Specific emotions were not specified: dialogs were divided into "problem-atic" vs. "non-problematic" dialogs. The classification algorithm is based on features derivable from ASR such as words, duration and number of words per utterance. Furthermore they investigated features such as dialog specific task-description (15 different), data coming from the dialog manager (e.g., prompt, prompt, confirmation, ...) and manually labeled features like words, age, or gender of the speaker. They achieved error rates of about 20 % for a binary decision (problematic vs. normal) after the first two utterances.

[Shafran et al. 2003] studied, beneath gender, age and dialect, the automatic classifi-cation of emotional expression again on a subset of AT&T's HMIHY database. After collapsing originally seven discrete emotion labels down to two (negative vs. posi-tive/neutral), a Hidden Markov Model (HMM)-based classifier resulted in an error rate of about 31 %, based on cepstral features, additional pitch information did not result in a significant increase.

[Yacoub et al. 2003] explored a database with about 2000 utterances performed by eight actors displaying 15 emotions uniformly distributed. They compared several classification approaches, namely ANNs, Support Vector Machines (SVMs), k-Nearest Neighbor (kNN), and CARTs. They investigated 39 prosodic and acoustic features, e.g., for pitch, energy and duration: minimum, maximum and mean, respectively, first derivative of the slope, jitter, shimmer, and ratio of audible vs. inaudible parts. The use of only the 19 best performing ones deteriorated the results by about 5 %. The results for a binary exploration(anger/neutral) showed that the ANN performed best with an error rate of about 10 % whereas CARTs and KNN resulted in a 20–30 % error rate. The SVM approach performed a little better under sparse data conditions.

[Lee, Narayanan 2005] investigated data coming from a flight-reservation application and looked at about 1200 dialogs with 7200 turns. Besides taking acoustic and prosodic features like F0, duration, energy and formants into account, they utilized a word content-based feature called "emotional salience". Furthermore features based on discourse got regarded for by the manual assignment of the turns to so-called "speech acts". A PCA reduces the feature set and therefore the complexity of the computation but does not result in an error reduction. A comparison between GMMs with KNN resulted in error rates of about 20 % for binary decision (negative vs. non-negative).

[Liscombe et al. 2005] also operated on a subset of the HMIHY data. They look at five different sets of features: 1) prosodic features like energy, pitch or duration based on voice/unvoiced frames, 2) lexical features like words and interjections that were manually labeled, 3) a semiautomatic extraction of phones and pauses, 4) manually labeled HMIHY Dialog Acts and 5), as context features the deviation from one to the next turn. A classifier called the BoosTexter (*boosting algorithm for combining results of weak learner decisions*) results in error rates of about 20 % when all features were combined and for a binary classification between negative and non-negative.

Training data

The compilation of training data for emotion classification mainly faces two problems. On the one hand, getting data that contains a sufficient quantity of emotional expression (recording), and on the other hand, deciding which emotion is expressed on the data (labeling). The state of emotion recognition in general still suffers from the prevalence of acted laboratory speech as the object of investigation. The high recognition rates of up to 100 % reported for such corpora cannot be transferred onto realistic, spontaneous data. For realistic databases, the performance for a two-class problem is typically < 80 %, for a four-class problem < 60 % as reported in the literature review of this chapter. Larger databases, i.e., more training data, seem to be a must but are difficult to obtain because the reference (ground truth, i.e., the phenomena that have to be recognized) cannot easily be obtained. For word recognition, a simple transliteration will suffice; for emotion recognition, manual annotation is normally necessary, time-consuming and costly, especially as emotion-related states are in no way a clearly defined issue.

The manual annotation is frequently done by a group of experienced listeners in order to avoid too much personal bias in the judgments. An important measure in this context is the inter-labeler agreement that tells how much the labelers are of the same opinion. As a way to measure the inter-labeler agreement the kappa-statistics K have often been used (e.g., [Lee, Narayanan 2005]); this sets the percentage of agreement in relation to the agreement expected by chance, as shown in (17.9),

$$K = \frac{P(A) - P(E)}{P(E)} \qquad (17.9)$$

where $P(A)$ stands for the average time the labelers agreed and $P(E)$ for the time they'd have agreed on a chance level. A value of 0 means no agreement, values between 0.4 and 0.7 are usually regarded as fair agreement and values above 0.7 denote excellent agreement. [Burkhardt et al. 2006] reported results based on three labelers. Two of them agreed nicely with a kappa value of 0.79 while the third disagreed quite often.

The agreement reported in [Burkhardt et al. 2006] was actually much higher then frequently reported in experiments dealing with emotional speech [Lee, Narayanan 2005], [Steidl et al. 2005] (about 0.45). The automatic classification ($K = 0.38$) resulted in a similarity comparable to the literature and to the human labelers, an outcome that was also reported in [Steidl et al. 2005].

In order to reuse data from different voice-portal applications, work on a set of standardized dialog tasks as well as a standardized way of emotional labeling is desirable.

17.4.2 Algorithm

The classifier used in [Burkhardt et al. 2005a] and [Burkhardt et al. 2006] is at its heart based on an acoustic-prosodic analysis of the speech signal, namely pitch-related features, energy features and duration. As a first step a voiced/unvoiced decision is used as a starting point for a frame-based pitch detection algorithm based on dynamic programming which is an advancement of the algorithm described in [Kompe 1989]. The pitch values are then transformed to semitones in order for the later comparisons to operate on relative intervals rather than absolute pitch values. The duration-related values are computed with respect to vowel vs. non-vowel phases in the speech. A small-scale phoneme recognizer is applied, which is based on a MFCC/HMM-based approach.

From these pitch, intensity and duration values, 31 prosodic features are extracted such as e.g., mean, minimum, standard deviation, regression coefficients etc., which are listed in detail in [Burkhardt et al. 2005a]. The feature vector is then classified into one of two classes using an algorithm based on Gaussian Mixture Models

(GMM). A score for every class gets calculated, which is the minimum of all negative logarithms from the evaluation of the corresponding densities. Then, those two scores are normalized as shown in (17.10).

$$S'_{\text{not angry}} = \frac{\min(scores)}{S_{\text{not angry}}}$$

$$S'_{\text{angry}} = \frac{\min(scores)}{S_{\text{angry}}}$$

$$\min(scores) = \min(S_{\text{not angry}}, S_{\text{angry}}). \tag{17.10}$$

As a result, one of the two scores $S'_{\text{not angry}}$ and S'_{angry} always has the value 1.0. Afterwards, the easiest way to classify is to decide the class that belongs to the score with the value 1.0, but, as we end up with separate probabilities for angry and non-angry speech, on a downstream stage a threshold filter can be applied, i.e., the caller only gets classified as angry if the non-anger probability is higher than a specified threshold. This is very important for online voice portal applications in which the dialog strategy should be more conservative in nature.

17.4.3 Results

Collecting training data

The emotion detection technology was used in several pilot voice portal implementations. Starting with a classifier trained on acted anger a group of about 50 researchers was instructed to call a voice portal and get angry with it. In a later phase, students were paid to call a fake hot-line that often failed in order to provoke anger, but of course the situation was still very artificial. Later, recordings from a pilot voice-portal with real customers were used, which provided for 18500 turns in 2300 dialogs, about 22 hours of data altogether. As this amount of data could no longer be labeled manually, it was pre-classified based on a training set of "faked anger" data gained in the above mentioned earlier phases of the project [Burkhardt et al. 2006] and only a subset containing 2232 turns in 167 dialogs was labeled manually.

Interestingly, most mis-classifications occurred because the classifier tended to mis-classify the neutral turns. This is probably caused by the fact that the faked data was performed under good audio conditions and contained clearly distinguishable emotional expression, while the real data was highly distorted and differences between anger and non-anger are often very small. It shows once more that training sets from laboratory data are not easily applicable for real world problems.

Working with thresholds

To adjust thresholds, several experiments with different training- and test sets from the pilot voice portal with real customer dialogs were performed. The following results are based on a disjunct test- and training set based on the decisions of one labeler alone, containing (randomly selected) 10 minutes anger out of 48 in the training and 6.5 minutes anger out of 28 in the test set. Because the distinction between low and high anger did not work well, the classifier was trained with only two classes: anger and non-anger. As reported [Burkhardt et al. 2005b], the trade-off between false acceptance and false rejection was controlled by means of thresholds, i.e., if one wants to avoid situations where users are accused of being angry although they were not, the anger value was disregarded in favor of anger only, if the non-anger value was lower than a certain threshold.

Recall and precision values for non-anger and anger detection as well as the overall accuracy (the total percentage of correctly identified cases) depending on the threshold for non-anger (left hand side) and anger (right hand side) are shown in Fig. 17.4.

Note that because the values are normalized it makes no sense to display results for both thresholds at the same time, as one value will always be 1. The anger recall rises with the increase of the non-angry threshold, as less and less samples get classified

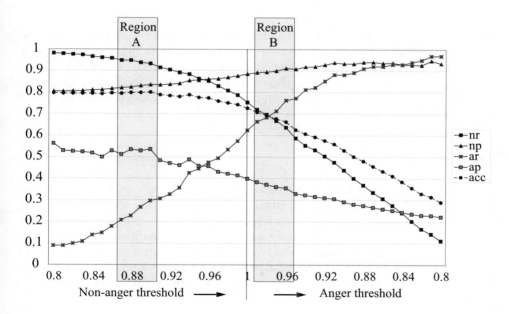

Figure 17.4: Recall and precision values in dependence of thresholds (see text), results from real data. n: non-angry, a: angry, r: recall, p: precision, acc: accuracy

as non-angry. As soon as the non-anger threshold reaches its limit and the anger threshold is lowered, the anger recall keeps on rising until it reaches its maximum of 1, at which point the decision will always be for "anger", irrespective of the classifier's outcome. The rise is monotonous, because the less the turns are classified as non-angry the more get classified as angry. At the same time, of course, the recall value for the non-angry turns drops, because more and more of them are misclassified as anger. The non-angry precision rises with the neutral-threshold because the less the turns get classified as non-angry the higher the percentage of correctly identified ones. The angry precision in contrast does not depend on the neutral threshold and therefore the curve is not monotonous in the left hand side.

All these statements get reversed on the right hand side of the figure, which displays recall and precision depending on the anger-threshold. The fact that the overall accuracy falls is a result of the by far greater number of non-angry turns, i.e., the accuracy is influenced mostly by the non-angry recall.

The optimal threshold to be used in a specific application depends, as mentioned earlier, on the scenario. If false accusations of being angry are to be avoided as is the case with a classifier that influences the dialog course of a customer voice portal, one will want to use a low threshold for the non-anger decision. This would be the case in region A of Fig. 17.4. On the other hand, someone who is primarily interested in identifying all the angry turns, e.g., with an offline statistical evaluation in mind, might opt for a lower anger-threshold like that given in region B of Fig. 17.4.

Experiments were also conducted on the so called "delta-features". This means that we not only compared the test sample with the models gained from the training database, but calculated the deviation of the features from the first turn, where the caller is assumed to be in a non-angry state. Taken alone, these delta features yielded worse results than the "world model", but taken in addition they resulted in slightly better results although the enhancement was not significant.

Conclusion

Anger detection via speech analysis is far from being an easy task, as was shown. Beginning with the collection of data, labeling it and training the classifier and implementing adequate dialog strategies many problems have to be solved. Although the acoustic classifier performed significantly worse under real conditions than with "laboratory" data, it still gives results well above the level of chance. As anger detection from short command-style utterance under low audio quality conditions will always be a problem and the occurrence of false alarms can not be excluded, the resulting dialog strategies will have to be conservative in nature.

The availability of high quality training data is an important issue. Therefore, in order to reuse data from different voice-portal applications work on a set of standardized dialog tasks as well as a standard way of emotional labeling would be desirable.

17.5 Applications in IVR Systems

This section envisages some applications that utilize speaker classification in a telecommunication scenario. Typical application fields are:

- dispatching of callers to trained agents

- market analysis of target groups

- call center quality management

- adaptive voice dialogs in IVR systems

- gaming and entertainment applications

Some of them have been repeatedly described in the scientific literature, others have already been mentioned in the public media, and few have even been deployed as real-world applications. [Burkhardt et al. 2005a] discusses several applications based on emotion recognition that stand for a family of related ideas.

[Batliner et al. 2006] introduces a taxonomy for emotion-aware applications, which can also be applied to speaker characterization in general. This taxonomy categorizes applications on the basis of the following four functional criteria.

Online/offline means the difference between whether the speaker classification is performed directly while the interaction is happening, as, for example, in an age detection system to enable age-specific dialog strategies, or later, as is the case with a statistical evaluation tool, for market research.

Mirroring/non-mirroring differentiates between applications where the classification is primarily used to give the user feedback, as exemplified by a language acquisition software that monitors the user's accent, in contrast to applications where the user is not directly aware of the classification process.

Impersonating/non-impersonating, originally called "emotional" vs. "non-emotional", classifies systems that simulate the characteristics of a certain speaker group, e.g., a dialog system that uses a youthful persona design because it detected a young customer calling.

Critical/non-critical means the differentiation between applications that depend strongly on a correct classification for each single instance, as, for example, in anger detection that triggers pacification strategies, versus a statistical reporting where the sum of classification results suffices for a general trend statement.

Of course the distinction based on that taxonomy is often not sharp, but it can be very useful to create new possible applications that utilize speaker classification by the inclusion or exclusion of certain features.

In the remaining part of this section, adaptive voice-dialogs in IVR systems are discussed in more detail.

17.5.1 Adaptive Voice-Dialogs

In automated voice portals dialog design becomes an important issue. State-of-the-art technology in language understanding and artificial intelligence does not yet allow for totally free and open dialogs in human–computer interaction. Dialog design comprises the way that the call flow is designed, i.e., which grammars are activated, which prompts will be played, which choices can be made by the user, and which feedback strategies are implemented.

In the case of static speaker classification such as age or gender, one possible application in this context would be to implement several designs and activate the one that best fits the current user profile. This might consist of very subtle changes, so that a misclassification would not lead to a perceptible difficulty for the callers. On the other hand a dynamic speaker classification such as anger detection could be used to adapt the dialog dynamically to a change in the user's state.

For the selection of appropriate dialog strategies it is important to reflect the interaction capabilities of the callers in order to avoid overcharge. The most important design criteria in this case are:

- balance between mixed initiative and directed dialogs
- explicit or implicit feedback strategies of the system
- design of menu trees (depth and width)
- design of audio prompts (volume, speed and pauses)
- usage of keypad input as an alternative to speech input

In a second step the designers of voice-dialogs have to care about the fulfillment of the expectations and needs of the callers. This is especially important for customer satisfaction and user acceptance of the service. The dialog adaptation should then be based on the following design parameters:

- design of the system's voice (persona design)
- order of presenting the menu entries
- usage of technical terms and colloquial speech in audio prompts
- introduction of music
- presentation of teasers and advertisements (e.g., in a waiting queue)
- offering of assistance from a human operator

Detailed information about the design of voice-dialogs can be found in [Balentine, Morgan 1999]. In this section, the feasibility and the advantages of adaptive voice-dialogs in two prototype voice applications of Deutsche Telekom are investigated. In the following these applications are described in detail and the results of corresponding usability and acceptability tests are shown.

17.5.2 A Voice Portal Based on Age/Gender Detection

Based on a prototype implementation of an adaptive voice portal the feasibility and advantages of adaptive voice-dialogs based on age and gender detection were analyzed. The basic application was similar to a voice portal that was deployed at Deutsche Telekom for the pre-qualifying of customers' requests (in German). After dialing the number, the caller enters a voice-dialog system that offers assistance for specific questions about, for example, the caller's subscriber line, tariffs, devices, and bills. In a first step the caller's requests are specified within the voice-dialog system. After that, the call is transferred to the next available human operator with matching skills.

This basic voice portal was extended by the integration of adaptive voice-dialogs based on automatic age and gender detection. The algorithms used are described in detail in Sec. 17.3, but to limit complexity in dialog design the number of classes was reduced by simply collapsing the classifier's output. The following four classes were used:

1. class C: Children / juveniles (age 0–19), male and female

2. class AM: Adults male (age 20–64)

3. class AF: Adults female (age 20–64)

4. class S: Seniors (age 65+), male and female

For each class, a specific voice-dialog was designed. The dialog designs differed from each other in the following parameters:

- persona design (age, gender and wording of prompts)

- speed of playing out prompts

- ASR timeout (the amount of time the ASR is waiting for speech input)

- the use of technical terms in prompts

- the level of detail for pre-qualifying the customer's request (depth of menu tree)

- assumptions on callers' preferences for telecommunication products

- escalation strategies: number of dialog loops with re-prompting after the ASR has detected the first invalid user input (after that, the call is transferred to a human agent)

- advertisement and teasers in waiting queue

- background music in waiting queue

Table 17.4: Proposed dialog design parameters for different classes

	AM	AF	C	S
Persona design	AM	AF	C (female)	S (male)
Speed of prompts	Normal	Normal	Normal	Slow
ASR timeouts	Normal	Normal	Normal	Long
Technical terms	Used	Avoided	Avoided	Avoided
Details of pre-qualifying	Additional information	Necessary information	Necessary information	Necessary information
Assumptions on preferences	Internet products	Mobile phones	Mobile phones	Fixed network
Escalation strategies	2–3 loops	2–3 loops	1 loop	1 loop
Advertisement in waiting queue	DSL products	Mobile phones	Mobile tariffs	Fixed network
Music	Rock	Pop	Hip-Hop	Big band

After the system's welcome prompt the caller was asked for his or her specific request. In this first prompt the dialog design for class AM was used. Then, the first utterance of the caller was analyzed by the classifier and an assignment to one of the four classes was made. After that, the system continued with the corresponding dialog design. The details of the respective dialog designs are listed in Table 17.4.

In order to assess the proposed dialog designs for the four selected classes a usability test was performed in cooperation with Siemens AG [Hempel 2006].

For this usability test 25 native German participants were recruited. They were divided into five groups: children (male and female), adults male, adults female, seniors male and seniors female, each group consisting of five people. All the participants had to perform a set of tasks in the voice-dialog system and had to answer a questionnaire. In principal, the test was designed to find answers to the following two main questions.

1. How do users value the usability of adaptive dialogs within an information portal designed for their target group?

2. Which additional preferences do different user groups have regarding wording style, gender and age of persona, background music and interaction style?

In general, the test results proved many of the proposed dialog design rules defined in Table 17.4. However, there were some findings that will lead to changes in future designs. The main results of the usability test are summarized in the following overview.

- The assumptions from Table 17.4 for the persona design were not fully confirmed. According to the test results the system's voice should not be younger than 20 and not be older than 60 years. Therefore, it is recommended to use persona AF also for class C and persona AM also for class S.

- For most of the dialog design criteria (dialog strategies, feedback strategies, escalation strategies, the structuring of dialog trees, the degree of automation, the wording and the presentation of content) the age is the dominating parameter, especially a rough differentiation between young and old people.

- But there is a significant difference in the assessments between male and female seniors. In general, female seniors showed much more problems interacting with a voice-dialog system than all other groups.

- In this context, the proposed differentiation into three age classes seems to be sufficient. However, it may be advantageous to introduce an additional differentiation between male and female seniors. Thus, the five classes C, AM, AF, SM, SF should be considered in future.

- Background music and jingles are a polarizing factor, primarily depending on the age of the user. In order to meet the preferences of the user it is necessary to make a finer differentiation or to make more discrete selections.

- Besides the user's age the usage frequency is the most important factor for dialog design criteria such as dialog strategies and feedback strategies.

- It was not possible to derive a valuable statement for preferred products in the waiting queue for the different classes because all groups refused advertisements. In order to find out preferences for specific product groups market surveys have to be made.

In general, it is important to notice that the context of the application, e.g., telecommunication, banking or entertainment, is also a dominating factor for an appropriate dialog design. The recommendations given in this section should therefore not be transferred to other application fields without validation.

17.5.3 Customer Self-Service Based on Anger Detection

In a second step, the feasibility and possible advantages of anger detection for optimizing voice-controlled customer self service applications were evaluated. The prototype application implemented for this evaluation was an automated voice-dialog (in German) for selecting a new mobile phone according to the customer's personal preferences.

After the system's welcome prompt the customer can say the product name of the desired mobile phone or can ask for assistance in selecting an appropriate model. For example, it is possible to say a combination of desired features like MP3 player, radio, camera, or to specify a price limit, and the system offers a set of suitable products for further selection. After making a decision for a particular model the user can receive detailed information about the product by fax or email (if an email address has been pre-specified with the user's account).

In order to detect potential problems that arise from an unsatisfactory course of interaction, automatic anger detection was introduced into the system. The objective was to help the customer by either offering the assistance of a human operator or trying to react with appropriate dialog strategies. The algorithm used for anger detection is described in Sec. 17.4. After an angry user utterance had been detected the system randomly chose one of the following two alternative approaches.

1. The wording of the prompts is slightly changed to calm down the user in order to continue the voice-dialog.

2. The system offers the transfer of the call to a human operator to complete the task.

This voice application was evaluated at T-Systems in an acceptability test with 200 test users (52 % female, 48 % male, aged in the range 18 to 65 years, average age 40 years). For most of them this was a first experience of voice controlled systems. The test users had to call the system in order to complete a set of tasks. One of the tasks was assigned to the assessment of the system's reaction in the case of detected anger. For this task the system was manipulated in order to give incorrect answers after the first user input in order to induce anger.

59 % of the test users admitted that they got angry after this incorrect system reaction, 56 % of them confirmed that they had really answered with an angry voice. 60 % of these callers (in total 40 people) noticed a corresponding system reaction to their angry utterance. These test users then had to answer the following question:

• "How did you experience the system's reaction to your angry utterance?"

Figure 17.5 shows the results of the assessment of the test users for the system's reaction to angry utterances. Around 70 % perceived the system's reaction as appropriate while only 56 % confirmed that it was helpful to complete the task. The difference

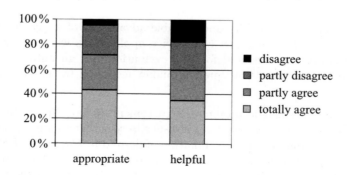

Figure 17.5: User experience of system reaction in case of detected anger

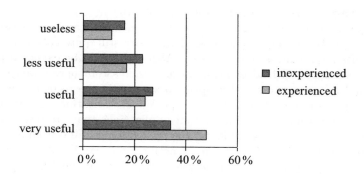

Figure 17.6: General assessment of anger detection in voice applications

between these results can be explained after looking at the different dialog strategies that were applied in the respective voice-dialogs. The test users, who were offered to be transferred to a human agent, were more successful at completing the task than those who continued the voice-dialog with a slightly changed, de-escalating wording of the prompts.

In a second step all participants in the acceptability test were asked to assess the introduction of anger detection in voice-controlled customer self service applications in general. The following question was asked.

- "How do you assess the introduction of anger detection in voice applications?"

Figure 17.6 shows the results of this general assessment of anger detection. The results differentiate between the users who perceived the system's reaction to angry utterances and those who did not have this experience. 72 % of the users with and 60 % of the users without experience confirmed that the introduction of anger detection in voice applications is very useful or useful. It is remarkable that the share of users that had the experience with the system and found it very useful is about 15 % higher than the share of those without such experience.

17.6 Discussion and Conclusion

As discussed in the previous sections, there is a growing interest in determining non-verbal information from the speech signal. In the absence of certain information about the speaker's identity, the goal is to detect side information about the speaker from the speech signal in order to improve automatic voice-dialog systems, for example by adapting the dialog design to pre-specified target groups. The algorithms for speaker classification described in this chapter focus on two types of parameters: age and gender recognition, and the detection of anger.

In the case of age and gender recognition it was shown that state-of-the-art algorithms can reach an accuracy that is high above the chance level and gets close to human performance. However, the age intervals defined are still quite coarse and the classification results are far from being reliable. Although it can be expected that further algorithmic research will lead to an additional increase in performance in the near future, significant error rates with respect to automatic age recognition will remain. Thus, the applications that benefit from speaker classification will be restricted to "non-secure" applications. This means that applications that are based on a reliable check of the caller's age, for example, for the prevention of fraud from minors, can not be considered in this case. However, most applications can be improved on a statistical basis, e.g., the increase of average user satisfaction, if the following two criteria are reflected in the corresponding application designs.

- The proposed dialog-design parameters for the pre-defined age and gender classes should be validated and adapted within the application itself. This means that the behavior of the callers in the application should be monitored and evaluated permanently and lead to an update of relevant design parameters within regular time intervals. Additionally, usability tests should accompany and justify these system updates.

- The negative effects of a misclassification should be minimized within the application. Thus, the dialog design assigned to a specific target group should not annoy or frustrate callers of another target group that have been assigned to the wrong group.

In the case of anger detection the situation is similar, but some significant differences have to be considered. The perception of anger in a caller's voice is a subjective impression of the listener within a specific application context. Thus, there is no reliable reference or set of training data that can be used for optimization of the anger-detection algorithm and substantial error rates have to be considered. It is therefore proposed to tune the thresholds of the anger-detection algorithm in a way that only the utterances where most of the labelers agreed on angry speech are assigned to the class "angry" with respective dialog strategies. The advantage is that the angriest customers are filtered out while non-angry customers are not disturbed during the interaction with the system. Another important issue for systems based on anger detection is that the classifier, usually trained on laboratory data, should be optimized based on speech data from the application itself.

In an industrial real-world deployment of an application based on speaker classification, a set of additional requirements have to be considered.

- The algorithm for speaker classification must integrate into the existing architecture of the overall voice platform.

- The delay caused by the processing of the classifier must not obstruct the dialog flow.

- The algorithm must be able to work on short one-word commands and poor audio conditions.

Acknowledgements

The authors would like to thank Deutsche Telekom Laboratories for funding the work described in this chapter. They are especially grateful to Dr. Udo Bub, Dr. Roman Englert, and Katja Henke for their continuous support during the execution of the related projects. Many researchers from other organizations have contributed to this work. The authors would especially like to thank Dr. Richard Huber, Dr. Bernt Andrassy, Dr. Christian Müller, and Dr. Thomas Hempel, for their valuable contributions corresponding to the algorithms for speaker classification and usability testing. This chapter has also benefited from careful reading and helpful comments from Frank Oberle.

Bibliography

Ang, J.; Dhillon, R.; Krupski, A.; Shriberg, E.; Stolcke, A. (2002). Prosody-based Automatic Detection of Annoyance and Frustration in Human–computer Dialog, *Proceeding of International Conference on Spoken Language Processing (ICSLP)*, Denver, CO, USA.

Baken, R.; Orlikoff, R. (2000). *Clinical Measurement of Speech and Voice*, 2nd edn, Singular Publishing Group, San Diego, CA, USA.

Balentine, B.; Morgan, D. P. (1999). *How to Build a Speech Recognition Application. A Style Guide for Telephony Dialogues*, Enterprise Integration Group, Inc., San Ramon.

Batliner, A.; Burkhardt, F.; van Ballegooy, M.; Nöth, E. (2006). A Taxonomy of Applications that Utilize Emotional Awareness, *Proceedings of Fifth Slovenian and First International Language Technologies Conference (SLTS)*, Ljubljana, Slovenia.

Board for Professional and Occupational Regulation (2003). Study of the Utility and Validity of Voice Stress Analyzers, *Technical report*, Department of Professional and Occupational Regulation.

Braun, A.; Cerrato, L. (1999). Estimating Speaker Age across Languages, *Proceedings of International Congress of Phonetic Sciences (ICPhS)*, San Francisco, CA, USA, vol. 2, pp. 1369–1372.

Breiman, L.; Friedman, J.; Olshen, R.; Stone, C. (1984). *Classification and Regression Trees*, Chapman & Hall, New York, NY, USA.

Burkhardt, F.; Ajmera, J.; Englert, R.; Burleson, W.; Stegmann, J. (2006). Detecting Anger in Automated Voice Portal Dialogs, *Proceedings of Interspeech*, Pittsburgh, PA, USA.

Burkhardt, F.; van Ballegooy, M.; Englert, R.; Huber, R. (2005a). An Emotion-Aware Voice Portal, *Proceedings of Conference for Electronic Speech Signal Processing (ESSP)*, Prague, Czech Republic.

Burkhardt, F.; van Ballegooy, M.; Stegmann, J. (2005b). A Voiceportal Enhanced by Semantic Processing and Affect Awareness, *Proceedings of INFORMATIK 2005*, Gesellschaft für Informatik, Bonn, Germany, vol. 2.

Cerrato, L.; Falcone, M.; Paoloni, A. (2000). Subjective Age Estimation of Telephonic Voices, *Speech Communication*, vol. 31, no. 2–3, pp. 107–102.

Devillers, L.; Lamel, L.; Vasilescu, I. (2002). Annotation and Detection of Emotion in a Task-oriented Human–Human Dialog Corpus, *Proceedings on ISLE Workshop on Dialogue Tagging*.

Domingos, P.; Pazzani, M. J. (1997). On the Optimality of the Simple Bayesian Classifier under Zero-One Loss, *Machine Learning*, vol. 29, no. 2-3, pp. 103–130.

Duda, R. O.; Hart, P. E.; Stork, D. G. (2000). *Pattern Classification*, 2nd edn, Wiley Interscience.

Ferrand, C. T. (2002). Harmonics-to-Noise-Ratio: an Index of Vocal Aging, *Journal of Voice*, vol. 16, no. 4, pp. 480–487.

Fukunaga, K. (1990). *Statistical Pattern Recognition*, 2nd edn, Academic Press, San Diego, CA, USA.

Hempel, T. (2006). Usability of a Telephone-Based Speech Dialogue System as Experienced by User Groups of Different Age and Background, *Proceedings of 2nd ISCA/DEGA Tutorial and Research Workshop on Perceptual Quality of Systems*, Berlin, Germany.

Höge, H.; Draxler, C.; van den Heuvel, H.; Johansen, F. T.; Sanders, E.; Tropf, H. S. (1999). SpeechDat Multilingual Speech Databases for Teleservices: Across the Finish Line, *Proceedings of Eurospeech 1999*, ISCA, Budapest, Hungary. http://www.speechdat.org/.

Jelinek, F. (1998). *Statistical Methods for Speech Recognition*, MIT Press, Boston.

Kienast, M.; Paeschke, A.; Sendlmeier, W. F. (1999). Articulatory Reduction in Emotional Speech, *Proceedings of Eurospeech 99 Budapest*, pp. 117–120.

Kompe, R. (1989). *Ein Mehrkanalverfahren zur Berechnung der Grundfrequenzkontur unter Einsatz der dynamischen Programmierung*, Master's thesis, Universität Erlangen-Nürnberg.

Lee, C. M.; Narayanan, S. S. (2005). Toward Detecting Emotions in Spoken Dialogs, *IEEE Transactions on Speech and Audio Processing*, vol. 13, no. 2, pp. 293–303.

Linville, S. E. (2001). *Vocal Aging*, Singular Publishing Group, San Diego, CA, USA.

Liscombe, J.; Riccardi, G.; Hakkani-Tür, D. (2005). Using Context to Improve Emotion Detection in Spoken Dialog Systems, *Proceedings of Interspeech*, Lisbon, Portugal.

Metze, F.; Ajmera, J.; Englert, R.; Bub, U.; Burkhardt, F.; Stegmann, J.; Müller, C.; Huber, R.; Andrassy, B.; Bauer, J. G.; Littel, B. (2007). Comparison of Four Approaches to Age and Gender Recognition for Telephone Applications, *Proceedings of International Conference on Acoustics, Speech, and Signal Processing (ICASSP)*, Honolulu, Hawaii.

Minematsu, N.; Sekiguchi, M.; Hirose, K. (2002). Automatic Estimation of One's Age with His/Her Speech Based Upon Acoustic Modeling Techniques of Speakers, *Proceedings of International Conference on Acoustics, Speech, and Signal Processing (ICASSP)*, Orlando, FL, USA.

Müller, C. (2005). *Zweistufige kontextsensitive Sprecherklassifikation am Beispiel von Alter und Geschlecht*, PhD thesis, Computer Science Institute; Universität des Saarlandes; Germany.

Müller, C.; Schötz, S. (eds.) (2007). *Speaker Classification*, Springer, New York – Berlin.

Müller, C.; Wittig, F.; Baus, J. (2003). Exploiting Speech for Recognizing Elderly Users to Respond to their Special Needs, *Proceedings of Interspeech (Eurospeech)*, ISCA, Geneva, Switzerland.

Mysak, E. D. (1959). Pitch Duration Characteristics of Older Males, *Journal of Speech and Hearing Research*, vol. 2, pp. 46–54.

Petrushin, V. (1999). Emotion in Speech: Recognition and Application to Call Centers, *Conference on Artificial Neural Networks In Engineering (ANNIE)*, St Louis, USA.

Rabiner, L. R. (1978). *Digital Processing of Speech Signals*, Prentice-Hall.

Rabiner, L. R. (1989). A Tutorial on Hidden Markov Models and Selected Applications in Speech Recognition, *Proceedings of the IEEE*, vol. 77, no. 2, pp. 257–286.

Redner, R. A.; Walker, H. F. (1984). Mixture Densities, Maximum Likelihood and the EM Algorithm, *SIAM Review*, vol. 26, no. 2, pp. 195–239.

Reynolds, D. A. (2002). An Overview of Automatic Speaker Recognition Technology, *Proceedings of International Conference on Acoustics, Speech, and Signal Processing (ICASSP)*, Orlando, FL, USA, vol. 4, pp. 4072–4075.

Reynolds, D.; Campbell, J.; Campbell, B.; Dunn, B.; Gleason, T.; Jones, D.; Quatieri, T.; Quillen, C.; Sturim, D.; Torres-Carrasquillo, P. (2003). Beyond Cepstra: Exploiting High-Level Information in Speaker Recognition, *Workshop on Multimodal User Authentication*.

Schölkopf, B.; Smola, A. (2002). *Learning with Kernels: Support Vector Machines, Regularization, Optimization, and Beyond (Adaptive Computation and Machine Learning)*, MIT Press, Cambridge, MA, USA.

Schötz, S. (2004). Automatic Prediction of Speaker Age Using CART, http://www.ling.lu.se/persons/Suzi/downloads/RF_paper_SusanneS2004.pdf. Term paper for course in Forensic Phonetics, Göteborg University.

Schultz, T.; Kirchhoff, K. (eds.) (2006). *Multilingual Speech Processing*, Academic Press.

Shafran, I.; Riley, M.; Mohri, M. (2003). Voice Signatures, *Proceedings of The 8th IEEE Automatic Speech Recognition and Understanding Workshop (ASRU)*, IEEE, U.S. Virgin Islands.

Steidl, S.; Levit, M.; Batliner, A.; Nöth, E.; Niemann, H. (2005). Of all Things the Measure is Man – Classification of Emotions and Inter-Labeler Consistency, *Proceedings of International Conference on Acoustics, Speech, and Signal Processing (ICASSP)*.

Tanner, D. C.; Tanner, M. E. (2004). *Forensic Aspects of Speech Patterns: Voice Prints, Speaker Profiling, Lie and Intoxication Detection*, Lawyers & Judges Publishing Company.

van Rijsbergen, C. J. (1979). *Information Retrieval*, Butterworths, London.

Walker, M.; Langkilde-Geary, I.; Wright, H.; Wright, J.; Gorin, A. (2002). Automatically Training a Problematic Dialogue Predictor for a Spoken Dialogue System, *Journal of Artificial Intelligence Research*, vol. 16, pp. 293–319.

Yacoub, S.; Simske, S.; Lin, X.; Burns, J. (2003). Recognition of Emotions in Interactive Voice Response Systems, *Proceedings of Interspeech (Eurospeech)*, Geneva, Switzerland.

Zweig, G.; Russell, S. (1998). Speech Recognition with Dynamic Bayesian Networks, *Proceedings of Fifteenth National Conference on Artificial Intelligence (AAAI)*, Madison, WI.

Index

Permissions List

Reproduced with permission of Elsevier

Fig. 7.5, reprinted from Antweiler, C.; Antweiler, M. (1995). System Identification with Perfect Sequences Based on the NLMS Algorithm, *AEÜ (International Journal of Electronics and Communication)*, vol. 49, no. 3, pp. 129–134.

Fig. 4.2, 4.3, and 4.4, reprinted from Enzner, G.; Vary, P. (2006). Frequency-Domain Adaptive Kalman Filter for Acoustic Echo Control in Hands-Free Telephones, *Signal Processing*, vol. 86, no. 6, pp. 1140–1156.

Reproduced with permission of Hindawi

Figs. 14.9, 14.10, 14.11, 14.12, 14.13, and 14.14, reprinted from Lotter, T.; Vary, P. (2006). Dual-Channel Speech Enhancement by Superdirective Beamforming, *EURASIP Journal on Applied Signal Processing*, Article ID 63297.

Reproduced with permission of IEEE

Figs. 3.4 and 3.5, reprinted from Werner, M.; Kamps, K.; Tuisel, U.; Beerends, J. G.; Vary, P. (2003). Parameter-based Speech Quality Measures for GSM, *Proceedings of the IEEE International Symposium on Personal, Indoor and Mobile Radio Communiocations (PIMRC)*, Beijing, pp. 2611–2615.

Figs. 3.9 and 3.10, reprinted from Werner, M.; Junge, T.; Vary, P. (2004). Quality Control for AMR Speech Channsl in GSM Networks, *Proceedings of the IEEE International Conference on Acoustics, Speech and Signal Processing (ICASSP)*, Montreal, pp. 1067–1079.

Fig. 7.13, reprinted from Antweiler, C.; Symanzik, H.-G. (1995). Simulation of Time Variant Room Impulse Responses, *Proceedings of the IEEE International Conference on Acoustics, Speech and Signal Processing (ICASSP)*, Detroit, pp. 3031–3034.

Fig. 8.5, reprinted from Riskin, E. A.; Ladner, R.; Wang, R.-Y.; Atlas, L. E. (1994). Index Assignment for Progressive Transmission of Full-Search Vector Quantization, *IEEE Transactions on Image Processing*, vol. 3, no. 3.

Figs. 8.10 and 8.11, reprinted from Ragot S.; et al. (2007). ITU-T G.729.1: An 8-32 kbit/s Scalable Coder Interoperable with G.729 for Wideband Telephony and Voice over IP, *Proceedings of the IEEE International Conference on Acoustics, Speech and Signal Processing (ICASSP)*, Honolulu, pp. 529–532.

Figs. 9.2, 9.3, 9.4, 9.5, 9.6, and portions reprinted from Jax, P.; Vary, P. (2006). Bandwidth Extension of Speech Signals: A Catalyst for the Introduction of Wideband Speech Coding?, *IEEE Communications Magazine*, vol. 44, no. 5, pp. 106–111.

Figs. 9.7 and 9.9, reprinted from Jax, P.; Geiser, B.; Schandl, S.; Taddei, H.; Vary, P. (2006). An Embedded Scalable Wideband Codec Based on the GSM EFR Codec, *Proceedings of the IEEE International Conference on Acoustics, Speech and Signal Processing (ICASSP)*, Toulouse, pp. 5–8.

Figs. 12.2, 12.3, and 12.4, reprinted from Heinen S.; Vary, P. (2005). Source-Optimized Channel Coding for Digital Transmission Channels, *IEEE Transactions on Communications*, vol. 53, no. 4, pp. 592–600.

Fig. 12.7, reprinted from Fingscheidt, T.; Hindelang, T.; Cox, R. V.; Seshadri, N. (2002). Joint Source-Channel (De)Coding for Mobile Communications, *IEEE Transactions on Communications*, vol. 50, no. 2, pp. 200–212
and from Hindelang, T.; Fingscheidt, T.; Seshadri, N.; Cox, R. V. (2000). Combined Source/ Channel (De)Coding: Can A Priori Information Be Used Twice?, *Proceedings of the IEEE International Conference on Communications (ICC)*, New Orleans, pp. 744–748.

Reproduced with permission of ISCA

Figs. 3.4 and 3.5, reprinted from Werner, M.; Kamps, K.; Tuisel, U.; Beerends, J. G.; Vary, P. (2003). Parameter-Based Speech Quality Measures for GSM, *Proceedings of First ISCA ITRW on Auditory Quality of Systems*, pp. 29–34.

Figs. 3.8 and 3.11, reprinted from Werner, M.; Vary, P. (2005). Quality Control for UMTS-AMR Speech Channels, *Proceedings of the European Conference on Speech Communication and Technology (Interspeech)*.

Fig. 17.4, reprinted from Burkhardt, F.; Ajmera, J.; Englert, R.; Stegmann, J.; Burleson, W. (2006). Detecting Anger in Automated Voice Portal Dialogs, *Proceedings of Interspeech (ICSLP)*.

Reproduced with permission of VDE-Verlag

Fig. 12.7, reprinted from Hindelang, T.; Hagenauer J.; Heinen, S. (2000). Source-Controlled Channel Decoding: Estimation of Correlated Parameters, *Proceedings of 3rd ITG Conference "Source and Channel Coding"*, Munich, Germany.

Reproduced with permission of VDI-Verlag

Figs. 12.6, 12.7, 12.12, 12.13, and 12.14, reprinted from Hindelang, T. (2002). *Source-Controlled Channel Coding*, VDI-Verlag, Reihe 10, Nr. 695, ISBN 3-18-369510-3.

Reproduced with permission of John Wiley & Sons, Ltd

Fig. 14.3, reprinted from Hamacher, V.; Raß, U. (2006). Hightech im Ohr: Physikalische und technische Grundlagen moderner Hörgeräte, *Physik in unserer Zeit*, vol. 37, no. 2, pp. 90–96.